CHEMISTRY
in Today's World

Modern Chemistry Series

Under the supervisory editorship of D. J. Waddington, B Sc. ARCS, DIC, Ph D, Professor of Chemical Education, University of York, this series is specially designed to meet the demands of the new syllabuses for sixth form, introductory degree and technical college courses. It consists of self-contained major texts in the three principal divisions of the subject, supplemented by short readers and practical books.

Major Texts

Modern Inorganic Chemistry
G. F. Liptrot MA, Ph D

Modern Organic Chemistry
R. O. C. Norman MA, D Sc, C Chem, FRIC, FRS and
D. J. Waddington, B Sc, ARCS, DIC, Ph D

Readers

Investigation of Molecular Structure
B. C. Gilbert MA. D Phil

Organic Chemistry: a problem-solving approach
M. J. Tomlinson B Sc, C Chem, MRIC
M. C. V. Cane B Sc, Ph D, C Chem, MRIC

Mechanisms in Organic Chemistry: Case Studies
R. O. C. Norman, MA, D Sc, C Chem, FRIC, FRS
M. J. Tomlinson B Sc, C Chem, MRIC
D. J. Waddington B Sc, ARCS, DIC, Ph D

Practical Books

Organic Chemistry Through Experiment
D. J. Waddington B Sc, ARCS, DIC, Ph D, and
H. S. Finlay B Sc

Inorganic Chemistry Through Experiment
G. F. Liptrot MA, Ph D
Also by G. F. Liptrot and J. S. F. Pode (not in the above series)
A Basic Course in Chemistry
Exploring Chemistry

CHEMISTRY
in Today's World

D. Ainley, B.Sc., M.Sc., B.Phil., C.Chem., M.R.I.C.

Lecturer, Department of Educational Studies,
University of Hull,
formerly Co-ordinator of the Faculty of Sciences, Lincoln Christ's Hospital School,
Lincoln

J. N. Lazonby, B.Sc., C.Chem., M.R.I.C.

Lecturer, Department of Education, University of York,
formerly Senior Teacher and Head of the Science Department,
Archbishop Holgate's Grammar School, York.

A. J. Masson, B.Sc., Ph.D.

Department of Chemistry,
Bradford Grammar School, Bradford.

Mills & Boon Limited
London

First published in Great Britain 1980
by Mills & Boon Limited, 15–17 Brooks' Mews, London W1

ISBN 0 263 06417 4

Filmset and printed in Great Britain by
BAS Printers Limited, Over Wallop, Hampshire
and bound by Hunter & Foulis Ltd., Edinburgh

Preface

The book is designed for the last two or three years of chemistry courses which lead to examinations at 16 + . Whereas the topics in the book are those which are required by the syllabuses of the various examination boards, we believe it is essential that the study of these topics should also be justified in other ways. This has been achieved by showing how certain lines of enquiry help to develop our understanding of the behaviour of substances and also by emphasising the parts played by individual scientists in the development of the subject. The other main reason for including topics is that they are important to our present and future lives. The applications of chemistry are considered, not only because of their economic importance, but also from the point of view of their effects on our environment and the need to conserve the world's resources. Many of the photographs in the text illustrate the part played by chemistry in Today's World.

While writing the book, we have assumed and indeed believe, that teachers prefer to devise their own courses and do not wish to provide a book for their students which places too many constraints on the teaching sequence or teaching methods employed. Thus we have written a single-volume book and divided each chapter into two distinct though related parts—investigations and text. This separation is intended to allow greater flexibility in use than is possible with a course book. It is not an attempt to divorce practical work from theoretical progress and thus within the text, wherever theories are outlined, they are presented as possible explanations of observable behaviour.

Instructions for investigations are given at the beginning of most chapters and those which are more appropriate as demonstrations are described in the text. The instructions are written in such a way that the investigations may be used as an integral part of a problem-solving type of approach. Each investigation is followed by questions which encourage the student to think about the observations before they are discussed more thoroughly in class. The investigations may be used at the discretion of the teacher, whereas students are able to work independently with the text without necessarily having carried out all of the investigations. The division of each chapter into numbered sections, the extensive use of cross-referencing within the text and the comprehensive index, facilitate the retrieval of information, and students will find the book easy to use for revision. The summaries at the end of each chapter provide the student with an overview of the topic rather than a condensed version of the content.

The book is supported by a data section and a set of questions. The data section includes tables of physical properties of selected elements and compounds. The questions cover the main topic areas in the book and each question in the second set of revision questions consists of several parts, with each successive part making increased demands on the students and so catering for the need to assess the knowledge and understanding of students who represent a wide range of ability.

A teachers' guide, which provides further details on the investigations and references to other sources of information and teaching aids may be obtained by sending a stamped addressed A5 envelope to J. N. Lazonby, Department of Education, University of York, Heslington, York Y01 5DD.

Acknowledgements

We thank Mr J. D. Haden for reading the manuscript and for his valuable advice, and Dr. J. McIntyre and Mr. D. J. M. Rowe for their helpful discussions on industrial chemistry. In particular we thank Professor D. J. Waddington for his encouragement, guidance and help at all stages in the production of the book.

Among the many sources of information consulted during the preparation of the book we particularly acknowledge the use we have made of the I.C.I. booklets, Fertilisers and Inorganic Chemicals.

The sources of photographs are acknowledged in the text. We thank the companies and organisations who have allowed us to use their photographs and we thank Mr. R. Edwards and Mr. J. Olive for those which were taken especially for the book.

We are grateful to the following examination boards for granting permission to use questions from past examination papers: East Midlands Regional Examinations Board, Joint Matriculation Board, Oxford and Cambridge Schools Examination Board, Oxford Delegacy of Local Examinations, South Western Examinations Board, University of Cambridge Local Examinations Syndicate, University of London University Entrance and Schools Examinations Council, Welsh Joint Education Committee, West Midlands Examinations Board and Yorkshire Regional Examinations Board.

Finally we thank our families for their patience and for their encouragement during the preparation of the book.

D.A.
J.N.L.
A.J.M.

Contents

Chapter 1

What are substances made of?

1.1 What does a chemist do?

St. Paul's Cathedral in London was designed by Sir Christopher Wren. On his tomb inside the cathedral is written 'If you seek my memorial, look about you'. In other words, the cathedral is evidence of his work.

Similarly, if we wish to see evidence of what a chemist does, we just have to look around us. Many of the materials which we use in our everyday lives involve the work of chemists at some stage in their production. The chemist's products include plastics, paints, textiles, dyes, medicines, fertilisers, insecticides and detergents. Most of these products are made from a small number of raw materials, such as coal, oil, air, salt and limestone, which are present in or around our Earth. It is the job of chemists to convert these raw materials into the substances we use.

Fig. 1.1 Some of the main sources of raw material for the chemical industry (a) inside the salt mine at Winsford, Cheshire. A lorry load of rock salt is being transported to another section of the mine. (Courtesy I.C.I. Ltd, Mond Division. Photo: Photo Graphics, Merseyside)

(b) A general view of the Swindon Limestone Quarry in Yorkshire, showing the rock face and the lime-kilns. (Courtesy Tilling Construction Industries Ltd. Photo: J. Holmes)

(c) An oil production platform in the North Sea

Chemists are also employed in medical work where, for example, the results of chemical tests on a sample of blood or urine can help a doctor to diagnose a patient's illness. Chemical techniques are also important in forensic science (the use of science for investigating crimes), checking the purity of water and food supplies, and many other areas which influence our daily lives.

Fig. 1.2 The materials used to make this motorbike, such as the metals for the engine and frame, the plastic for the seat and cable covering, the rubber for the tyres and the glass for the lights, are all produced by chemical processes. (Photo: Russell Edwards, B.Sc)

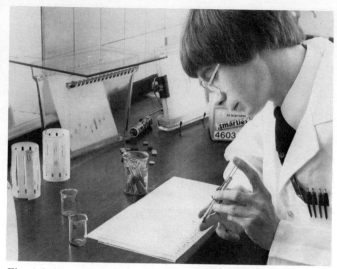

Fig. 1.3 A use of chemical techniques in the food industry—this photograph shows thin layer and paper chromatography being used to separate and identify the colouring materials in Smarties. (Courtesy Rowntree Mackintosh Ltd)

1.2 Chemical and physical changes

Most raw materials consist of mixtures of substances. The chemist must first find ways of separating the useful substances from these mixtures and then converting them into the materials which are required.

In order to carry out these steps, chemists perform two types of changes on substances. One type of change is more drastic than the other. The differences can be illustrated by thinking about the ways in which candle wax can be changed.

When candle wax is heated it melts to form a colourless liquid, but if the liquid is cooled, it changes back to solid wax again. The melting of candle wax has not changed it into a different substance. It is still candle wax even when it is a liquid.

On the other hand, if candle wax is held in a flame, it will burn with a yellow flame, giving off smoke and gases. In this example, the candle wax has been changed into different substances. It would be very difficult to change these substances back into candle wax.

Fig. 1.4 A lighted candle involves both chemical and physical changes. (Photo: Russell Edwards, BSc)

Melting is a physical change

Burning is a chemical change

A change in which different substances are not formed (e.g. melting wax) is called a **physical change**, as we have only changed the physical form of the substance.

A change in which different substances are formed (e.g. burning wax) is called a **chemical change**, as new chemical substances have been formed.

Chemical changes can be called chemical reactions. This name is particularly appropriate if the chemical change occurs when two substances are put together. Clearly, in this type of change the two substances can be thought of as reacting together to form different substances. For example, the change which occurs when a piece of wood is burned, can be thought of as a reaction between wood and air. Ash, smoke and gases are formed by this reaction.

This particular chemical reaction needs some heat to start it off, but once started, the wood burns and gives out more heat. Many reactions need heat to start them off, but some occur by simply mixing the substances. For example, a chemical reaction occurs between two components of health salts when they are dissolved in water. Some types of glue work by mixing two components together; a chemical reaction occurs while the glue is setting. The resin which is used with fibreglass to repair damaged motor cars is made by mixing two substances which spontaneously react together.

Sometimes it is obvious that new substances have been formed, but if it is not, what other indications might there be that a chemical change has occurred? One sign might be that there is an obvious energy change. For example, when a substance burns energy in the forms of heat and light is given out. When the components of a glue or resin are mixed no light is given out (i.e. there is not a flame) but the mixture does become hot which indicates that a chemical change is occurring.

During a chemical change different substances are formed and it is usually difficult to convert these substances back to the original substances. This means that a chemical change is not easily reversible. However, you must take care when using this indication of a chemical change. If, for example, you break a glass bottle, it is not easy to reverse the change, but this is not a chemical change as no new substances have been formed. The broken pieces of bottle are still made of glass. A physical change has occurred because you have changed the physical form of the bottle. To reverse the change, it would be necessary to melt the glass and reshape it. This would not be easy for most of us, but it would still be a physical change.

The differences between physical and chemical changes are summarised in Table 1.1.

TABLE 1.1. *Physical and chemical changes*

	PHYSICAL CHANGES	CHEMICAL CHANGES
Differences	No new substances are formed. The product is chemically identical to the starting material.	New substances are formed in the change.
Possible signs	1. The change can often be simply reversed. 2. The energy changes are usually small.	Usually the product can only be converted back to the original material with great difficulty. Often accompanied by obvious energy changes.

1.3 What substances consist of

Chemists have investigated materials by subjecting them to physical and chemical changes. They have found that every material can be broken down into one or more of about ninety substances which cannot themselves be broken down into anything simpler. These substances, from which everything is made, are called **elements**.

3

For example, if electricity is passed through molten salt, the salt is broken down into two substances. One of these substances is a silvery-grey metal called sodium and the other is a pale green gas called chlorine. Try as hard as we might, we cannot break down sodium and chlorine into anything simpler. Sodium and chlorine are therefore elements.

Water can be split up into hydrogen and oxygen, but hydrogen and oxygen cannot be broken down any further. Therefore, hydrogen and oxygen are elements, but water is not.

Other common elements which you are likely to have heard of are iron, aluminium and copper. If you have a piece of copper, then it is just copper—you cannot get anything else out of it. On the other hand blue copper sulphate is not an element. By carrying out chemical changes on the crystals it is possible (although not easy) to obtain from them four different elements: copper, sulphur, hydrogen and oxygen.

A list of elements is given in the Data Section. From what is known about the elements, we are able to say that nowhere in the whole Universe are we likely to find any more than the ninety or so elements which can be found in the Earth or its atmosphere.

1.4
The elements

There is considerable variety in the elements which exist. The most obvious way in which they differ is in their physical states. Many are solids, some are gases and two are liquids at room temperature and pressure.

Another way of classifying elements is to divide them into **metals** and **non-metals** as in Table 1.2. All of the gases, one of the liquids (bromine) and a small number of solids are non-metals. All of the other solids and the other liquid (mercury) are metals.

TABLE 1.2. *Some common elements*

SOLIDS AT ROOM TEMPERATURE		LIQUIDS AT ROOM TEMPERATURE		GASES AT ROOM TEMPERATURE	
METALS	NON-METALS	METALS	NON-METALS	METALS	NON-METALS
aluminium calcium chromium copper gold iron lead magnesium platinum radium silver sodium tin uranium zinc	carbon phosphorus sulphur	mercury	bromine		chlorine fluorine helium hydrogen neon nitrogen oxygen

If you were to obtain a sample of a solid element from a material, how would you know whether it was a metal or a non-metal? One possible way of deciding would be simply to look at it and to handle the solid. Metals are usually shiny and most are silvery-grey in colour (copper and gold are obvious exceptions). They are generally hard and strong, and they feel cold to the touch because they easily conduct heat away from your hand. Metals can usually be beaten or rolled into strips or sheets, and drawn into wires.

Solid non-metallic elements are not as uniform in their appearances and natures as metals. They show a variety of colours. Sulphur is yellow, phosphorus is either red or light yellow and carbon is black or grey. They are usually hard and brittle (although some forms of carbon and phosphorus are not). They are poor conductors of electricity (except carbon), and they are poor conductors of heat, so they do not feel as cold to the touch as metals. Generally non-metals are less dense than metals.

The physical differences between metallic and non-metallic elements are summarised in Table 1.3.

TABLE 1.3. *The physical properties of metallic and solid non-metallic elements*

METALLIC ELEMENTS	SOLID NON-METALLIC ELEMENTS
1. Shiny, silvery-grey colour	1. Variously coloured
2. Generally hard and strong	2. No uniformity in hardness and strength
3. Generally high densities	3. Generally low densities
4. Good conductors of heat and electricity	4. Generally poor conductors of heat and electricity

All this table does is to show the general properties of the two groups of elements. There are exceptions to most of these properties. A more definite way of deciding if an element is metallic or non-metallic is to examine the chemical properties of the elements. This is done in 11.19.

1.5 Putting elements together

Iron and sulphur are two solid elements and the effects of putting them together in different ways are easily studied. Iron, when alone, is attracted to a magnet and goes rusty when exposed to moist air. When dilute sulphuric acid is added to it, bubbles of gas are steadily given off and the gas can be shown to be hydrogen.

If powdered sulphur is stirred with iron filings, so that the two are thoroughly mixed, the iron filings in the mixture will still be attracted to a magnet, will still go rusty if the mixture is exposed to moist air and will still react with dilute sulphuric acid, causing hydrogen to be given off. The presence of the sulphur in the mixture has not changed the behaviour, or properties, of the iron.

If the mixture of iron filings and sulphur is heated it glows red-hot, and the glow persists for a while even when the flame is removed. This shows that heat is being given out by the elements and suggests that they are reacting with each other. The iron becomes joined to, or combined with, the sulphur to form a black **compound** called iron(II) sulphide.

If there was sufficient sulphur in the mixture to combine with all the iron, the product would not be attracted to a magnet and would show no sign of rusting when exposed to moist air. If dilute sulphuric acid is added to the black solid a gas is given off, but this time, instead of hydrogen, it is a foul-smelling gas called hydrogen sulphide. The properties of the iron, when it is joined to (combined with) the sulphur, are quite different to those which it shows when alone or when simply mixed with the sulphur.

Every substance in this world which contains more than one element is either a mixture of elements or a compound in which the elements have been joined together. In a mixture the elements are able to show the same properties as they do when alone, but a compound has its own properties which are usually different from those of the elements in it.

Air is a **mixture** of gases, the most abundant ones being oxygen and nitrogen and it will allow things to burn in it and animals to breathe in it, just as pure oxygen will. These processes are slower in air than in pure oxygen since the oxygen in air is diluted by the nitrogen and not so much of it is available in one place as would be the case in pure oxygen.

Nitrogen dioxide is a **compound** of nitrogen and oxygen and is a brown gas which is acidic (air is not acidic), dissolves readily in water (not much of air dissolves in water) and will allow only very hot substances to burn in it. The properties of oxygen have been much changed by combining it with nitrogen.

Another difference between a mixture and a compound becomes obvious when we see how the two things can be separated into the elements. Iron can be extracted from a mixture of iron and sulphur simply by holding a magnet near it or by dissolving away the sulphur in a liquid called trichloroethane. If the iron and sulphur have combined to form iron(II) sulphide, the extraction of the iron is not so easy. A magnet will not attract the iron in it nor will trichloroethane dissolve the sulphur from it.

Usually the separation of mixtures is quite easy; the separation of a compound into its elements is much harder. The separation of a mixture usually involves physical changes, while chemical changes have to be used with compounds.

1.6
What is a pure substance?

A pure substance is one single chemical element or compound. For example, to a chemist pure water consists solely of the compound water and nothing else. A swimming bath supervisor may consider that the water which is put into the swimming pool is pure, but it is pure only in the sense that all the bacteria have been removed and it is safe to swim in. Swimming bath water contains many other chemical substances dissolved in it, including the chlorine which has been added to kill the bacteria, and therefore it is far from chemically pure.

When a substance is pure it has a constant melting point and boiling point. An impurity lowers the melting point and raises the boiling point of a substance. Pure water boils at $100\,^{\circ}\mathrm{C}$ and freezes at $0\,^{\circ}\mathrm{C}$ but swimming bath water will boil at a temperature slightly above $100\,^{\circ}\mathrm{C}$ and freeze at a temperature slightly below $0\,^{\circ}\mathrm{C}$.

An indication as to whether or not a sample of a substance is pure can be obtained by finding its melting point or boiling point and comparing them to the accepted values for the pure substance. However, for a compound, the most reliable way of deciding if it is pure is to analyse it. All pure samples of any particular compound always consists of the same elements combined together in the same proportions by mass. The constant composition by mass of compounds is discussed in more detail in 4.7.

1.7
Summary

1. It is the chemist's job to find how to make useful substances from raw materials which are found in or around the Earth. To do this he must first find what the raw materials are made of.
2. In order to find what substances are made of, they must be changed. The changes are of two sorts—physical changes and chemical changes.
3. All substances in our Universe consist of one or more of about ninety substances which cannot be broken down into anything simpler. These substances are called elements.
4. Elements can be roughly classified (according to their physical properties) as metals or non-metals.
5. Elements can be put together in two ways. They can either be simply mixed or they can be made to combine together to give a compound.
6. A pure substance consists of one single chemical element or compound.

Chapter 2

How can pure substances be obtained from mixtures?

Is green ink a pure substance or a mixture?

Green writing ink could consist of a pure green liquid or it could consist of a solution of a dye in a solvent. If the ink is a pure liquid and it is heated, it will all evaporate and will leave behind no residue. If, however, it is a solution of a dye, evaporation will leave behind the dye.

You will need a beaker and a watchglass of such a size that it will rest on top of the beaker.

Pour water into a beaker up to a depth of about 4 cm. Put about 2 cm³ of green ink on a watchglass and then place the watchglass on top of the beaker.
  Heat the beaker with a Bunsen burner, gently at first and then more strongly. The water in the beaker will boil and the steam will heat the watchglass, evaporating the ink. Continue heating until there is no further change on the watchglass.

Questions

1 Is there a residue on the watchglass?
2 Is ink a single substance?

What is the liquid in green ink?

If a solution of a solid in a liquid is heated, the liquid boils and evaporates and, if the vapour is led away and cooled, it condenses to reform the liquid. The solid remains behind after the heating. This process is called distillation.
  If a thermometer is placed in the vapour above the boiling liquid, it will record the boiling point of the liquid and this can be used to identify the liquid.

You will need the apparatus shown in Fig. 2.1.

Pour green ink into a conical flask to depth of about 2 cm. To make sure that the liquid boils evenly, add to it two or three anti-bumping granules or a very small quantity of pumice powder.
  Set up the apparatus and heat the flask, gently at first and then more strongly, until the liquid boils. Do not let the ink froth too far up the flask.
  As the ink is boiling note the temperature shown by the thermometer. Continue boiling until you have collected liquid in the test-tube to a depth of 2 to 3 cm.

7

Fig. 2.1

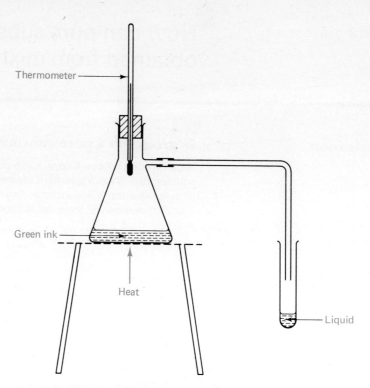

Thermometer

Green ink

Heat

Liquid

Questions

1 What is the colour of the liquid collected in the test-tube?
2 What is the boiling point of the liquid?
3 What is the liquid in the ink?

Investigation 2.3

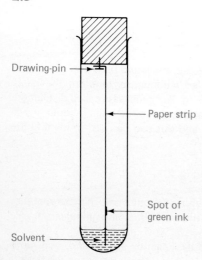

Drawing-pin

Paper strip

Spot of green ink

Solvent

Fig. 2.2

Questions

Does green ink contain more than one dye?

Chromatography is a method of separating two or more dyes in a solution, involving washing the dyes at different rates along a piece of paper with a suitable solvent.

You will need a strip of chromatography paper which is about 3 cm longer than a boiling-tube and about $1\frac{1}{2}$ cm wide, a cork for the boiling-tube, a drawing pin, a fine glass tube (a melting point tube is ideal for this) and a supply of a solvent containing butanol, ethanol and 2M ammonia solution, mixed in the proportion by volume of 3:1:1.

Fold over the strip about $1\frac{1}{2}$ cm from one end and pin this flap to the bottom of the cork. Now, holding the cork and paper strip by the side of the boiling-tube, cut the strip so that it is just long enough to touch the bottom of the tube when the cork is in the tube.
    Dip the fine glass tube into the green ink and touch it on to the paper about $2\frac{1}{2}$ cm from the bottom of the strip to get a spot no more than $\frac{1}{2}$ cm across. Add the solvent to the boiling-tube to a depth of about 1 cm and carefully lower the strip into the boiling-tube. The apparatus will now look like that in Fig. 2.2. Let the tube stand undisturbed until the liquid has risen almost to the top of the paper.

1 How many dyes are there in green writing ink?

8

2 If you were given another sample of green ink, how would you find if it was the same sort as you have been investigating here?

3 A spot of an orange dye and a spot of a green dye were added to a single strip of paper and then a solvent is allowed to run through the paper. The result then looks like Fig. 2.3. What does this tell you about the composition of the two dyes?

Fig. 2.3

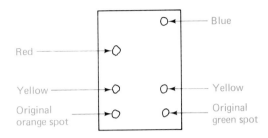

Blue

Red

Yellow — Yellow

Original orange spot — Original green spot

What happens when a hot saturated solution is cooled?

In this experiment a plastic teaspoon makes a useful measure of the amount of solid used to make a solution.

Using a measuring cylinder, transfer 70 cm³ of water to a beaker and add potassium nitrate, a spoonful at a time, stirring the whole time with a glass rod and not adding another portion until the previous one has dissolved. Continue until a portion of the solid will not dissolve, even after prolonged stirring.

Now warm and stir the mixture until the remaining solid dissolves and then add five more spoonfuls of potassium nitrate. Continue to warm and stir until this solid also dissolves.

Pour portions of this solution into two boiling-tubes and set one of them on one side to cool slowly. Cool the second quickly under a running cold water tap.

Questions

1 What sort of solution has been formed when no more potassium nitrate will dissolve in the water?

2 What happens to the solubility of potassium nitrate as the temperature of the water rises?

3 Why, therefore, does a hot concentrated solution of potassium nitrate deposit crystals when it is cooled?

4 Examine the crystals formed in the two boiling-tubes. How are the two samples similar? How are they different?

How can a small quantity of a soluble impurity be removed from soluble potassium nitrate?

The impurity, mixed with the white potassium nitrate, is blue copper(II) sulphate. Although both substances are soluble in water, there could be a big enough difference in the quantities of each substance present in the mixture to allow us to separate the two substances by crystallising a solution of the mixture. We would then be separating them by fractional crystallisation.

9

Put the mixture of solids into a boiling-tube to a depth of about 2 cm and then add about 20 cm³ of water. Warm the boiling-tube in a beaker of water on a tripod and gauze over a Bunsen burner, stirring the mixture with a glass rod or shaking the tube until all the solid dissolves.

Cool the boiling-tube in a beaker of cold water and, when it is at room temperature, filter off the crystals which have formed, collecting the filtrate in another boiling-tube. Wash the crystals in the filter paper with a small volume of cold water. This must be done carefully—if you do not use enough, you will not be able to see the true colour of the crystals, and, if you use too much, the crystals will all dissolve.

When the washings have drained through the filter paper, examine carefully the crystals which remain.

Questions

1   What colour and shape are the crystals in the filter paper?
2   What is the substance whose crystals are in the filter paper?
3   What colour is the filtrate from which the crystals were separated?
4   What must the filtrate contain to give it this colour?
5   Why could you not have used this method to obtain some pure potassium nitrate if there had been much more copper(II) sulphate than potassium nitrate in the original mixture?

2.6 Solutions

Most natural substances are not chemically pure (1.6); that is, they consist of a mixture of two or more components. An everyday problem for the chemist is that of separating mixtures so that their individual components can be studied, or made use of, separately. Much of this chapter is concerned with the particular ways by which pure substances can be obtained from solutions.

A **solution** is a special type of mixture. If salt is dissolved in water to give salt solution and a drop of the solution is put on a glass slide under a microscope, it is impossible to see any difference between the salt and the water. If, however, a mixture of salt and sand is examined under the microscope, the crystals of salt and the grains of sand are clearly seen and it is easy to say which is which. A solution is a mixture in which it is impossible to see any difference between the constituents, even under close examination.

Usually a solution is made by dissolving a solid in a liquid, but whisky is a solution of, amongst other things, a liquid (alcohol) in another liquid (water). Brass is a solid solution where the two constituents are both solids and where, in the mixture, one cannot be distinguished from the other. The constituent of the solution which does the dissolving (e.g. the water in sea water) is called the **solvent**, while the substance which is dissolved is called the **solute**. When it is difficult to tell which component has done the dissolving, the one present to the larger extent is called the solvent and the one present to the smaller extent is then the solute. In brass, which contains 62% of copper and 38% of zinc, the copper is the solvent and the zinc the solute.

The most common solutions we meet are those in which water is the solvent. Because of the elements in it and the way they are joined together, water has the power to dissolve many substances—solids, other liquids and even gases. For example, fish breathe in the air which is dissolved in the water in which they are swimming. Although a very large proportion of the surface of the Earth is covered by water, it is impossible to find any natural water which is chemically pure, simply because the water must have come into contact with something which will dissolve in it. We are able to put this ability of water to dissolve many substances to countless everyday uses—you have only to remember the difficulties which you meet at home when the water supply is cut off, even for a short time, to realise how important to us is the solvent action of water.

2.7
Filtration and evaporation

As we have mentioned previously, mixtures are usually separated into their constituents by methods involving physical changes and, since solutions are mixtures, they can be separated using physical changes or methods, rather than by chemical changes or reactions. The most common form of solution is the one where a solid has been dissolved in a liquid. The solid usually has a much higher boiling point than the liquid and, if the solution is heated, Fig. 2.4, the liquid will evaporate and boil at much lower temperatures than the solid. The vapour will then escape from the solution and eventually all the liquid evaporates, leaving behind the solid. In this case the solute has been separated from the solution by **evaporation**.

Fig. 2.4 Evaporation of a solution

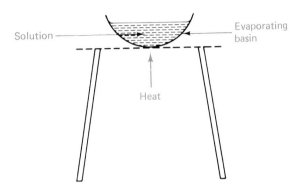

Evaporation is an important stage in the separation of pure salt from rock salt. The impure salt which can be found in the form of solid rock in certain parts of the world, contains earthy impurities which do not dissolve in water and are said to be **insoluble** in water. The first stage of the separation is to add water to the impure salt. The salt dissolves in the water and the resulting mixture is dirty salt solution. The insoluble impurities are removed by the process of **filtration** which, in the laboratory, involves pouring the mixture into a filter paper in a filter funnel as in Fig. 2.5.

Fig. 2.5 Filtration

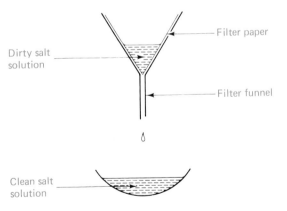

The filter paper has lots of fine holes through it and acts like a sieve, allowing the salt solution to pass through but not the insoluble impurities. The evaporating basin in which the clean salt solution has been collected is then heated. The water evaporates off leaving the salt behind.

This technique of separating two solids by dissolving one of the solids in a suitable solvent, filtering and then evaporating is used in the extraction of sugar from sugar beet, where again the solvent is water. There are other separations where the most suitable solvent is not water. For example, if you wish to prepare pure crystals of naphthalene from an impure sample, you could dissolve the naphthalene in ethanol and then filter off the impurities before evaporating the ethanol to recover the pure naphthalene.

Obtaining the colouring matters from plants uses a similar process. Propanone is poured on to the plants, which are then crushed so that the dyes dissolve in the propanone. The remainder of the plant material is filtered off and the solution is evaporated to obtain the dyes.

Fig. 2.6 (a) (above) Sugar beet arriving at the factory

(b) (above right) Evaporators used for removing some of the water from the sugar solution

(c) (right) Bags of sugar leaving the production line. (All three photos courtesy British Sugar Corporation)

2.8 Distillation

If we want to extract the solvent from the solution, the vapour of the boiling liquid can be led away and cooled so that it **condenses**. This can be done by heating the solution in a flask which is connected to a Liebig condenser, Fig. 2.7. The condenser has a tube surrounded by a wider tube, through which passes a continuous flow of cold water.

The vapour passes into the central tube and is cooled and condensed. The solute remains in the flask and the solvent which drips from the end of the condenser is usually pure. The process of evaporating a liquid and then cooling the vapour to condense it to give the liquid again is called **distillation**.

In distillation the solution is boiled to evaporate the solvent. A thermometer, placed in the vapour above the boiling liquid, gives the boiling point of the solvent. The boiling point can be used to identify the solvent or sometimes to check whether it is pure.

Fig. 2.7 Distillation of a solution

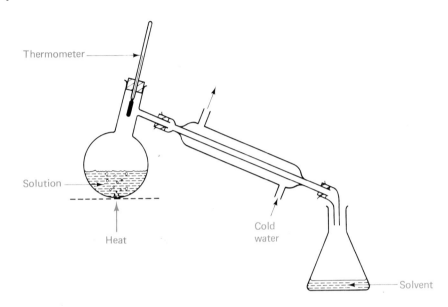

2.9 Fractional distillation

The process of distillation can be used to separate mixtures of liquids where the temperatures at which the two liquids boil are sufficiently different. If such a mixture is heated, the liquid with the lower boiling point will tend to evaporate first and that with the higher boiling point should boil off after the other liquid has all gone. In practice it is not as simple as this. A single distillation is rarely sufficient to separate the liquids and in some cases complete separation is impossible.

Separation of liquids by distillation is particularly important in that branch of the chemical industry which is concerned with converting crude oil into useful products. Crude oil is a mixture of many different liquids and many distillations would be required to separate completely even a small number of the liquids. Instead of attempting complete separations, the oil is distilled to produce a series of **fractions** and the process by which this is done is called **fractional distillation**. Each fraction contains liquids which boil over a certain range of temperatures. This can be demonstrated in the laboratory by using the apparatus in Fig. 2.8. The oil is poured on to fibre called rocksil, which allows the oil to be heated more evenly and reduces the risk of fire.

As in the previous distillation, a thermometer is placed in the vapour to record the boiling point of the liquid whose vapour is passing over into the condenser. The oil is heated, gently at first and then more strongly, and the test-tube which collects the condensed liquid (called the distillate) is replaced by another each time the temperature reaches the end of one of the chosen boiling ranges. For example, the first test-tube will contain the fraction which boils between room temperature and 70 °C and then the second fraction will boil between 70 °C and 120 °C and the third between 120 °C and 170 °C and so on.

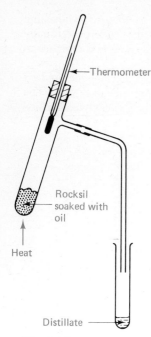

Fig. 2.8 Fractional
distillation of crude oil

Fig. 2.9 Copper pot-stills
which are used to produce
whisky in a distillery in
Ross-shire, Scotland. The
fermented liquid is boiled
in the still and the spirit
vapour passes out at the
top and is condensed to
form whisky. (Courtesy
The Distillers Company
Limited)

2.10 Saturated solution

Another important example of fractional distillation is the production of spirits such as whisky and gin, from a less concentrated solution of alcohol which has been produced by fermentation (32.5). The mixture of alcohol and water, plus all the other substances present which give each drink its special flavour, is distilled, but again only a partial separation is required. It is possible to use fractional distillation to produce an alcohol-water mixture containing 96% of alcohol, but this, as well as having lost the original flavour of the drink, is now very poisonous. The spirit is therefore usually concentrated until it contains about 40% to 50% by mass of alcohol.

Both of the above examples of fractional distillation are discussed in more detail later in the book, 31.5 and 32.5.

In the case of a solution of a solid in a liquid, it is not possible to go on dissolving the solute in the solvent indefinitely. If the solvent is at a constant temperature, only a certain mass of solute will dissolve in a given mass of solvent and the solution which is formed when no more solute will dissolve is called a **saturated solution**.

Often, the substances which we refer to as 'insoluble' do dissolve very slightly but only a minute trace is required to produce a saturated solution. For example, 1 dm³ of water will dissolve only 0.0014 g of silver chloride at room temperature. This quantity is so small that, for most purposes, silver chloride is regarded as being insoluble in water.

When you state whether a substance is soluble or insoluble, it is important that you also name the solvent being used because some substances are insoluble in one solvent but soluble in another. Sulphur is regarded as being insoluble in water but it is very soluble in another colourless liquid, trichloroethane.

2.11
Miscible and immiscible liquids

For mixtures of liquids the terms **miscible** and **immiscible** are frequently used, rather than soluble and insoluble. If some alcohol is added to water, the two form one liquid and therefore these liquids are said to be miscible. The process of fractional distillation will be required to separate them.

If oil or petrol is added to water, they will form a mixture which, no matter how long it is shaken, will always separate into two layers when allowed to stand. Water and oil are said to be immiscible and can be separated easily by putting them into a separating funnel, Fig. 2.10, which has a tap in its stem. The more dense liquid will settle to the bottom of the funnel and, by opening the tap, it can be run off into one container. When all the lower layer has been run off, the tap is closed and the first container removed. The remaining liquid can then be run into a second container.

2.12
Solubility and solubility curves

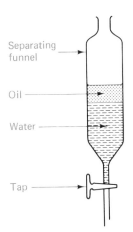

Separating funnel

Oil

Water

Tap

Fig. 2.10 Separation of immiscible liquids

If the mass of the solvent in a saturated solution is 100 g, the mass of the solute required to saturate it is called the **solubility** of the solute at that temperature. Usually the solubility of a solute in a solvent increases as the temperature is raised and this can be shown graphically by plotting a solubility curve, an example of which is shown in Fig. 2.11.

This curve shows that the solubility of potassium nitrate in water at 60 °C is 110 g per 100 g, while at 30 °C it is only 45 g per 100 g. If we had, therefore, a saturated solution of potassium nitrate at 60 °C, made up in 100 g of water, 110 g of the solute would be dissolved, but, if the solution was then cooled to 30 °C, only 45 g could remain in solution. The other 65 g of potassium nitrate has to come out of solution and this would be in the form of crystals. The more slowly the solution is cooled, the larger the crystals would be. One way of preparing crystals, therefore, is to cool a hot saturated solution of the substance in a suitable solvent.

Sometimes it is necessary to separate two solutes in a solution. For example, you may have some sodium nitrate, contaminated with a small amount of salt. If you prepare a hot solution which is saturated as far as the sodium nitrate is concerned, it is unlikely to be saturated as far as the salt is concerned. If this solution is cooled, crystals of sodium nitrate will form, while the sodium chloride remains in solution. This is the basis of the technique known as **recrystallisation** which is used for the purification of many solid compounds.

If the two substances in a mixture are present in comparable amounts, they can still be separated by forming crystals of one of them, but this time it is more difficult to do. Sodium chlorate is a compound which is used as a weedkiller and in matches and is made by reacting a solution of sodium hydroxide with the gas, chlorine. The reaction, however, also forms sodium chloride as well as the sodium chlorate. A hot concentrated solution of the mixture would be likely to be saturated as far as both substances are concerned, and, if it was cooled, crystals of both substances would be formed. If, however, the solubility curves of the two compounds are examined, Fig. 2.12, the problem of the separation of the sodium chlorate can be solved.

The curves show that at a temperature of about 100 °C sodium chloride is much less soluble than sodium chlorate and, if a solution containing the two is crystallised at about this temperature, it will be sodium chloride which forms the crystals. These can be removed and eventually there will be more sodium chlorate in the solution than sodium chloride. If the solution is now cooled to 20 °C (room temperature), the crystals which form will be those of sodium chlorate since it is now much more concentrated. The method used in this case to separate the two solutes is called **fractional crystallisation** and can be employed when one of the two has a significantly greater solubility than the other at a particular temperature.

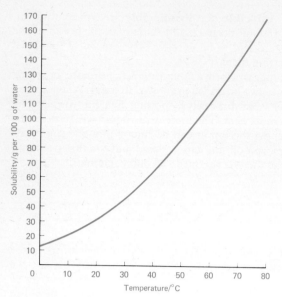

Fig. 2.11 Solubility curve for potassium nitrate

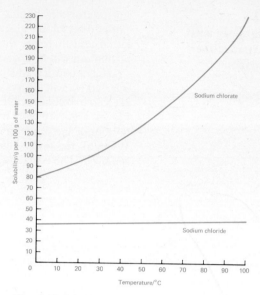

Fig. 2.12 Solubility curves of sodium chlorate and sodium chloride

2.13 Chromatography

Another method of separating two solutes in a solution uses a technique called **chromatography**. This relies on the fact that the components in a mixture can be washed at different speeds through a substance like paper or powdered chalk with a suitable solvent.

It can easily be demonstrated using a solution of a green substance called screened methyl orange. A drop of the solution is placed on the centre of a piece of filter paper and allowed to soak in. The spot will have a blue ring around it, showing that separation has started, but it can be improved by adding drops of water to the centre of the spot with a teat pipette, allowing each drop to soak in before the next one is added. When a few drops of water have been added, the paper will look like that in Fig. 2.13.

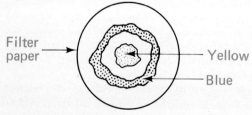

Filter paper — Yellow — Blue

Fig. 2.13 Separation of screened methyl orange on filter paper

Fig. 2.14 Separation of screened methyl orange on a paper strip

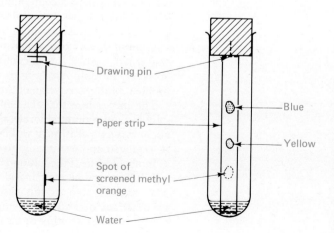

Drawing pin

Paper strip

Spot of screened methyl orange

Water

Blue

Yellow

16

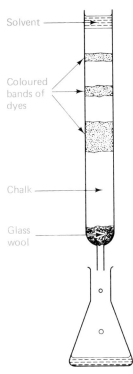

Solvent

Coloured
bands of
dyes

Chalk

Glass
wool

*Fig. 2.15 Separation of
dyes by column
chromatography*

A better separation can be obtained with the apparatus shown in Fig. 2.14. As the water is drawn upwards through the paper, the dyes separate as before to give two separate spots on the paper. The separated components on the piece of paper is called a **chromatogram**.

The separation of a mixture by paper chromatography depends on the components of the mixture having different tendencies to be adsorbed (stick to the surface) on to the paper and on the different solubilities of the components in the solvent. If a component tends either to be strongly adsorbed on to the paper or to be not so soluble in the solvent, it will move along the paper slowly (e.g. the yellow dye in methyl orange). On the other hand, if a component tends either not to be adsorbed so well on to the paper or to be rather soluble in the solvent, it will move along the paper more rapidly (e.g. the blue dye in methyl orange).

While paper chromatography is useful for determining whether substances are mixtures or even identifying the substances in a mixture, it is not very suitable for actually obtaining samples of the components of the mixture. To do this, column chromatography has to be used. This was the first form of the technique to be used, being first described in 1916 by a Russian botanist called Tswett who was interested in separating the coloured substances in the leaves of plants.

He did this by dissolving out the dyes from the plants with a suitable solvent and then pouring the solution into the top of a glass tube, packed with powdered chalk, Fig. 2.15. The dyes were adsorbed into the first few centimetres of chalk and then more solvent was poured through the column. The dyes with considerable liking (affinity) for the chalk, or with the lower solubility in the solvent, stayed near the top of the column, while those with lower affinity for the chalk, or higher solubility in the solvent, were washed quickly through the chalk, so that coloured bands were formed in the column. If enough solvent is added to the top of the column, each coloured band can be washed out of the column in turn into separate containers.

2.14
Gas-liquid
chromatography

A more recent development of the technique of chromatography uses a gas such as nitrogen instead of a solvent. Instead of the paper or powder it uses a powder which has been coated with a very thin layer of a liquid. The liquid-coated powder is put into a long glass or metal tube (often 2 m long but coiled to save space). A small quantity of the mixture to be analysed is introduced into one end of the tube, while the gas passes through the tube.

The different components of the mixture, depending on their solubilities in the layer of liquid, take different times to pass through the tube. A detector records when each component passes out of the tube. By comparing the time for each component to pass through the tube to the times taken by known pure substances, the components of the mixture can be identified.

This technique, which is called **gas-liquid chromatography**, is particularly useful as a very small sample of the mixture (as little as 0·001 g) can be analysed (its components identified) in a few minutes. When the technique was introduced, chemists were immediately able to carry out analyses which had been previously either not possible or very time-consuming. Gas-liquid chromatography is now used, for example, in investigations in hospitals and police laboratories where a minute trace of blood can be analysed, in the food industry where the chemist can rapidly check the purity of the ingredients and in the oil industry for controlling the composition of products such as petrol.

2.15 Summary

1. A solution is a special type of mixture in which it is impossible to see the two parts even on very close inspection.
2. In a solution of a solid in a liquid, the liquid is called the solvent and the solid is called the solute.
3. The solute and solvent are separated by physical methods. The solute could be obtained by evaporation and the solvent by distillation.
4. A solid which is soluble in a particular solvent can be separated from a solid which is not soluble in the solvent by adding the mixture to the solvent, filtering off the insoluble solid and evaporating off the solvent.
5. Mixtures of miscible liquids can be separated by fractional distillation, whereas a separating funnel can be used for immiscible liquids.
6. A solution in which no more solute will dissolve is called a saturated solution. More solute can usually be made to dissolve by raising the temperature of the solution.
7. Two solutes in one solution can be separated by fractional crystallisation or by chromatography.

Fig. 2.16 A gas chromatograph which can be used for analysing very small samples of substances such as blood. The main parts of the chromatograph have been labelled. (Photo: J. Olive)

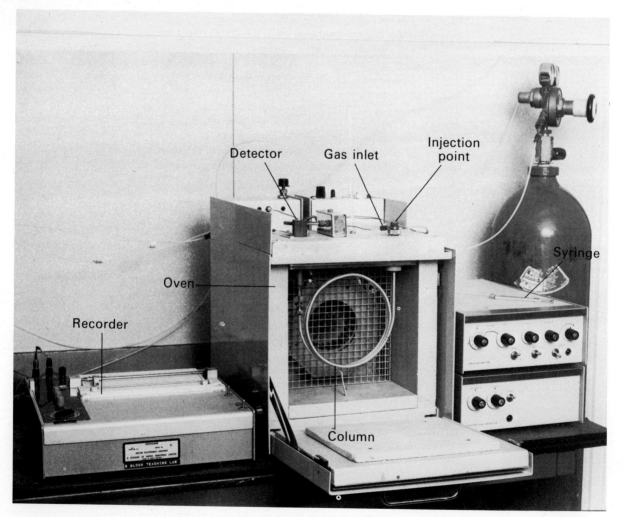

Chapter 3

Particles of substances

**Investigation
3.1**

How big are the particles making up a crystal?

You will need a large plastic container such as an aquarium or bucket, and a 250 cm³ measuring cylinder.

From a bottle of potassium manganate(VII), pick out a small crystal and add it to 250 cm³ of water in a beaker. Stir with a glass rod until the crystal is dissolved.

Now pour the solution into a large plastic container.

Add more water from a measuring cylinder, 250 cm³ at a time, stirring after each addition and counting the number of additions made, until the pink colour of the solution is only just visible. From the number of additions made, calculate the volume of the solution when it reaches this stage.

We shall assume that the pink colour can just be seen when there is one particle in one drop of the solution and we have to find the number of drops in the solution where one drop contains one particle. We now know the volume of this solution in cm³ and therefore we need to find the number of drops in 1 cm³.

Fill up a burette with water and clamp it vertically in a burette stand. Run water from the burette until the surface of the water in it is on one of the cm³ graduation marks.

Set the tap of the burette so that the water is coming out of it drop by drop and count the number of drops which are delivered as the level falls to the next cm³ graduation mark. This is the number of drops in 1 cm³.

Now calculate the number of drops in the solution of potassium manganate(VII); this will be the approximate number of particles in one crystal of the substance.

Questions

1 Approximately how many particles were there in the crystal?
2 Remembering the size of the crystal you started with, what is your estimate in mm of how big a particle in the crystal is?

**Investigation
3.2**

What happens when a soluble solid is left in contact with water?

Into the bottom of a boiling-tube, put either the largest crystal of copper(II) sulphate you can find in the bottle, or a layer of small crystals. Carefully, using a beaker, pour water into the tube so that it forms a layer about 7 cm deep on top of the solid. Leave the boiling-tube undisturbed for a few days.

Questions

1 What starts to happen to the solid in the boiling-tube?
2 What do you see immediately above the solid in the boiling-tube?
3 What happens to this as time goes by?
4 What does this suggest about the particles in the solution?
5 Would what you have seen take place slower or quicker if the room was warmer?

What is the length of a particle of stearic acid (octadecanoic acid)?

This measurement relies on the fact that, if you pour a little of a solution of stearic acid on to the surface of water, it will form a layer on the water which is just one particle thick.

The volume of the layer will be equal to its surface area multiplied by its thickness (which is the length of one particle). If we know the volume of stearic acid in the layer and can measure its surface area, we can calculate the thickness of the layer and hence the length of the particle.

You will need a large plastic tray, a very fine pipette, some powdered chalk, a ruler, a 10 cm³ measuring cylinder and a solution of stearic acid containing 0·1 g in 1000 cm³ of light petroleum.

Add water to the tray to give a depth of about 3 cm. Sprinkle powdered chalk on the surface of the water.

Using the fine pipette, add two drops of the stearic acid solution to the water near the centre of the tray. If the layer produced is only small, add a further drop of the solution to the centre of it. Measure the diameter of the layer to the nearest cm, using your ruler.

It is now necessary to find how many of the drops, delivered by the pipette, make up 1 cm³. Fill up the pipette with water and then add the water drop by drop to the small measuring cylinder, counting the number of drops required to raise the level of water from one cm³ graduation mark on the cylinder to the next one.

A specimen set of results and the calculation of the length of the particle of stearic acid are given later in this chapter (3.8). You can use these, if you wish, to help in answering the following questions.

Questions

1   What is the surface area of the layer? (You can assume that it is circular with the diameter you have measured and you should remember that the area of a circle is given by the formula πr^2.)
2   What volume of stearic acid is in 1000 cm³ of your solution? (Assume the density of stearic acid is 1 g cm⁻³.)
3   What volume of stearic acid is in 1 cm³ of your solution?
4   What volume of stearic acid is in 2 (or 3) drops of your solution?
5   What is the thickness of the layer? (Remember that the volume of the layer of stearic acid is equal to its area multiplied by its thickness.)
6   What, therefore, is the length of a particle?
7   How many of these particles, laid end to end, would be needed to make a length of 1 cm?

3.4 Evidence for the existence of particles

Most of the crystals in the bottle of potassium manganate(VII) (potassium permanganate) in your laboratory will be no more than 1 mm long and 0·5 mm wide. If you take one of these small crystals and dissolve it in, say, 100 cm³ of water, a deep purple solution is formed. What there was in the crystal has now been spread out through 100 cm³ of water and this is easiest to picture and understand if we think of the crystal as a collection of lots of bits of potassium manganate(VII) (it is easier to spread out sawdust than a block of wood). When the crystal is dissolved in water, the water gets between these bits, or **particles**, and so spreads them out. If we add more water to the solution, the colour becomes lighter, but it still persists. The extra water has simply caused the particles to be spread out further, there now being more water between them than before.

If our idea about the crystal consisting of particles is correct, then how big are these particles? A very rough idea can be obtained by pouring the solution of potassium manganate(VII) into a large tank (a plastic aquarium will do nicely for this) and then adding measured volumes of water until, after stirring, the pink colour of the solution can only just be seen. Now the particles must be really spread out, yet there must be at least one particle in each drop of the solution for its colour still to be visible. Usually the colour given to its solution by a small crystal of potassium manganate(VII) will be almost invisible when the crystal is dissolved in about 15 dm³ of water (15 000 cm³). Each cm³ of water consists of about 20 drops (you can use a burette to check this, if you wish) and therefore, if there is at least one potassium manganate(VII) particle in each

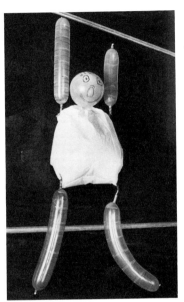

Fig. 3.1 (a) Liquids
have no definite shape, so
although all these pints of
beer have the same volume,
they take the shapes of the
different containers.
(Photo: Russell
Edwards, B.Sc)

(b) This Balloon man
shows that gases take on the
shape and volume of the
container in which they are
placed. (Photo: Russell
Edwards B.Sc)

(c) This sculpture made
from old cog-wheels shows
that solids have fixed shapes
as well as fixed volumes (at
constant temperature).
(Photo: Russell
Edwards, B.Sc)

drop, there must be at least 300 000 particles in the 15 dm³ of solution. All these particles must have come from the single crystal and therefore there must have been at least 300 000 particles in that crystal, which was no more than 1 mm long. Clearly the particles must be very small indeed.

In the case of potassium manganate(VII) the fact that the compound could be 'spread out' by dissolving it in water, suggests that the crystal consists of particles. The same idea can be applied to other crystalline solids which dissolve in water.

Unlike solids, liquids do not have fixed shapes. If you spill a glass of water on a flat table top, the water will change its shape and form a thin layer on the surface of the table. Again, as with a solid, this is best understood if it is assumed that water is made up of particles. The particles are able to move in all directions so that when the water is spilt, it changes shape, but the total volume of the liquid remains constant. Similarly, if a liquid is poured from a glass into a cup, it will change shape but its volume will not change. In contrast, a piece of solid will keep both its volume and its shape when placed in a different container.

Gases are similar to liquids in that they change shape and are therefore likely to consist of particles, but gases do behave differently in some ways and these differences can tell us more about the particles. Ethoxyethane is the liquid often known as ether (it was one of the first anaesthetics to be used) and it evaporates easily to form its vapour which is a gas. If you open a bottle of ethoxyethane in one corner of your laboratory, it won't be long before someone, sitting in another corner, notices the smell of the vapour. This shows that some of the gas has travelled quickly from one corner of the room to the other and the particles must therefore be moving very quickly. The same sort of behaviour is shown by other gases, which leads us to conclude that all gases are likely to consist of particles which are constantly moving with high speeds. The smell of the gas can eventually be detected in all corners of the room and this shows that the particles of the gas are moving in all directions and are only restricted by the boundaries of the container—in this example the room. Thus there is a difference between a liquid and a gas. The total space taken up by a liquid does not alter when the liquid changes shape whereas, when a gas has been released and has filled a container, its volume will have increased to the volume of the container.

If the particles of a gas are moving with high speeds, they will have high energies, even if the particles are heavy. If a small volume of the deep red liquid, bromine, is poured into a gas jar and a second gas jar is then inverted over the first, as in Fig. 3.2, traces of reddish-brown bromine vapour soon appear in the upper jar, and, within a short time, there is as much vapour in the upper jar as in the lower. We now know that bromine particles have a relative mass of 160 (about five times the mass of those in air) and we might expect them to be held in the lower jar by the large force of gravity on them. Such are their energies at room temperature, however, that their speeds are still high enough to carry them against gravity into the upper jar. This quick spontaneous movement of a gas to fill up all the available space is called **diffusion**.

Fig. 3.2 Diffusion of bromine vapour

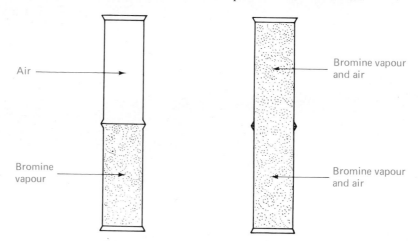

Air

Bromine vapour

Bromine vapour and air

Bromine vapour and air

The speeds of gas particles can be calculated. Oxygen particles, when the gas is at room temperature, move at about 1000 miles per hour, while for hydrogen, whose particles are lighter, the speed is about 3600 miles per hour. It is interesting to compare these speeds with that of a supersonic aeroplane like Concorde, which can fly at 1350 miles per hour.

The fact that the speeds of gas particles depend on their masses (when compared at the same temperature), can be shown by another diffusion experiment. When ammonia gas and the fumes from concentrated hydrochloric acid (hydrogen chloride) mix, they form dense white fumes which consist of tiny particles of a white solid called ammonium chloride. Fig. 3.3 shows a long glass tube with a plug of cotton-wool at each end, one soaked in concentrated ammonia solution and the other in concentrated hydrochloric acid. The two plugs must be put into the tube at the same time. A cork is put into each end of the tube, so that the plugs are completely enclosed in the apparatus. The ammonia solution gives ammonia particles which, being particles of a gas, tend to move away quickly. Because of the corks, the only way the gas can move is along the tube. The same applies to the hydrogen chloride particles at the other end of the tube. Where the two sorts of particles meet, white fumes of ammonium chloride form. The white fumes are seen to form further away from the starting point of the ammonia than the starting point of the hydrogen chloride, showing that the ammonia particles move more quickly than the hydrogen chloride particles. The mass of an ammonia particle is almost half that of a hydrogen chloride particle and so the investigation indicates that lighter particles move more quickly than heavier ones.

Fig. 3.3 Diffusion of ammonia and hydrogen chloride

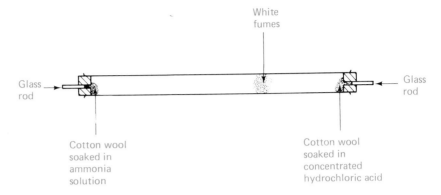

Further evidence about the speed of the particles depending on their masses is provided by an experiment, using the apparatus shown in Fig. 3.4. The walls of the porous pot have fine holes in them, so that gases can pass slowly through the pot. The U-tube contains a coloured liquid and is a manometer, used for detecting and measuring gas pressures. If the porous pot is held upside down, as in (a), and a beaker, also upside down and containing hydrogen (which is less dense than air), is held over it, the level of the coloured liquid in arm B of the manometer rises, showing that there is a greater pressure inside the porous pot than outside it. The particles of hydrogen are very light, while those in air, which was originally in the porous pot, are about fifteen times heavier. The hydrogen particles move much quicker than the particles in air and will therefore get into the porous pot faster than the air particles can get out. A few seconds after the beaker has been put over the porous pot, there will be more gas inside the pot than at the start and the pressure inside it will be greater. This extra pressure pushes down the liquid in arm A and up in arm B of the manometer.

Fig. 3.4 Diffusion of hydrogen and carbon dioxide

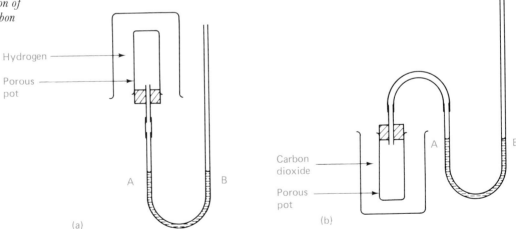

If the porous pot in Fig. 3.4 is the right way up (as in (b)) and a beaker of carbon dioxide, also the right way up, is held round it, the level of liquid in arm A of the manometer goes up. The carbon dioxide particles have masses about $1\frac{1}{2}$ times that of the air particles in the pot and therefore move more slowly. This time the air will diffuse out faster than the carbon dioxide gets in and there will be less gas and less pressure in

23

the pot than at the start. The atmosphere outside the apparatus will then push the liquid down in arm B of the manometer and up in arm A.

3.5 How close together are the particles?

One of the most obvious differences between a gas and the other two states of matter is a difference in density. The densities of a substance in the three states gives us an indication of how close together the particles of that substance are packed in each state. For example, the density of water is 1.00 g cm^{-3} at $4\,^{\circ}$C, the density of ice is almost the same, 0.97 g cm^{-3}, whereas the density of steam at $110\,^{\circ}$C is about 0.0006 g cm^{-3}. The particles in steam must therefore be widely separated whereas those in ice and water must be more closely packed.

3.6 The movement of particles and the effect of temperature

Gases

Another thing which decides how fast the particles of a gas move is its temperature. If the diffusion experiments with bromine and with ammonia and hydrogen chloride, shown in Fig. 3.2 and Fig. 3.3, are done on a hot day, the effects occur more rapidly than on a cold day. The particles of the gases move faster when the gases are warmer. At $0\,^{\circ}$C, the average speed of oxygen molecules is 461 metres per second (1021 miles per hour); at $25\,^{\circ}$C it has risen to 482 metres per second (1068 miles per hour) and at $100\,^{\circ}$C to 539 metres per second (1194 miles per hour).

Liquids

In 1827 Robert Brown, a botanist, while examining pollen grains in water under a microscope, noticed that the grains were not stationary but moved about in a zig-zag manner over small distances (Fig. 3.5). This behaviour is now called Brownian motion and can be observed with a number of materials such as chalk, magnesium oxide, graphite and even toothpaste when they are suspended in a liquid such as water. It provides direct evidence that the particles of a liquid are constantly moving. The grains of the solid move because they are hit by the particles of the liquid. The grains are being bombarded by these particles on all sides the whole time, but occasionally a grain will be hit by a lot more particles on one side than on the others and so it will move. This is why the movements are haphazard, it being impossible to predict that a particular particle will move at a particular time.

Fig. 3.5 Brownian motion

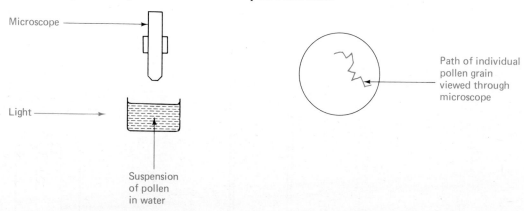

Just as the movement of gas particles will show itself by the diffusion of the gas, the movement of liquid particles can give rise to the diffusion of liquids. If some phenylamine (aniline) is poured into a test-tube and then, with great care, a little ethoxyethane (ether) is poured on top of this liquid, the two will form separate layers. If the test tube is now corked and left undisturbed for some time, the two layers gradually become mixed until a uniform mixture is obtained. As with gases, the diffusion of these liquids is faster at a higher temperature and we can therefore conclude that the particles move faster when the liquids are warmer.

24

Solids

We have now decided that the particles in a gas and in a liquid are constantly moving—do those in a solid move in the same way? Before we can answer this question, we must pause to reconsider the differences between solids and the other two states of matter. Solids are much stronger than liquids and gases. It takes a karate expert to drive his hand through a wooden plank, but anyone can drive their hand through air or water. Also, as previously mentioned, solids unlike liquids and gases do not readily change shape. These particular characteristics of solids indicate that the forces between the particles in a solid are able to hold the particles in fixed positions and prevent them from being pulled apart or from being able to move about from place to place within the solid. Nevertheless there is some movement. The particles do vibrate within the solid. There can be some stretches and contractions within the particle itself. As the solid becomes hotter, the speeds of these movements become greater and, in the case of the vibrations, not only will they become faster but they will take up more room. This is why the solid 'takes up more room', or expands, when it is heated.

On rare occasions we can see the diffusion of solid materials. A blackboard, on which something has been written in chalk, is hard to clean if the writing is left on the board over a school holiday. This is probably due to the diffusion of the chalk into the surface of the blackboard. If two different metals are bound tightly together and left for a long time, it is possible to detect traces of one metal in the surface of the other. Again the process of diffusion is evidence for some sort of movement of the particles in the two solids which enables particles of one to cross over to the other.

3.7 Changes of state

When a solid is heated, it usually, at a particular temperature, melts to form a liquid. As the heating of the solid takes place, the vibrations of the particles increase, but the forces between the particles are still strong enough to hold the particles together. Eventually, however, a point is reached where the vibrations become so great that the particles can no longer be held in position by the forces of attraction and they start to move, the solid being virtually shaken to pieces. When the particles are able to move around within the substance a liquid has been formed. The temperature at which the vibrations become just too great for the forces of attraction to hold the particles in position is the **melting point** of the solid.

If a liquid is heated, it will, at some temperature which depends on what the liquid is, begin to boil. As the liquid is heated, its particles rush about with increasing speed within the liquid. The liquid behaves as if it has, like rice pudding or custard, a skin

Fig. 3.6 The attractive forces which hold the particles of water on the surface together (surface tension) prevent these needles from sinking. (Photo: Russell Edwards, B.Sc)

over its surface (caused by the particles at the surface attracting each other—surface tension) and, if a particle is going to escape, it must break through the skin. If a rugby player is trying to get over his opponents' try line in a particular place and has in front of him enough defenders to stop him, his only chance is to run as fast as he can at the defence in the hope that his momentum might carry him through. In the same way, only when the particles of a liquid are moving fast enough can they break through the surface skin or 'defence' of the liquid and so escape. The **boiling point** of the liquid is the temperature at which all the particles are moving fast enough to escape in this way. When they have escaped, they will be further apart than in the liquid, with virtually no forces holding them together, and so will have formed a gas, which is the vapour of the substance.

When the vapour is cooled, its particles slow down, so that forces between them can start to pull them together to form the liquid again. When the liquid is cooled further, the slowing down continues and the forces can really take over to pull the particles into an orderly arrangement which is the solid.

3.8 How big are the particles?

We have already seen that the particles in a crystal are very small, but exactly how big are they? What is the mass of a particle and what is its length? Stearic acid (octadecanoic acid) is a substance whose compounds are found in beef fat and whose particles are comparatively heavy. As will be seen in Chapter 4, we now know that 6×10^{23} particles of this substance have a mass of 284 g, and therefore the mass of one particle is $4 \cdot 7 \times 10^{-22}$ g or $0 \cdot 000\,000\,000\,000\,000\,000\,000\,47$ g. If you remember that one gram is about 1/1000th of the mass of a bag of sugar, you will realise how very small is the mass of a stearic acid particle.

Stearic acid is one substance with which we can do quite a simple experiment to find an approximate value for the length of the particle. When a small volume of this substance is added to water in a tray, it will spread out as far as possible over the surface of the liquid forming a layer which, for substances such as stearic acid, will be just one particle thick. The volume of the layer will be equal to its surface area multiplied by its thickness (i.e. the length of one particle) (Fig. 3.7), and, if we know the volume of stearic acid added to the water and can determine the area of the layer, the length of a particle can be calculated.

Fig. 3.7 Layer of stearic acid

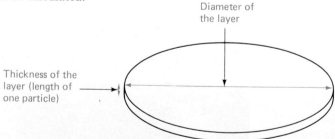

Diameter of the layer

Thickness of the layer (length of one particle)

Usually the experiment is done in a large tray or bowl with a fine powder on the surface of the water so that the boundary of the layer can be easily seen and its diameter measured. A very small volume of stearic acid, even less than a normal-sized drop, will give a big layer when spread out to one particle thick, so it is usually made into a solution in a volatile solvent called light petroleum. Light petroleum evaporates very quickly from the thin layer, leaving only stearic acid. A small number of drops of the solution will contain sufficient stearic acid to leave a reasonably-sized layer after the light petroleum has evaporated.

The layer is produced by adding the solution to the water with a dropping pipette and the volume of the solution added can be calculated, as described below, from the number of drops added.

The following are the sort of results which would be obtained in this experiment.

Mass of stearic acid in 1000 cm³ of solution $= 0\cdot1$ g
Density of stearic acid $= 0\cdot94$ g cm⁻³
Volume of stearic acid in 1000 cm³ of solution $= 0\cdot105$ cm³
This is very nearly equal to $0\cdot1$ cm³.

Two drops of this solution were found to be spread out into a circular layer with a diameter of 8 cm and therefore a radius of 4 cm.

1 cm³ of the solution of stearic acid was found to contain 20 drops.

Area of the layer $= \pi \times (\text{radius})^2 = \pi \times 4^2 = 50$ cm²

Volume of stearic acid making up the layer $= \dfrac{0\cdot1}{1000} \times \dfrac{2}{20}$

$= 10^{-5}$ cm³

Volume of the layer = area of the layer × thickness of the layer (Fig. 3.7)

Length of one particle = thickness of the layer $= \dfrac{\text{volume}}{\text{area}}$

$= \dfrac{10^{-5}}{50}$

$= 2 \times 10^{-7}$ cm

or 2 millionths of a millimetre.

It is hardly necessary to say that this is very small indeed. If similar experiments could be carried out with other substances, we should find that their particles are also about this size.

There is a lot of other evidence, in addition to that discussed in this chapter, which indicates that everything in the world is made up of very small particles. The particles which make up a particular substance might be atoms, molecules or ions. In later chapters you will find out which substances contain which type of particle and what the differences are between the three types of particles.

3.9 Summary

1. All substances, whether solids, liquids or gases, consist of very small and very light particles.
2. These particles are spread out when the substance dissolves in a solvent or evaporates. The spontaneous spreading out of the particles of a gas or a liquid is called diffusion.
3. The particles are constantly moving. In a gas the particles move from place to place and fill the container. In a liquid the particles move from place to place within the liquid. In a solid they vibrate about fixed positions.
4. The speed of movement depends on the mass of the particles and on the temperature of the substance. The lighter the particles, the faster do they move. The higher the temperature of the substance, the faster do the particles move.
5. The low densities of gases, compared to those of liquids and solids, indicate that the particles in a gas are widely spaced.
6. There are forces of attraction between the particles and the closer the particles, the more effective are these forces.
7. When a solid melts, the vibrations of the particles overcome the forces holding them together, and the particles become disordered.
8. When a liquid boils, the particles are moving with sufficient speed to break through the surface skin of the liquid and so escape and form a gas in which the particles are well separated.

27

Chapter 4

Chemical combination and amounts of substances

Investigation 4.1

Is there any change in the total mass when a chemical reaction occurs?

The following pairs of solutions are suitable for this experiment. When they are mixed there are obvious signs that chemical reactions have occurred. There is no need at this stage to know what the products of the reactions are.

A		B
copper(II) sulphate	and	sodium carbonate
lead(II) nitrate	and	sodium chloride
barium chloride	and	sodium sulphate

You will need a conical flask with a rubber bung to fit, and an ignition-tube with a small length of cotton attached to it.

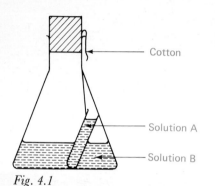

Fig. 4.1

Place a small quantity of solution B in the conical flask and half fill the ignition-tube with solution A. Suspend the ignition-tube inside the conical flask by means of the cotton as shown in Fig. 4.1 and put the bung in the flask.

Make sure the outside of the flask is dry and then weigh the flask and its contents.

Tilt the flask so that the solutions mix and shake it gently. Now weigh the flask and its contents again.

Questions

1 What evidence is there that a chemical reaction has occurred?
2 Why is a bung kept in the flask?
3 Is there any change in the total mass?

Investigation 4.2

Using some small objects to weigh some larger objects

Your teacher will provide a collection of objects (such as screws or coins of different sizes, or blocks of different sizes (made from wood or centicubes).

Weigh one of the largest objects. Then find out how many of the smallest objects need to be placed on the balance pan to give almost the same reading as obtained with the large object.

Question

1 If a small object is given a relative mass of 1 unit, what would be the relative mass of the large object in these units?

Repeat the procedure with the other objects and in each case find the relative mass of each object on this new scale.

Questions

2 If a similar procedure was adopted for atoms of elements, what would be the obvious element to choose as the standard with a relative mass of 1 unit?
3 Why would you choose this element?

Investigation 4.3	**Counting a set of objects by finding their total mass**

Your teacher will provide one opaque bag containing 10 small objects and three opaque bags containing different numbers of larger objects. Each of the large objects has approximately twice the mass of a small object.

Weigh each bag and estimate the numbers of objects in each of the three bags containing the larger objects.

Question

Bromine atoms have twice the mass of calcium atoms. 40 g of calcium contain approximately 6×10^{23} atoms of calcium. What mass of bromine would contain the same number of bromine atoms?

Investigation 4.4

What are the combining masses of copper and oxygen in black copper oxide?

The gas which is used for Bunsen burners will remove the oxygen from the copper oxide and leave a residue of copper.* The combining masses found from the experiment may be used to find the empirical formula of black copper oxide.

You will need the apparatus shown in Fig. 4.2. The hard glass test-tube has a small hole in the closed end.

Fig. 4.2 Removing oxygen from copper oxide

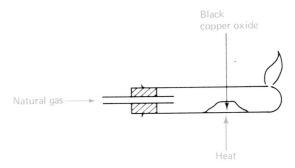

Weigh the tube empty and then with about two spatula measures of pure dry black copper oxide placed halfway down the tube as shown.

Clamp the tube and connect it to the gas supply, taking care not to disturb the copper oxide.

Pass a small stream of gas through the tube, and when you are sure that all the air has been swept from the tube (needs about 10 seconds), light the gas as it emerges from the small hole, taking care to keep your face well back from the hole. Adjust the gas pressure so that the flame is about 3 cm high.

Heat the test-tube in the region of the copper oxide with a small non-luminous flame until all of the copper oxide appears to have been changed to copper.

Stop heating the tube but keep the gas passing until the tube is cool. Then weigh the tube and its contents.

If you have time, connect the tube to the gas supply again and reheat the solid for 5 minutes. Allow the tube to cool with the gas passing over the solid

---

* The efficiency of natural gas in this reaction can be improved by first passing the gas through a U-tube containing cotton wool soaked in alcohol, but your teacher may prefer to demonstrate this.

and then reweigh the tube. Continue this procedure until two consecutive weighings are the same.

Make a table of your results and calculate the mass of copper and the mass of oxygen in your sample of copper oxide.

Look up the relative atomic masses of copper and oxygen in the Data Section and use them to calculate the number of moles of copper and oxygen in your sample. Convert these numbers to the simplest ratio and hence find the empirical formula for black copper oxide. The worked examples (4.11) will help you with this calculation.

Questions

1 Why is it necessary to keep passing the gas until the tube and its contents are cold?

2 Why will your result be more accurate if you keep repeating the procedure until a constant mass is obtained?

Investigation 4.5

What are the combining masses of magnesium and oxygen in magnesium oxide?

A known mass of magnesium is converted to magnesium oxide by carefully heating it in air, and then the mass of the magnesium oxide is found. The combining masses found from the experiment may be used to calculate the empirical formula of magnesium oxide. You will need a crucible and lid, a pair of tongs and a pipeclay triangle.

It is necessary to work with considerable care and skill in this experiment in order to obtain a reasonably accurate result.

Weigh the crucible and lid. Place about 30 cm of clean, loosely coiled magnesium ribbon into the crucible, replace the lid and reweigh.

Place the crucible on a pipeclay triangle on a tripod and heat it gently at first and then more strongly (Fig. 4.3).

Fig. 4.3

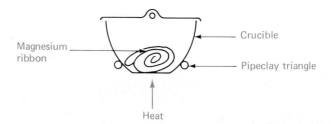

Magnesium ribbon — Crucible — Pipeclay triangle — Heat

Every two or three minutes it is necessary to allow more air to enter the crucible. This is done by removing the Bunsen burner and lifting the crucible lid for a very short time with a pair of tongs, taking care not to allow any smoke to escape.

When the reaction appears to be complete, allow the crucible to cool and weigh the crucible, lid and contents.

Make a table of your results and calculate the mass of oxygen which has combined with your known mass of magnesium.

Look up the relative atomic masses of magnesium and oxygen in the Data Section and use them to calculate the number of moles of magnesium and oxygen which combined together in your experiment. Convert these numbers to the simplest ratio and hence find the empirical formula of magnesium oxide. The worked examples (4.11) will help you with this calculation.

30

1 How could you check that all the magnesium has reacted?
2 What is the most likely source of error in this experiment?

What are the combining masses of zinc and iodine?

Zinc and iodine react together in the presence of alcohol. In this experiment a known mass of iodine is reacted with excess zinc of known mass.

The mass of zinc reacting is found by recovering the unreacted zinc and subtracting its mass from the original mass of zinc.

You will need a hard glass test-tube, preferably one which will fit a centrifuge.

Weigh the test-tube. Put about 0·5 g of zinc powder into the tube and reweigh it.

Add about 1·0 g of iodine (taking care not to allow it to touch your hands) and again reweigh the test-tube.

Although the mass of the zinc present is less than that of the iodine, the zinc is in excess as far as the numbers of atoms present are concerned.

Stand the test-tube in a rack and slowly, drop by drop, add about 2 cm³ of alcohol. When the reaction between zinc and iodine appears to have slowed down, shake the tube until the brown-yellow colour of the iodine disappears.

Centrifuge the mixture or allow it to stand until the excess zinc settles. Pour off the liquid from above the zinc. This liquid is a solution of the product of the reaction (zinc iodide) in alcohol.

Wash the residue of zinc by shaking with about 0·5 cm³ of alcohol. Centrifuge or allow to stand and then pour off the alcohol.

Dry the zinc by laying the tube on a gauze which has been previously heated. Keep the Bunsen flame away from the mouth of the test-tube as alcohol is flammable. When the zinc is dry, reweigh the tube and zinc.

Make a list of your results and calculate the mass of zinc which has reacted with the mass of iodine which you used.

Look up the relative atomic masses of zinc and iodine in the Data Section and use them to calculate the numbers of moles of zinc and iodine reacting in your experiment. Convert these numbers to the simplest ratio and hence find the empirical formula of zinc iodide. If you need help with this calculation, refer to the worked examples (4.11).

What can be deduced from the fact that there is some zinc left when all of the iodine colour has disappeared?

By the end of the eighteenth century, scientists recognised the differences between elements and compounds and naturally they became interested in finding out more about what was happening when two elements combined to form a compound. One way of beginning this investigation was to find out how much of one element combined with a certain mass of another element.

A scientist who made an important contribution to this work was Joseph Louis Proust. In 1799 he found that 100 g of calcium carbonate, whether it was in the form of chalk or marble, always contained 40 g of calcium, 12 g of carbon and 48 g of oxygen. In a similar manner Investigation 4.4 would enable you to find the mass of oxygen which combines with a particular mass of copper.

Fig. 4.4 John Dalton (1766–1844). This portrait shows, on the table, his drawings of symbols for atoms. (Courtesy The Science Museum)

From results of experiments such as these it became obvious that, however a compound is made, the mass of one element combining with a certain mass of another element is always the same. In other words compounds always have the same composition by mass.

This pattern of behaviour is known as the **Law of Constant Composition** (sometimes called the **Law of Definite Proportions**). Having accepted this fact, scientists tried to think of reasons why compounds should have constant compositions. In 1808 John Dalton suggested the theory that all elements consist of very small particles for which he used the name atoms (derived from the Greek word atomos meaning indivisible) and that all the atoms of each particular element are identical in all respects including mass. He was then able to explain the Law of Constant Composition by saying that when two elements combine to form a compound, the atoms of the elements must always combine in fixed proportions and hence the ratio of the mass of one element to the mass of the other will always be constant.

For example, if atoms of element A, each with a mass of a grams, combine with atoms of element B, each with a mass of b grams, in the fixed ratio of 1 atom of A to 2 atoms of B, then the compound formed will have a constant composition with the elements always being present in the proportions $a:2b$ by mass.

The explanation proposed by Dalton is known as **Dalton's Atomic Theory** and, although it has been shown in more recent years that all the atoms of an element are not necessarily exactly identical in mass, the picture of the atoms of two elements combining in fixed proportions to form a compound is still accepted and is of fundamental importance to the understanding of chemistry.

Dalton's Atomic Theory also provided an explanation of the **Law of Conservation of Mass** for which the experiments of Hans Landolt in 1806 had provided evidence. Landolt found that the total mass of the products of a chemical reaction is equal to the total mass of the substances reacting together. Dalton's Theory suggests that when a reaction occurs, none of the atoms of the elements involved are destroyed and no new atoms are formed. After the reaction the atoms are simply combined in a different way and hence the total mass is not changed. Investigation 4.1 provides evidence for this law.

An understanding of this law should provide you with what may be a new way of looking at a common chemical reaction which you must have seen many times. That is, when a piece of coal or wood burns it is combining with the oxygen part of the air and appears to leave little else other than a small amount of ash. However, if you were to

32

find the total mass of the ash plus all the soot, smoke and gas produced, it would be equal to the mass of the coal or wood plus the mass of the oxygen used and hence there has been no change in the total mass during the reaction.

4.8 Multiple proportions

Dalton also put forward the idea that the atoms of some elements might be able to combine in two or more definite proportions, for example, 1:1 and 1:2. If this is the case then it ought to be possible to find some elements which combine together in more than one proportion by mass, but one of the proportions would always be a simple multiple (usually 2 or 3 times) the other.

For example, under different conditions lead and oxygen combine to form three different oxides, one being yellow, one being dark brown and one red. The oxygen can be removed from samples of these oxides by passing hydrogen or natural gas over the heated oxides in an apparatus such as that shown in Fig. 4.5.

Fig. 4.5 Removing oxygen from two oxides of lead

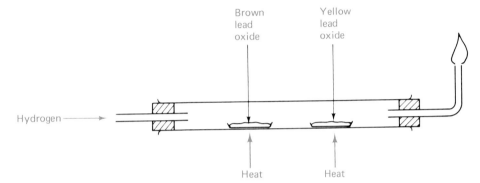

The following are the results of such an analysis of two of the oxides:

	Yellow oxide	Brown oxide
Mass of boat empty	= 4·82 g	4·56 g
Mass of boat + lead oxide	= 10·34 g	10·15 g
Mass of boat + lead	= 9·95 g	9·39 g
Mass of the lead oxide	= 5·52 g	5·59 g
Mass of lead in the oxide	= 5·13 g	4·83 g
Mass of oxygen in the oxide	= 0·39 g	0·76 g

In the yellow oxide:
5·13 g of lead combined with 0·39 g of oxygen
that is, 1 g of lead combined with 0·076 g of oxygen.

In the brown oxide:
4·83 g of lead combined with 0·76 g of oxygen
that is, 1 g of lead combined with 0·157 g of oxygen

The experiment shows that the masses of oxygen combining with 1 g of lead in the two oxides are in the ratio of 0·076 to 0·157 which within the limits of experimental error is a ratio of 1:2.

This pattern of behaviour in which two elements combine in two or more fixed proportions which are simple multiples of each other is known as the **Law of Multiple Proportions**.

4.9 Relative atomic masses

Every substance in the Universe is made from one or more of the elements. The investigations in Chapter 3 indicated that in compounds the elements are combined together to form very small particles.

Dalton put forward the theory that the elements themselves consist of particles and hence these particles, which he called atoms, must be extremely small—even smaller than the particles measured in Chapter 3. It is for this reason that scientists prefer to use relative masses of atoms for most of their work rather than actual masses. That is, if we decide to give hydrogen, which has lighter atoms than any other element, an atomic mass of 1, then as oxygen atoms are 16 times as heavy as hydrogen atoms, oxygen will have a **relative atomic mass** (compared to hydrogen) of 16. Similarly,

1 atom of sulphur is 32 times as heavy as 1 atom of hydrogen and hence the relative atomic mass of sulphur is 32,

1 atom of iron is 56 times as heavy as 1 atom of hydrogen and hence the relative atomic mass of iron is 56.

Hydrogen, being the element with atoms of the smallest mass, is the obvious choice as a standard with which to compare the masses of atoms of other elements.

However, as the methods of determining relative masses have changed during the last hundred years it has been necessary to use different standards. The present internationally accepted standard is the atom of carbon which has been given a relative atomic mass of 12.* Therefore the masses of atoms of other elements are compared to one twelfth of the mass of the carbon atom.

Relative atomic masses are found now using a mass spectrometer. Atoms of the element, in the form of a stream of gas, become electrically charged and are then deflected by electric and magnetic fields. The heavier the atom is the more difficult it is to deflect it and hence by using the angle of deflection and the strengths of the fields it is possible to work out the mass of the atoms.

The changing of the standard has resulted in different values for the relative atomic masses of some elements, but these are only important when very accurate values are required. For most purposes the list of approximate relative atomic masses (which differs little from that based on hydrogen as the standard) given in the Data Section is adequate.

4.10 Numbers of atoms

Chemists think of chemical reactions between elements in terms of atoms of the elements combining in some simple whole number ratio. Thus chemists, although interested in the masses of substances reacting together, are more interested in the numbers of atoms involved in reactions. The fact that atoms are so very small does create problems. For example, if 1 g of a mixture of iron and sulphur is heated in a test-tube, a reaction occurs which involves about 14 000 000 000 000 000 000 000 atoms (1.4×10^{22}). To avoid having to use such large numbers in calculations, chemists have created an easier unit amount of substance called the **mole**. We shall see below that this is the amount of substance which contains a particular number of atoms. This is a similar idea to file paper being ordered by a school in reams (each containing 500 sheets) rather than in individual sheets.

The number of atoms chosen for a mole of a substance is the number of atoms in one relative atomic mass (expressed in grams) of hydrogen. (Strictly speaking using the modern standard this should be 12 g of carbon-12.)

The relative atomic mass of hydrogen is 1, therefore the unit amount of substance, called the **mole, is the amount of substance which contains the same number of particles as there are atoms in 1 g of hydrogen (or 12 g of carbon-12)**. This number is approximately 600 000 000 000 000 000 000 000 which is more conveniently written as 6×10^{23} and is called the **Avogadro constant** after the nineteenth-century Italian chemist Amadeo Avogadro whose important work on gases is discussed in Chapter 5.

* ^{12}C is the most common isotope of carbon (13.5).

Thus 1 mole of hydrogen atoms has a mass of 1 g and contains 6×10^{23} atoms. The mass of oxygen containing 6×10^{23} atoms of oxygen must be 16 times larger as each oxygen atom is 16 times as heavy as each hydrogen atom. Therefore 1 mole of oxygen atoms has a mass of 16 g which is in fact the relative atomic mass of oxygen expressed in grams. Similarly, the relative atomic mass (expressed in grams) of iron is 56 g and this amount of iron must contain the same number of atoms of iron as 1 g of hydrogen contains atoms of hydrogen, as each atom of iron is 56 times as heavy as each atom of hydrogen. By applying similar arguments to all elements it can be seen that 1 relative atomic mass (expressed in grams) of any element must contain 6×10^{23} atoms of that element and is therefore 1 mole of that element.

4.11 Empirical formulae

It is quicker for a cashier in a bank to find out the number of coins in a bag by finding the mass of the coins, rather than by emptying the coins out and counting them. For example, if a bag containing 2p coins has a mass of about 360 g then the bag contains 50 coins.

The mass of a 1p coin is about half that of a 2p coin and therefore for a bag to contain 50 1p coins it must have a mass of about 180 g.

Fig. 4.6 A bank cashier counts coins by weighing them. In a similar manner a chemist counts atoms by weighing substances. (Courtesy Midland Bank Limited)

In a similar manner, using the knowledge that 1 relative atomic mass of any element expressed in grams contains the same number of atoms, it is possible to count atoms by finding the mass of the element. However, as explained earlier, because of the large numbers involved, it is easier if moles of atoms are used rather than actual numbers. The following example illustrates the usefulness of this method of counting atoms.

The results of an experiment to find the combining masses of magnesium and oxygen were:

6 g of magnesium combined with 4 g of oxygen to form the compound magnesium oxide.

The number of moles of atoms of each element taking part in the reaction is found by dividing the masses reacting by the mass of 1 mole of each element (that is, by the relative atomic mass of each element, which for magnesium is 24 and for oxygen is 16).

$\frac{6}{24}$ moles of magnesium atoms combined with $\frac{4}{16}$ moles of oxygen atoms or

$\frac{1}{4}$ moles of magnesium atoms combine with $\frac{1}{4}$ moles of oxygen atoms or

1 mole of magnesium atoms combine with 1 mole of oxygen atoms.

As compounds have a constant composition by mass, we can conclude from this result that magnesium and oxygen will always combine to form the compound magnesium oxide in the ratio of 1 mole of magnesium atoms to 1 mole of oxygen atoms, which in turn means that they must be combining in the ratio of 1 atom of magnesium to 1 atom of oxygen. This useful information is usually written down in a shorthand form known as the **formula** of the compound. Thus the formula of magnesium oxide is MgO where

Mg is the symbol for magnesium and represents 1 mole of atoms (or 1 atom) of the element,

and O is the symbol for oxygen and represents 1 mole of atoms (or 1 atom) of the element.

MgO is the simplest formula which gives the ratio of the number of atoms of each element combining to form the compound and is known as the **empirical formula** of the compound.

In water the ratio of hydrogen atoms to oxygen atoms in the empirical formula is $2:1$ and the formula, as you probably know, is written as H_2O.

A list of the symbols used for the elements is given in the Data Section. The general rules for the construction of formulae from these symbols are given in Chapter 7 but another example of the determination of an empirical formula of a compound from experimental results is the following.

In an experiment to find the combining masses of mercury and chlorine, 4·00 g of the compound of mercury and chlorine were found to contain 2·95 g of mercury.

Therefore the mass of chlorine in the compound $= 4{\cdot}00 - 2{\cdot}95$ g

$$= 1{\cdot}05 \text{ g}$$

That is, 2·95 g of mercury combine with 1·05 g of chlorine.

The relative atomic mass of mercury is 201 and that of chlorine is 35·5. Therefore,

$\dfrac{2{\cdot}95}{201}$ moles of mercury atoms combine with $\dfrac{1{\cdot}05}{35{\cdot}5}$ moles of chlorine atoms.

That is,

0·0147 moles of atoms of mercury combine with 0·0296 moles of atoms of chlorine.

To obtain this mole ratio in its simplest form it is necessary to divide both numbers by the smallest number.

That is,

$$\dfrac{0{\cdot}0147}{0{\cdot}0147} \qquad\qquad \dfrac{0{\cdot}0296}{0{\cdot}0147}$$

1·00 mole of atoms combines with   2·01 moles of atoms.

This result indicates that mercury and chlorine combine in the ratio of 1 atom of mercury to 2·01 atoms of chlorine, but the atomic theory requires that atoms only combine in whole numbers and hence the ratio must be $1:2$.

The symbol for mercury is Hg and that for chlorine is Cl and the empirical formula for the compound of mercury and chlorine is therefore $HgCl_2$.

**4.12
Summary**

1. Compounds have a fixed composition by mass and this is due to atoms of the elements present combining in fixed whole number proportions.
2. Actual masses of atoms are very small and therefore relative atomic masses are used.
3. We are interested in the numbers of particles involved in reactions, but as atoms are so very small, the numbers of particles involved in reactions are very large, and hence we use the mole as the unit amount of substance containing approximately 6×10^{23} particles (the same number of particles as there are atoms in 12 g of carbon—12).
4. The empirical formula of a compound is that which represents the simplest ratio of the number of moles (and hence atoms) of the elements present in the compound.

Chapter 5

What can we learn from the behaviour of gases?

5.1 Physical behaviour of gases

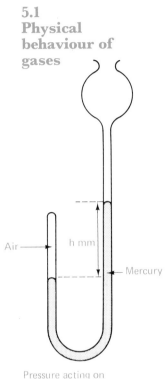

Pressure acting on
trapped air =
(atmospheric + h) mmHg

Fig. 5.1 Investigating the effect of changing the pressure on the volume of a fixed mass of air

Many gases are different from each other in very obvious ways. For example, the active part of air (oxygen) supports combustion whereas the inactive part (nitrogen) does not. Some gases are coloured, chlorine for example being pale green. Others, such as hydrogen, are explosively flammable, some such as carbon monoxide are very poisonous, and some such as hydrogen sulphide are very smelly.

Despite all these obvious differences, gases do have some similar characteristics. For example, if a gas is heated, it expands (its volume increases), and if a gas is subjected to a greater pressure it contracts (its volume decreases).

Early scientists, having made observations such as these, investigated the changes more closely to see if there were any recognisable patterns in the way they occurred. Robert Boyle in 1662 trapped some air in a J-tube as in Fig. 5.1. He varied the pressure by either increasing or decreasing the amount of mercury in the tube. He found that when the pressure was doubled, the volume of the gas approximately halved, and when the pressure was halved the volume doubled.

However, what was more interesting was that when the experiment was repeated using other gases, their pressures and volumes varied in the same way. These experimental results gave rise to a law which describes this physical behaviour of all gases, even though the gases have different chemical properties.

A relationship in which one variable decreases regularly as the other increases regularly, is called an inverse relationship. **Boyles Law** states that **the pressure of a fixed mass of gas at a constant temperature is inversely proportional to its volume**.

In 1787 Jacques Charles carried the investigation a stage further by keeping the pressure constant and seeing what effect changes in temperature had on the volume of a gas. He found in this case that there is a direct relationship between the variables, that is, as one variable increases regularly the other also increases regularly. **Charles' Law** states that **the volume of a fixed mass of gas at a constant pressure is directly proportional to the temperature measured on the Kelvin scale**.

Fig. 5.2 Variation of volume of a gas with temperature (it is necessary to extrapolate the graph below the temperature at which the gas changes into a liquid)

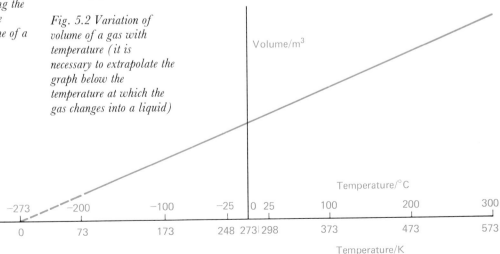

The Kelvin scale is that which takes as its zero the temperature which is often referred to as the absolute zero. When a gas is cooled, its volume decreases and if it does not change into a liquid or solid, we can predict that it would have zero volume at a temperature of $-273°$ Celsius which is 273 degrees below the freezing point of water, $0°C$. Fig. 5.2 shows how, if the volume-temperature graph is extrapolated back until the volume is zero, the temperature would be $-273°C$. The temperature on the Celsius scale is given above the horizontal axis and the equivalent on the Kelvin scale is given below the axis. It can be seen from this graph that temperature on the Kelvin scale = temperature in $°C + 273$.

For example,
$$0°C = 273 \text{ K},$$
$$25°C = 298 \text{ K}$$
$$\text{and } -25°C = 248 \text{ K}$$

A third gas law, which is not usually attributed to a particular scientist, describes the variation of pressure with temperature when the volume is kept constant and therefore is sometimes known as the **Constant Volume Law**. The law states that **the pressure of a fixed mass of gas at a constant volume is directly proportional to its temperature measured on the Kelvin scale**.

Really accurate experimental work does reveal that gases do not obey the laws exactly over large ranges of temperature and pressure, but nevertheless the fact that gases with widely differing properties all approximately obey the laws requires some explanation.

5.2 Kinetic theory of gases

In Chapter 3 the differences between gases and the other physical states of matter, solids and liquids, were accounted for by the theory that they consist of particles which are well separated from each other and are moving around at high speeds in all directions within the boundaries of the container. It was also pointed out that the speeds of the molecules increase with temperature. This theory, because it is concerned with the motion of the particles, is known as the **Kinetic Theory of Gases**. It can be used to provide a qualitative explanation of the changes in pressure of a gas which result from changes in its volume and temperature, and if you study chemistry or physics later on, you will develop the theory further to provide a quantitative explanation of the gas laws.

The theory assumes that the pressure exerted by a gas on the walls of its container is due to the force exerted by the particles when they collide with the walls. Changes in the pressure, which are described more precisely by the gas laws, can be explained qualitatively as follows.

When the volume of a gas is decreased (at constant temperature), the particles have less space to move around in, therefore there will be more collisions with the walls of the container and the pressure exerted by the gas will increase.

When the temperature of a gas is increased (at constant volume), the speeds of the molecules increase, hence the number of collisions with the walls and the force exerted by each collision will increase and so the pressure exerted by the gas will increase.

Thus the theory is able to explain the observed changes in pressure with volume and temperature.

5.3 Calculations involving the gas laws

Using $\propto$ to represent 'proportional to' and p, V and T to represent respectively pressure, volume and temperature, the three gas laws may be written in mathematical terms as follows:

Boyle's Law $\qquad p \propto \dfrac{1}{V}$ for a constant mass at a constant temperature

Charles' Law $\qquad V \propto T$ for a constant mass at a constant pressure

Constant Volume Law $\quad p \propto T$ for a constant mass at a constant volume

In each case, $\propto$ can be replaced by an equals sign and a constant (k_1, k_2 and k_3).

$$p = \frac{k_1}{V} \quad \text{or} \quad pV = k_1$$

$$V = k_2 T \quad \text{or} \quad \frac{V}{T} = k_2$$

$$p = k_3 T \quad \text{or} \quad \frac{p}{T} = k_3$$

These three equations can be incorporated into one equation which is known as the **Combined Gas Equation**.

$$\frac{pV}{T} = \text{a constant}$$

You can check that this is correct by first letting T be a constant and including it in the constant at the right-hand side. The resulting equation is Boyle's Law. Similarly the other two laws can be obtained by first letting p and then V be constants.

The combined form of the gas laws is particularly useful for calculating what volume a fixed mass of gas would occupy at different temperatures and pressures. For example, if a gas occupies a volume V_1 at temperature T_1 and pressure p_1, the volume it would occupy at a temperature T_2 and pressure p_2 can be represented by V_2 and in each case,

$$\frac{pV}{T} = \text{a constant}$$

Thus, $\qquad \dfrac{p_1 V_1}{T_1} = \text{a constant} = \dfrac{p_2 V_2}{T_2}$

or $\qquad \dfrac{p_1 V_1}{T_1} = \dfrac{p_2 V_2}{T_2}$

As mentioned earlier, temperature must always be measured on the Kelvin scale, but the units used for volume and pressure can vary according to the circumstances. The systematic unit for pressure is the Pascal (Pa) where one Pascal means a pressure of 1 N m^{-2} (or one kiloPascal (kPa) $= 1000 \text{ N m}^{-2}$). Frequently it will be more convenient to use other units for pressure, for example mm of mercury (mmHg) which indicate the height of a column of mercury that the pressure would support. Occasionally it may be more convenient to use atmospheres, when, for example, a pressure of 10 atmos would mean a pressure ten times greater than the normal atmospheric pressure. Details of the volume are usually given in dm^3 or cm^3. Whatever units are selected for the pressure and the volume they must be used on both sides of the Combined Gas Equation.

Example: A chemical engineer knows that the process for which he is designing a chemical plant will produce 2000 dm^3 of a gas, measured at 500 kPa and 150 °C during every hour of operation. He wants to store the gas at a pressure of 100 kPa and a temperature of 30 °C. What volume would the gas occupy at this new pressure and temperature?

$$p_1 = 500 \text{ kPa} \qquad\qquad p_2 = 100 \text{ kPa}$$
$$V_1 = 2000 \text{ dm}^3 \qquad\qquad V_2 \text{ is unknown}$$
$$T_1 = 273 + 150 = 423 \text{ K} \qquad T_2 = 273 + 30 = 303 \text{ K}$$

Substituting into the Combined Gas Equation:

$$\frac{p_1 V_1}{T_1} = \frac{p_2 V_2}{T_2}$$

$$\frac{500 \times 2000}{423} = \frac{100 \times V_2}{303}$$

$$V_2 = \frac{500 \times 2000 \times 303}{423 \times 100}$$

$$= 7163 \text{ dm}^3$$

The gas would occupy 7163 dm³ at the new temperature and pressure.

5.4 Standard temperature and pressure

Clearly there is little point in recording volumes of gases unless the temperature and pressure at which they were measured is known. Similarly, it is pointless comparing volumes occupied by gases unless they are all measured at the same temperature and pressure. When the volumes have been measured at different temperatures and pressures, the combined gas equation can be used to calculate what the volumes would be at the same temperature and pressure. Any temperature and pressure could be selected, but it is more sensible if most people tend to use the same temperature and pressure when comparing volumes of gases. For this reason, it is customary for volumes of gases to be quoted for a **standard temperature and pressure** (s.t.p.). The standards which have been selected are:

standard temperature = 273 K (0 °C, the freezing point of water)
standard pressure   = 101 kPa (760 mmHg, the normal atmospheric pressure)

Example: In an experiment to compare the densities of gases at s.t.p. the mass of 1000 cm³ of carbon dioxide was found at 22 °C and 768 mmHg. What volume would this mass of gas occupy at s.t.p.?

$$p_1 = 768 \text{ mmHg} \qquad\qquad p_2 = 760 \text{ mmHg}$$
$$V_1 = 1000 \text{ cm}^3 \qquad\qquad V_2 \text{ is unknown}$$
$$T_1 = 273 + 22 = 295 \text{ K} \qquad T_2 = 273 \text{ K}$$

Substituting into the Combined Gas Equation:

$$\frac{p_1 V_1}{T_1} = \frac{p_2 V_2}{T_2}$$

$$\frac{768 \times 1000}{295} = \frac{760 \times V_2}{273}$$

$$V_2 = \frac{768 \times 1000 \times 273}{295 \times 760}$$

$$= 935 \text{ cm}^3$$

The carbon dioxide would occupy 935 cm³ at s.t.p.

Fig. 5.3 Joseph Gay-Lussac (1778–1850) who did some important experiments on the combining volumes of gases. (Courtesy Palais de la Découverte, Paris)

5.5 Combining volumes of gases

Armed with the knowledge that it is only sensible to compare volumes of gases when they are measured at the same temperature and pressure, Joseph Louis Gay-Lussac, who was a professor of chemistry in Paris, investigated the volumes of gases taking part in chemical reactions. Fig. 5.4 shows a modern apparatus which can be used to investigate the reaction between nitrogen monoxide and oxygen. These two gases react on contact at room temperature to produce a product which occupies a smaller volume.

Fig. 5.4 Reacting volumes of nitrogen monoxide and oxygen

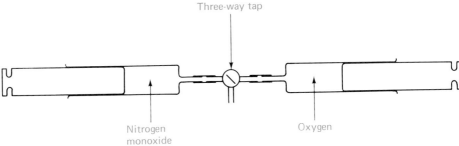

One syringe contains 50 cm³ of nitrogen monoxide and the other contains 50 cm³ of oxygen. The oxygen is pushed into the nitrogen monoxide in 5 cm³ portions and after each addition, the total volume of gas in both syringes is recorded. As the reaction proceeds the total volume decreases until the reaction is complete.

Typical results are given in the graph, Fig. 5.5, which shows that the reaction is complete when approximately 25 cm³ of oxygen have been added to the 50 cm³ of nitrogen monoxide. Provided a little time has been allowed for the gases to cool after each addition, all of the volumes have been measured at the same temperature and pressure.

Fig. 5.5 Reaction of nitrogen monoxide with oxygen

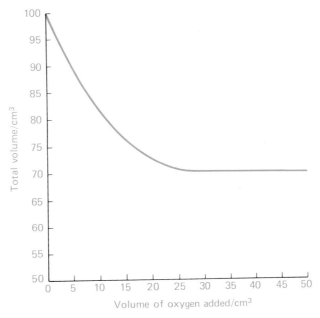

Twice as much nitrogen monoxide as oxygen reacts and therefore the result can be expressed in the following way:

$$\text{nitrogen monoxide} \quad + \quad \text{oxygen} \quad \rightarrow \quad \text{products}$$
$$\text{2 volumes} \qquad\qquad \text{1 volume}$$

41

Results which could be obtained for some of the reactions investigated by Gay-Lussac, expressed in the same way, are:

hydrogen   +   oxygen   →   steam
2 vol   1 vol   2 vol

hydrogen   +   chlorine   →   hydrogen chloride
1 vol   1 vol   2 vol

Gay-Lussac described his experimental results in the law which states that **the volumes of gases which react together and the volumes of the products, if they are gases, are in simple whole number ratios to one another, all the volumes having been measured at, or converted to, the same temperature and pressure**.

The remarkable thing is that it is always approximately a small whole number ratio. Clearly, the next problem facing the scientists of that time was to develop a theory which would account for this experimental fact. As with the laws of Boyle and Charles, it seemed likely that the explanation would rely on the idea that gases consist of particles, but in this case it would be the numbers of particles which would be important.

Gay-Lussac published his results in 1808 which was about the same time as Dalton put forward his Atomic Theory (4.7). If we attempt to explain Gay-Lussac's Law by using Dalton's theory of all matter consisting of small particles called atoms, then we find the explanation will only work if it is possible to divide atoms. For example, experiment shows that:

hydrogen   +   chlorine   →   hydrogen chloride
1 vol   1 vol   2 vol

If we assume that equal volumes of gases at the same temperature and pressure contain equal numbers of atoms (and let this be n in this case),

	n atoms of hydrogen	+	n atoms of chlorine	→	$2n$ compound atoms of hydrogen chloride
or,	1 atom	+	1 atom	→	2 compound atoms
and	$\frac{1}{2}$ atom	+	$\frac{1}{2}$ atom	→	1 compound atom

Dalton stressed in his theory that atoms could not be divided and hence it is not possible to use his theory to explain Gay-Lussac's observations. Dalton, realising this, suggested that Gay-Lussac was mistaken and that his observations were inaccurate. This inflexible attitude of Dalton and others led to the value of Gay-Lussac's work not being recognised for many years.

5.6 Avogadro's hypothesis

It was a hypothesis put forward by an Italian scientist, Amadeo Avogadro, in 1811 which eventually provided the answer to the problem. He suggested that some elements, instead of being made up of individual separate atoms of the element, actually exist as larger particles each of which contains a small number of atoms of the element combined together. He called these larger particles **molecules**. This led him to put forward his hypothesis in which he stated that: **equal volumes of all gases at the same temperature and pressure contain equal numbers of molecules**.

This hypothesis, which is now known to be a fact, is more appropriately referred to as Avogadro's Law.

It was not until almost fifty years later (1858) that the full significance of the hypothesis was realised by another Italian, Stanislao Cannizzaro. If we apply it to Gay-Lussac's results we can see how it provides an explanation.

$$
\begin{array}{cccc}
\text{hydrogen} & + & \text{chlorine} & \rightarrow & \text{hydrogen chloride} \\
\text{1 vol} & & \text{1 vol} & & \text{2 vol}
\end{array}
$$

Applying Avogadro's Law,

$$
\begin{array}{cccc}
n \text{ molecules} & + & n \text{ molecules} & \rightarrow & 2n \text{ molecules} \\
\text{of hydrogen} & & \text{of chlorine} & & \text{of hydrogen chloride}
\end{array}
$$

or $\quad$ 1 molecule $\quad + \quad$ 1 molecule $\quad \rightarrow \quad$ 2 molecules

and $\quad \frac{1}{2}$ molecule $\quad + \quad \frac{1}{2}$ molecule $\quad \rightarrow \quad$ 1 molecule

If one molecule of hydrogen and 1 molecule of chlorine each contain an even number of atoms, then it is possible to have half of a molecule of each element without contradicting Dalton's Atomic Theory. Thus Avogadro's Law enables the experimental results of Gay-Lussac to be reconciled with the theory of Dalton. Except for the group of elements known as the noble gases (e.g. helium, neon and argon) most of the gaseous elements are diatomic, that is, there are two atoms of the element in each molecule.

5.7 Relative molecular masses

Because, as Avogadro stated, equal volumes of gases at the same temperature and pressure contain equal numbers of molecules, the ratio of the masses of equal volumes is equal to the ratio of the molecular masses of the gases. That is,

$$
\frac{\text{mass of 1 vol. of gas A}}{\text{mass of an equal vol. of gas B}} = \frac{\text{mass of } n \text{ molecules of gas A}}{\text{mass of } n \text{ molecules of gas B}}
$$

$$
= \frac{\text{mass of 1 molecule of gas A}}{\text{mass of 1 molecule of gas B}}
$$

Therefore, by comparing the masses of equal volumes of gases (which is the same as comparing densities), it is possible to compare the molecular masses of gases.

Relative atomic masses were obtained originally by comparison with hydrogen whose relative atomic mass was given a value of 1 (4.9). It is possible to calculate the **relative molecular mass** of a gas on the hydrogen scale by comparing the mass of a volume of the gas to the mass of an equal volume of hydrogen. In 1858 Cannizzaro used Avogadro's hypothesis to reason this out. This is shown in the following steps:

$$
\frac{\text{mass of 1 vol. of gas}}{\text{mass of equal vol. of hydrogen}} = \frac{\text{mass of } n \text{ molecules of gas}}{\text{mass of } n \text{ molecules of hydrogen}}
$$

$$
= \frac{\text{mass of 1 molecule of gas}}{\text{mass of 1 molecule of hydrogen}}
$$

But 1 molecule of hydrogen contains 2 atoms,

$$
= \frac{\text{mass of 1 molecule of gas}}{\text{mass of 2 atoms of hydrogen}}
$$

$$
2 \times \frac{\text{mass of 1 vol. of gas}}{\text{mass of equal vol. hydrogen}} = \frac{\text{mass of 1 molecule of gas}}{\text{mass of 1 atom of hydrogen}}
$$

$$
= \text{relative molecular mass of the gas}
$$

The $\dfrac{\text{mass of 1 vol. of gas}}{\text{mass of equal vol. of hydrogen}}$ is known as the **Relative Vapour Density** of the gas, which means the density of the gas relative to hydrogen at the same temperature and pressure.

The more usual way of writing the above relationship is:

the relative molecular mass of a gas = 2 × its relative vapour density.

Hence the link made by Avogadro between volumes of gases and numbers of molecules has provided a method for calculating the relative molecular mass of a gas from its relative vapour density. If higher temperatures are used, the method may also be used to find the relative molecular masses of volatile liquids (liquids which easily change to a vapour on heating).

5.8 Numbers of molecules

In the previous section we saw that the ratio of the masses of equal volumes of two gases (at the same temperature and pressure) must always be equal to the ratio of the relative molecular masses of the gases. In the case of hydrogen and chlorine, which have relative molecular masses of 2 and 71 respectively,

$$\frac{\text{mass of 1 vol. of hydrogen}}{\text{mass of 1 vol. of chlorine}} = \frac{\text{mass of } n \text{ molecules of hydrogen}}{\text{mass of } n \text{ molecules of chlorine}}$$

$$= \frac{2}{71}$$

If we actually use 2 g of hydrogen and 71 g of chlorine, that is their relative molecular masses expressed in grams, then these masses will also occupy equal volumes and the equal volumes will contain equal numbers of molecules. This number is approximately 6×10^{23} which as previously mentioned in 4.10 is called the Avogadro constant. The amount of gas (its relative molecular mass expressed in grams) which contains this number of molecules is referred to as 1 mole of molecules of the gas.

The volume occupied by 1 mole of a gas is called the molar volume and when measured at s.t.p., it is very nearly $22 \cdot 4$ dm³.

This relationship provides an alternative method of finding the relative molecular mass of a gas or volatile liquid.

Example: Propanone is a liquid at room temperature but its boiling point (56 °C) is very low and so it is easy to convert it to a gas and to measure the volume occupied by a certain mass of the gas. When $0 \cdot 165$ g of propanone was injected into a heated glass syringe, the gas produced was found to occupy $85 \cdot 0$ cm³ when measured at 100 °C and 758 mmHg pressure.

Substituting into the Combined Gas Equation to find the volume the gas would occupy if it could be measured at s.t.p.:

$$\frac{758 \times 85 \cdot 0}{373} = \frac{760 \times V_2}{273}$$

$$V_2 = 62 \cdot 0 \text{ cm}^3$$

$0 \cdot 165$ g of propanone vapour would occupy $62 \cdot 0$ cm³ at s.t.p.

1 cm³ of propanone vapour at s.t.p. would have a mass of $\dfrac{0 \cdot 165}{62 \cdot 0}$ g.

But 1 mole of any gas occupies 22 400 cm³ at s.t.p.

22 400 cm³ of propanone vapour at s.t.p. would have a mass of $\dfrac{0 \cdot 165}{62 \cdot 0} \times 22\,400$ g

$$= 59 \cdot 6 \text{ g}$$

Hence the value obtained for the relative molecular mass of propanone is 59.6. As explained in 7.13, the molar volume of gases may also be used to calculate the volumes of gases involved in chemical reactions.

44

The combining masses of the elements which form a compound can be used to calculate the empirical formula (4.11) of the compound. This formula gives the simplest ratio of the number of atoms of each element present in the compound. Some compounds exist in the form of molecules and the **molecular formula** of such a compound gives the number of atoms of each element present in a molecule and this may or may not be the same as the empirical formula of the compound. The following example shows how the relative molecular mass of a compound can be used to find the molecular formula for the compound when its empirical formula is known.

Example: Analysis of the compound called benzene shows that it contains carbon and hydrogen in the ratio of 1 mole of carbon atoms to 1 mole of hydrogen atoms and hence its empirical formula is:

$$CH$$

The relative vapour density of benzene is approximately 39 and so its relative molecular mass is approximately 78.

The relative atomic masses of carbon and hydrogen are 12 and 1 respectively. The molecular formula of benzene could not be CH as this would give a relative molecular mass of 13. The molecular formula must be C_6H_6 which, without changing the empirical formula from CH, gives a relative molecular mass of 78.

The fact that the value for the relative molecular mass found from relative vapour density measurements is only approximate does not matter as it is only necessary to decide which whole number multiple of the empirical formula is required.

Alternatively the relative molecular mass could have been calculated from the molar volume, as in the previous section, rather than from the relative vapour density. The example in the previous section gave a value of 59·6 for the relative molecular mass of the compound called propanone. The empirical formula of the compound is C_3H_6O. If the molecular formula of the compound is C_3H_6O, its relative molecular mass would be $(3 \times 12) + (6 \times 1) + (16) = 58$ which is very close to the value of 59·6 and therefore C_3H_6O must also be the molecular formula of the compound.

1. Boyle's Law, Charles' Law and the Constant Volume Law are examples of how gases, even though they may appear to be very different, do have some similar physical properties.
2. The three gas laws can be combined into a more convenient form:

$$\frac{p_1 V_1}{T_1} = \frac{p_2 V_2}{T_2}$$

3. Standard temperature and pressure, s.t.p., are 273 K and 101 kPa (760 mmHg).
4. Gay-Lussac's experimental law of combining volumes was eventually explained by a hypothesis put forward by Avogadro.
5. Avogadro's hypothesis also led to a method for calculating relative molecular masses of gases and volatile liquids from their relative vapour densities.
6. The relative molecular mass of a gas, expressed in grams, is 1 mole of molecules of the gas and occupies $22 \cdot 4 \, dm^3$ at s.t.p. It contains approximately 6×10^{23} molecules.
7. The relative molecular mass of a compound can be used to obtain its molecular formula from its empirical formula.

Chapter 6

What can we learn about substances by passing electricity through them?

Investigation 6.1

Which pure elements conduct electricity?

In order to test whether a substance is a conductor or non-conductor of electricity, it is necessary to put the substance in a circuit with a source of electricity, such as a battery. You will also need some means of detecting when a current is flowing through the circuit; a bulb or an ammeter would be suitable for this.

You will need a 6 volt battery, a 6 volt bulb in a holder (or other suitable matched battery and bulb), some connecting wires fitted with crocodile clips; and small pieces (not powder) of each of the following elements: copper, lead, sulphur, zinc, iron and carbon (graphite).

Connect up the circuit as in Fig. 6.1 which shows that two of the wires each have one free end. Test the conductance of each element by connecting them between the two free wires. Record which elements are good conductors and which appear not to be.

Fig. 6.1

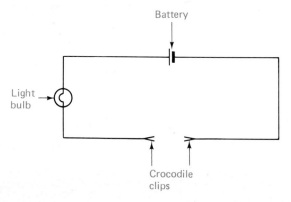

Questions

1 Excluding graphite, to which class of elements do all of the conductors belong?

2 It is difficult for you to test the conductance of many elements because they are either gases at room temperature (e.g. oxygen, nitrogen and chlorine) or they are too dangerous (e.g. bromine, phosphorus, sodium, potassium and mercury). Table 6.1 shows the results which are obtained with these elements, using the liquid forms of the gases as this is a fairer test.

Using your results (but again excluding graphite) and the results in Table 6.1, decide to which class of elements the non-conductors belong.

TABLE 6.1. *The conductance of some elements*

ELEMENT	CONDUCTOR	NON-CONDUCTOR
oxygen (l)		✓
nitrogen (l)		✓
chlorine (l)		✓
bromine (l)		✓
phosphorus (s)		✓
sodium (s)	✓	
potassium (s)	✓	
mercury (l)	✓	

Investigation 6.2

Which pure compounds conduct electricity?

The compounds will either be in liquid or powder (small crystals) form and so in this case it is better to connect them to the circuit by means of two carbon rods called electrodes which are attached to the free ends of the two wires.

You will need the apparatus used in Investigation 6.1 plus two carbon electrodes and a selection of compounds such as sodium chloride, copper(II) sulphate, lead(II) bromide, potassium iodide, pure water, alcohol and sugar.

1. Connect up the circuit as in Fig. 6.2, using about 1 cm depth of each compound in a small beaker test its conductance. It would be wasteful to throw away the samples of some of these compounds, so be very careful not to contaminate them.

Fig. 6.2

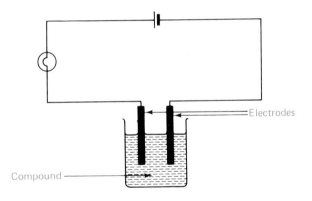

Electrodes

Compound

It could be argued that a fairer test of the solid compounds would be to use a solid lump of the compound rather than the powder. If there is one available, test the conductance of a large crystal of copper(II) sulphate.

2. Heat the lead(II) bromide, either in a small beaker on a tripod and gauze, or in a boiling tube supported by a clamp and stand, until it melts. TAKE GREAT CARE THAT THE MOLTEN LEAD(II) BROMIDE DOES NOT COME INTO CONTACT WITH YOUR SKIN—IT WILL CAUSE A VERY PAINFUL BURN.

Test the conductance of the molten lead(II) bromide by dipping the carbon electrodes into it (Fig. 6.3). The cold electrodes may cause some of the liquid to solidify, so keep heating it for a short while until you are sure that it is all molten. Then watch the bulb and allow the lead(II) bromide to cool until it is solid again.

Fig. 6.3

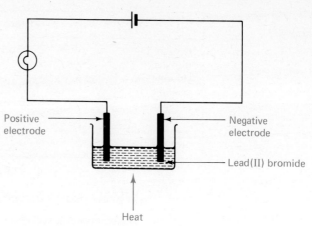

Your teacher may demonstrate to you whether a similar result is obtained for some other molten compounds, for example, lead(II) iodide, potassium iodide and lead(II) chloride.

Questions

1 It appears that compounds in the solid state never conduct electricity, but when in the liquid state, some compounds do conduct electricity and some do not. Thinking back to what you know about the differences between solids and liquids, what happens to the particles of a solid when it changes to a liquid?

2 All compounds contain at least two elements. Examine the names of those compounds which conduct electricity when molten and decide to which classes of elements the components of these compounds belong.

Investigation 6.3

Which compounds conduct electricity when they are dissolved in water?

In this experiment you are going to test the conductance of mixtures of compounds which, on their own at room temperature, do not conduct electricity.

You will need, in addition to the apparatus used in Investigation 6.2, either samples of the following compounds from which you can make aqueous solutions, or previously prepared solutions of the compounds sodium chloride, copper(II) sulphate, potassium iodide, alcohol and sugar.

1. Connect up the same circuit as in Fig. 6.2 and test the conductance of each aqueous solution.

Question

By examining the names of the compounds which do conduct electricity when dissolved in water, decide to which two classes of elements the components of these compounds belong.

2. Repeat the experiment using dilute solutions of sulphuric acid, hydrochloric acid and nitric acid. (It is too dangerous for you to test the conductance of these compounds when pure, or for you to make up your own aqueous solutions of them.)

Question

Which element is present in all acids?

In which direction do the particles in potassium manganate(VII) move when a direct electric current is passed through a solution of the compound?

The manganate(VII) part of the compound is purple in colour, whereas the potassium does not contribute to the colour of the compound.

Rather than make up a solution of potassium manganate(VII) and then pass a current through it, a crystal of the compound is placed on a piece of damp filter paper. In this way it is possible to see what happens to the purple part when the compound dissolves in the water on the filter paper and comes under the influence of the electric current.

You will need two microscope slides, two strips of filter paper cut to the exact size of the slides, and a 20 volt source of direct current.

Moisten the filter papers with tap water and place one on each slide. Place one crystal of potassium manganate(VII) on the centre of each piece of filter paper.

Connect one piece of filter paper by means of crocodile clips to the source of direct current (Fig. 6.4), but do not switch on the power until all of the connections have been made. After checking that the microscope slide is horizontal, switch on the power supply and adjust it to 20 volt d.c.

Leave the current passing for about 5 minutes and after this time compare what has happened to the purple manganate(VII) part of the compound on each slide.

Fig. 6.4

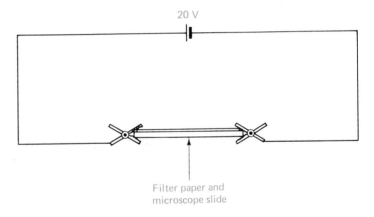

20 V

Filter paper and
microscope slide

Questions

1 Is the manganate(VII) part of the compound positively or negatively charged?

2 Overall the compound is neutral, therefore, is the charge on the potassium part likely to be positive or negative?

3 The potassium particles are colourless and so it is not possible to see them move, but towards which electrode would you expect them to move?

4 Predict which part of the sodium chloride is likely to be positively charged and which negatively charged. Towards which electrode would you expect the sodium and chloride parts to move when a direct current is passed through a solution of the compound?

Investigation 6.5

How much copper and silver are produced by passing a known quantity of electricity through a solution of a copper compound and a solution of a silver compound?

Electricity is the flow of charge through a substance. The quantity of charge (and hence electricity) which passes through a conductor is measured in coulombs where:

number of coulombs = current in amperes × time in seconds (6.8)

When a direct current is passed through aqueous solutions of copper(II) sulphate and silver nitrate with copper and silver electrodes respectively, in each case the metal passes into solution at the positive electrode (the anode) and is deposited at the negative electrode (the cathode). Thus, by weighing the cathodes before and after passing the current, it is possible to find the masses of the two metals which have been deposited.

Also, if the two solutions are connected in series in the circuit, Fig. 6.5, the same current and hence the same quantity of electricity will pass through both solutions.

Fig. 6.5

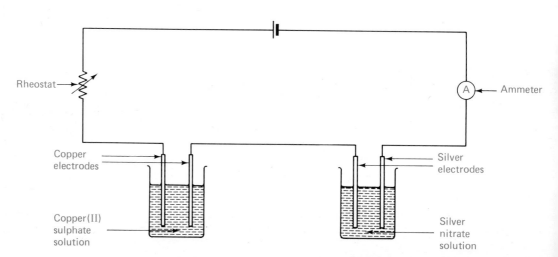

You will need a 6 volt battery or alternative source of electric current, a rheostat, a milliammeter, approximately 0·05M solutions of copper(II) sulphate and silver nitrate and two copper electrodes and two silver electrodes.

As the masses of copper and silver deposited are small, it is necessary to work very carefully so that experimental errors are kept to a minimum.

Connect up the circuit as shown in Fig. 6.5, but do not dip the electrodes into the solutions. Weigh each cathode after checking that it is clean and dry.

Place the electrodes into the solutions, make a note of the time and quickly adjust the current by means of the rheostat so that it is 100 milliamperes. Keep the current at this value for about 30 to 40 minutes. Record the time when the current is stopped.

Carefully remove the cathodes and wash them by gently dipping them into a beaker of distilled water and then into another beaker of propanone. When they are dry, weigh each cathode.

The silver in particular tends to fall off the cathode. If this does occur,

50

collect the pieces by filtering and then after washing and drying find their mass and add it to the increase in mass of the silver cathode.

Calculate the quantity of electricity passed in coulombs and the mass of copper and the mass of silver deposited.

Chemists are more interested in numbers of particles involved in reactions (4.10) rather than masses. Look up the relative atomic masses of copper and silver in the Data Section and use them to calculate the number of moles of copper atoms and silver atoms deposited by this quantity of electricity.

Then calculate the quantity of electricity which is required to neutralise 1 mole of copper ions to form 1 mole of copper atoms and the quantity of electricity required to neutralise 1 mole of silver ions to form 1 mole of silver atoms.

Questions

1 Which ion, the copper or the silver, requires the more electricity to neutralise it? Therefore, which of the ions has the greater charge?

2 The charges on the copper and silver ions are in a small whole number ratio to one another. Examine your results and deduce what this ratio is. (This will only be possible if you have obtained reasonably accurate results.)

6.6 Conductors and non-conductors

In our homes we are all familiar with substances which conduct electricity (conductors) and substances which do not conduct electricity (non-conductors or insulators). For example, the flex leading to a reading lamp or a television has metal wires inside which conduct electricity and a plastic covering which does not conduct electricity.

Investigations 6.1, 6.2 and 6.3 use a battery as a safe low-voltage supply of direct current for testing the conductance of various substances. An appropriately small bulb is used to indicate whether or not the substance being tested conducts electricity. Several patterns emerge from investigations such as these.

1. All metals conduct electricity and all non-metals, except carbon in the form of graphite, are non-conductors.
2. Compounds in the solid state do not conduct electricity.
3. Some compounds in the liquid state do conduct electricity, e.g. lead(II) bromide, lead(II) iodide, sodium chloride. All of these compounds have both a metallic part and a non-metallic part.
4. Some compounds when mixed with water conduct electricity, e.g. aqueous solutions of sulphuric acid, hydrochloric acid, sodium chloride, copper(II) sulphate, potassium iodide and sodium hydroxide. All of these compounds except the acids have both a metallic part and a non-metallic part. All acids contain hydrogen and another non-metallic part.

6.7 Conduction of electricity

Conduction in metals and in graphite is due to the movement of very small charged particles called electrons. Electrons and the part they play in conduction are discussed in more detail in Chapters 19 and 20. This chapter is concerned primarily with the conduction of electricity through compounds and what we can learn from this about the composition of these compounds.

The patterns listed in the previous section indicate that compounds only conduct when they are in the liquid form and, if they are soluble, when they are in aqueous solution. This suggests that conduction only occurs when the compounds are in a mobile form.

Investigation 6.4 shows that the purple part of potassium manganate(VII) (permanganate) moves towards the positive terminal of the electricity supply.

Further evidence for conduction involving the movement of particles is provided by passing a current through copper(II) chromate(VI) solution. If a supply of direct electric current is connected by means of two carbon rods (called electrodes) to a solution of copper(II) chromate(VI), in dilute sulphuric acid, in a U-tube, Fig. 6.6, the solution conducts electricity and the original green colour of the copper(II) chromate(VI) solution changes. Around the negative electrode the solution becomes blue and around the positive electrode it becomes yellow. This shows that one part of the copper(II) chromate(VI) is moving towards the positive electrode and, as this is blue, it is likely to be the copper. Another part of the copper(II) chromate(VI) is moving towards the positive electrode and, as this is yellow, it is likely to be the chromate. Thus copper(II) chromate(VI) consists of a positive part, copper, which is attracted to the negative electrode and a negative part, chromate, which is attracted towards the positive electrode.

Fig. 6.6 The conduction of electricity by copper(II) chromate(VI) solution

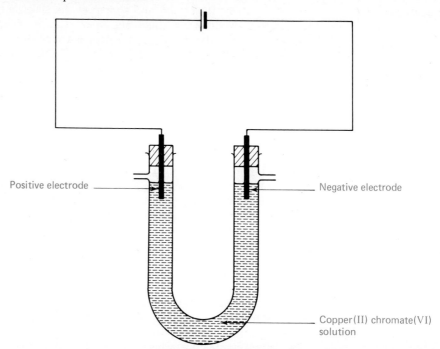

Positive electrode

Negative electrode

Copper(II) chromate(VI) solution

When a direct electric current is passed through molten lead(II) bromide, with the electrodes far enough apart to be able to distinguish what happens at each electrode, Fig. 6.3, a small globule of lead is seen to collect under the negative electrode and faint brownish fumes of bromine are given off near the positive electrode. Again this provides evidence that a compound which conducts electricity has a positive part, in this case lead, and a negative part, in this case bromine.

It can be concluded that conduction by a compound depends on the presence of positively and negatively charged particles (that is, particles which are attracted to the negative and positive electrodes respectively) and the freedom of movement of the charged particles. The converse must be true. If a substance does not conduct electricity, then either it does not contain charged particles or the particles are unable to move.

When the elements lead and bromine are not combined with other elements, they are not charged. The atoms of lead and bromine are said to be neutral. Bromine exists as diatomic molecules (5.6) rather than single atoms, and the molecules are also neutral. The charged forms of these elements which are present in the compound lead(II) bromide are called **ions**.

52

When a direct electric current is passed through molten lead(II) bromide, positively charged lead ions are attracted to the negative electrode where they are converted to neutral lead atoms, and negatively charged bromide ions are attracted to the positive electrode where they are converted into neutral bromine molecules. During this process the ions of lead and bromine have lost their charge and therefore have been neutralised. The next section is concerned with the quantity of electricity which is required to neutralise ions.

6.8 Quantity of electricity

An electric current is the movement of charge through a conductor. In the case of metals and graphite the charge is carried by electrons. In the case of compounds the charge is carried by ions. The quantity of electricity passing through a substance is thus the quantity of charge passing through it.

Charge is measured in units called coulombs, but it is normal to measure the flow of electricity by the current, the units for which are called amperes. The current, in amperes, is the rate of flow of charge in coulombs per second.

For example, if a current of 2 amperes is passing through a conductor, charge is flowing through it at the rate of 2 coulombs per second. Therefore, if this current is allowed to flow for 30 seconds, the total quantity of electricity passing through the substance in this time is:

$$2 \times 30 = 60 \text{ coulombs}$$

6.9 Neutralisation of ions

By recording the current and the time, and finding the mass of lead produced, it is possible to find the mass of lead ions which are neutralised by a certain quantity of electricity.

For example, if on passing a current of 4 amperes through molten lead(II) bromide for 10 minutes 2·5 g of lead are produced, we can say that the quantity of electricity used

$$= 4 \times 10 \times 60$$
$$= 2\,400 \text{ coulombs}$$

Thus 2·5 g of lead are formed by neutralising lead ions with 2 400 coulombs of electricity.

It is more useful to express the quantity of lead in moles of atoms rather than in grams.

$$1 \text{ mole of lead atoms} = 207 \text{ g}$$

$$\text{hence } 2·5 \text{ g of lead} = \frac{2·5}{207} \text{ moles of atoms}$$

$$= 0·0121$$

0·0121 moles of lead atoms are formed by 2400 coulombs of electricity.

1 mole of lead atoms are formed by 198 000 coulombs of electricity. 1 mole of lead atoms contains approximately 6×10^{23} atoms (4.10). Hence 6×10^{23} ions of lead are neutralised to form 6×10^{23} atoms of lead by 198 000 coulombs of electricity.

A small change in the mass of lead produced will result in a considerable change in the final quantity of electricity required to neutralise 1 mole of ions. It is therefore more sensible when using the results of this experiment to acknowledge this limitation by not claiming a high degree of accuracy in the final answer and approximating it to 200 000 coulombs, which is more conveniently written as 2×10^5 coulombs.

Table 6.2 shows the approximate results from these two experiments and some other experiments which are too difficult or dangerous to perform in a school laboratory.

TABLE 6.2. *Quantity of electricity required to neutralise 1 mole of ions of a selection of elements*

ION	COULOMBS MOLE^{-1}
silver	1×10^5
copper	2×10^5
lead	2×10^5
sodium	1×10^5
calcium	2×10^5
aluminium	3×10^5
chloride	1×10^5
bromide	1×10^5
hydrogen	1×10^5

6.10
Charges on ions

The quantity of electricity required to neutralise 1 mole of ions will depend on the size of the charge on the ions. Inspection of the results in Table 6.2 shows that there is a recognisable pattern, namely, that all of the ions require either approximately 10^5 coulombs per mole or a simple multiple of this quantity. None of the ions requires less than 10^5 coulombs per mole.

1 mole of copper ions (6×10^{23} ions) requires twice as much electricity to neutralise them as is required by 1 mole of silver ions (6×10^{23} ions) and therefore, the charge on a copper ion is twice that on a silver ion. The actual charge on one silver ion is very small, it is approximately:

$$\frac{10^5}{6 \times 10^{23}} = 1 \cdot 7 \times 10^{-19} \text{ coulombs}$$

Therefore, as in the case of atomic masses (4.9) it is more convenient to use relative charges than actual charges. If the smallest charge found on any positive ion (e.g. silver) is said to be $1+$, then the charge on a copper ion, which requires twice as much electricity, will be $2+$. Similarly, negative ions require simple multiples of 10^5 coulombs for neutralisation of 1 mole of ions and the relative negative charges on the ions are represented by $1-$ and $2-$.

Table 6.3 shows the symbols used to represent 1 mole, and the relative charges, of some common positive and negative ions. Note that the 1 in front of the single positive and negative charge is usually omitted.

TABLE 6.3. *Formulae and charges of some common ions*

POSITIVE IONS		NEGATIVE IONS	
NAME	FORMULAE	NAME	FORMULAE
sodium	Na^+	chloride	Cl^-
potassium	K^+	bromide	Br^-
silver	Ag^+	iodide	I^-
calcium	Ca^{2+}	oxide	O^{2-}
copper	Cu^{2+}		
lead	Pb^{2+}		
aluminium	Al^{3+}		
hydrogen	H^+		

Compounds have no net charge and hence are neutral. If the relative charges of two ions which form a compound are known, the empirical formula (4.11) of the compound is obtained by using the minimum number of positive and negative ions which are required to produce a neutral compound. For example, there must be twice as many chloride ions (Cl^-) as calcium ions (Ca^{2+}) in the compound calcium chloride

54

and hence its empirical formula is $CaCl_2$. Calcium oxide will contain equal numbers of calcium ions (Ca^{2+}) and oxide ions (O^{2-}) and hence its formula will be CaO.

Now using this procedure try to predict the empirical formulae of the following compounds: potassium bromide, silver oxide, aluminium chloride and copper oxide.

The construction of formulae from combining powers, derived from conduction experiments and by other means, will be discussed in more detail in Chapter 7.

6.11 Michael Faraday

The quantity of electricity required to neutralise 1 mole of singly charged positive or negative ions, that is approximately 10^5 coulombs (more accurately 96 500 coulombs) is known as 1 Faraday after the famous scientist Michael Faraday (1791–1867).

Fig. 6.7 Michael Faraday working in his laboratory at the Royal Institution. (Courtesy The Royal Institution, London)

Faraday was the son of a blacksmith and he educated himself while serving an apprenticeship as a bookbinder. He brought himself to the attention of Sir Humphry Davy who was at the Royal Institution in London by attending some of his lectures and sending him a bound copy of the notes he had made of the lectures. Faraday joined the staff of the Royal Institution and eventually became the director of its laboratory. While there he made many important discoveries in both physics and chemistry. He summarised the results of his experiments on the chemical effects of electricity in two laws.

Faraday's first law states that
the quantities of substances liberated by an electric current are directly proportional to the quantity of electricity used.

Faraday's second law can be stated, in modern form, as
the quantity of electricity required to neutralise 1 mole of ions to form 1 mole of atoms of an element is 1 Faraday (96 500 coulombs) or a simple multiple of this quantity.

Thus as sodium ions have a relative charge of 1 +, 1 mole of atoms will be formed by

1 Faraday, whereas calcium ions have a relative charge of 2 + and 1 mole of atoms will be formed by 2 Faradays of electricity. Thus if the relative charge on ion is known, it is possible to use Faraday's second law to predict the mass of the element which will be formed by a certain quantity of electricity.

Example: What mass of copper will be deposited at the negative electrode when a current of 2 amperes is passed through a solution of copper(II) sulphate for one hour?

The quantity of electricity used $= 2 \times 1 \times 60 \times 60$

$$= 7\,200 \text{ coulombs}$$

The copper ions in copper(II) sulphate have a relative charge of 2 +. Thus 2 Faradays of electricity will be required to form 1 mole of copper atoms.

That is $2 \times 96\,500$ coulombs will form 63·5 g of copper

$$\text{1 coulomb will form} \qquad \frac{63 \cdot 5}{2 \times 96\,500} \text{ g of copper}$$

$$\text{7200 coulombs will form} \qquad \frac{63 \cdot 5 \times 7200}{2 \times 96\,500} \text{ g}$$

$$= 2 \cdot 37 \text{ g of copper}$$

The mass of copper deposited will be 2·37 g.

Faraday was also the first person to use the names which are still commonly used when describing the chemical effects of electricity.

Electrolyte and **non-electrolyte** are used to describe compounds which, respectively, do and do not conduct electricity when either in the liquid form or in aqueous solution.

Electrolysis is the name given to the process by which a compound is broken down by passing a direct electric current through it. The materials dipping into the electrolyte during electrolysis are called electrodes. The positive electrode is called the **anode** (from the Greek meaning the way up) as this was the electrode by which electricity was thought to enter the apparatus, and the negative electrode is called the **cathode** (from the Greek meaning the way down).

The ions which are attracted to the anode, that is the negative ions, are called **anions** and those attracted to the cathode, that is the positive ions, are called **cations**.

You will meet these terms again when the products of electrolysis and the production of electricity are discussed in Chapter 19 and Chapter 20 respectively.

6.12 Summary

1. For a substance to be able to conduct electricity it must contain charged particles which are free to move through the substance.
2. For compounds which conduct electricity, the charged particles are called ions.
3. The ions in a compound are able to move only when the compound is in the liquid state and (if it is soluble) when it is dissolved in water.
4. When a molten compound or an aqueous solution of a compound conducts electricity, the positive ions are attracted to the negative electrode where they are neutralised and the negative ions are attracted to the positive electrode where they are neutralised.
5. The quantities of electricity required to neutralise 1 mole of ions of different elements are related to the relative charges on the ions.
6. The relative charges on ions can be used to predict the empirical formulae of compounds.
7. Faraday was the first person to use the specialised terms electrolyte, electrolysis, anode and cathode when describing the chemical effects of electricity.
8. The mass of a substance liberated during electrolysis depends on the quantity of electricity used and on the relative charge on the ions which are discharged.

Chapter 7

Representing substances by formulae, and chemical reactions by equations

How many moles of copper are formed when 1 mole of iron reacts with excess copper(II) sulphate solution?

When iron is added to copper(II) sulphate solution, the iron dissolves and copper is displaced and left as a precipitate. If excess copper(II) sulphate solution is used, all of the iron will dissolve and the mass of copper formed from a certain mass of iron can be found.

These masses can be used to find the number of moles of copper formed when 1 mole of iron is reacted with the copper(II) sulphate solution, which is the first step in finding the equation for the reaction.

You will need clean iron filings, a concentrated solution of copper(II) sulphate and a small volume of propanone.

Weigh a dry test-tube. Place about 0·5 g of iron filings in the test-tube and reweigh the tube.

Carefully warm about half a test-tube full of copper(II) sulphate solution. When it is hot, pour it, in about 1 cm³ portions, on to the iron filings. After each addition, shake the test-tube to ensure good mixing.

When the reaction appears to be complete and the blue colour of excess copper(II) sulphate solution can be seen, allow the copper to settle. Carefully remove as much as possible of the excess copper(II) sulphate solution by means of a teat pipette.

Add about 2 cm³ of pure water, shake the tube, allow the copper to settle and remove the liquid again. Repeat this procedure using propanone instead of water.

The remaining traces of propanone can be removed, leaving dry copper, by placing the test-tube in a beaker of boiling water for a few minutes.

Weigh the dry test-tube and copper.

Make a list of your results and calculate the mass of iron used and the mass of copper produced. Look up the relative atomic masses of iron and copper and use them to calculate the number of moles of copper produced when 1 mole of iron reacts with copper(II) sulphate solution.

The formula of copper(II) sulphate is $CuSO_4$. Predict the formula of the other product of the reaction besides copper and write a complete balanced equation for the reaction.

Question

Select the appropriate symbols from the list (s), (l), (g) and (aq) (see 7.11) and indicate the physical states of the reactants and products in the equation.

How many moles of hydrogen are produced when 1 mole of magnesium reacts with excess hydrochloric acid?

If a known mass of magnesium is added to an excess of hydrochloric acid, all the magnesium will react and the hydrogen produced can be collected in a syringe and its volume measured.

Thus the volume of hydrogen produced by a certain mass of magnesium will be known, from which it is possible to calculate the number of moles of hydrogen molecules produced by 1 mole of magnesium atoms, which is the first step in finding the equation for the reaction.

You will need a boiling-tube and syringe connected together as shown in Fig. 7.1, some clean magnesium ribbon and dilute (2M) hydrochloric acid.

Fig. 7.1

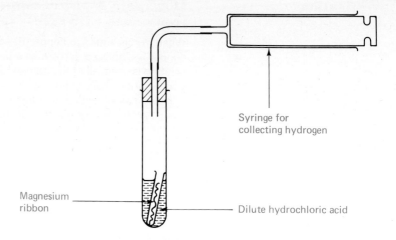

Syringe for collecting hydrogen

Magnesium ribbon

Dilute hydrochloric acid

Pour dilute hydrochloric acid into the boiling-tube to a depth of about 3 cm. Weigh about 8 cm of clean magnesium ribbon.

Put the ribbon into an ignition-tube and stand the tube vertically in the acid, taking care that the magnesium does not come into contact with the acid.

Connect the boiling-tube to the syringe and note the reading on the syringe.

Allow the acid to come into contact with the magnesium by tilting the boiling-tube. Make sure the piston of the syringe does not stick by rotating it gently as the gas is being given off. When all the magnesium has reacted, record the new reading on the syringe and calculate the volume of hydrogen produced.

Look up the relative atomic mass of magnesium in the Data Section and use it to calculate the number of moles of magnesium atoms reacting.

Using the fact that 1 mole of molecules of a gas will occupy approximately 24 dm^3 at room temperature and pressure, calculate the number of moles of molecules of hydrogen produced. Then calculate (to the nearest whole number) the number of moles of hydrogen molecules produced by 1 mole of magnesium atoms.

The formula of 1 mole of magnesium atoms is Mg.

The formula of 1 mole of hydrogen molecules is H_2.

The formula of 1 mole of hydrochloric acid is HCl.

Predict the formula of the other product besides hydrogen and write the balanced equation for the reaction.

Question

What are the physical states of the reactants and products? Select the appropriate symbols for these states (7.11) and include them in the equation.

58

How many moles of potassium iodide react with 1 mole of lead(II) nitrate?

When a solution of lead(II) nitrate is mixed with a solution of potassium iodide, a bright yellow precipitate is formed. All potassium compounds are soluble and hence the precipitate is likely to be lead(II) iodide.

The first stage of the experiment involves taking several equal portions of potassium iodide solution and adding to each successive portion a larger volume of lead(II) nitrate solution in order to find out the volume of lead(II) nitrate solution which is needed to use up all of the potassium iodide present. When the reaction is complete no more precipitate will be formed, and if the precipitate is allowed to settle, a constant height of precipitate will indicate the 'end point' of the reaction.

You will need six test-tubes with equal internal diameters, 0·5M solutions of potassium iodide and lead(II) nitrate, and two burettes or plastic syringes for measuring out up to 5·0 cm³ of the solutions.

Place the six test-tubes into a test-tube rack and put 5·0 cm³ of the potassium iodide solution into each test-tube (Fig. 7.2).

Fig. 7.2

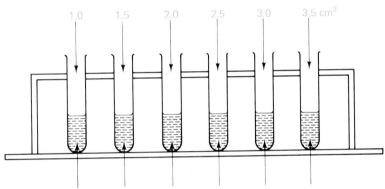

Volume of lead(II) nitrate solution to be added

1.0   1.5   2.0   2.5   3.0   3.5 cm³

5.0 cm³ of potassium iodide solution in each test-tube

Add 1·0 cm³ of lead(II) nitrate solution to the first test-tube, 1·5 cm³ to the second, 2·0 cm³ to the third, 2·5 cm³ to the fourth, 3·0 cm³ to the fifth and 3·5 cm³ to the sixth. Stir each test-tube an equal number of times and allow the precipitates to settle.

Measure the heights of the precipitates and plot a graph of height of precipitate against volume of lead(II) nitrate solution added.

From the graph, estimate the volume of lead(II) nitrate solution which has to be added to first give the final constant height of precipitate. This volume is the minimum quantity of the lead(II) nitrate solution which is required to react completely with the 5·0 cm³ of potassium iodide solution.

Use these two volumes and the molarities (7·9) of the solutions to calculate the number of moles of each substance taking part in your reaction. Then calculate the number of moles of potassium iodide which reacts completely with 1 mole of lead(II) nitrate.

The formula of potassium iodide is KI and that of lead(II) nitrate is $Pb(NO_3)_2$. Putting the appropriate numbers in front of these formulae, write down the left-hand side of the equation for the reaction. Now predict the complete balanced equation for the reaction.

1   What are the physical states of the reactants and products of this reaction? Indicate these physical states by including the appropriate symbols (7.11) in the equation.

2   Why is it necessary to use test-tubes with similar internal diameters?

7.4
Formulae and combining ratios

When atoms of one element combine with the atoms of another, they always do so in a small whole number ratio. The combining masses of the two elements can be found by experiment and used to calculate the ratio of the number of moles of each element combining and hence the ratio of the number of atoms combining (4.11). The usual way of writing down the result is to represent 1 mole of atoms of each element by a symbol and the ratio by small numbers written after each symbol. Thus in the case of water the ratio is:

2 moles of hydrogen atoms to 1 mole of oxygen atoms. This can be represented by H_2O which is called the formula for water (or more strictly the empirical formula, 4.11).

When the ratio involves the number 1, it is usual not to write the 1 in the formula, that is water is written as H_2O rather than as H_2O_1.

Thus it is possible to find the formulae of compounds from experimental results. Table 7.1 lists some simple compounds and their formulae which have been found in this way.

TABLE 7.1. *The formulae of some simple compounds*

NAME	FORMULA
hydrogen chloride	HCl
sodium chloride	NaCl
hydrogen bromide	HBr
sodium bromide	NaBr
hydrogen iodide	HI
sodium iodide	NaI
magnesium chloride	$MgCl_2$
water	H_2O

There do appear to be some patterns in these formulae. For example, when a halogen (chlorine, bromine or iodine) combines with sodium, it is always one mole of halogen atoms which combines with one mole of sodium atoms rather than two or three. Also, one mole of hydrogen atoms combines with one mole of atoms of each halogen. The existence of a pattern becomes even more convincing when the formula of the compound between sodium and hydrogen (sodium hydride, NaH) is included:

1 mole of hydrogen atoms combines with 1 mole of chlorine atoms,
1 mole of sodium atoms combines with 1 mole of chlorine atoms and
1 mole of sodium atoms combines with 1 mole of hydrogen atoms.

This pattern can be summarised by saying that hydrogen, sodium and chlorine in these compounds each have a **combining power** of 1. The formula of magnesium chloride, $MgCl_2$, indicates that there are always two moles of chlorine required for every mole of magnesium. Thus magnesium, compared to chlorine, has a combining power of 2. You should now be able to predict the formula of magnesium bromide and of magnesium iodide.

Magnesium has a combining power of 2 and bromine has a combining power of 1, therefore two moles of bromine are needed for every mole of magnesium and the formula of the compound is $MgBr_2$. Similarly, the formula of magnesium iodide is MgI_2.

The formula of water, H_2O, indicates that there are always two moles of hydrogen

60

atoms for every mole of oxygen atoms and as hydrogen has a combining power of 1, the oxygen must have a combining power of 2.

You should now be able to predict the formula of a compound of sodium and oxygen, and the formula of a compound of magnesium and oxygen.

It does appear from an examination of the formulae of compounds that elements do have particular combining powers in several compounds.

An alternative name which is often used for combining power is **valency**.

7.5
Charges on ions and valencies

A clue as to why some metallic elements and non-metallic elements have particular valencies is provided by the experiments, described in Chapter 6, during which electricity is passed through compounds either in their liquid state or dissolved in water. These experiments show that compounds of metals and non-metals frequently contain charged atoms (called ions). The metallic ions (and the hydrogen ion) are positively charged and the non-metallic ions are negatively charged.

The formula of a compound which contains ions is the smallest ratio of ions of each element which will produce a neutral compound. For example,

the relative charge on a magnesium ion is $2+$ (Mg^{2+}),
the relative charge on a bromide ion is $1-$ (Br^-).
Thus the formula of magnesium bromide is $MgBr_2$.

We can see that magnesium has a valency of 2 because its ion has a relative charge of $2+$ and bromine has a valency of 1 because its ion has a relative charge of $1-$.

To find more fundamental reasons for why particular elements have particular valencies and ions particular charges we need to find out more about the structures of atoms. Chapters 14 and 15 are concerned with atomic structure and why and how atoms combine together.

7.6
Construction of formulae

The previous sections and the previous three chapters, show how it is possible to find the formulae of compounds from experimental results, but obviously we cannot do an experiment every time we want to use a formula. We must remember the formula, or look at a list of formulae which have been found by other people, or construct the formulae by using the valencies of the elements or the charges on their ions.

Table 7.2 shows the charges on some common positive ions and hence their valencies.

TABLE 7.2. *Charges on some positive ions*

VALENCIES					
1		2		3	
NAME	ION	NAME	ION	NAME	ION
sodium	Na^+	magnesium	Mg^{2+}	aluminium	Al^{3+}
potassium	K^+	calcium	Ca^{2+}	iron(III)	Fe^{3+}
silver	Ag^+	zinc	Zn^{2+}		
copper(I)	Cu^+	copper(II)	Cu^{2+}		
hydrogen	H^+	iron(II)	Fe^{2+}		
ammonium	NH_4^+	lead(II)	Pb^{2+}		

Notice that all of the ions in Table 7.2 (except hydrogen and ammonium) are ions of metallic elements.

It can be seen from Table 7.2 that some metals form more than one type of ion. There exist, for example, two compounds in which copper and oxygen are combined:

Cu_2O, called copper(I) oxide (the (I) indicating that the compound contains copper which is using a valency of 1),

CuO, called copper(II) oxide (the (II) indicating that the compound contains copper which is using a valency of 2).

The most important negative ions are those which are derived from the common acids. Table 7.3 shows the charges on these ions together with, where appropriate, the acids with which they are associated.

ACID		ION		
NAME	FORMULA	NAME	FORMULA	VALENCY
hydrochloric	HCl	chloride	Cl^-	1
nitric	HNO_3	nitrate	NO_3^-	1
sulphuric	H_2SO_4	sulphate	SO_4^{2-}	2
		hydrogensulphate (bisulphate)	HSO_4^-	1
sulphurous	H_2SO_3	sulphite	SO_3^{2-}	2
		hydrogensulphite (bisulphite)	HSO_3^-	1
carbonic	H_2CO_3	carbonate	CO_3^{2-}	2
		hydrogencarbonate (bicarbonate)	HCO_3^-	1
		oxide	O^{2-}	2
		hydroxide	OH^-	1

Notice that all of the ions in Table 7.3 are made up of one or more non-metallic elements.

It is important to remember that not all compounds conduct electricity when molten or dissolved in water and therefore not all compounds contain ions. Some common non-metals seldom, if ever, form ions. For example, carbon which has a valency of 4 and nitrogen which has a valency of 3 only rarely form ions.

When using the valencies from Tables 7.2 and 7.3 to construct formulae, several rules are applied:

1. The valencies (or charges on the ions) must balance. For example, the atoms of an element with a valency of 2, such as zinc, will combine with twice as many atoms of an element, such as chlorine, which has a valency of 1. Thus the formula of zinc chloride is $ZnCl_2$.

2. Where one of the components consists of atoms of more than one element bonded together, e.g. nitrate, NO_3^-, and two or more of these ions are required in the formula, the ion is enclosed in brackets before writing the number to indicate the ratio. For example, in magnesium nitrate there will be twice as many nitrates as magnesiums, therefore, the formula is $Mg(NO_3)_2$ rather than $MgNO_{32}$, which would mean something completely different and obviously would be incorrect. Similarly, the formula MgN_2O_6 does not enable us to recognise easily that the compound is a nitrate.

3. It is usual to put the more metallic part of the compound first in the name and the formula. For example, it is normal to write sodium chloride which is NaCl, rather than chloride sodium and ClNa.

For the moment it is important that you treat these tables as aids to help you construct formulae. Do not be concerned about the seemingly large number of elements and valencies; these will become more meaningful when you have studied the elements themselves and know more about the structures of atoms. Remember that it would be possible to use the tables to construct the formula of a compound that is not known to exist, e.g. aluminium carbonate. You must have good evidence for the existence of a compound before constructing and using its formula.

In Chapter 4 one mole of atoms of an element was defined as the amount of the element that contains the same number of atoms as in 1 g of hydrogen (or 12g of carbon-12), i.e. 6×10^{23} atoms. The amount of an element which contains this number of atoms is one relative atomic mass of the element expressed in grams. The mole is also useful for substances which do not exist in the form of single atoms. In these cases the mole is the amount of substance which contains the same number of particles as there are atoms in 1 g of hydrogen (or 12 g of carbon-12). For example, some non-metallic elements exist in the form of diatomic molecules (5.6),

e.g. hydrogen, H_2; oxygen, O_2; nitrogen, N_2; chlorine, Cl_2.

In the case of oxygen, its relative atomic mass, expressed in grams, 16 g, contains 6×10^{23} atoms combined together to form 3×10^{23} molecules.

One relative molecular mass of oxygen, expressed in grams, 32 g, contains 12×10^{23} atoms combined together to form 6×10^{23} molecules.

For compounds, you may not always know whether they exist as separate molecules, e.g. carbon dioxide as CO_2 molecules, or as collections of oppositely charged ions, e.g. sodium chloride, Na^+Cl^-, but whatever the case, 1 mole of the compound is always 1 relative formula mass of the compound expressed in grams. For example:

1 mole of iron (56 g) contains 6×10^{23} atoms of iron,

1 mole of sulphur (32 g) contains 6×10^{23} atoms of sulphur.

When these amounts combine to form iron sulphide, 88 g of the compound will be formed which will contain the atoms of iron and sulphur combined together to form 6×10^{23} iron-sulphur pairs.

i.e. 6×10^{23} atoms of Fe $+ 6 \times 10^{23}$ atoms of S $\rightarrow 6 \times 10^{23}$ FeS units.

Table 7.4 shows how the unit amount of substance (the mole) can be used for elements and in Table 7.5 for compounds.

TABLE 7.4. *Moles of elements*

ELEMENT	MASS OF 1 MOLE OF ATOMS	MASS OF 1 MOLE OF MOLECULES
carbon	C = 12 g	
sodium	Na = 23 g	
hydrogen	H = 1 g	H_2 = 2 g
oxygen	O = 16 g	O_2 = 32 g
nitrogen	N = 14 g	N_2 = 28 g
chlorine	Cl = 35·5 g	Cl_2 = 71 g

TABLE 7.5. *Moles of compounds*

COMPOUND	FORMULA	MASS OF 1 MOLE	
iron(II) sulphide	FeS	56 + 32	= 88g
sodium chloride	NaCl	23 + 35·5	= 58·5 g
water	H_2O	$(2 \times 1) + 16$	= 18 g
carbon dioxide	CO_2	$12 + (2 \times 16)$	= 44 g
calcium hydroxide	$Ca(OH)_2$	$40 + 2(16 + 1)$	= 74 g
magnesium sulphate	$MgSO_4$	$24 + 32 + (4 \times 16)$	= 120 g

7.8
Representing chemical reactions

When a chemical reaction occurs, and the identities of the starting substances (reactants) and the products are known, it is possible to represent the reaction by a **word equation**,

e.g.   iron + sulphur → iron sulphide

This word equation tells us only the names of the reactants and products, but it does not tell us the ratio of the numbers of atoms of iron and sulphur which combine together, and the equation could not be used to calculate the mass of iron sulphide which would be formed from a certain mass of iron, and yet these are the sorts of questions that a chemist needs to be able to answer.

In the previous section, we saw that:

1 mole of iron (Fe) combines with 1 mole of sulphur (S) to form 1 mole of iron(II) sulphide.

This enables us to write an equation using formulae rather than words:

$$Fe + S \rightarrow FeS$$

This **equation** is more useful than the word equation because anyone reading it will know that atoms of iron and sulphur combine in a 1:1 ratio. Later in this chapter you will see how it could be used to predict how much iron(II) sulphide would be formed from a certain mass of iron.

Wherever possible, chemists represent chemical reactions by equations of this type as they convey so much useful information.

7.9
Determination of equations

The equation for a chemical reaction can be worked out from experimental results or when the type of reaction occurring is known, the equation can be predicted.

Equations from reacting masses
Example: 6 g of magnesium (Mg) on heating in oxygen (O_2) produced 10 g of magnesium oxide (MgO).

Thus   6 g of Mg + 4 g of O_2 → 10 g MgO

The relative atomic masses of Mg and O are 24 and 16 respectively.

$$\frac{6}{24} \text{ moles of Mg} + \frac{4}{32} \text{ moles of } O_2 \rightarrow \frac{10}{40} \text{ moles of MgO}$$

$$\frac{1}{4} \text{ moles of Mg} + \frac{1}{8} \text{ moles of } O_2 \rightarrow \frac{1}{4} \text{ moles of MgO}$$

Converting this to a simple whole number ratio:

$$2 \text{ moles of Mg} + 1 \text{ mole of } O_2 \rightarrow 2 \text{ moles of MgO}$$

Therefore this reaction may be represented by the formula equation:

$$2Mg + O_2 \rightarrow 2MgO$$

Equations from volumes of solutions reacting together
Chemical reactions are frequently carried out by firstly dissolving the reactants in water and then reacting them together. The concentration of a solution can be expressed in grams per cubic decimetre (or grams per litre) which is abbreviated to g dm^{-3} or more usefully as the number of moles per cubic decimetre, i.e. mol dm^{-3}. The concentration of a solution expressed in mol dm^{-3} is known as the **molarity** of the solution.

For example, 1 mole of sodium hydroxide (NaOH) has a mass of 40 g, a solution containing this amount of sodium hydroxide in 1 dm³ of solution has a concentration of 1 mol dm⁻³ and is called a 1M (or molar) solution.

A solution containing 4 g of sodium hydroxide in 1 dm³ of solution has a concentration of 0·1 mol dm⁻³ and is called a 0·1M solution. If the molarities of two solutions reacting together are known and there is some means of detecting when the reaction is complete, and hence what volumes react together, it is possible to find the ratio of the number of moles of the substances which react together. This ratio is the first stage in writing the equation.

Example: Sulphuric acid solution reacts with sodium hydroxide solution. The end of the reaction can be detected by an indicator which is a substance which is a different colour in an acid than it is in an alkali (11.13). In an experiment to find the ratio of the number of moles of the compounds which react together, 12·5 cm³ of 0·1M H_2SO_4 solution are found to react with 25·0 cm³ of 0·1M NaOH solution.

1000 cm³ of 0·1M H_2SO_4 solution contains 0·1 mole of H_2SO_4

12·5 cm³ contains $\dfrac{0·1}{1000} \times 12·5 = 0·00125$ moles of H_2SO_4

1000 cm³ of 0·1M NaOH solution contains 0·1 mole of NaOH

25·0 cm³ contains $\dfrac{0·1}{1000} \times 25·0 = 0·0025$ moles of NaOH

Thus 0·0025 moles of NaOH react with 0·00125 moles of H_2SO_4 .

Dividing through by 0·00125 to obtain a simple whole number ratio,

2 moles of NaOH react with 1 mole of H_2SO_4

At this stage the equation for the reaction can be written as:

$$2NaOH + H_2SO_4 \rightarrow \text{products}$$

To complete the equation it is necessary either to analyse the resulting solution in order to identify the products and the number of moles of each product formed or to predict the products from our knowledge that acids react with alkalis to form salts (11.9) and water. In this reaction the products are sodium sulphate and water.

At this stage we can write:

$$2NaOH + H_2SO_4 \rightarrow Na_2SO_4 + H_2O$$

To complete the equation we must use our knowledge that:

All the atoms which appear on the left-hand side of the equation must appear on the right-hand side (Law of Conservation of Mass, 4.7).

In order to satisfy this condition it is necessary to place a 2 in front of the H_2O which indicates that two moles of water will be formed whenever two moles of sodium hydroxide react with one mole of sulphuric acid. The complete equation is:

$$2NaOH + H_2SO_4 \rightarrow Na_2SO_4 + 2H_2O$$

An equation which satisfies this second condition is said to be **balanced**. Only when an equation is balanced will the correct ratio of moles of reactants and products be indicated.

It must be emphasised that when balancing equations numbers must be placed in front of formulae where necessary, but the formulae themselves must not be changed. In the above example, it was correct to write $2H_2O$ but it would have been incorrect to write H_4O_2 as this would no longer represent water, indeed there is no substance with a formula H_4O_2.

7.10
Predicting equations

We cannot carry out experiments every time we wish to write down the equation for a reaction and so it is useful to be able to remember a lot of different types of equations which will enable you to predict the identity of the products of the reaction. Then using your knowledge of the valencies (or charges on ions) you can work out the formulae for the reactants and products.

Example: One general type of reaction is that all carbonates react with acids to form carbon dioxide, a salt and water (28.15). This enables you to predict that in the case of calcium carbonate with hydrochloric acid, the products of the reaction will be carbon dioxide, calcium chloride and water. Using the valencies in 7.5, the formulae of the reactants and products can be predicted to be:

calcium carbonate	$CaCO_3$
hydrochloric acid	HCl
carbon dioxide	CO_2
calcium chloride	$CaCl_2$
water	H_2O

The equation, before checking whether it is balanced, is:

$$CaCO_3 + HCl \rightarrow CO_2 + CaCl_2 + H_2O$$

This equation is not balanced because there are two moles of chlorine atoms and two moles of hydrogen atoms on the right-hand side whereas there is only one mole of each element on the left-hand side. To balance the equation it is necessary to write a 2 in front of HCl. The equation becomes:

$$CaCO_3 + 2HCl \rightarrow CO_2 + CaCl_2 + H_2O$$

This equation is now balanced and shows that 1 mole of calcium carbonate will react with 2 moles of hydrochloric acid to form 1 mole of carbon dioxide, 1 mole of calcium chloride and 1 mole of water. This information can be used to predict the amount of a product that can be formed from a known amount of a reactant.

7.11
States of reactants and products

Chemical equations often include additional information which indicates the physical state of each substance. The equation shows whether the substance is in its solid, liquid or gas form at the temperature at which the reaction is performed, or if it is present in the form of a solution in water. For example the above equation can be written as,

$$CaCO_3(s) + 2HCl(aq) \rightarrow CO_2(g) + CaCl_2(aq) + H_2O(l)$$

where: (s) indicates that calcium carbonate is a solid,
(aq) indicates that hydrochloric acid and calcium chloride are in aqueous solution,
(g) indicates that carbon dioxide is a gas and is therefore given off as bubbles of gas,
(l) indicates that water is in its liquid form.

Where it is helpful to do so, the physical states of reactants and products are included in the equations used in the remaining chapters of this book.

7.12
The masses of reactants and products

If the balanced equation for a reaction is known, it is possible to use it to predict the mass of one reactant which is required to react with a certain mass of the other reactant and also the masses of the products which will be formed.

Example: The first stage in the conversion of galena, which is an important ore of lead, into the metal, is to heat it in air.

66

Galena, lead(II) sulphide, is converted to lead(II) oxide. The equation for the change is:

$$2PbS(s) + 3O_2(g) \rightarrow 2PbO(s) + 2SO_2(g)$$

The next stage, which can be carried out easily in a school laboratory, converts the lead(II) oxide to lead by heating it with carbon:

$$PbO(s) + C(s) \rightarrow Pb(l) + CO(g)$$

This equation is balanced and can be used to predict the mass of lead formed from a certain mass of lead(II) oxide.

For example, what mass of lead can be obtained from 10 g of lead(II) oxide? The balanced equation shows that:

$$1 \text{ mole of lead(II) oxide} \rightarrow 1 \text{ mole of lead}$$

The relative atomic masses of lead and oxygen are 207 and 16 respectively. Thus 1 mole of lead(II) oxide, PbO, has a mass of $207 + 16 = 223$ g and 1 mole of lead, Pb, a mass of 207 g.

Therefore, 223 g of lead(II) oxide $\rightarrow$ 207 g of lead
Divide both sides by 223,

$$1 \text{ g of lead(II) oxide} \quad \rightarrow \quad \frac{207}{223} \text{ g of lead}$$

$$10 \text{ g of lead(II) oxide} \quad \rightarrow \quad 10 \times \frac{207}{223} = 9 \cdot 28 \text{ g of lead}$$

It can be predicted that complete conversion of 10 g of lead(II) oxide will produce 9·3 g of lead.

You should now be able to take the process a stage further back and predict the mass of pure lead(II) sulphide which you would need to start with in order to end up with this mass of lead.

Example: The usual method of preparing ammonia, NH_3, in the laboratory is to heat ammonium chloride with calcium hydroxide (33.14). The balanced equation for the change is:

$$2NH_4Cl(s) + Ca(OH)_2(s) \rightarrow CaCl_2(s) + 2NH_3(g) + 2H_2O(g)$$

This equation can be used to calculate the minimum mass of calcium hydroxide needed to convert a certain mass of ammonium chloride (e.g. 40 g) into ammonia.

From the balanced equation:

$$2 \text{ moles of } NH_4Cl \text{ react with 1 mole of } Ca(OH)_2$$

The relative atomic masses of the elements are:

$N = 14, H = 1, Cl = 35 \cdot 5, Ca = 40, O = 16$
1 mole of NH_4Cl has a mass of $14 + (4 \times 1) + 35 \cdot 5 = 53 \cdot 5$ g
1 mole of $Ca(OH)_2$ has a mass of $40 + 2(16 + 1) = 74$ g

Thus 107 g of ammonium chloride reacts with 74 g of calcium hydroxide.
Divide both sides by 107:

$$1 \text{ g of ammonium chloride reacts with } \frac{74}{107} \text{ g of calcium hydroxide}$$

Therefore,

$$40 \text{ g of ammonium chloride reacts with } 40 \times \frac{74}{107} \text{ g of calcium hydroxide}$$

$$= 27 \cdot 7 \text{ g}$$

The minimum mass of calcium hydroxide required to convert completely 40 g of ammonium chloride into ammonia is 27·7 g.

7.13 Volumes of reactants and products

When gases are used or produced in chemical reactions, it is often more important to know the volumes of the gases rather than their masses. In Chapter 5 (5.8) we saw that 1 mole of any gas must occupy the same volume if measured at the same temperature and pressure. This volume when measured at s.t.p. (273 K and 101 kPa (760 mmHg)) is 22·4 dm³ and at room temperature and pressure (293 K and 101 kPa (760 mmHg)) it is 24·0 dm³. This relationship between moles of gases and their volumes can be used to predict the volumes of gases involved in reactions.

Example: In an important industrial process (Solvay Process, 29.5) sodium chloride is converted into sodium hydrogencarbonate, which is later converted into sodium carbonate. The balanced equation for the second stage of the process is:

$$2NaHCO_3(s) \rightarrow Na_2CO_3(s) + CO_2(g) + H_2O(g)$$

This equation can be used to calculate the volume of carbon dioxide produced by heating a certain mass (e.g. 1 kg) of sodium hydrogencarbonate.

The balanced equation shows that,

2 moles of $NaHCO_3$ will produce 1 mole of CO_2

The relative atomic masses of the elements are:

Na = 23, H = 1, C = 12, O = 16.

1 mole of $NaHCO_3$ has a mass of $23 + 1 + 12 + (3 \times 16) = 84$ g
1 mole of CO_2 will occupy 22·4 dm³ at s.t.p.

Thus 2×84 g of sodium hydrogencarbonate will produce 22·4 dm³ of carbon dioxide at s.t.p.

Divide both sides by 168,

1 g of sodium hydrogencarbonate will produce $\dfrac{22·4}{168}$ dm³ of carbon dioxide

1000 g of sodium hydrogencarbonate will produce $1000 \times \dfrac{22·4}{168}$ dm³

$$= 133·3 \text{ dm}^3 \text{ of carbon dioxide at s.t.p.}$$

Thus for every 1 kg of sodium hydrogencarbonate which is heated 133·3 dm³ (measured at s.t.p.) of carbon dioxide will be produced. Clearly this is useful information for the chemical engineer who is responsible for designing the equipment for this process.

7.14 Summary

1. Elements and ions, including those which are made up of combinations of elements (e.g. SO_4^{2-}), appear to have particular valencies. The valencies used in the formation of certain compounds can be related to the charges on the ions present.
2. The valencies or charges on ions can be used to predict empirical formulae of compounds.
3. It is useful to represent a chemical reaction by a balanced equation which gives the formulae of each reactant and product and the ratio of the number of moles of reactants and products involved in the reaction.
4. It is common to indicate in an equation for a reaction the physical states of the reactants and products by using the appropriate symbols: (s), (l), (g), (aq).
5. Balanced equations can be used to predict the masses and, when gases are involved, the volumes of substances taking part or produced in chemical reactions.

Chapter 8

Air and oxygen

Investigation 8.1

Is there any change in mass when magnesium burns?

You will need a crucible and lid, a pipeclay triangle and a pair of tongs.

Weigh a crucible and lid. Place about 10 cm of loosely coiled magnesium ribbon into the crucible, replace the lid and find the total mass.

Place the crucible on a pipeclay triangle on a tripod and heat it gently at first and then more strongly.

Every two or three minutes remove the Bunsen burner and lift the crucible lid for a short time with a pair of tongs, taking care not to allow any smoke to escape.

When the burning appears to be complete, allow the crucible to cool and weigh the crucible, lid and contents.

Make a list of your results and note how the mass of the ash which has been produced compares with the original mass of the magnesium.

Questions

1 One theory, put forward in the early eighteenth century, suggested that combustion (burning) is a type of decomposition. According to this theory magnesium, on burning, would split up into simpler substances, one being left behind as ash while the other is given off as a gas. Do your results on the mass change in this experiment suggest whether this theory is likely to be true or not? Explain your answer carefully.

2 Another theory, put forward in the late eighteenth century, suggested that combustion involves combination with something from the air. Do your results fit this theory?

What did you notice during the experiment, in particular when you lifted the lid, which might suggest that air is involved in the process?

Investigation 8.2

Is there any change in mass when magnesium is heated under sand?

Place about 10 cm of loosely coiled magnesium ribbon into a crucible. Now fill the crucible to about two thirds of its depth with fine dry sand and tap it gently so that the sand settles. Weigh the crucible and its contents.

Support the crucible in a pipeclay triangle and heat it strongly for about 10 minutes. Allow the crucible to cool and reweigh it.

Record your results in a table and note if there was a change in mass similar to that in Investigation 8.1.

Finally empty out the sand and examine the magnesium and decide whether or not it has burned.

Question

Why do the results of this Investigation differ from those of Investigation 8.1?

What happens to copper when it is heated?

1. Take a piece of clean copper foil, about 3 cm square, and weigh it. Now hold it with tongs and heat it in a hot Bunsen flame until it goes black all over. Allow the foil to cool (this will only take a minute or so) and weigh it again. Compare the mass of the black-coated copper with that of the original foil.

Question

1 At least three theories can be proposed for what happens to copper on heating. One theory says that it decomposes (splits up) on heating. Another theory suggests that soot from the flame is deposited on the copper. A third theory suggests that the copper combines with air to form a new substance. Which of these theories could account for your result? Explain your answer carefully.

2. Rub the foil gently with fine sandpaper to remove the black coat and expose the clean copper surface. Now weigh the foil for the third time. How does its mass now compare with that of the original foil?

Questions

First fold   Second fold

Press tightly
together

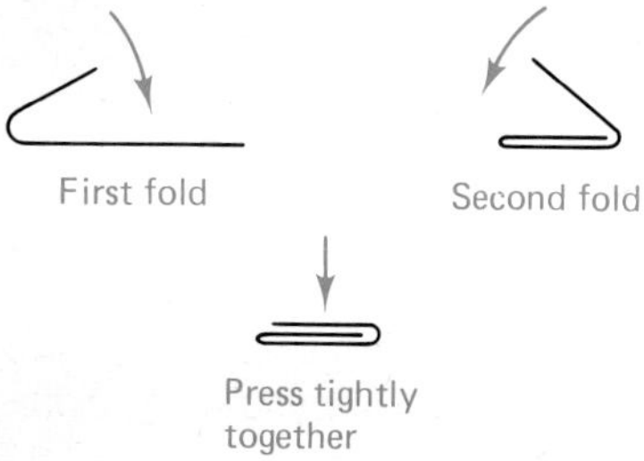

Fig. 8.1 Folding copper foil

2 Can the result of this part of the investigation be explained by the soot theory, which suggests that something is deposited on the surface of the copper and then rubbed off again? Give reasons for your answer.
3 Can the result be explained by the theory that the copper has combined with air and some of the copper itself is part of the black coat? Give reasons for your answer.

3. Finally take the foil and fold it over as shown in Fig. 8.1. Press it flat with a suitable hard object, so that it is folded tightly. Heat the folded foil until the outside is black, then allow it to cool.
   When it is cool, unfold it and examine the surface of the copper which was on the inside.

Question

4 Did the copper on the inside go black? Does this result show that air is involved when copper turns black?

**8.4
Air**

Air is a substance which we tend to take for granted. It is all around us and our life depends upon it, but we cannot see it. We live at the bottom of an ocean, called the Earth's atmosphere, and this ocean of invisible gas exerts a pressure of about 1 kg on each square centimetre of the Earth's surface, yet we are not aware of any burden on our shoulders. We can feel the presence of the air when the wind blows, or when we ride a bicycle fast, but on the whole the air is an unobtrusive substance. Its importance in chemical reactions did not become clear until the late eighteenth century, when scientists turned their attention to a precise study of something with which they were all familiar, that is, **combustion** (burning). The discovery of the chemical composition of air was the turning point in the development of chemistry, and chemistry as an exact science really began at this point.

**8.5
Combustion:
early theory**

The use of fire has always been of vital importance in the development of mankind. Early man found that he could keep himself warm with fire, and cook food to make it more pleasant to eat. Later it was found that by the use of fire, metals could be extracted from minerals and later still fire was used in the production of glass. It was known that fire could be used to bring about chemical changes, and yet the nature of fire itself was not understood.

In the earlier part of the eighteenth century, a theory emerged which said that when any substance was converted into an ash by heating, something was given off—presumably some kind of gas—leaving the ash behind. In modern language, we would call this process 'decomposition' and summarise it as:

$$\text{substance} \xrightarrow{\quad \text{burning} \quad} \text{'ash'} + \text{'something given off'}$$

Now, prior to the eighteenth century, the sciences of physics and astronomy had made enormous progress because of the development of measurement as an aid in scientific investigations. The early alchemists had not thought it worthwhile to make measurements (e.g. of changes in mass or changes in volume) and had relied entirely on what they could see happening, but in the eighteenth century the idea of measuring things in chemistry caught on. As a result certain facts emerged which were hard to explain by the idea of burning as decomposition. The decomposition theory suggested that the ash should have a smaller mass than the original substance and also that on burning in a closed container an increase in volume should be observed (measuring the volume at the original pressure) owing to a gas being given off. When measurements were made, the facts were found to contradict both of these suggestions.

8.6 Combustion: the facts

If magnesium ribbon is heated in a crucible with a loosely-fitting lid (Investigation 8.1) it burns to a white ash. This ash is found to have a greater mass than the original magnesium. Copper, on heating in air, turns black on the surface (Investigation 8.3). In this case there are no flames to indicate that the copper is burning but again an increase in mass is found.

Experiments can be done in which these, and other, elements are burned or heated in a confined volume of air. Two of these experiments will be described. First, magnesium can be burned inside a bell jar over water (Fig. 8.2). A coil of magnesium ribbon floats in a basin inside the bell jar. Its end is ignited by a taper inserted through the neck of the jar, then the stopper is quickly replaced.

Fig. 8.2 Burning magnesium in a bell jar

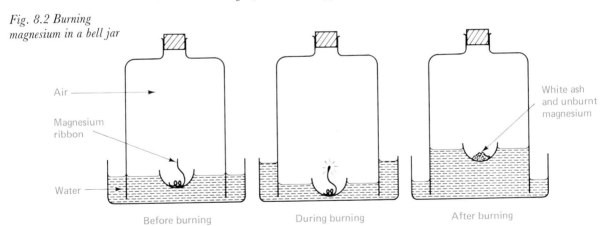

Air

Magnesium ribbon

Water

White ash and unburnt magnesium

Before burning   During burning   After burning

During the burning the water level inside the jar falls, owing to the fact that the intense heat generated by the burning magnesium causes the air in the bell jar to expand. After a short time the magnesium stops burning and the water level inside the jar rises, showing that the air has been used up during the combustion.

The extent of the rise in water level corresponds to about one fifth of the original volume of air inside the jar. Note that some unburnt magnesium still remains in the basin, showing that the combustion did not stop because the supply of magnesium ran out.

If the stopper is removed, a burning splint inserted through the neck is immediately put out, showing that the 'air' remaining in the jar will not support combustion. Thus the magnesium went out because something in the air was used up.

A second experiment, with copper, can be done using gas syringes (Fig. 8.3). A hard glass tube contains clean copper turnings or very small pieces of copper wire, held in place by pieces of glass rod which fit loosely inside each end of the tube. One of the syringes contains a known volume of air, say 100 cm³.

Fig. 8.3 Heating copper in air

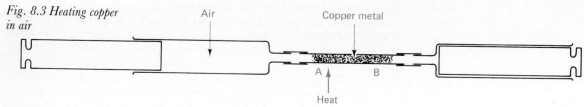

The copper is heated strongly at point A while the syringes are pushed backwards and forwards a few times. The copper in the region of A turns black on its surface and the volume of air in the apparatus decreases. After cooling, it is found that the volume of 'air' remaining is about 80 cm³. If the copper is heated again, at point B, it does not go black in that region and no further decrease of 'air' volume occurs when the syringes are pushed backwards and forwards. This experiment shows that the blackening of copper occurs until something in the air (about one fifth of the original volume) is used up. Thus even though the copper does not appear to be burning it is using up the same proportion of the air as the magnesium does when it burns.

It might be argued that these results are all very well when the substance turns to an ash on heating in air, but what about when a substance 'burns away completely', like a candle? At first sight, it does not seem to make sense to claim that a candle gains mass when it burns, but experiments may easily be done to show that this is the case.

For example, when a candle is burned over water in a bell jar the water level inside the jar drops slightly, owing to the expansion of the air caused by the heat produced. After a while the candle goes out, but the water level inside the jar rises only slightly above the original level. It can also be observed that a mist or condensation has collected on the upper part of the wall of the jar. A burning splint inserted through the neck of the jar is put out showing that the remaining 'air' will not support combustion. This shows that the something in air has been used up, even though hardly any reduction in volume has been observed. The answer to this problem is that the candle, on burning, produces a gas as one of its combustion products, and this gas virtually takes the place of the part of the air which is used up when the candle burns.

Candle wax is a hydrocarbon (a compound of carbon and hydrogen) and on combustion the products are carbon dioxide and steam (hence the condensation observed in the bell jar). When a candle is burned in the open air both these products escape into the atmosphere, but if it can be arranged so that these products are trapped then the increase in mass can be demonstrated. Fig. 8.4 shows an apparatus which can be used for this purpose. A glass chimney, with a candle mounted in the bottom, contains lumps of calcium oxide to absorb the water vapour and sticks of sodium hydroxide to absorb the carbon dioxide gas. The whole apparatus is put on one side of a balance and weights added to the other side until it is balanced. After lighting the candle, the balance gradually tilts so as to show an increase in mass on the candle side. This result indicates that the burning candle does involve combination with air, just as the burning of magnesium involves combination with air.

It can be shown that when any substance burns in air there is an increase in mass, provided the products (which may be gases) are collected. It can also be shown that combustion always stops when about one fifth of the air available has been used up, no matter what substance is being burned.

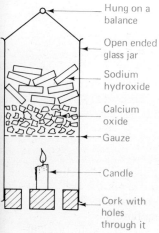

Fig. 8.4 Showing that there is an increase in mass when a candle burns

The nature of the 'something used up' in combustion was first discovered by experiments carried out near the end of the eighteenth century, as described in the next section.

8.7 The discovery of oxygen

The key figure in this story is Joseph Priestley who was born in 1733 at Fieldhead, near Leeds. After attending Batley Grammar School, he studied theology and became a minister of the church. Like many clerical gentlemen in those days he became interested in science, and in 1767, on moving back to Leeds, he found himself living next to a brewery. He noticed that during the process of fermentation a gas was given off (we know this to be carbon dioxide) and this aroused a particular interest in gases as a result of which, incidentally, he invented soda-water (a solution of carbon dioxide in water, which with added flavouring and sugar is lemonade). Priestley did not have much money and so shortly after doing the experiments with carbon dioxide he was sponsored in his researches by the Earl of Shelburne—later to become Prime Minister.

It was in Shelburne's country mansion near Calne, in Wiltshire, that Priestley carried out the experiment which sealed his name for ever in the history of chemistry.

Fig. 8.5 Joseph Priestley (1733–1804). Courtesy The Science Museum)

Fig. 8.6 Some of the apparatus which was used by Joseph Priestley. (Courtesy The Science Museum)

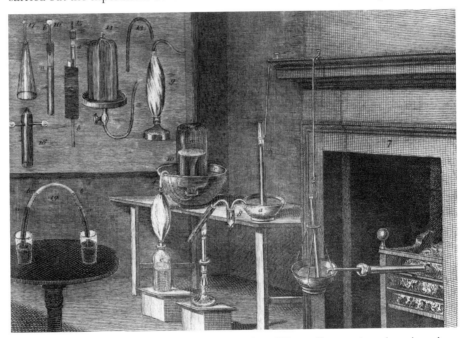

In 1774 Priestley was given a large convex lens (30 cm diameter), or burning glass, by means of which the sun's rays could be focused to produce intense heat (the Bunsen burner had not yet been invented), and with this device he started to examine the action of heat on many substances. One of these substances was an orange powder, calcined mercury* (or, as it is called now, mercury(II) oxide). On August 1st, 1774, he heated this substance with his burning glass and obtained from it a gas. In Priestley's own words, 'What surprised me more than I can well express, was, that a candle burned in this air' (as he called it) 'with a remarkably vigorous flame . . . I was utterly at a loss how to account for it'. Later he tried breathing the air and noted that his 'breast felt peculiarly light and easy for some time afterwards'. He recommended its use in medicine—as, indeed, it is now used in the treatment of pneumonia, for example.

* 'Mercurius calcinatus per se' meaning mercury heated by itself.

73

Fig. 8.7 *A portrait of Lavoisier (Courtesy Mary Evans Picture Library)*

The gas which Priestley discovered was **oxygen**, the component of air which combines with substances when they burn and which is therefore present, in combined form, in ashes (oxides). Priestley's experiment involved the decomposition of mercury oxide into its constituent elements on heating:

$$\text{mercury oxide} \quad \rightarrow \quad \text{mercury} \quad + \quad \text{oxygen}$$
$$2HgO(s) \quad \rightarrow \quad 2Hg(l) \quad + \quad O_2(g)$$

This experiment is easily demonstrated, but it must be done in a fume cupboard as mercury vapour is very poisonous. If a little mercury(II) oxide (orange powder) is heated in a test-tube, it first darkens and then silvery-grey globules of liquid mercury start to condense on the upper part of the test-tube. If, at this point, a glowing wooden splint is inserted into the test-tube it is rekindled and burns briefly with a very bright flame. The rekindling of a glowing splint is the usual test for oxygen gas, which is a far better supporter of combustion than ordinary air.

It should be noted that, although Priestley is credited with the discovery of oxygen, he did not realise the full implications of these results. The credit for the interpretation of Priestley's experiment must go to the brilliant French chemist, Antoine Lavoisier. Karl Scheele in Sweden had prepared several gases including oxygen before 1773, but he did not publish his results until 1777.

Lavoisier was born in Paris in 1743, the son of a wealthy aristocrat. In 1774 Priestley visited Paris with Shelburne and had dinner with Lavoisier. Priestley described his experiment whereupon, in Priestley's own words, 'all the company, and Mr and Mrs Lavoisier as much as any, expressed great surprise'. Lavoisier was quick to see the implications of Priestley's result and repeated the experiment for himself. Eventually, in 1777, he presented to the Academy in Paris a memoir describing the experiment for which he is most famous; this is the experiment involving the reversible oxidation of mercury.

Fig. 8.8 *Lavoisier's apparatus*

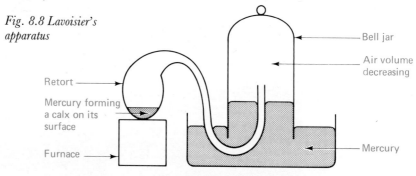

Bell jar

Air volume decreasing

Retort

Mercury forming a calx on its surface

Furnace

Mercury

Fig. 8.9 *Lavoisier's apparatus for heating mercury. (Courtesy The Science Museum)*

Lavoisier carried out the experiment using the apparatus shown, in simplified form, in Fig. 8.8. A known mass of mercury was placed in a retort (mounted on a furnace) whose neck led into a known volume of air trapped in a bell jar over mercury.* The retort was heated with the furnace to just below the boiling point of mercury. As the reaction proceeded a layer of calx (mercury oxide) gradually accumulated on the surface. After twelve days heating the layer no longer increased in size; the apparatus was allowed to cool and the volume of air in the bell jar was found to have decreased by nearly one fifth. The remaining air in the bell jar would not support combustion.

The layer of calx (mercury oxide) was collected and then heated strongly. It turned back into mercury (as in Priestley's experiment), giving off a gas whose volume was equal to the volume of air absorbed during the original heating. Lavoisier confirmed Priestley's comments, noting that the gas was much more capable of supporting respiration and combustion than ordinary air. He named it oxygen which was derived from the Greek words for 'sour' and 'I produce', as he observed that the products of burning substances in the gas were often acidic.

The air remaining in the bell jar he called azote (meaning that it would not support life) but later it was given its present name, nitrogen. Lavoisier also noted that when the oxygen evolved by heating the mercury calx was mixed with the azote remaining in the bell jar, the resulting gas had all the properties of ordinary air. The clear conclusion was that air is a mixture of two gases, one of which (oxygen) is used up in combustion, while the other gas (nitrogen) takes no part in the process.

8.8
The composition of air

Experiments such as those using copper and magnesium which were described in 8.6, show that the proportion of oxygen in the air is about one fifth.

The normal composition of dry air (i.e. air from which water vapour has been removed) is by volume:

oxygen	21%
nitrogen	78%
other gases	1%

The other gases are, like nitrogen, unreactive; the major part of the 1% consists of argon, one of the noble gases. Small amounts of other noble gases are present and also a little carbon dioxide. The composition of air is fairly constant from place to place, though it does vary with altitude.

A question which is sometimes asked is—'Is air just a mixture of these gases or is it a compound of them?' The evidence is clear that air is simply a mixture, on the following grounds:

(a) The composition of air does vary slightly from place to place, and with altitude.

(b) The composition of air, expressed in moles (4.10 and 5.8), does not correspond to any simple chemical formula.

(c) The component gases of air can be mixed together in the appropriate proportions to form a mixture which has all the properties of ordinary air. During this mixing there is no evidence of chemical action, e.g. volume change or evolution of heat.

(d) The components of air can be separated from one another by purely physical processes, such as the distillation of liquid air (see the next section).

8.9
Industrial preparation of oxygen

Pure oxygen is an important industrial material. It is prepared directly from the air, using the principle that the substances present in the air have different boiling points. The boiling points (at atmospheric pressure) of oxygen and nitrogen are $-183\,°C$ and $-196\,°C$ respectively.

*The mercury in the trough played no part in the chemical reaction; in those days chemists often collected gases over mercury rather than over water. This would be too expensive today!

75

Fig. 8.10 A road tanker
filling up with liquid
oxygen. (Courtesy British
Oxygen Company Ltd)

Fig. 8.11 Liquid nitrogen
(b.p. − 196°C) is used
for rapid food freezing.
The photograph shows
prepared food being put
into a nitrogen food
freezing cabinet.
(Courtesy British Oxygen
Company Ltd)

Fig. 8.12 An important
use of oxygen is in
resuscitation. (Courtesy
British Oxygen Company
Ltd)

Air, after being freed from dust and water vapour, is liquefied by cooling under pressure (over 100 atmospheres) and then the liquid air is allowed to evaporate in a carefully controlled way so that the nitrogen and oxygen are separated. Nitrogen, having the lower boiling point, evaporates first. The separated gases are stored under pressure in steel cylinders, although some oxygen and nitrogen are transported as liquids in refrigerated tankers.

Liquid nitrogen is particularly useful in the food industry where it can be sprayed directly on to food. This results in very rapid freezing which does less damage to the food and preserves its flavour and texture. Excess nitrogen evaporates into the atmosphere and because it is chemically inert there are no health or fire hazards.

Pure oxygen has many uses, but by far the largest is in the steel industry. The pig iron produced by reduction of the iron ore in the blast furnace (24.9) contains too much carbon and other impurities which make the iron brittle. These impurities are removed by burning them out of the molten iron with a controlled blast of oxygen. Oxygen is also used in welding and cutting metals (e.g. the oxy-acetylene torch, 31.11), in medicine to help patients breathe, in rocket fuels and as a starting material in many industrial chemical processes. In all these, it behaves as it would in air, but it is in more concentrated form and hence it is more effective.

8.10 Laboratory preparation of oxygen

The preparation of oxygen by heating mercury(II) oxide (as done by Priestley and Lavoisier) is not convenient because it is very expensive and the mercury vapour which is also produced is very poisonous.

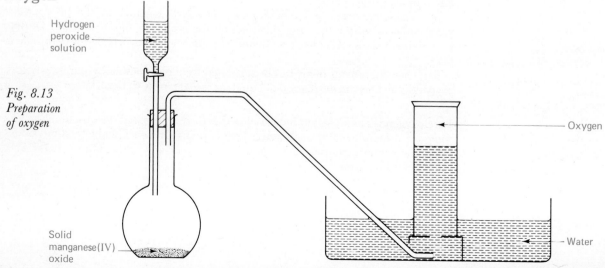

Hydrogen
peroxide
solution

Fig. 8.13
Preparation
of oxygen

Oxygen

Water

Solid
manganese(IV)
oxide

The safest way of preparing a reasonable quantity of oxygen in a laboratory is by the catalytic decomposition of hydrogen peroxide solution. The apparatus is shown in Fig. 8.13. Hydrogen peroxide decomposes according to the equation:

$$2H_2O_2(aq) \rightarrow 2H_2O(l) + O_2(g)$$

The aqueous solution is fairly stable at room temperature, but the decomposition can be speeded up by the use of a catalyst; this is a substance which increases the rate of a chemical reaction but can be recovered unchanged at the end. Manganese(IV) oxide is a suitable catalyst for this reaction. A few grams of this solid (the amount is not critical) is placed in the flask and hydrogen peroxide solution is dropped on to it from the tap funnel. Immediate effervescence occurs, and the rate of oxygen production is controlled by the rate of addition of the hydrogen peroxide solution. The hydrogen peroxide solution usually used is of 20 volume concentration which means that 1 volume of the solution, on decomposition, produces about 20 volumes of oxygen gas measured at s.t.p. Thus about 1 dm³ of oxygen is obtained from 50 cm³ of 20 vol hydrogen peroxide solution. The catalyst, manganese(IV) oxide, is not used up in the reaction, and so addition of hydrogen peroxide solution can be continued indefinitely.

8.11 Properties of oxygen

As you cannot see or smell air, it follows that oxygen is a colourless odourless gas. As previously mentioned, it forms about one fifth of the volume of air and it has about the same density as air.

It is neutral to moist litmus and not very soluble in water (1 dm³ of water dissolves about 40 cm³ of oxygen at room temperature and pressure).

Pure oxygen is an excellent supporter of combustion and will rekindle a glowing wood splint, this being the usual laboratory test for the gas. Most elements will combine directly with oxygen (though not usually at room temperature) to form oxides; exceptions include the halogens and unreactive metals such as gold, but oxides of even these elements can be made indirectly.

The behaviour of oxides when treated with water is important, since it forms part of the basis for the classification of elements as metals or non-metals, about which more is said in Chapter 11. Briefly, the metal oxides which are soluble form solutions which turn litmus blue and are therefore alkaline (e.g. sodium, potassium and calcium). The non-metal oxides which are soluble form solutions which turn litmus red and are therefore acidic (e.g. carbon, sulphur and phosphorus).

It should be noted, however, that many metal oxides are insoluble in water and do not react with it to form an alkali, but most of them, nevertheless, are basic in character (11.10).

8.12 Summary

1. When substances burn in air they gain in mass and about one fifth of the air is used up. The remaining 'air' will not support combustion.
2. Air consists of about one fifth oxygen, which combines with substances when they burn, and about four fifths nitrogen (and traces of other gases) which does not take part in combustion. Air is a mixture, but its composition is fairly constant from place to place.
3. Oxygen is prepared industrially from liquid air. Its main use is in the steel industry.
4. Oxygen may conveniently be prepared in the laboratory by the catalytic decomposition of hydrogen peroxide solution. It is an excellent supporter of combustion and will rekindle a glowing splint.
5. Oxides of metals often show basic properties and oxides of non-metals often show acidic properties.

Chapter 9

Air: burning, breathing and rusting

Investigation 9.1

Is there a change in mass when iron rusts?

Place about one spatula-full of iron filings on a watchglass or in a small basin and then weigh the container with the filings. Add enough water to wet the filings thoroughly and then leave them to stand for several days.

When the iron has rusted, and any excess water has evaporated, weigh the container and filings again.

Record your results and note how the mass of the rusted filings compares to that of the clean iron filings at the beginning.

Question

Does the result of your investigation indicate that rusting is a decomposition (splitting up) process or a process in which iron combines with something?

Investigation 9.2

Under what conditions does iron rust?

The aim of this investigation is to find out if rusting is caused by water alone, or air alone, or if both air and water are required.

You will need four clean iron nails, four clean test-tubes (one of them dry), and a cork to fit the dry tube.

Fig. 9.1 The rusting of iron

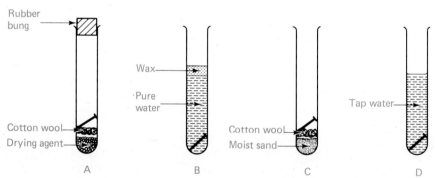

Set up the tubes as shown in Fig. 9.1. Put a few lumps of anhydrous calcium chloride (a good drying agent which absorbs water vapour from the atmosphere) in the bottom of tube A (the dry one), then a layer of cotton wool, then the nail and then close the tube with the cork.

Take some pure water in tube B and boil it for one minute to drive off any dissolved air. Then drop in the nail and seal the water surface with wax to keep the air out. Vaseline can be used or a piece of candle wax can be floated on the surface of the water and heated at the side of the tube with a low Bunsen flame until it melts. When the tube cools, the wax solidifies.

Put about 2 cm depth of sand in tube C, add enough water to soak it thoroughly, then add a layer of cotton wool and drop the nail on top.

Put some ordinary tap water in tube D and just drop the nail in. Leave all the tubes for several days and then examine them to see which nails have rusted.

Set out the results in a table:

	Air present	Water present	Does iron rust?
Tube A			
Tube B			
Tube C			
Tube D			

Questions

1 What substances must be present for iron to rust?
2 Why can iron be prevented from rusting by painting it?

9.3
Fuels and the control of fire

Fire is both our friend and our enemy. In the right place, under control, it warms our houses, cooks our food, drives generators to produce electricity and runs motor vehicles and other machines. In the wrong place, out of control, it destroys land and property and kills people.

The fuels which we commonly use for producing heat by controlled combustion are wood, coal, coal gas or natural gas, and the various petroleum products such as oil, petrol and paraffin. These substances are all of organic origin (i.e. derived from living matter) and are mainly hydrocarbons. Some further discussion of these fuels can be found in Chapters 28 and 31. They burn to produce carbon dioxide (gas) and water vapour:

$$\begin{matrix} \text{hydrocarbon} \\ \text{(fuel)} \end{matrix} + \begin{matrix} \text{oxygen} \\ \text{(from air)} \end{matrix} \rightarrow \text{carbon dioxide} + \text{water}$$

The production of carbon dioxide and water when a hydrocarbon fuel burns is easily shown by holding an inverted gas jar for a few seconds over a burning candle, or a Bunsen burner with a small flame. Condensation appears in the jar, showing the formation of water. If a little lime water is then added to the jar and shaken up it turns milky, showing the presence of carbon dioxide (carbon dioxide is the only gas which turns lime water milky, 28.13).

Much energy, as heat, is liberated during burning and in order to use this energy most efficiently we must achieve complete and rapid combustion of the fuel, so that a high temperature is produced. To do this we need a good supply of oxygen, and this is where careful design of the burner is important. The principle involved is conveniently illustrated by considering the construction of the familiar piece of laboratory apparatus, the Bunsen burner.

The Bunsen burner, developed around 1855 by the German chemist Robert Wilhelm Bunsen and almost unchanged in design since then, is shown in Fig. 9.2. The chimney has holes near the base and a rotating collar (or sleeve) which has matching holes so that by rotation of the collar the entry of the air at the base of the chimney can be controlled. If we light the burner with the air holes closed, we get a yellow, luminous flame which is not very hot and deposits particles of unburnt carbon as soot on a cold surface held in the flame. Under these conditions we are simply burning the gas at the end of a tube, and clearly the combustion is inefficient and does not produce a high temperature.

If we now open the air holes, the character of the flame changes considerably. It becomes blue, non-luminous and much hotter, and it will in fact burn off the soot produced by the luminous flame.

What causes the difference between the two flames? It is simply the fact that with the air holes open, the gas rushing through the narrow jet at the base of the chimney draws air in through the air holes. What emerges at the top of the chimney is a mixture of gas

Fig. 9.2 Bunsen burner

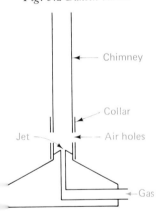

Fig. 9.3 Bunsen flame

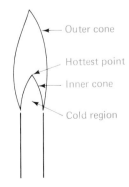

79

Fig. 9.4

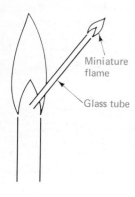

Miniature flame

Glass tube

and air instead of pure gas. Combustion is then much faster (which is why the flame becomes smaller with the air holes open) and a much higher temperature is produced.

The overall appearance of the blue, non-luminous flame is shown in Fig. 9.3. The pale blue line separating the inner and outer cones is the flame-front, where combustion begins. Below the flame-front, inside the inner cone, is a cold region containing unburnt gas and air. If the lower end of a glass tube is held inside the inner cone, as shown in Fig. 9.4, and a light is applied to the upper end of the tube, a perfect miniature version of the blue flame burns at the end of the tube, as shown in the diagram. The hottest point in the Bunsen flame is the point just above the inner cone, close to the tip of the flame-front. The temperature at this point is greater than 1100°C.

All domestic gas-burning appliances, such as gas fires and gas cookers, have air-inlet holes in the gas pipe leading to the burners. When a gas cooker is partly dismantled for cleaning it may be possible to see the air holes, somewhere between the control-knob and the burner. If the air holes were not there the gas would burn with a yellow flame which would deposit soot and would not produce a high temperature.

In a device such as a motor-car engine the petrol must be mixed with air before it enters the cylinders, otherwise combustion would not be possible. The carburettor in the motor-car engine is equivalent to the air hole in the Bunsen burner or gas cooker.

In the carburettor, petrol (a volatile liquid) is forced through a narrow jet, so that it becomes mixed with air while, at the same time, vaporizing. The resulting mixture of air and petrol vapour then passes into the cylinders where it is ignited by sparks from the sparking plugs.

We have discussed so far how fire can be harnessed usefully; what about the other side of the coin? Fire can occur where it is not wanted, and then it is disastrous. As pointed out earlier, organic substances based on the elements carbon and hydrogen are used as fuels. However, vegetation is also obviously organic, and substances which we use for furnishing and decorating our homes—wood, carpets, curtains and chair coverings—are frequently organic in origin; even plastics are made up mainly of carbon and hydrogen. All these substances are combustible and can become fuels by accident. If fire occurs where it is not wanted, then the problem is how to put it out.

We have seen that in order to achieve efficient combustion we need a good supply of oxygen from the air and thus, in order to stop unwanted combustion, we must restrict the air supply as far as possible. If a person gets his clothes on fire, the usual remedy is to roll him up in a blanket or rug, and, if a chip-pan catches fire, you cover it with a damp towel. These methods prevent air getting to the fire and it is extinguished. The carbon

Fig. 9.5 The two steps in putting out a chip pan fire (a) removing the source of heat by turning off the gas supply, and

(b) cutting off the supply of air by placing a lid on the pan. (Both photos courtesy the London Fire Brigade Photographic Service)

dioxide fire extinguisher works in a similar way, displacing air and replacing it by a gas which will not support combustion. Obviously, if a whole building is on fire, a different method has to be used. In this case the standard procedure is to spray water on the fire. This has two effects; first the water turns to steam, thus displacing air from the fire, and, secondly, the temperature of the combustible materials is lowered. If the temperature drops below a certain point, the fire cannot continue, since most substances fortunately will only combine with oxygen at a fairly high temperature. When firemen refer to a fire as having been brought under control they do not mean that it is actually put out; they mean that they have reduced the temperature to the point where the fire is no longer growing. The fire is then no longer capable of maintaining itself and it is only a matter of time before it is extinguished.

A lot of research has been, and is still being, done on fire-proofing materials so that they will not readily burn. This type of research is very important, since it is obviously better to prevent fire than to have to put it out once it has started.

9.4 Respiration and fuels for living

Human beings, and animals in general, have to eat food to live. Food is the fuel on which the body runs. The overall complex series of processes by which food is consumed and used by the body is called **metabolism**. Food is required to maintain growth and replace body tissue, and to provide energy which enables the muscles to work and energy (as heat) to maintain the body temperature at about 37 °C (for human beings), a temperature considerably higher than the normal surrounding temperature.

The first stage of metabolism is digestion of the food in the stomach. As a result of this process, the main energy-source in food—carbohydrate—is converted into glucose, a type of sugar, which passes round the blood stream and is deposited in the body tissues. This glucose provides the body's energy—but how? It was Lavoisier who first suggested the answer. He showed, as described in the previous chapter, that combustion of substances involves their combination with oxygen from the air. At the time it was already well known that human beings and other animals require air to live, and Lavoisier suggested that life itself is a chemical process in which oxygen taken in by the process of breathing is used to oxidise food and thus produce energy.

The process by which oxygen is used by the body is known as **respiration**, and Lavoisier was the first person to make measurements on the respiration in human beings. During the process, oxygen is taken in by the lungs, circulates in the blood-stream through the body, and combines with the glucose in the body tissues. The glucose is oxidised to carbon dioxide and water, energy being released during the process:

$$C_6H_{12}O_6 + 6O_2 \rightarrow 6CO_2 + 6H_2O + \text{energy}$$
glucose

As a result of this the air which we breathe out (exhale) contains less oxygen, and more carbon dioxide, than the air we breathe in (inhale), Table 9.1.

TABLE 9.1. *Typical composition (by volume) of inhaled and exhaled air*

	INHALED (NORMAL) AIR	EXHALED AIR
nitrogen	78%	78%
oxygen	21%	16%
carbon dioxide	0·03%	4%

It can easily be shown that exhaled air contains more carbon dioxide. The test for this gas is that it turns lime water milky (28.13). If you breathe gently into a test-tube

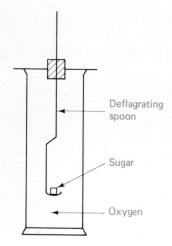

Fig. 9.6 Burning a food-substance

Fig. 9.7 An area of Middlesbrough before and after the introduction of smoke control. (Courtesy Dept. of Planning, Cleveland County Council)

containing a little lime water and then close the end of the tube and shake it vigorously, the lime water soon goes milky. Lime water can be shaken with ordinary air for several minutes without going milky.

The behaviour of food as a fuel may be shown by actually burning it. A gas jar (Fig. 9.6) is filled with oxygen, then a little sugar is heated on a deflagrating spoon until it catches fire and lowered into the jar. The sugar burns well, producing condensation (water) on the wall of the jar. If a little lime water is then added to the jar and shaken it turns milky, showing that carbon dioxide has been formed. The experiment can also be done using other foods such as bread, which is rich in carbohydrate.

It has been shown experimentally that although the oxidation of food in the body is a very slow and steady process compared to actually burning it (as above), the total quantity of energy liberated from a given mass of carbohydrate is almost the same in both processes. Consequently the energy content, or calorific value, of foods can be determined in the laboratory by burning them in a calorimeter. The results of such experiments are very important in the study of nutrition, in which man's food requirements are investigated. In these days, when many people in the world are starving while others are dying from diseases caused by over-eating, the study of nutrition is of great social importance.

It is of interest to note the quantity of energy required by a human being. A moderately active young person requires a daily energy intake of something like 12 000 kJ (1 kiloJoule is roughly the work done in lifting a 100 kg (220 lb) mass 1 metre off the ground). This quantity is equivalent to the energy given off when about 700 g of sugar is burned. This energy would keep a 1 kW (1-bar) electric fire burning for about 3 hours, or, if converted completely into mechanical work, would lift a 100 ton mass (a large railway engine) about 10 metres off the ground.

It should be pointed out that, although food is a fuel, not all fuels can be used as foods. It must be remembered that before the energy in food can be used in the body the food must first be digested. Wood, for example, is a good fuel, but we could not live by eating sawdust because the carbohydrate in wood is mainly in the form of cellulose, a substance which cannot be digested by the human stomach.

It might appear that, because of the carbon dioxide liberated during the respiration of humans and animals, the carbon dioxide content of the atmosphere would steadily increase. This does not happen because the carbon dioxide is used by plants in their production of carbohydrates by photosynthesis. This process is discussed in 23.14 and 28.8.

9.5
Pollution of air

If we consider our bodies to be machines which require air in order to work, then they will only work smoothly when the air is pure and contains nothing which spoils the machine. In modern life, however, we insist on pouring into the air substances which make it impure and which may hinder the smooth working of our bodies.

Air pollution is principally caused by the fuels we burn to set free energy. Every year the atmosphere over Britain is contaminated by 1·5 million tonnes of smoke which consists of fine solid particles, drifting about in the air and finally being deposited somewhere. If they settle on buildings, the stonework is turned black and ugly, and, if they pass into the lungs, they can cause bronchitis. If the smoke particles become trapped in droplets of water in a fog, a smog is formed and the atmosphere becomes harmful to people with chest diseases. In 1956 the British Government introduced an Act of Parliament to control the amount of smoke in the air and at the present time the quantity has fallen to half what it was in 1954. The Act allows only smokeless fuels to be burnt in certain areas and by the middle of the 1980s smoke pollution will have been largely eliminated. Industry has also played its part by the careful design of chimneys so that fuels are burnt more efficiently, and the replacement of steam locomotives on the railways by diesel and electric engines has removed another major source of smoke.

Another major cause of air pollution does not, however, alter with a change of fuel. Whenever we burn coal, coke or fuel oil, we pour into the atmosphere sulphur dioxide which is made from the sulphur in the fuel. On average, $5\frac{1}{2}$ million tonnes are discharged into the air over the United Kingdom every year. The world figure is about 150 million tonnes per year, two thirds of which results from the burning of coal (Fig. 9.8). The amount of sulphur dioxide escaping from chemical processes such as the Contact Process for making sulphuric acid (27.20) has been reduced by up to 75% in recent years by technical improvements.

Sulphur dioxide dissolves in water forming an acidic solution, and, together with smoke, it can cause diseases of the lungs. Its solution in rain water corrodes stonework, particularly in old buildings, but the main effect of the gas is that it is poisonous to plants (27.17). The problem of removing the sulphur dioxide from the gases formed by burning fuels has not been fully solved. Some of it is sometimes washed from the gases before they pass through the furnace chimney. The main method of reducing the effect of the rest is to discharge it into the air through tall chimneys so that the gases are spread over a large area. Probably the only satisfactory way of removing this major pollutant would be a changeover from the burning of fossil fuels to nuclear fuels as the main source of energy, but that could bring other problems.

Internal combustion engines, like those of motor cars, cause air pollution in a number of ways and motor vehicles are now responsible for 60 per cent of the impurities in the air. Car engines do not burn petrol or diesel fuel completely and the partly burnt gases are given out through the exhausts. Complete combustion of the hydrocarbons in the fuel would give carbon dioxide,

e.g. $$C_8H_{18}(g) \ + 12\tfrac{1}{2}O_2(g) \ \rightarrow \ 8CO_2(g) \ + \ 9H_2O(g)$$

but partial oxidation produces carbon monoxide which is poisonous because it prevents the blood carrying oxygen. On the open road the carbon monoxide will rapidly diffuse and will cause no harm, but in congested streets in heavy traffic the level of the gas can increase towards the danger limit. Because of the limited efficiency of the motor-car engine which causes the formation of the carbon monoxide, some of the hydrocarbons in the fuel escape combustion altogether and are also contained in the exhaust gases. Such compounds are not harmful to health, but they are a nuisance because of their smell.

Car engines have sparking plugs which ignite the mixture of air and petrol vapour, and surprisingly they are also a cause of impurities in the air. The air contains nitrogen and oxygen and the action of an electrical spark causes them to combine together to form nitrogen monoxide:

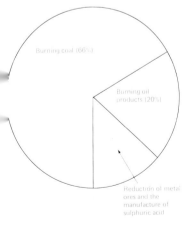

Fig. 9.8 The major sources of man-made sulphur dioxide which escapes into the atmosphere

Burning coal (66%)

Burning oil products (20%)

Reduction of metal ores and the manufacture of sulphuric acid

$$N_2(g) + O_2(g) \rightarrow 2NO(g)$$

When this is discharged into the air, it combines with oxygen, forming nitrogen dioxide:

$$2NO(g) + O_2(g) \rightarrow 2NO_2(g)$$

This gas can cause damage to the lungs, but usually the amounts produced are small and the compound either diffuses away or is dissolved in rainwater before it can do much harm. Experiments have been carried out to make the nitrogen monoxide and the carbon monoxide in the exhaust gases react in the presence of a catalyst in the exhaust system, forming non-polluting carbon dioxide and nitrogen,

$$CO(g) + NO(g) \rightleftharpoons N_2(g) + CO_2(g)$$

but this has not yet proved of practical use.

A much more dangerous pollutant comes from the petrol. In order to prevent the engine 'knocking' (in which the mixture of air and petrol vapour explodes on compression before the spark is passed, causing the engine to work less efficiently) a compound of lead, tetraethyl lead, is added to the petrol. This, by itself, would cause lead to be deposited on the walls of the cylinders and so another compound, dibromoethane, is also added, the effect of this being to cause the lead to be lost from the engine in the exhaust fumes. Once it has been taken into the body, lead remains in the tissues until the level rises to the danger point when it can cause brain damage and even death, and, although the amounts passed into the air by petrol exhausts are small, there have been reports that they are causing some harm, particularly to young children. Certainly lead compounds from this source have been shown to cause the death of plants on roadside verges.

Just as it is possible to judge how polluted a sample of water is by seeing what creatures live in it, the level of air pollution can be estimated by studying the occurrence of plants called lichens. These crust-like growths look more dead than alive, although they often have pleasing colours and were used as sources of dyes. They will only exist when the air pollution is low and so they are more frequently found in country districts where the air is purer.

9.6
Respiration of water creatures

When life first emerged on earth, millions of years ago, it did so in the sea; the mammals, living on land, developed much later. Marine and river life is still of fundamental importance in the balance of nature, and water creatures, like land creatures, need oxygen to convert their food into energy and so live. Whereas land creatures absorb oxygen from the air using their lungs, water creatures absorb dissolved oxygen from the water using gills.

The presence of dissolved air in a sample of water obtained either from a tap or from some natural source such as a river, can be shown experimentally, using the apparatus in Fig. 9.9. The round flask is filled with water by connecting the rubber tube (shown clipped in the diagram) to the tap. Water is allowed to run into the flask until both the flask and the delivery tube are completely full, and the rubber tube is then closed with a clip. A graduated tube is filled with water and inverted over the delivery tube. When the flask is heated, air is expelled as the temperature of the water approaches its boiling point. When no more air collects in the graduated tube, heating is stopped.

If a burning wood splint is put into the 'boiled out' air obtained from the water, it will burn more brightly for an instant, showing that air dissolved in water is richer in oxygen than ordinary air. This is because the solubility of oxygen in water is about twice that of nitrogen.

The oxygen dissolved in natural water is vital for fish. It enters the water at the surface, where air is present, and is also given off by water plants during the process of photosynthesis (28.8). In an indoor fish-tank, where there is little sunlight for photosynthesis and where the surface area of the water is rather small relative to the volume of water, it may be necessary to increase the oxygen content by using an aerator which is a small electrical pump for bubbling air through the water.

Fig. 9.9 Showing that air is dissolved in tap water

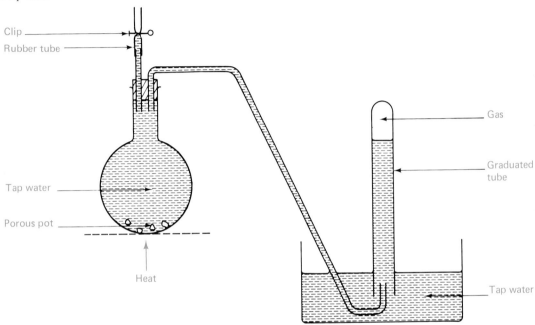

9.7 Air and the rusting of iron

Iron, usually in the form of steel, is by far the most widely used metal. Its excellent strength and mechanical properties, together with its relatively high availability, make it a most suitable material for building ships, bridges, motor vehicles, etc., and it would be hard to imagine our industrialised society without it. What a pity it is that its otherwise excellent qualities are marred by its tendency to rust. The process of rusting, which, if left unchecked, results in the eventual disintegration of the metal, costs the world hundreds of millions of pounds per year.

Let us consider what happens when iron rusts. If we did not know that iron was an element, we might think that rusting is a kind of decomposition. The use of the balance helps us here, just as it helped in sorting out what happens during combustion (8.6). In Investigation 9.1 some iron filings were weighed, moistened with water (since we know that rusting occurs most readily in damp conditions) and then left to rust. When thoroughly rusted (and dry) they were weighed again, and an increase in mass was found. Thus rusting involves combination—but with what? Water is an obvious possibility, but as air is present it is another. In order to sort this out we must do a set of controlled experiments as in Investigation 9.2.

Each of the four test-tubes, Fig. 9.1, contains a clean iron nail. In tube A the iron is in contact with dry air, in tube B it is in contact with air-free water, in tube C it is in contact with damp air and in tube D it is in water which contains dissolved air. The four tubes are left for a few days and the result is that the nails in A and B do not rust, while the nails in C and D do rust. This shows that both air and water together are necessary for rusting to occur.

85

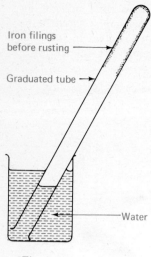

Iron filings before rusting

Graduated tube

Water

Fig. 9.10.

Since it is now clear that air is involved in rusting, we should now do an experiment to see if air is used up when iron rusts. A suitable apparatus for this investigation is shown in Fig. 9.10. Some iron filings are sprinkled on the inside of a wet graduated tube, which is then inverted in a beaker of water. The volume of air inside the tube at the start of the investigation is noted. Over a period of a few days the water inside the tube rises, as the iron filings rust, showing that air is used up. The rise of the water level stops when the volume of the air in the tube has decreased by about one fifth, and if a burning splint is put into the remaining air, it is extinguished. These observations show that when iron rusts oxygen is used up and that rusting must be a kind of slow combustion which is aided by the presence of water. The brown substance, rust, is hydrated iron(III) oxide: $Fe_2O_3.xH_2O$ (where x is variable).

Since rusting is an oxidation process, it is clear that in order to prevent rusting the surface of the iron must be protected from the air, and there are many ways of doing this. Tools and machinery can be smeared with oil or grease, and steel structures, such as bridges, can be painted. Sheet steel is often coated with a metal which is more resistant to corrosion. For example, it can be coated with a thin layer of zinc (galvanised) and it can be tin plated, as in the so-called tin cans used for storing foods. A more detailed treatment of the protection of iron by this method is given in 20.9. Another method which is sometimes used on mesh fencing and underground pipes is to coat the iron with plastic.

A different approach is to blend the iron with other metals, chosen such that the resulting alloy (24.11) is resistant to corrosion. The stainless steel which is used for cutlery and other kitchen utensils is an alloy of iron with chromium. Unfortunately, chromium is far too expensive for this method to be used on a large scale.

9.8 Summary

1. Naturally occurring gas, oil, coal, wood and other substances are used as fuels. These substances are mainly hydrocarbons, and burn to form carbon dioxide and water.
2. Fuel-burning devices such as the Bunsen burner must be designed so that a good air supply is available to oxidise the fuel efficiently and produce maximum heat.
3. Unwanted fire may be put out by excluding air from the region of combustion and/or by lowering the temperature of the fuel.
4. Food is the fuel on which the body runs. Oxygen is used up during respiration (breathing). The body's energy is obtained from the oxidation of glucose produced by the digestion of food.
5. The blood transports both glucose and oxygen around the body, and also transports carbon dioxide back to the lungs where it is exhaled.
6. The main substances which cause air pollution are smoke, sulphur dioxide, carbon monoxide, oxides of nitrogen and lead compounds.
7. Oxygen is slightly soluble in water (more so than nitrogen) and is essential for fish life.
8. The rusting of iron is an oxidation process which occurs when both air and water are present. Rusting may be prevented by excluding air from the surface of the iron, e.g. by painting or galvanising.

Chapter 10   # Water

Investigation 10.1

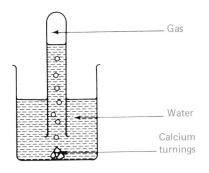

Fig. 10.1

The action of calcium on water

About half fill a beaker with water and add three calcium turnings.

Completely fill a test-tube with water and, with your thumb over the mouth, invert the tube and open it with its mouth under the water in the beaker. Hold the tube over the calcium turnings (Fig. 10.1) until it is filled with gas.

Place your thumb over the end of the tube and open it near a Bunsen flame and note what happens.

When all the calcium has reacted with the water, stir the residue with a glass rod and then filter the mixture.

Now using a drinking straw, blow into the filtrate.

Questions

1 Calcium is an element and so the gas which is given off could not have come from it. From which substance must the gas have come?

2 In view of what happens when carbon dioxide is blown through the filtrate, what compound has been formed in solution?

3 Which elements are contained in this compound?

4 Therefore which elements do you think are contained in the water?

Investigation 10.2

What happens when an electric current is passed through water?

You will need an electrolysis cell, with carbon electrodes, such as that shown in Fig. 10.2(a).

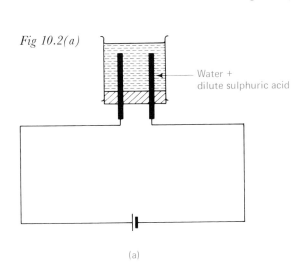

Fig 10.2(a)

(a)

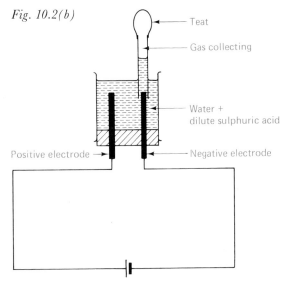

Fig. 10.2(b)

(b)

Pour some tap water into the cell and connect the carbon rods to a direct current power supply (about 6 volts).

Then, without switching off the current, add some dilute sulphuric acid to the water and carefully observe what happens.

Switch off the current. By means of a rubber teat completely fill a piece of glass tubing with water and place it over the carbon rod connected to the negative terminal of the electricity supply. Support the tube so that it covers about the top 1 cm of the rod (Fig. 10.2(b)).

Switch on the current and when the tube is full of gas, remove it from the cell and test the gas with a lighted splint.

In the same way collect a tube of the gas given off at the other carbon electrode (positive) and then insert into it a glowing splint.

Questions

1 What gas is liberated at the cathode (negative electrode)?
2 What gas is liberated at the anode (positive electrode)?
3 The mass of sulphuric acid in the cell does not change as the electric current passes. What, therefore, does this experiment tell us about which elements are present in water?

Investigation 10.3

What number of moles of water is combined with one mole of copper(II) sulphate in crystals of the compound?

When hydrated copper(II) sulphate is heated, the water of crystallisation evaporates and the anhydrous compound remains. By determining the loss of mass in a sample of crystals, the number of moles of water driven off from one mole of anhydrous copper(II) sulphate can be found.

Weigh a crucible, first empty and then about three-quarters full of copper(II) sulphate crystals.

Place a pipeclay triangle on top of a tripod and then put the crucible into the pipeclay triangle. Heat the crucible with a Bunsen flame, gently at first and then a little more strongly.

When there appears to be no further change in the copper(II) sulphate, allow the crucible to cool, and, when it is cool enough for you to bear it on the back of your hand, weigh it again.

Place the crucible back in the pipeclay triangle and reheat it for a further five minutes. Allow it to cool and weigh it again. If there has been a change in mass since the previous occasion continue the process of reheating and cooling until two consecutive masses are within 0·01 g of each other.

Calculate the mass of the anhydrous compound and hence the mass of the water which has been driven off from the crystals.

Using the relative atomic masses given in the Data Section, calculate the number of moles of copper(II) sulphate ($CuSO_4$) and the number of moles of water (H_2O) in the sample of crystals.

Questions

1 How many moles of water (to the nearest whole number) are combined with 1 mole of copper(II) sulphate in the crystals?
2 What, therefore, is the formula of the crystals?

Investigation 10.4

What happens to certain compounds when they are exposed to the air?

You will need three watchglasses and samples of anhydrous calcium chloride, sodium carbonate decahydrate crystals (washing soda), and copper(II) oxide which has been strongly heated and then kept in a desiccator.

Weigh a watchglass, first empty and then containing a small quantity of calcium chloride. Place the watchglass in some place where it is open to the air and can remain undisturbed for a few days. Repeat the process with the dry copper(II) oxide on the second watchglass, and sodium carbonate decahydrate crystals on the third.

A few days later, weigh each watchglass again, and note any changes in mass and appearances of the solids.

Questions

1 What substance has been taken from the air to bring about the change in the calcium chloride?

2 Copper(II) oxide also takes this substance from the air. What is the difference between the two compounds as far as the absorption is concerned?

3 How could the change in mass of the hydrated sodium carbonate crystals be explained?

Investigation 10.5

Is tap water harder than pure water (distilled water)?

Hard water will not easily form a lather with soap, and the hardness of samples of water can be compared by finding the volume of a soap solution which is required to give a 'permanent lather' in the water. A permanent lather is a lather which lasts for at least one minute after the mixture has been shaken.

1. Using a measuring cylinder, measure 10 cm³ of distilled water and pour it into a boiling-tube.

By means of a teat-pipette, add one drop of soap solution to the water, fit the tube with a bung and then shake the tube vigorously. If a permanent lather is not obtained, repeat the addition of one-drop portions of the soap solution, shaking after each addition, until such a lather is produced. Note the number of drops needed to produce a permanent lather.

Repeat the procedure using 10 cm³ of tap water instead of distilled water and note the number of drops of soap solution needed.

Questions

1 Is tap water as hard as, or harder than distilled water?

2 Describe what else, other than a lather, is formed when soap solution is added to tap water.

2. Boil some tap water for about 10 minutes in a beaker and then allow it to cool to room temperature. Measure 10 cm³ of the water into a boiling-tube and repeat the dropwise addition of soap solution until a permanent lather is obtained, and note the number of drops required.

Questions

3 Is boiled tap water harder, softer or as hard as unboiled tap water?

4 Do you think that boiling has removed all, some or none of the impurities from tap water?

3. Measure a further 10 cm³ portion of tap water into a boiling-tube and add a small quantity of sodium carbonate crystals. Shake the tube until the solid

dissolves, and then repeat the addition of drops of soap solution until a permanent lather is obtained, and note the number of drops required.

Questions

5 Does adding sodium carbonate remove more or less impurities than boiling the water?

6 Bath salts contain sodium carbonate crystals. What is one reason for adding the salts to bath water?

Fig. 10.3

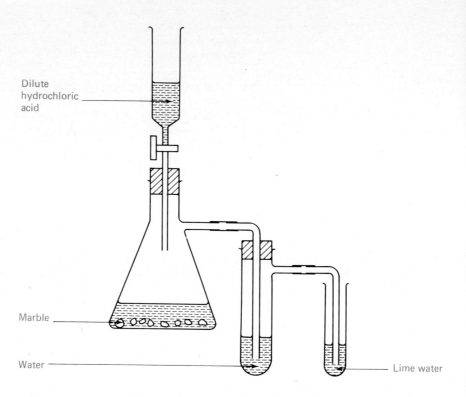

Dilute hydrochloric acid

Marble

Water

Lime water

Investigation 10.6

How is calcium hydrogencarbonate produced in water, and what sort of hardness is given to the water by this substance?

You will need the apparatus shown in Fig. 10.3.

When the hydrochloric acid is added from the tap-funnel to the marble in the filter-tube, carbon dioxide is given off. The water in the filter-tube removes droplets of hydrochloric acid from the stream of gas, so that they will not neutralise the lime water.

Half fill a test-tube with lime water (calcium hydroxide solution) and put it over the end of the delivery tube. Add some dilute hydrochloric acid from the tap-funnel to the marble and pass the carbon dioxide through the lime water. The lime water soon turns milky, and there is then in the test-tube a mixture of calcium carbonate and water.

Continue to pass the carbon dioxide into the mixture until all the milkiness disappears. The extra carbon dioxide has reacted with the calcium carbonate and water to produce a solution of calcium hydrogencarbonate.

Divide the solution into two equal portions. To the first, using a teat-pipette, add soap solution, drop by drop, shaking after each addition, until a permanent lather is produced (Investigation 10.5).

Boil the second portion until an obvious change takes place. Cool the solution and then repeat the dropwise addition of soap solution until a permanent lather is produced.

Questions

1 Write an equation for the reaction which takes place when carbon dioxide is passed into a mixture of calcium carbonate and water.

2 If this reaction explains how some samples of tap water contain calcium hydrogencarbonate, where is the carbon dioxide which takes part in the reaction likely to come from and where is the calcium carbonate likely to come from?

3 What effect does boiling have on the hardness of the water and what do you see in the solution after it has been boiled?

4 Apart from the difficulty in producing a lather with soap, what do you think is likely to be a disadvantage of the presence of calcium hydrogen-carbonate in the water supply?

10.7
Water and the elements it contains

Water is one of the most common compounds, but there are parts of the world where there are severe shortages of water for human consumption and for land irrigation (Fig. 10.4). Also in many areas seasonal fluctuations in rainfall (Fig. 10.5) mean that it is essential to devise satisfactory systems of storing, treating and distributing water.

Water is an essential constituent of all living things and this is based on the fact that it is the best solvent known, dissolving many more things than any other liquid. Clearly, since it is such an important liquid, water has been widely studied and we now have a considerable knowledge about its constitution and the reasons for its useful properties.

If two or three calcium turnings are dropped into a beaker of water, they sink to the bottom of the beaker and give off a steady stream of bubbles of gas. The gas can be collected, as shown in Fig. 10.1, by allowing it to bubble into an inverted test-tube full of water.

When the tube is full of the gas, and it is opened with its mouth close to a Bunsen flame, there is a 'pop', showing the gas to be hydrogen. Since calcium is an element and cannot be split up into other substances, the hydrogen given off in this reaction must have come from the water and water must be a compound of hydrogen.

Some clue as to what is combined with the hydrogen in water is given when we examine what is left when the calcium has reacted with water. As the reaction takes place, a white solid separates from the solution and, if this is stirred with the water when the reaction is complete, a saturated solution is produced. If the surplus solid is filtered off and then a stream of carbon dioxide is blown into the colourless filtrate, the liquid turns milky as a white precipitate forms. A solution which reacts like this with carbon dioxide is lime water (calcium hydroxide solution) and it seems as if this has been formed by the action of the calcium on water. Calcium hydroxide contains the elements calcium, hydrogen and oxygen, and the oxygen must, like the hydrogen, have come from the water. Water, therefore, is likely to be a compound of **hydrogen** and **oxygen**.

Other metallic elements, potassium and sodium, react with cold water, forming the metal hydroxide and setting free hydrogen (17.4), thus confirming what the reaction with calcium suggests.

91

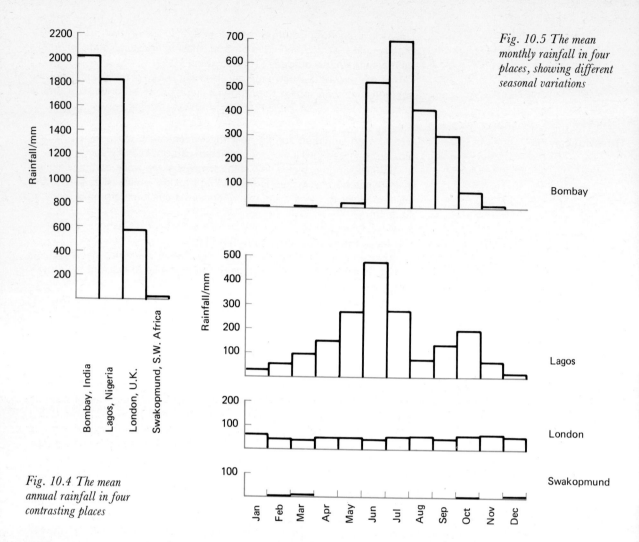

Fig. 10.5 The mean monthly rainfall in four places, showing different seasonal variations

Bombay

Lagos

London

Swakopmund

Fig. 10.4 The mean annual rainfall in four contrasting places

10.8 Synthesis of water

The synthesis of a compound consists of the building up of the compound from its elements or from simpler compounds. If water is a compound of hydrogen and oxygen, it should be possible to form water by causing these two elements to combine. This can, in fact, be done using the apparatus shown in Fig. 10.6.

Fig. 10.6 Burning hydrogen in air

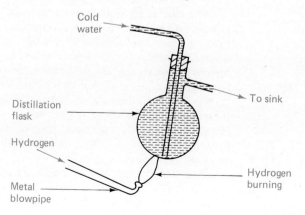

As the flame of the burning hydrogen plays on the cooled flask, drops of a colourless liquid collect on the flask. The liquid gives positive results in two simple chemical tests for water. It turns white anhydrous copper(II) sulphate blue and anhydrous cobalt chloride pink (10.18). If sufficient of the liquid could be collected, it could be shown to have a freezing point of $0\,°C$ and a boiling point of $100\,°C$, confirming it to be water.

As has been described in Chapter 8, when an element burns in oxygen, the oxide of the element is formed. If water is formed by burning hydrogen in the oxygen of the air, the water must be the oxide of hydrogen and this confirms what the action of metals, like calcium, has suggested.

10.9 Electrolysis of water

The fact that water is a compound of hydrogen and oxygen is also confirmed by the action of an electric current on water. Pure water hardly conducts a current but, if some sulphuric acid is added, the conductance of the water increases considerably and the water is split into hydrogen and oxygen.

The reactions which occur at each electrode and result in water being split up into hydrogen and oxygen are explained in 19.9.

The volume of hydrogen produced in a given time is twice that of the oxygen, suggesting that there are twice as many moles of hydrogen as oxygen in the water which is decomposed in the electrolysis. It is the water which is being electrolysed, and not the sulphuric acid, as the mass of the sulphuric acid does not change during the electrolysis.

10.10 The formula of water

Another way of forming water from hydrogen and oxygen, which does not involve burning one gas in the other, consists of passing hydrogen over hot copper(II) oxide. Because copper is lower than hydrogen in the reactivity series (17.9), hydrogen is able to take oxygen from the copper to form the water and copper is left as the solid product of the reaction. The copper(II) oxide is reduced to copper by the hydrogen which is oxidised to water. Since the masses of the starting materials and products are easily found in this reaction, the reaction can be the basis of a simple method for finding the formula of water. A suitable apparatus for this experiment is shown in Fig. 10.7.

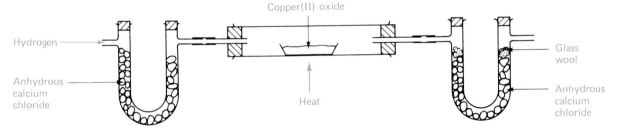

Fig. 10.7 Reducing copper(II) oxide with hydrogen

The anhydrous calcium chloride in the U-tube, placed after the copper(II) oxide, will absorb every bit of water formed in the reaction, and, if the mass of the tube is found before and after passing hydrogen gas, the mass of water formed can be determined. The first U-tube also contains anhydrous calcium chloride and is to ensure that any water collected by the second U-tube must have come from the reaction and was not brought in by the hydrogen.

If the mass of the porcelain boat and copper(II) oxide is found before heating then when the reaction is obviously complete, the difference in mass between the two results is the mass of oxygen lost by the oxide. This is also the mass of oxygen in the water collected in the second U-tube. Subtracting the mass of oxygen from the mass of water gives the mass of hydrogen in the water.

If the number of moles of hydrogen atoms and oxygen atoms in the collected water are then calculated, the formula of the water can be found.

A typical set of results from this experiment is given below.

Mass of porcelain boat + copper(II) oxide before heating	=	26·32 g
Mass of porcelain boat + copper after heating	=	23·52 g
Mass of oxygen lost	=	2·80 g
Mass of oxygen in the water	=	2·80 g
Mass of second U-tube at the start	=	156·48 g
Mass of second U-tube after absorption of water	=	159·63 g
Mass of water produced in the reaction,	=	3·15 g
and mass of hydrogen in the water = 3·15 − 2·80	=	0·35 g
Mass of one mole of hydrogen atoms	=	1·00 g
No. of moles of hydrogen atoms in 0·35 g of hydrogen	=	0·35
Mass of one mole of oxygen atoms	=	16·00 g
No. of moles of oxygen atoms in 2·80 g of oxygen = $\dfrac{2\cdot80}{16}$	=	0·175

0·35 moles of hydrogen atoms combine with 0·175 moles of oxygen atoms.
2 moles of hydrogen atoms combine with 1 mole of oxygen atoms.

The empirical formula of water is H_2O.

The molecular formula of water could be H_2O or a multiple of this such as H_4O_2 or H_6O_3. To see which is the correct molecular formula we need to do an experiment to find the relative molecular mass of the water. This proves to be 18, which agrees with the formula H_2O, and not with H_4O_2 for which the relative molecular mass would need to be 36, nor with H_6O_3 for which it would need to be 54.

Now we know that the **molecular formula** of water is **H_2O**, we can write equations to describe more fully the action of water on certain metals. For example, as has already been described, if calcium turnings are added to cold water, the metal sinks to the bottom and there is a steady effervescence of hydrogen, a white precipitate of calcium hydroxide forming as the metal reacts with the water:

$$Ca(s) + 2H_2O(l) \rightarrow Ca(OH)_2(s/aq) + H_2(g)$$

10.11
The water cycle

About 70% of the Earth's surface is covered by sea which contains 95% of all the water on this planet. The other 30% of the surface consists of land masses on which live a wide variety of plants and animals, all requiring water for their survival and development. The acquisition of this necessary water by living things from the large bulk stored in the sea, is part of the Natural Water Cycle, which not only provides what is needed but also ensures that the water which is not used is put back into store again for future use.

Although water has a boiling point of 100°C at normal atmospheric pressures, it does evaporate at temperatures well below that, particularly if there is a large surface area of it exposed to the air. This is the case with the sea. Water in the sea vaporizes at quite a high rate and the water vapour diffuses through the atmosphere, being joined as it does so by the vapour formed from water on the land. The vapour forms clouds in the upper atmosphere and, under certain conditions, the amount of water in a cloud becomes too great and the water returns to the surface of the Earth as rain.

The proportion of rainfall which is lost by evaporation before it can serve any useful purpose will depend on the climate, but even in the U.K., where the climate is temperate, over half of the annual rainfall is lost in this way.

Rain-water which is not immediately absorbed by the soil or by plants, is then likely to run over or through rocks, or through the soil, and collect to make a spring or a stream. When a number of these join together, a river may be produced, and this then carves out a suitable path for itself, so that, enlarged by the entry of many other streams and springs, it can make its way to the sea.

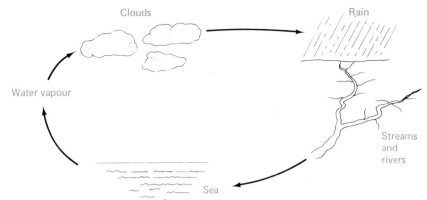

Fig. 10.8 The Natural Water Cycle

Clouds

Rain

Water vapour

Streams and rivers

Sea

This constant movement of water from sea to sky, from sky to land, and from land to sea again is called the **Natural Water Cycle** and it is represented in Fig. 10.8.

The Natural Water cycle has been (and still is) important in changing the land on which we live. As the rain-water runs over and through rocks on its way to form a stream or river, it can either dissolve away constituents of the rock, or it can wear away the rock and carry the debris with it to deposit it at another place, where it may form a constituent of soil. The wearing away, or erosion, of the rocks can give rise to various features of our landscape, such as deep valleys, caves, pot-holes and the estuaries of rivers.

The water may also dissolve out of the rock certain minerals and these will eventually find their way to the sea. A significant example of this is the dissolving of calcium compounds which are used by both fresh-water and salt-water shell fish for the making of the shells. The dissolving of calcium salts in the Earth by rain water will be discussed later in 10.22.

If man is to be able to have sufficient water to satisfy his needs, he must disturb the Natural Water Cycle so that he can collect the water. This is usually done in two ways. The most usual is to make a large resorvoir, which is simply an artificial lake, collecting water from a large number of streams or springs or by natural seepage from the region around it. This lake is usually made by constructing a dam across the narrow part of a valley with a large stream flowing through it, and this water, which would quickly run to the sea, is held back and stored for man's use. An alternative method consists of sinking wells or boreholes into porous rocks such as chalk, limestone or sandstone. Rain-water falling on these rocks infiltrates into the rock and remains in the pores or cracks in just the same way as water will stay in a large sponge. The water is then usually pumped out of the rock through the wells or boreholes to the surface.

10.12 Air in water

If someone holds your head under water in a lake or a river, you very soon begin to doubt, as you struggle for breath, that there is air dissolved in the water. Yet fish, whose bodies require oxygen from the air in order to produce from food the energy they need to stay alive, can live in deep water for a large part of their lives. This suggests that there is air dissolved in water, and that, while we have no natural mechanism for taking that air out of the water into our bodies, the fish are able to extract it and use it for their needs.

The presence of air dissolved in water can be shown using the experiment described in 9·6.

10.13 Pollution of water

Water normally contains about 10 parts per million by mass of dissolved oxygen, although this varies slightly between summer and winter due to the temperature difference. This dissolved oxygen is essential for fish and other aquatic animals, and, if this oxygen is removed from the water faster than it is replaced, these creatures are unable to live in that water. This is what happens when a stretch of water is polluted in certain ways. If **sewage or some industrial waste** is released into a river, it provides food for oxygen-consuming bacteria and they may use up the oxygen faster than it is dissolved from the air in contact with the water. The water is then said to have a high **BOD (biochemical oxygen demand)** and will not support the aquatic animals which need a lot of oxygen.

The BOD of water can be measured by adding to the water a small volume of a solution of potassium manganate(VII) and recording the time for its purple colour to disappear. The greater the demand for oxygen, the faster will oxygen be taken from the potassium manganate(VII) and hence the faster will it be decolourised.

The BOD can also be assessed by examination of the animals living in the water. For example, mayflies and stoneflies require well-oxygenated water, whilst the bloodworm can survive in a river which has very low oxygen levels. By using the knowledge of which animals and plants can survive under certain conditions, regular checks on the plant and animal life can indicate the general condition of the water in a river.

Plants, sown in the soil in fields, need to take in nitrogen and phosphorus in the form of soluble nitrates and phosphates if they are to grow healthily, and, if these compounds are missing from the soil, they have to be supplied in the form of **fertilisers** (33.10). If the fields are close to a stream or river and the fertiliser is applied in a rainy time, it may be washed out of the soil into the river before it is taken up by the plants, and the concentrations of nitrate and phosphate in the river water increase. As well as being food for land plants, these compounds feed plants which grow in water, including the primitive ones without leaves, stems or roots, called algae. When the water is warm in the summer and when the concentrations of nitrate and phosphate are high, the algae grow very rapidly, forming what is called an algal bloom in the water, and, as they grow, they use up dissolved oxygen in the water, thus making life impossible for those creatures which need a lot of oxygen. The water has thus become polluted by the fertiliser.

Fig. 10.9 (a) Measuring the dissolved oxygen content of polluted river water

(b) treating polluted water with oxygen. (Both photos courtesy British Oxygen Company)

Pollution in river water may also be caused by **detergents**. Before synthetic detergents became widely used, the soap, made by the action of alkali on vegetable oils, was easily broken down by sewage bacteria and, when the water was discharged into a river, it did not pollute the water. When synthetic detergents were introduced, they were not broken down in the same way and were still contained in the water when it entered the river. The detergent reduces the amount of oxygen dissolved from the air at the surface of the water, and so the water contains less oxygen than it should, again making it impossible for some creatures, like trout, to live in it.

The detergent may cause two other problems. When a detergent solution is agitated, it produces a lather or foam. River water containing detergent may flow over a weir or be stirred up in some other way, and it will then produce a foam which can then blow about in the wind, contaminating the neighbourhood. The presence of these clouds of foam, blown about near a river, sometimes called 'detergent swans', indicate that there is detergent in the river and therefore the river authorities can expect the water to be deoxygenated. A second difficulty is that some detergents contain phosphates which, as was described earlier, can encourage the growth of algae.

Synthetic detergents have been improved since they were first introduced. It was discovered that, if a molecule of a detergent has a chain of carbon atoms which is branched (Fig. 32.12), it cannot easily be broken down, but, if the chain is straight, a very large proportion of it (up to 99%) is broken down in the sewage works before it enters the river and the pollution is very much reduced.

10.14
The treatment of water

It is necessary to treat water before it is used and also after it has been used. Water which is to be used for domestic purposes must obviously be safe to drink, whereas water which is to be used in industry must not contain anything which will reduce the efficiency of the processes on which the industry depends. The power industry, for example, needs water which is more pure (chemically) than drinking water, as the boilers which convert the water into steam will only remain efficient if the water contains no more than $0 \cdot 00002$ g of dissolved solids per dm^3.

Water which has been used, such as domestic waste (sewage) and water leaving factories (industrial effluent), must be treated before it is allowed to enter natural waterways such as rivers or lakes.

Fig. 10.10 A water treatment plant showing the reaction tanks in which river water is clarified by adding iron(II) sulphate and softened by adding calcium hydroxide and sodium carbonate. (Courtesy the Yorkshire Water Authority)

The main steps in water treatment are clarification, disinfection and softening.

Clarification involves removing the suspended solids from the water. The water is treated with chemicals to encourage the solids to coagulate (stick together). The coagulated solids are then removed by filtration.

97

Disinfection (sterilisation) of water supplies in the U.K. is done by treating it with chlorine which kills the bacteria (25.10). In some countries ozone (O_3) is used instead of chlorine. The disadvantage of ozone is that it does not have any residual sterilising effect. Thus if the sterilised water becomes contaminated again before it is used, there will be no ozone left in the water to remove the contamination. This is not the case with chlorine, as it is more stable and some of it would still be present to remove the contamination. The disadvantage of chlorine, as some of you will probably know, is that if it is necessary to use a lot of chlorine it will cause the water to have an unpleasant taste.

The third stage, known as **water softening**, is more important for industrial supplies and is discussed in detail in 10.25.

The techniques and the standards required by law for the treatment of domestic and industrial effluent have improved considerably during the last 20 years. In the U.K., for example, the miles of rivers which are badly polluted have been reduced from about 6% to 3% of the total.

10.15 Water in the air

If you add some ice to a cold drink in a glass and then leave it for a short time, the outside of the glass becomes cloudy as a coating of droplets of a colourless liquid forms on it. If we could collect some of this liquid, we could show by its freezing point and its boiling point that it is water. The water is present as vapour in the air surrounding the glass and, when the air is cooled by the cold mixture in the glass, the vapour condenses to liquid which collects on the surface of the glass.

The major source of water vapour in the air is the evaporation of water on the surface of the Earth in seas, lakes and rivers (10.11). Other sources include animals which produce it as a by-product in the conversion of food into energy (9.4), and the combustion of hydrocarbon fuels (9.3). These other sources produce an approximately constant amount of water vapour each day. Any variation in the amount of water vapour in the air is due to changes in the climatic conditions. When the temperature is high the rate of evaporation increases, but if the temperature decreases sufficiently, such as it might do overnight, some of the water vapour can condense to water droplets which can take the form of dew or, under certain conditions, fog. If it is very cold overnight, water vapour can be converted into tiny particles of ice, which appear as white hoar frost.

10.16 Hygroscopy and deliquescence

When you buy a new camera or tape-recorder, you may find in the package a small packet of a substance called silica gel. This compound has the property of absorbing up to about 70% of its own mass of water without becoming wet, and it is included in the packing to absorb the water vapour from the air, so that it cannot condense on the equipment. A substance which can absorb water from the atmosphere in this way is called a **hygroscopic substance**.

Copper(II) oxide and table salt are other hygroscopic substances. This explains why occasionally you cannot pour table salt on to your food from a salt cellar. Magnesium carbonate is usually added to table salt to absorb the water which would be taken in by the salt, and so prevent the salt crystals sticking together.

Some of the substances which absorb water from the atmosphere are soluble in water and will take in so much water that they are able to dissolve in it to form a solution. Such a substance is called a **deliquescent substance**.

Sodium hydroxide, anhydrous iron(III) chloride, copper(II) nitrate and anhydrous calcium chloride are examples of deliquescent substances. Anhydrous calcium chloride is used for drying gases and in **desiccators** which are used in science laboratories for storing substances in a dry atmosphere. The anhydrous calcium chloride removes all the water from the air in the air-tight vessel (Fig. 10.12) and so eliminates the chance of water passing from the air to a substance stored in the desiccator.

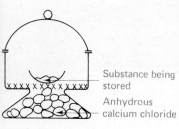

Substance being stored

Anhydrous calcium chloride

Fig. 10.11 A desiccator

10.17
Water in crystals

Crystals of copper(II) sulphate are blue in colour, but, if they are heated (Investigation 10.3), the solid turns white and steam is evolved. Although the original crystals feel perfectly dry, there is locked up in each of them a certain amount of water, joined chemically to the particles making up the crystal, and the blue variety of copper(II) sulphate is therefore called **hydrated copper(II) sulphate**.

When the crystals are heated, the bonds holding the water molecules to the particles in the crystals (which are copper(II) ions and sulphate ions), are broken and the water molecules are able to escape in the form of water vapour. The white solid which remains is therefore copper(II) sulphate without water, which is called **anhydrous copper(II) sulphate**. The water present in the original crystals and driven off by heating, is called **water of crystallisation**.

The breaking of the bonds between the ions in the crystal and the water molecules requires energy to be supplied to the crystals, and therefore, when water is added to the anhydrous solid so that the crystals can reform, energy will be liberated in the form of heat. If a few drops of water are added to white anhydrous copper(II) sulphate around the bulb of a thermometer, the solid rapidly turns blue and the temperature rises quickly to round about $100\,°C$.

10.18
Detecting water

The blue colour of copper(II) sulphate crystals is due to the presence of hydrated copper(II) ions. One way of testing for the presence of water in a liquid is based on this fact and consists of adding the liquid to anhydrous copper(II) sulphate. If the white solid turns blue, there is water present. A similar test involves anhydrous cobalt(II) chloride which is blue and turns pink when water is added, as the hydrated salt is formed. If filter paper is dipped into a solution of cobalt(II) chloride and is then heated, it turns blue and, when this touches anything with water in it, the colour changes to pink. This cobalt chloride paper provides a very simple way of detecting the presence of water.

10.19
The formulae of crystals

Because each ion in the blue crystals of copper(II) sulphate is linked to a small whole number of water molecules, one mole of the salt (consisting of 6×10^{23} copper(II) ions and 6×10^{23} sulphate ions (4.10)) is joined to a small whole number of moles of water. This number can be found by heating a known mass of the crystals in a crucible until the mass of the residue does not alter on further heating (this is called heating to constant mass).

The following is a typical set of results for this experiment.

Mass of crucible	$= 23\cdot61$ g
Mass of crucible + copper(II) sulphate crystals	$= 28\cdot59$ g
Mass of crucible + anhydrous copper(II) sulphate:	
(after first heating)	$= 26\cdot85$ g
(after second heating)	$= 26\cdot79$ g
(after third heating)	$= 26\cdot79$ g

The number of moles of water combined with one mole of copper(II) sulphate can be calculated from these results as follows.

Mass of crystals used	$= 4\cdot98$ g
Mass of anhydrous copper(II) sulphate formed	$= 3\cdot18$ g
Mass of water driven from the crystals	$= 1\cdot80$ g

$3\cdot18$ g of anhydrous copper(II) sulphate combine with $1\cdot80$ g of water.

One mole of anhydrous copper(II) sulphate ($CuSO_4$) has a mass of

$$63 \cdot 5 + 32 + (4 \times 16) = 159 \cdot 5 \text{ g}$$

One mole of the anhydrous salt combines with $\dfrac{1 \cdot 80 \times 159 \cdot 5}{3 \cdot 18} = 90$ g of water.

Since one mole of water has a mass of 18 g, one mole of anhydrous copper(II) sulphate combines with $\frac{90}{18} = 5$ moles of water and the formula of the hydrated copper(II) sulphate is $CuSO_4.5H_2O$.

The equation for the action of heat on the crystals is therefore:

$$CuSO_4.5H_2O(s) \rightarrow CuSO_4(s) + 5H_2O(g)$$

The numbers of moles of water joined to one mole of a compound in its hydrated form differs from compound to compound.

10.20
Efflorescence

If a sample of crystals of sodium carbonate decahydrate (washing soda—$Na_2CO_3.10H_2O$) is left exposed to the air, the clear colourless crystals quickly become covered with an opaque white powder. Some of the water of crystallisation in the crystals has spontaneously evaporated and the white solid formed is sodium carbonate monohydrate ($Na_2CO_3.H_2O$). A compound which spontaneously loses some or all of its water of crystallisation to the air in this way is said to be **efflorescent**. Sodium sulphate decahydrate (Glauber's salt) is also efflorescent and when left in the atmosphere changes to the anhydrous compound.

10.21
Hard water

If you wash your hands in tap water using solid soap, you may notice in the water a white or grey scum and some difficulty in forming a lather with the soap. The amount of scum and the difficulty in forming the lather will depend on the part of the country in which the water has been collected. Water which does not form a lather with soap is called **hard water**. The hardness is due to impurities in the water (10.5). Tap water is pure in the sense that it is safe to drink, but it is chemically impure in that it contains substances other than water. The impurities which cause hardness are usually calcium or magnesium ions (or often both).

Soap consists of the sodium salts of organic acids—e.g. stearic acid (octadecanoic acid). These acids have relatively large molecular masses and hence sodium stearate will be represented here by NaSt. Most metal stearates (except those of sodium and potassium) are insoluble in water, so if soap is dissolved in hard water they are precipitated as an insoluble scum:

$$Ca^{2+}(aq) + 2St^-(aq) \rightarrow CaSt_2(s)$$

The soap is used up in this way and it is difficult to form a lather. The manufacture of soap and how it works when being used to clean fabrics are explained in 32.10.

10.22
Permanent and temporary hardness

The substances present in natural water vary from place to place; their identities depend upon the type of rocks and soil through which the water has flowed on its way to the reservoirs. One important type of rock is gypsum (calcium sulphate) which is slightly soluble in water. Thus if water flows through gypsum rocks it will become hard by dissolving a little calcium sulphate.

Another important type of rock is limestone or chalk (calcium carbonate). If the water to be supplied to our taps is collected in a limestone area it is likely to be hard. Limestone, the main constituent of which is calcium carbonate, does not dissolve in pure water, and so how can calcium ions from the limestone get into the water to make it hard?

As we have already seen, the source of our tap water, whether it is collected in a reservoir or pumped out from a borehole, is rain, and as the rain falls through the air it dissolves some of the carbon dioxide in the air, forming carbonic acid:

$$H_2O(l) + CO_2(g) \rightarrow H_2CO_3(aq)$$

carbonic acid

When the solution of carbon dioxide then falls on or runs through limestone, it dissolves the calcium carbonate, forming a solution of calcium hydrogencarbonate:

$$CaCO_3(s) + H_2CO_3(aq) \rightarrow Ca(HCO_3)_2(aq)$$

This solution contains calcium ions and, if it forms part of the water supply for a particular place, the water supplied to that place is hard. The slow dissolving of limestone by rain-water, in the way just described, is responsible for many landscape features which are found in limestone or chalk areas, e.g. caves, pot-holes, 'water sinks' (where a river suddenly disappears underground to reappear perhaps miles away) and steep-sided valleys.

Hardness due to calcium hydrogencarbonate is different from that due to calcium sulphate in that the former can be reduced or even removed by boiling the water. Calcium hydrogencarbonate (which only exists in solution) is decomposed on heating, liberating carbon dioxide and water and forming a precipitate of calcium carbonate. The calcium ions, originally in the solution of calcium hydrogencarbonate, become combined with carbonate ions in the precipitate:

$$Ca(HCO_3)_2(aq) \rightarrow CaCO_3(s) + H_2O(l) + CO_2(g)$$

The calcium ions are no longer available to react with the soap. Hardness which can be removed simply by boiling the water is called **temporary hardness**.

If the metals are present in salts such as sulphates, or anything else other than hydrogencarbonates, boiling will have no effect. Hardness which is not removed by boiling the water is called **permanent hardness**.

Natural water will often contain both types of hardness, so that boiling will soften it partially but not completely. Addition of washing soda (sodium carbonate) will, however, remove all hardness, e.g.,

$$CaSO_4(aq) + Na_2CO_3(aq) \rightarrow CaCO_3(s) + Na_2SO_4(aq)$$

and $Ca(HCO_3)_2(aq) + Na_2CO_3(aq) \rightarrow CaCO_3(s) + 2NaHCO_3(aq)$

Before the invention of soapless detergents it was usual to add sodium carbonate, in the form of washing soda crystals, to the water when washing clothes. Bath salts, which are simply sodium carbonate crystals with colouring and perfume added, are used to soften bath water.

10.23 Disadvantages of hard water

The problems associated with hardness of water go beyond difficulty in washing. Further problems, associated particularly with temporary hardness, arise from the fact that when water containing, say, calcium hydrogencarbonate is heated, precipitation of calcium carbonate occurs. If such water is fed directly into a heating system, in the home or industry, precipitation occurs within the boiler and the hot-water pipes. The precipitated solid (sometimes called fur) then builds up inside the pipes, making them effectively narrower and impeding both the water flow and the efficiency of heat transfer. Ultimately the whole system will need replacing, which is very expensive. In industry, particularly (since the quantity of water concerned is much greater than in the home), the water must be softened before it goes into the heating system. By far the cheapest way of doing this is to add just the right amount of slaked lime (calcium hydroxide) which precipitates the calcium hydrogencarbonate as calcium carbonate:

$$Ca(HCO_3)_2(aq) + Ca(OH)_2(aq) \rightarrow 2CaCO_3(s) + 2H_2O(l)$$

It is important to note that addition of lime will only remove temporary hardness. If it is necessary to remove all of the dissolved solids an ion exchange method is used (10.25).

In the home, it is often possible to see furring simply by looking at the inside of a kettle. You can show for yourself that the fur consists of carbonate deposits by adding a little vinegar (dilute ethanoic acid), when you will see effervescence owing to carbon dioxide being given off.

Fig. 10.12 Scale deposited inside (a) a water pipe and (b) a hot water tank. (Both photos courtesy Houseman (Burnham) Ltd)

10.24 Advantages of hard water

Despite these problems, hardness in tap water is not wholly a bad thing. The human body contains about 1000 to 1500 g of calcium in bones and teeth and in the blood (where it is necessary for clotting when the skin is damaged) and it needs a daily intake of this element to replace that lost by natural wastage. The recommended amount for an adult is $\frac{1}{2}$ a gram. Growing children, whose bones are still developing, require more, about $\frac{3}{4}$ of a gram. A large majority of the calcium is supplied by foods such as cheese and milk, but some of it is provided by tap water containing dissolved calcium salts. Another advantage of hard water appears in older houses where the water pipes are made of lead. Pure water dissolves a very small quantity of lead and, if someone drinks this water over a long period of time, the amount of lead in that person's body builds up until it reaches a harmful, or even fatal, level. Hard water does not dissolve lead and, although most water pipes are today made of copper with which the problem does not arise, tap water which is very soft is usually artificially hardened before it is supplied to prevent it dissolving lead.

The location of certain industries is influenced by the hardness of the water which is available. For example, a plentiful supply of hard water is needed for the brewing industry as it improves the quality of the beer, whereas the woollen industry needs a good supply of soft water for washing the wool.

**10.25
Other methods
of removing
hardness**

Distillation is an obvious way of softening water as this removes all dissolved solids and hence produces pure water. Generally this method, due to the energy needed to heat the water, is too expensive. However, if there is even a shortage of impure fresh water, as is the case in some parts of the world (10.7), distillation of sea water may be the best method of obtaining adequate supplies of water for domestic use. The use of solar energy, which is likely to be fairly continuously available in such countries, could be the ultimate solution to the problem where such severe water shortages exist.

An important method which is used for domestic, laboratory and industrial applications is the **ion-exchange method**. This is based on the use of an ion-exchange resin, or zeolite, which is an earthy material consisting of a complex sodium aluminium

Fig. 10.13 A deioniser producing pure water for laboratory use. (Courtesy Houseman (Burnham) Ltd)

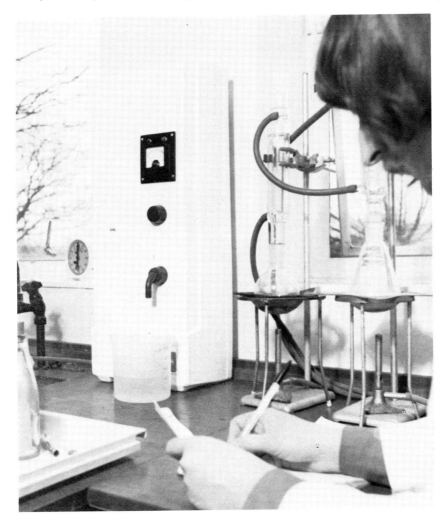

silicate, represented here simply as Na_2Z. As the hard water containing, say, Ca^{2+} ions, flows through the resin ion-exchange occurs:

$$Ca^{2+}(aq) + Na_2Z(s) \rightarrow CaZ(s) + 2Na^+(aq)$$

This leaves the water soft, but obviously the resin eventually becomes used up; it can be regenerated by running concentrated sodium chloride solution through it, whereupon the Ca^{2+} ions in the resin are again replaced by Na^+ ions. The resin is then ready for use again.

Another method used for softening water on a small scale is to use a substance known commercially as **Calgon**, which contains a complex sodium phosphate. (Calgon is a trade name and is derived from the phrase CALcium GONe.) When added to hard water, this combines with Ca^{2+} and Mg^{2+} ions; with the ions tied up as soluble complex phosphates, they are no longer free to cause hardness. (The use of phosphate additives in detergents has caused pollution problems—see 10.13.)

A completely different method, developed relatively recently, uses reverse osmosis. In this method, which is generally used in conjunction with an ion-exchange system (see above), the water is forced under high pressure through a semi-permeable membrane. This acts like an incredibly fine filter; so fine, in fact, that although water will pass through it, dissolved substances will not. The method has the merit that not only will it remove salts, but it will also remove dissolved organic matter, which cannot be removed by the ion-exchange method.

10.26 Modern detergents

Nowadays, hardness of water is much less important with regard to washing than it used to be, because modern washing powders and washing-up liquids are formulated in such a way that they will not form a scum in hard water. The essential feature of modern detergents is that, if they are ionic, the anion is one whose calcium and magnesium salts are soluble in water; consequently, the presence of Ca^{2+} or Mg^{2+} ions has no effect on the operation of the detergent.

(For a general discussion of detergents, see 32.12.)

10.27 Summary

1. Water is a compound of hydrogen and oxygen and it has the formula H_2O. It can be formed by burning hydrogen in oxygen.
2. Water can be decomposed by adding potassium, sodium or calcium, the products of the reaction in each case being the hydroxide of the metal and hydrogen.
3. Water can be split into hydrogen and oxygen by electrolysis, the addition of a compound like sulphuric acid being necessary to make the water conduct the electric current.
4. In the natural world water is involved in a cycle, evaporating from the sea, returning to the Earth as rain and then running back to the sea via springs, streams and rivers. In order to collect water, man has to disturb the Natural Water Cycle.
5. All samples of water, except those which have just been boiled, contain dissolved air. This air is necessary for plants and animals to live in the water.
6. There is usually water vapour in the air. A hygroscopic substance is able to absorb water from the air. A deliquescent substance absorbs water from the air to such an extent that it dissolves in the water.
7. Many crystalline compounds contain water of crystallisation. Water of crystallisation is linked chemically to the particles of which the crystals consist.
8. An efflorescent compound is one which loses water of crystallisation spontaneously by evaporation into the air.
9. Hard water forms a scum and does not easily form a lather with soap. Hardness is due to the presence of calcium ions and/or magnesium ions in solution in the water. A soapless detergent does not form a scum with hard water.
10. Calcium hydrogencarbonate is formed when rain-water, containing dissolved carbon dioxide, falls on rocks containing calcium carbonate and causes temporary hardness which can be reduced or removed by boiling the water. Other soluble calcium salts cause permanent hardness which cannot be removed by boiling.
11. Removal of hardness from water involves tying up the calcium and magnesium ions, either in a precipitate, or as a complex, or with an ion-exchange resin.

Chapter 11

Acids, bases and salts

Investigation 11.1

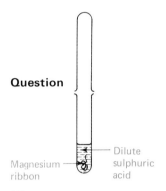

Magnesium ribbon — Dilute sulphuric acid

Fig. 11.1

Question

Questions

How do dilute acids react with metals?

1. Pour dilute sulphuric acid into a test-tube to a depth of about 3 cm.

Add about 3 cm of loosely coiled magnesium ribbon to the acid. Immediately invert a second test-tube over the first, as shown in Fig. 11.1.

When you think sufficient gas has been given off to fill the upper tube, quickly hold the mouth of the tube to a Bunsen flame and record what happens.

1 What is the gas given off when dilute sulphuric acid reacts with magnesium? Write an equation for the reaction.

2. Repeat the procedure using dilute hydrochloric acid instead of sulphuric acid.

Repeat the procedure again, using either of the acids with first granulated zinc, then iron powder and then copper turnings in place of the magnesium. If the reaction does not start, warm the mixture very carefully, but do not boil it.

2 Which of the metals does not react with either dilute sulphuric acid or dilute hydrochloric acid?
3 How do the rates of reaction of those metals which do react compare with that of magnesium?

Investigation 11.2

Questions

Question

How do acids react with bases?

1. Pour dilute sulphuric acid into a boiling-tube to a depth of about 3 cm.

Add a small amount (about one quarter of a spatula-full) of copper(II) oxide to the acid and gently warm the tube over a Bunsen flame until the solid dissolves. Note the colour of the solution.

1 What is the substance likely to be which is responsible for the colour of this solution? What type of substance is this? Write an equation for the reaction which has taken place.

2. Repeat the procedure using first magnesium oxide and then lead(II) oxide to see whether these solids will dissolve in the acid.

2 What could be a possible reason for one of these solids not dissolving in sulphuric acid?

3. Pour dilute sulphuric acid into a boiling-tube to a depth of about 2 cm.

Add a few drops of litmus solution to the acid and then add, in small volumes, sodium hydroxide solution. Stir the solution with a glass rod after each addition and continue adding the sodium hydroxide solution until there is an obvious change in the litmus solution.

4 What happens to the sulphuric acid when sodium hydroxide is added to it?
5 What type of substance is sodium hydroxide? One of the products of the reaction is water. What type of substance is the other product?

Investigation 11.3

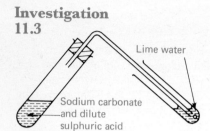
Lime water

Sodium carbonate and dilute sulphuric acid

Fig. 11.2

How do acids react with carbonates?

You will need a right-angled delivery tube in a bung which fits a test-tube.

1. Pour lime water (calcium hydroxide solution) into a test-tube to a depth of about 2 cm.
   Place a few crystals of sodium carbonate into a second test-tube. Then add dilute sulphuric acid to the crystals to a depth of about 3 cm. Quickly fit the test-tube with the bung and delivery tube and pass the gas which is given off through the lime water, Fig. 11.2. Note what happens to the lime water.

Question

1 Which gas has this effect on lime water?

2. Repeat the procedure using first copper(II) carbonate and then marble chips (calcium carbonate) instead of sodium carbonate.
   Repeat the procedure using dilute hydrochloric acid instead of sulphuric acid with each of the three carbonates.

Question

2 Why does the reaction between sulphuric acid and calcium carbonate stop after a short time?

Investigation 11.4

How is the pH of a solution determined and what happens to the pH of a solution as it becomes more acidic or alkaline?

The pH of a solution tells us how acidic or alkaline a solution is. Universal Indicator is a mixture of indicators whose colour depends on the pH of the solution to which it is added.

You will need six beakers—one 1000 cm³, one 500 cm³, one 250 cm³, two 100 cm³ and one 50 cm³.

Add Universal Indicator solution to about 2000 cm³ of tap water in a large bottle until the bluish-green colour of the solution is quite deep.
   Using a measuring cylinder, measure out 800 cm³ of this solution and pour it into the largest beaker. Similarly pour 400 cm³ of the solution into the 500 cm³ beaker, 200 cm³ into the 250 cm³, 100 cm³ into the first 100 cm³ beaker, 50 cm³ into the second 100 cm³ and 25 cm³ into the 50 cm³ beaker.
   Now add 8 cm³ of 0·04M ethanoic acid solution to each beaker and stir with a glass rod.
   Compare the colours in the beakers with those on the chart provided, or on the label of the bottle of Universal Indicator, and hence find the pH of the solution in each beaker.

Questions

1 In which of the beakers is the solution most acidic? Give the reason for your answer.
2 What is the pH of the solution in that beaker?

3 In which of the beakers is the solution least acidic? Give the reason for your answer.

4 What is the pH of the solution in this beaker?

5 What, therefore, happens to the pH of a solution as it becomes more acidic?

6 Which is the more acidic—solution A with a pH of 3 or solution B with a pH of 5?

Now wash out the beakers and repeat the experiment, this time using 8 cm³ of 0·04M sodium hydroxide solution instead of ethanoic acid.

7 In which of the beakers is the solution the most alkaline? Give the reason for your answer.

8 What is the pH of the solution in this beaker?

9 In which beaker is the solution least alkaline? Give the reason for your answer.

10 What is the pH of the solution in this beaker?

11 What, therefore, happens to the pH of a solution as it becomes more alkaline?

12 Which is the more alkaline—solution C with a pH of 9 or solution D with a pH of 13?

Investigation 11.5

The preparation of a pure dry sample of a salt by reacting an acid with an insoluble base.

In this Investigation you will use the reaction between copper(II) oxide (a base which is insoluble in water) and dilute sulphuric acid to produce a solution of copper(II) sulphate:

$$CuO(s) + H_2SO_4(aq) \rightarrow CuSO_4(aq) + H_2O(l)$$

The solution is then concentrated by evaporation until it is saturated. On cooling this solution crystals of the salt will be formed.

You will need a beaker, glass rod, filter funnel and an evaporating basin.

Pour 50 cm³ of dilute sulphuric acid into a beaker and warm the beaker on a tripod and gauze over a Bunsen flame.

Add copper(II) oxide in small portions, stirring the mixture with a glass rod and waiting until each portion dissolves before adding the next.

When no more of the base dissolves (even when the solution is boiling), filter the mixture through a filter paper in a funnel and add a few drops of dilute sulphuric acid to the filtrate.

Pour the solution into an evaporating basin and heat the basin gently on a tripod and gauze over a Bunsen flame.

From time to time pour a small portion of the solution into a test-tube and cool it under the tap, scratching the inside of the test-tube with a glass rod. When the crystals form in one of these samples, pour the remainder of the solution from the basin into a boiling-tube and cool it rapidly under the tap, scratching the tube as before.

When the solution is at room temperature and no more crystals separate, filter off the crystals, wash them with a small volume of distilled water and then spread out the filter paper on another filter paper so that the crystals can dry.

Why can this method not be used to prepare lead(II) sulphate from lead(II) oxide and sulphuric acid?

Investigation 11.6

The preparation of a pure dry sample of a salt by reacting an acid with a soluble base (an alkali).

In this method of preparing a salt an indicator is used to show that all of the acid has reacted with the alkali. The indicator is then removed from the solution of the salt by adsorbing it on charcoal.

You will need a supply of dilute nitric acid (2M) and a more concentrated solution of potassium hydroxide (4M). THE POTASSIUM HYDROXIDE SOLUTION IS VERY CORROSIVE AND MUST BE TREATED WITH GREAT CARE.

Pour 50 cm³ of dilute nitric acid into a conical flask, followed by a few drops of litmus solution. Add concentrated potassium hydroxide solution, 1 cm³ at a time, swirling the liquid in the flask after each addition until the litmus solution changes colour.

Pour the potassium nitrate solution into a beaker and add a small quantity of activated charcoal. Heat the solution to boiling and then filter it while still hot through a filter paper.

Pour the filtrate, which should now be colourless, into an evaporating basin and heat the basin to evaporate the solution to make it saturated.

From time to time pour a small portion of the solution into a test-tube and cool the tube under the cold water tap, scratching the inside of the tube with a glass rod as the tube is cooled. When crystals form in one of the test samples, pour the rest of the solution into a boiling-tube and cool it under the tap, scratching as before.

When the solution is at room temperature and no more crystals separate, filter off the crystals through a filter paper in a funnel and wash them with a small volume of distilled water. Open out the filter paper on to another filter paper and allow the crystals to dry.

Question

Why is it necessary to use a different method when a soluble base is being used than when an insoluble base is the starting substance?

Investigation 11.7

The preparation of a pure dry sample of lead(II) sulphate.

Lead(II) sulphate cannot be prepared satisfactorily by adding sulphuric acid to either lead(II) oxide or lead(II) carbonate. Both of these compounds are insoluble in water. Also the rates of the reactions of sulphuric acid with them are very slow, as the product, lead(II) sulphate, is insoluble and is deposited on the lead(II) oxide or lead(II) carbonate thus protecting them from further attack by the acid.

It is necessary to use a two-stage process, the first of which involves preparing a soluble compound of lead. In this Investigation lead(II) carbonate is used as the starting material.

Add 25 cm³ of dilute nitric acid to a 250 cm³ beaker and then add lead(II) carbonate in small portions, the mixture being stirred after each addition and each portion being allowed to dissolve before the next one is added. When eventually no more will dissolve, filter the mixture into a boiling-tube.

To the filtrate add dilute sulphuric acid until, when a small portion of the mixture is filtered and dilute sulphuric acid is added to the filtrate, no precipitate is produced.

Precipitation of lead(II) sulphate is now complete in the boiling-tube. Filter off the precipitate, wash it well with distilled water and then open out the filter paper on to another filter paper. The product can now be dried, either at room temperature or in a warm oven.

Questions

1 Which ions must have combined together to form the precipitate of lead(II) sulphate? Write an ionic equation for the formation of the precipitate.
2 When the precipitate is finally washed on the filter paper, what substances will be removed from it in the washings?

Investigation 11.8

Burning elements in oxygen and the behaviour of the oxides with water.

You will need either three gas jars or three boiling-tubes full of oxygen, and three deflagrating spoons, either full size or the miniature type to fit a boiling-tube.

Wind a short piece of magnesium ribbon round the stem of one spoon, leaving one end sticking out, so that the ribbon can be ignited in a Bunsen flame. Teachers may prefer to demonstrate the magnesium experiment.

Dip the bowl of a second spoon into powdered charcoal and the third into powdered iron dust.

Heat each spoon in turn in a Bunsen flame and, when the element is hot or burning, lower the spoon carefully into the oxygen.

Note the colours of the flames when the elements burn. ON NO ACCOUNT STARE AT THE MAGNESIUM WHILE IT IS BURNING.

When the jars or tubes are cool, add a small amount of water to each and shake the vessels thoroughly. Now add a few drops of Universal Indicator to each jar or tube. By comparing the colour of each solution with those on the chart or the label on the indicator bottle, find the pH of each solution.

Your teacher may show you the burning of other elements, such as sulphur, red phosphorus and calcium.

Questions

1 Which of the elements you burnt in oxygen are metallic?
2 Which of the elements you burnt in oxygen are non-metallic?
3 What sort of solutions are formed when the oxides of metals dissolve in water?
4 What sort of solutions are formed when the oxides of non-metallic elements dissolve in water?
5 Why does iron oxide appear to give a neutral solution in water?
6 What type of oxide would you expect to be formed by (i) chlorine, (ii) copper (copper(II) oxide does not dissolve in water)?

11.9 Acids

Let us suppose that you are an underwater archaeologist and that your job is to equip yourself with wet suit, face mask and air supply and to dive down to the bottom of the sea to seek for and explore the remains of ships, sunk a long time ago. Let us also suppose that on one trip, you brought to the surface a gold plate which had been under the sea for many years and had become covered with coral and barnacles and limpets.

One way of cleaning your prize would be to try to rub off the encrustation with a brush and detergent solution, but this would take a long time. A much easier method would be to put the plate in a tank containing dilute hydrochloric acid and simply leave it. The acid would dissolve the calcium carbonate in the deposits on the metal, which could then be cleaned easily. A similar sort of process can be used for cleaning off the scale or 'fur' which is formed when hard water is heated in a kettle. Also, when a piece of copper has been enamelled by a method in which the enamel is painted on to the copper and is then made solid by firing in a kiln, the firing produces a black deposit on the exposed copper, which can then be cleaned off by acid.

In all these examples, an acid acts faster than water, with something being eaten away by the acid more rapidly than it can be removed by water. The power of acids to 'eat away', or corrode, materials has interested chemists for a long time. As often happens, the practical uses of these substances led people to try to discover what acids have in common which results in them all having this property.

Before we can get an answer to this question about acids, we ought first to see exactly what will react with them and what products are formed in the reactions. The important reactions which most acids undergo, are the following.

1. Most acids react with certain **metals**, setting free hydrogen, the metal taking the place of the hydrogen,

$$\text{e.g.} \quad \underset{\text{zinc}}{\text{Zn(s)}} \quad + \quad \underset{\substack{\text{sulphuric} \\ \text{acid}}}{\text{H}_2\text{SO}_4\text{(aq)}} \quad \rightarrow \quad \underset{\substack{\text{zinc} \\ \text{sulphate}}}{\text{ZnSO}_4\text{(aq)}} \quad + \quad \underset{\text{hydrogen}}{\text{H}_2\text{(g)}}$$

2. Most acids react with a **metal oxide or a metal hydroxide**, forming water and a compound in which the metal in the oxide or hydroxide has taken the place of the hydrogen in the acid,

$$\text{e.g.} \quad \underset{\substack{\text{magnesium} \\ \text{oxide}}}{\text{MgO(s)}} \quad + \quad \underset{\substack{\text{hydrochloric} \\ \text{acid}}}{2\text{HCl(aq)}} \quad \rightarrow \quad \underset{\substack{\text{magnesium} \\ \text{chloride}}}{\text{MgCl}_2\text{(aq)}} \quad + \quad \underset{\text{water}}{\text{H}_2\text{O(l)}}$$

$$\underset{\substack{\text{sodium} \\ \text{hydroxide}}}{\text{NaOH(aq)}} \quad + \quad \underset{\substack{\text{nitric} \\ \text{acid}}}{\text{HNO}_3\text{(aq)}} \quad \rightarrow \quad \underset{\substack{\text{sodium} \\ \text{nitrate}}}{\text{NaNO}_3\text{(aq)}} \quad + \quad \underset{\text{water}}{\text{H}_2\text{O(l)}}$$

The metal oxide or hydroxide neutralises the acid and is called a base. If a base is soluble in water, it is known as an alkali.

3. Most acids react with **carbonates**, liberating carbon dioxide,

$$\text{e.g.} \quad \text{ZnCO}_3\text{(s)} + \text{H}_2\text{SO}_4\text{(aq)} \rightarrow \text{ZnSO}_4\text{(aq)} + \text{H}_2\text{O(l)} + \text{CO}_2\text{(g)}$$

In each of these reactions a compound is formed in which a metal has taken the place of the hydrogen in the acid,

$$\text{e.g.} \quad \underset{\substack{\text{sulphuric acid} \\ \text{or} \\ \text{hydrogen sulphate}}}{\text{H}_2\text{SO}_4} \quad \text{gives} \quad \underset{\text{zinc sulphate}}{\text{ZnSO}_4}$$

This leads to the simple description of an acid as being a substance containing hydrogen which can be replaced by a metal.

11.10
Bases and
alkalis

The compounds referred to earlier, in which the hydrogen of an acid has been replaced by a metal, are called salts and the simplest description of a base is a compound which will react with an acid forming a salt and water. Copper(II) oxide is a base because it reacts with sulphuric acid to form copper(II) sulphate and water, and sodium hydroxide is a base because it reacts with hydrochloric acid to form sodium chloride and water. Thus bases are metal oxides or hydroxides. One important exception to this is a solution of ammonia (33.15), which reacts as though it were ammonium hydroxide, NH_4OH. It reacts with, for example, hydrochloric acid forming ammonium chloride (a salt) and water:

$$NH_4OH(aq) + HCl(aq) \rightarrow NH_4Cl(aq) + H_2O(l)$$

The ammonium part clearly resembles the sodium part of the reaction mentioned above involving sodium hydroxide.

A base which is soluble in water is called an alkali. Sodium hydroxide and potassium hydroxide are two important alkalis. They dissolve readily in water, producing very corrosive solutions (called alkaline solutions) which react rapidly with acids. If a metal oxide dissolves it does so by reacting with the water to form the metal hydroxide which then dissolves,

e.g.
$$CaO(s) + H_2O(l) \rightarrow Ca(OH)_2(aq)$$

11.11 Acids and alkalis in terms of ions

Solutions of acids and alkalis readily conduct electricity which indicates that the solutions contain ions (6.7). Hydrogen is common to all acids and when an electric current is passed through a solution of an acid, the hydrogen is attracted to the negative electrode, suggesting that hydrogen is present in the form of positively charged ions. This leads to a more detailed description of an acid as being a compound which dissolves in water, liberating hydrogen ions into the solution.

This definition also had to be modified when it was discovered that a solution of an acid does not contain free hydrogen ions.

Hydrogen ions have a great liking for water and the water in the solution holds them in the form of another positive ion, called the hydroxonium ion (or the oxonium ion):

$$H^+ \quad + \quad H_2O \quad \rightarrow \quad \underset{\substack{\text{hydroxonium} \\ \text{ion}}}{H_3O^+}$$

The dissolving of other compounds to give acid solutions will be similar,

$$H_2SO_4(l) + H_2O(l) \rightarrow H_3O^+(aq) + HSO_4^-(aq)$$

which is followed by:

$$HSO_4^-(aq) + H_2O(l) \rightarrow H_3O^+(aq) + SO_4^{2-}(aq)$$

A definition of an acid in terms of ions would therefore be that an acid is a substance which dissolves in water, liberating hydroxonium ions. The formula H_3O^+ is rarely used in equations. For simplicity, it is usual to write the ion as $H^+(aq)$, showing it to be a hydrated hydrogen ion. For example a solution of hydrochloric acid can be represented by:

$$H^+(aq) + Cl^-(aq)$$

11.12 Neutralisation

When the neutralisation of an acid by an alkali takes place, the positive ion from the alkali and the negative ion from the acid remain unchanged in the solution and the mixture of the two in solution constitutes the solution of the salt, produced in the neutralisation. These ions, which play no part in the neutralisation, are sometimes called spectator ions:

e.g.
$$Na^+(aq) + OH^-(aq) + H^+(aq) + Cl^-(aq) \rightarrow \underbrace{Na^+(aq) + Cl^-(aq)}_{} + H_2O(l)$$

unchanged spectator ions—together make up a solution of sodium chloride

Just as the names of the spectators are not included in the press report of a football match, so the formulae of the spectator ions are not included in our report of the reaction, which is the equation, and so the equation for all neutralisations in aqueous solution is:

$$H^+(aq) + OH^-(aq) \rightarrow H_2O(l)$$

11.13 Recognizing acids

One way to detect the presence of an acid in a solution is to taste the solution. Like an 'acid drop' sweet, acids generally have sour tastes—indeed, the German word for acid is saure. Most of the sour-tasting substances we meet in our food contain acids. Vinegar contains acetic acid (now called ethanoic acid); oranges, lemons and grapefruits contain citric acid, and sour milk tastes as it does because, as the milk turns sour, it is attacked by bacteria which convert it to lactic acid.

Tasting, however, is not a safe way of detecting an acid. If you tasted prussic acid (hydrogen cyanide solution), you would hardly have time to report your findings before you turned blue and died in terrible agony. Ethanedioic acid (found in the leaves of the rhubarb plant) could also cause you a premature end, though not as quickly as with prussic acid. We want a safer test for an acid than tasting it.

It has been known for a long time that a colouring matter, obtained from a species of lichen, changes colour when it is put into an acid solution. The dye, called litmus, is now made from simpler compounds and dissolves in water to give a purple solution which turns red when an acid is added. A dye which changes colour when an acid is added is called an indicator, and clearly the safest way of finding whether a solution has an acid in it is to add such an indicator. Litmus also changes colour when alkalis are added to it. In this case it changes from purple to blue and so it can be used to detect the presence of alkalis as well as acids.

Nowadays we do not have to rely on plant material for indicators. We have a range of synthetic dyes which will detect acids and alkalis for us, and dyes such as methyl orange, phenolphthalein, methyl red and bromothymol blue are regularly used for this purpose. The colours which indicators show when added to acidic or alkaline solutions are listed in Table 11.1.

TABLE 11.1. *Indicators for acids and alkalis*

INDICATOR	COLOUR IN	
	ACID	ALKALI
litmus	red	blue
methyl orange	orange	yellow
phenolphthalein	colourless	red
methyl red	red	yellow
bromothymol blue	yellow	blue

The indicators mentioned above do not tell us how acidic a solution is. If 5 cm³ of dilute sulphuric acid are added to 5 cm³ of water, the mixture will turn purple litmus solution red. If 5 cm³ of dilute sulphuric acid are added to 100 cm³ of water, the solution will still turn purple litmus red, although it is obviously more dilute than the previous one and therefore not as acidic. So how can we determine how acidic a solution is, and how can we express our answer?

The figure the chemist uses to denote the level of acidity of a solution is the pH number of the solution. Any acidic solution will have a pH of less than 7 and, because the letter p in pH is concerned with the reciprocal of the concentration of acid (the hydrogen ions from the acid), the lower the pH, the more acidic is the solution. A

112

solution with a pH of 2 is more acidic than a solution with a pH of 4. Sour milk has a pH of about 5·5, and the fact that the pH is less than 7 tells us that it must have acid in it. Vinegar, on the other hand, has a pH of 2·4 and this therefore is more acidic than sour milk.

The idea of pH can also be used for alkaline solutions and will tell us how alkaline a solution is. Solutions of alkalis have pHs greater than 7 and in these cases the more alkaline the solution, the higher is the pH. A solution with a pH of 13 is more alkaline than one with a pH of 10.

A small number of water molecules in a pure sample of water are split up into hydrogen ions and hydroxide ions:

$$H_2O(l) \rightarrow H^+(aq) + OH^-(aq)$$

but as these are present in equal numbers, pure water is neutral. Its pH is 7 and this is the pH of any neutral solution. For example, sodium chloride solution has a pH of 7 and therefore is neutral, being neither acidic nor alkaline. The whole pH scale can be summarised as in Table 11.2.

TABLE 11.2.
The pH scale

1	2	3	4	5	6	7	8	9	10	11	12	13	14

ACIDIC SOLUTIONS

NEUTRAL SOLUTIONS

ALKALINE SOLUTIONS

ACIDITY INCREASING → | ALKALINITY INCREASING →

A dilute solution of sodium carbonate has a pH of about 11. This tells us that it is a moderately alkaline solution. A solution of carbon dioxide in water has a pH of 5 which indicates that the solution is acidic, but only slightly so.

One way of finding the pH of a solution is to use a mixture of indicators called Universal Indicator, which shows different colours at different pHs. For example, it is red when the pH is 4 or less, orange at pH 5, yellow at pH 6 and bluish-green in a neutral solution at pH 7. The maker of the indicator usually supplies a chart or a label on the bottle which shows the colour the indicator gives at a particular pH. To find the pH of a solution, it is only necessary to add a few drops of the indicator solution and then to compare the colour produced with those on the chart or label.

11.14
Weak acids and weak alkalis

If a solution of hydrochloric acid is diluted the pH of the solution increases, showing that the solution is less acidic because, as you would expect, the concentration of the hydrogen ions in the solution has decreased. However, when the pH of different acids in solutions of equivalent concentrations are measured, it is found that the pH value for some acids is higher than expected. For example, the pH of a solution of hydrochloric acid containing $0·1$ mol dm^{-3} is 1, whereas the pH of a solution of ethanoic acid of the same concentration is about 3, indicating that ethanoic acid is weaker than hydrochloric acid.

The conductance of a solution of a weak acid such as ethanoic acid is less than that of strong acid in a solution of equivalent concentration. This observation suggests that there are fewer ions present in the solution of the weak acid. The solution of ethanoic acid is less acidic than expected because not all of the acid is split up into hydrogen ions and ethanoate ions:

113

$$CH_3COOH(aq) \rightarrow H^+(aq) + CH_3COO^-(aq)$$
ethanoic
acid
ethanoate
ions

Similarly, some alkalis, for example ammonia solution, are weak alkalis because their solutions contain fewer hydroxide ions than you would expect.

11.15
Salts

As we have already said, a salt is the compound formed when the hydrogen of an acid is replaced by a metal. When one mole of the acid can give rise to one mole of hydrogen ions (or more correctly hydroxonium ions), as with hydrochloric acid and with nitric acid, replacement can only result in the formation of one salt of a particular metal. For example, the only sodium salt which can be formed from hydrochloric acid (HCl) is sodium chloride (NaCl), where all the hydrogen has been replaced by the metal. An acid of this sort is called a monobasic acid.

If, however, one mole of acid can form two moles of hydrogen ions, as with sulphuric acid (H_2SO_4), two salts of a particular metal can be produced. The first is formed when half the hydrogen ions have been removed, and the second when the rest have reacted. When sulphuric acid is neutralised with sodium hydroxide solution and the same number of moles of sodium hydroxide as there are of sulphuric acid have been added, the following reaction will have taken place:

$$H_2SO_4(aq) + NaOH(aq) \rightarrow NaHSO_4(aq) + H_2O(l)$$

The salt formed here is sodium hydrogensulphate. Because there is still some replaceable hydrogen in it, it is an acid as well as a salt, and so is known as an **acid salt**.

When another equimolar quantity of alkali is added, a second reaction takes place:

$$NaHSO_4(aq) + NaOH(aq) \rightarrow Na_2SO_4(aq) + H_2O(l)$$

the product this time being sodium sulphate, which, with all the hydrogen ions from the acid having been tied up in the form of water, is called a **normal salt**.

Because each mole of sulphuric acid can produce two moles of hydrogen ions, it is called a dibasic acid. Phosphoric acid (H_3PO_4) is a tribasic acid and can form three sodium salts, two (NaH_2PO_4 and Na_2HPO_4) being acid salts and one (Na_3PO_4) being a normal salt.

11.16
Preparations of salts

Whenever we are faced with the job of preparing a salt, the first question asked must be 'is it soluble or insoluble in water?'

For simple salts the answer to that question is given in Table 11.3, which contains an indication of those salts which are not soluble in water.

TABLE 11.3. *Common insoluble salts*

sulphates	lead(II) sulphate, barium sulphate, calcium sulphate
chlorides	lead(II) chloride, silver chloride
nitrates	none
carbonates	all except sodium carbonate, potassium carbonate, ammonium carbonate

If the salt to be prepared is insoluble it must be prepared by a method which does not result in the formation of a mixture of product and starting material. For example, if you try to prepare lead(II) sulphate from lead metal and sulphuric acid, the reaction will be very slow because the lead will become coated with insoluble lead(II) sulphate,

114

and you will end up with a mixture of lead(II) sulphate and unreacted lead rather than pure lead(II) sulphate.

If the salt is soluble it can be prepared by reacting the appropriate acid with the metal or its oxide, hydroxide or carbonate as would be expected from the reactions of acids met earlier in this chapter.

11.17 Preparations of soluble salts

The following methods are available.

1. **The action of an acid on a metal**.

This method would not be suitable for the preparation of salts of potassium, sodium and calcium, since the reactions of these metals with acids would be too violent, while the metals just above hydrogen in the reactivity series react too slowly for this to be the best method.

In those cases where it is convenient, the metal is allowed to react with the acid, warming if necessary, until gas is no longer given off. The excess metal is filtered off and the solution is concentrated by evaporation until a hot saturated solution is obtained. This will then form crystals on cooling, and these can be filtered off, washed with a little water and dried on the filter paper.

2. **The action of an acid on a base (a metal oxide or hydroxide)**.

If the base is insoluble in water, it is added in small portions to the warm acid with continuous stirring, until no more will dissolve. The undissolved solid is filtered off and the solution is again concentrated by evaporation before cooling to produce crystals of the product.

If the base is soluble in water (i.e. an alkali), the problem is then to know when the acid has just all been neutralised. This is solved by putting an indicator into the acid and then adding the alkali solution in small portions (preferably from a burette) until the indicator just changes colour. The indicator can be removed by boiling the solution of the salt for a few minutes with a pinch of charcoal, which has the ability to adsorb complicated molecules, like those of the indicator, on its surface. On filtering, the dye stays with the charcoal in the filter paper and the crystals can be obtained from the filtrate as in Method 1.

In the case of the neutralisation of sulphuric acid with sodium hydroxide solution, the indicator changes colour when all the hydrogen of the acid has been converted to water and a solution of the normal salt (sodium sulphate—Na_2SO_4) has been formed. If half the volume of sodium hydroxide solution required to change the colour of the indicator is added to the same volume of the sulphuric acid as above, only half the hydrogen of the acid is converted to water and the solution can be crystallised to give the acid salt (sodium hydrogensulphate—$NaHSO_4$).

3. **The action of an acid on a carbonate**.

As in the preparation from an insoluble base, an insoluble carbonate is added in small portions to the acid, warming if necessary, until there is no further effervescence and some of the solid remains undissolved. The mixture is filtered and the filtrate is crystallised as before.

If the carbonate is sodium, potassium or ammonium carbonate, it is soluble in water and the preparation of a salt, starting with it, would follow the same method, using an indicator, as a preparation from an alkali.

11.18 Preparation of insoluble salts

Insoluble salts have to be prepared by a method called **double decomposition**, in which, in two compounds, there is a 'changing of partners'. Lead(II) sulphate is, as indicated in Table 11.3, insoluble in water. If dilute sulphuric acid (hydrogen sulphate) is added to a solution of lead(II) nitrate, partners are changed and the products are lead(II) sulphate and nitric acid (hydrogen nitrate). The lead(II) sulphate, being insoluble, separates as a solid, called a precipitate:

115

$$\text{Pb(NO}_3)_2\text{(aq)} + \text{H}_2\text{SO}_4\text{(aq)} \rightarrow \text{PbSO}_4\text{(s)} + 2\text{HNO}_3\text{(aq)}$$

What actually happens here is that the solution of lead(II) nitrate contains lead(II) ions and the dilute sulphuric acid has in it sulphate ions. When the solutions are mixed, the lead(II) ions and the sulphate ions come together to form solid particles of lead(II) sulphate which appear as the precipitate:

$$\text{Pb}^{2+}\text{(aq)} + \text{SO}_4^{2-}\text{(aq)} \rightarrow \text{PbSO}_4\text{(s)}$$

Any insoluble salt is prepared by a method similar to this. All that it is necessary is to mix together a solution of salt containing the positive ion of the salt to be prepared, and a solution containing the negative ion.

The precipitate in each case can be separated by filtration, washed with water and then dried.

The coming together of the ions to form crystals of the precipitate will only take place in solution and, if the salt has to be prepared from the insoluble metal oxide, hydroxide or carbonate, a solution of the metal ions has first to be prepared by dissolving the base or carbonate in dilute nitric acid. The solution containing the appropriate anion can then be added to form the precipitate.

11.19 Metallic and non-metallic elements

It is not always easy to decide whether an element is a metal or a non-metal by examining its physical properties (1.4). There are some non-metallic elements, like carbon in the form of graphite, which are metallic in appearance and behave in some ways like metals, and there are some metals which have low densities and melting points, like non-metallic elements. The only satisfactory way of deciding arises from the chemical reactions of the two sorts of elements.

The usual method is to convert the elements to their oxides and then examine the properties of the oxides. In the laboratory, this is most conveniently illustrated with elements which burn in oxygen. The apparatus in Fig. 11.3 can be used to burn some solid elements. The colours of the flames and the names of the oxides are given in Table 11.4.

Some metal oxides can be distinguished from the oxides of the non-metallic elements by adding water and then Universal Indicator to the jar after the element has been burned. The oxides of the non-metallic elements cause the indicator to turn red, orange or yellow, showing that their solutions are acidic. In each case the oxide combines with the water to form an acid, sulphur dioxide forming sulphurous acid, carbon dioxide carbonic acid, and phosphorus(V) oxide phosphoric acid. Non-metallic elements can therefore be said to form acidic oxides. Some oxide solutions are more acidic than others, and this is why the pHs of the solutions, and hence the colours produced with the Universal Indicator, vary.

The oxides of the metals, calcium, sodium, potassium and to a very slight extent, magnesium combine with the water, forming the metal hydroxides which dissolve in the water and are therefore alkalis. The Universal Indicator will turn blue or purple,

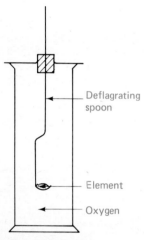

Fig. 11.3 Burning elements in oxygen

TABLE 11.4. *The burning of some metallic and some non-metallic elements in oxygen*

ELEMENT	COLOUR OF FLAME	NAME OF OXIDE
calcium	red	calcium oxide
sodium	yellow	sodium peroxide
potassium	light purple	potassium superoxide
iron (powder)	yellow sparks	iron oxide
magnesium	bright white	magnesium oxide
sulphur	blue	sulphur dioxide
phosphorus	yellow	phosphorus(V) oxide
carbon	red glow	carbon dioxide

116

showing that the pHs of the resulting solutions are greater than 7. Iron oxide does not dissolve in water and so is not an alkali, the pH of its mixture with water being 7 (the pH of the water). Nevertheless, the iron oxide is like the other metal oxides in that it will neutralise acids, and so, while not an alkali, it is a base. Metallic elements, therefore, form basic oxides.

11.20 Amphoteric oxides and hydroxides

Some metal oxides react with both acids and alkalis. For example, zinc oxide reacts with hydrochloric acid,

$$ZnO(s) + 2HCl(aq) \rightarrow ZnCl_2(aq) + H_2O(l)$$

and with sodium hydroxide solution:

$$ZnO(s) + 2NaOH(aq) + H_2O(l) \rightarrow Na_2Zn(OH)_4(aq)$$
$$\text{sodium zincate}$$

In the first reaction zinc oxide is acting as a basic oxide and in the second it is acting as an acidic oxide. An oxide which has both basic and acidic properties is called an **amphoteric oxide**. Aluminium oxide and lead(II) oxide are also amphoteric.

If the hydroxides of these three metals are prepared by adding a few drops of sodium hydroxide solution to solutions of salts of the metals, they all appear as white precipitates,

e.g.
$$Pb(NO_3)_2(aq) + 2NaOH(aq) \rightarrow Pb(OH)_2(s) + 2NaNO_3(aq)$$

or
$$Pb^{2+}(aq) + 2OH^-(aq) \rightarrow Pb(OH)_2(s)$$

When an excess of sodium hydroxide solution is added to each of these precipitates, they dissolve to form clear solutions. The metal hydroxides have reacted with the alkali,

e.g.
$$Pb(OH)_2(s) + 2NaOH(aq) \rightarrow Na_2Pb(OH)_4(aq)$$

$$Al(OH)_3(s) + NaOH(aq) \rightarrow NaAl(OH)_4(aq)$$

Thus in addition to the expected basic properties these hydroxides show acidic properties and are therefore amphoteric.

11.21 Summary

1. Acids usually react with metals forming salts, react with bases forming salts and water, and react with carbonates setting free carbon dioxide.
2. An acid is a substance which dissolves in water, liberating hydroxonium ions (usually written as hydrated hydrogen ions, $H^+(aq)$).
3. A base is a compound which neutralises an acid, forming a salt. It is usually the oxide or hydroxide of a metal. An alkali is a soluble base.
4. A salt is the compound formed when the hydrogen of an acid is replaced by a metal. If all the hydrogen is replaced, a normal salt is formed. If only part is replaced, an acid salt is formed.
5. Acid solutions can be recognized by their ability to change the colours of certain dyes called indicators.
6. The pH of a solution tells us how acidic or alkaline a solution is. As acidity rises, the pH falls. The more alkaline a solution, the higher its pH. A neutral solution has a pH of 7.
7. Soluble salts can be prepared by the action of the appropriate acid on the appropriate metal, metal oxide, metal hydroxide or metal carbonate. Insoluble salts have to be prepared by double decomposition and are precipitated on mixing two solutions containing the appropriate ions.
8. Metallic elements form basic oxides; non-metallic elements form acidic oxides.
9. Aluminium, lead and zinc form amphoteric oxides and hydroxides.

Chapter 12

Families of elements

What patterns are there in the melting points and boiling points of elements?

The melting points and boiling points of the elements are given in the Data Section. For the elements with atomic numbers from 1 to 38, plot a graph of melting point against atomic number and on a second sheet of graph paper, plot a graph of boiling point against atomic number. In both graphs leave a gap between elements 20 and 31 (i.e. omit elements 21 to 30).

Questions

1 Which elements correspond to the minimum points on the graphs?
2 In what way are these elements chemically similar?
3 Which elements correspond to the maximum points on the graphs?
4 Do you expect the properties of these elements to be fairly similar or quite different?
5 Lithium, sodium and potassium have similar properties. Look at the graphs to see where these elements lie in relation to the maximum points on the graphs. What other groups of elements can you find from the graphs, where the members of a group have similar properties?

What pattern is there in the atomic volumes of the elements?

The atomic volume of an element is the volume which would be occupied by 1 mole of atoms of the element if it were in the solid state.

By looking up the required values in the Data Section, plot a graph of atomic volume against atomic number for the elements with atomic numbers from 3 to 38 but leaving a gap between elements 20 and 31 as before.

Questions

1 Look at your graph and decide whether it is better to describe the way in which atomic volume varies with atomic number as regular or periodic. A regular variation is one in which one value increases steadily as the other increases. A periodic variation is one in which a certain type of variation is repeated at intervals.
2 Which elements correspond to the maximum points on the graph?
3 The elements which you listed in your answer to Question 2 are also very similar in their chemical behaviour. From your graph, suggest elements (one in each case) which are likely to behave in a similar manner to each of the following:
    (a) chlorine
    (b) magnesium
    (c) argon

To find a more convenient and useful arrangement of the elements.

Write the symbols of the elements 3 to 38 (but omitting elements 21 to 30) in one continuous row. Now rewrite the list, except this time arrange the elements in a series of horizontal rows so that elements with similar characteristics are placed vertically under each other. Start with lithium (Li) and then start a new row each time a similar element occurs. Compare your table of the elements with the relevant part of the Table on p. 382.

Questions

1 Are the elements in the first vertical column of the table metals or non-metals?

2 Are the elements in the next to the last vertical column in the table metals or non-metals?

3 Look at the names of the elements between 20 and 31. Are those that you recognise metals or non-metals?

12.4
The elements

In this world of ours it is possible to find and to extract about ninety substances which, except under very rare conditions, cannot be split into other substances. These substances, as we have said before, are the **elements** and within the list of the elements is a wide variety of substances. There are many metals, like magnesium, iron, copper and gold; there are some non-metallic solids, like carbon, sulphur and phosphorus; there are many gases, like oxygen, nitrogen and neon and there are even two, mercury and bromine, which are liquids. Why is there such variety among the elements?

In order to try to answer this question, we have to carry out a deeper study of the elements. If you were given a pile of coins of different sorts and you wanted to study them, the chances are that your first move would be to make groups of similar coins and so **classify** the coins. This is what chemists set out to do with the elements—to divide them into groups containing those which are similar in some way, and so classify them.

Chlorine, bromine and iodine are elements which, at first sight, look very different. Chlorine is a pale green gas, bromine a deep red liquid and iodine a shiny black solid, but the three elements do react in similar ways with the same substances. Each will combine directly with hydrogen, forming a gas which dissolves in water to give an acidic solution:

$$H_2(g) \quad + \quad Cl_2(g) \quad \rightarrow \quad 2HCl(g)$$
$$H_2(g) \quad + \quad Br_2(g) \quad \rightarrow \quad 2HBr(g)$$
$$H_2(g) \quad + \quad I_2(g) \quad \rightarrow \quad 2HI(g)$$

Heated metals such as zinc will combine with all three of them, forming similar salts:

$$Zn(s) \quad + \quad Cl_2(g) \quad \rightarrow \quad ZnCl_2(s)$$
$$Zn(s) \quad + \quad Br_2(g) \quad \rightarrow \quad ZnBr_2(s)$$
$$Zn(s) \quad + \quad I_2(g) \quad \rightarrow \quad ZnI_2(s)$$

Clearly these three elements can be grouped together into a family on the basis of their reactions. The same thing is true of the three metals, lithium, sodium and potassium, which show close similarities in their reactions with water, air and chlorine. Again they obviously constitute a group or family. Here, therefore, we have two families of elements (called the **halogens** and the **alkali metals** respectively). What

119

relationship is there between these two groups and other groups of elements whose members react with the same substances in the same ways? Into what sort of framework do these groups fit?

12.5 The search for a classification

The search for the framework, in which the elements are classified, occupied many years and it is only comparatively recently that a complete classification has been found.

In 1817 Johann Döbereiner, who was a professor at the University of Jena, noticed that three elements with very similar chemical behaviour, calcium, strontium and barium, had relative atomic masses which fitted a simple relationship. The relative atomic mass of strontium (88) is almost midway between that of calcium (40) and that of barium (137) and this led him to call this group of elements a 'triad'. By 1829 two other Döbereiner triads had appeared, and these were, in fact, the two groups of elements which we met earlier in the chapter—the halogens (chlorine, bromine and iodine) and the alkali metals (lithium, sodium and potassium). Not only were the chemical behaviours of the three elements similar in each case, but the relative atomic mass of the middle one fell halfway between the other two. Döbereiner's triads have been swamped in the modern classification of the elements, but his efforts represented significant progress in the search for the classification since he was the first to suggest relative atomic mass as a basis for grouping the elements.

12.6 Periodicity

At the time of Döbereiner, relative atomic masses were not known with great certainty and it was another forty years before the work of Cannizzaro and others led to a list of accurate relative atomic masses. The values proposed by Cannizzaro were the ones used by a London industrial chemist, John Newlands. In 1863 he produced his *Arrangement of Elements* which was to show up an important phenomenon on which the modern classification is based. Newlands listed the elements in the order of increasing relative atomic mass and noticed that every eighth element appeared to show similar chemical behaviour. About his idea he wrote 'the eighth element, starting from a given one, is a kind of repetition of the first, like the eighth note in an octave of music' and this he called his Law of Octaves.

By this 'law', Newlands had made two significant contributions to the development of a classification of the elements. Firstly he assigned a number to each element, this being the number of the place an element occupied when they were arranged in order of increasing relative atomic mass. (A similar system of numbers, known as atomic numbers, is used in the modern classification of the elements.) Secondly, and more important, he showed the existence of a **periodicity** within the list of elements. A periodic event is something which occurs at regular intervals, like a new issue of a magazine appearing on the bookstalls on a particular day each week. Newlands was suggesting that similar properties occurred at regular intervals in the list of elements.

Unfortunately, this periodic relationship held good for the first sixteen elements but fell down after the seventeenth, and this made scientists reluctant to accept Newlands' ideas. Indeed, when he lectured on them at a meeting of the Chemical Society in London, he was received with considerable scorn and it was twenty years before his contribution was recognised by the award of the Davy Medal by the Royal Society.

12.7 The first periodic tables

In 1869 the periodicity, first spotted by Newlands, was confirmed by the publication of two **Periodic Tables** in which the elements showing similar chemical behaviour were grouped together. The second of the two Tables was published in December of that year in Germany by Julius Lothar Meyer, who based his arrangement of the elements on the curve obtained by plotting the atomic volumes of the elements against their relative atomic masses. The atomic volume of an element is the volume which would be occupied by 1 mole of atoms of the element if it were a solid, and the curve which Lothar Meyer obtained is shown in Fig. 12.1.

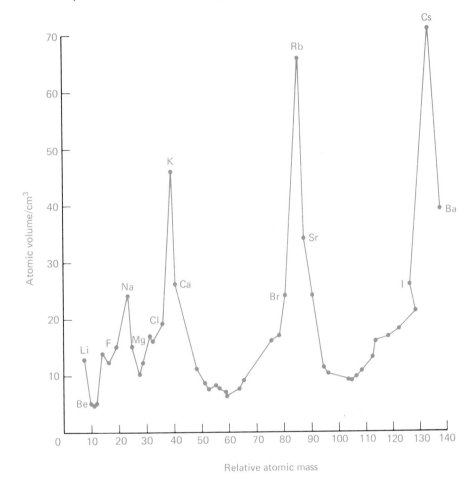

Fig. 12.1 Lothar Meyer's atomic volume curve

The highest points on the left-hand side of the curve are those for lithium, sodium and potassium, which, as we have seen, are similar elements, and the corresponding points on the right-hand side of the curve are those for rubidium and caesium, suggesting that these are also alkali metals.

Indeed, an examination of the properties of rubidium and caesium shows that they are similar to the other alkali metals and hence should be incorporated in the same family of elements. The regular spacing of the highest points confirms the idea of periodicity which Newlands suggested.

Notice how the points for the members of the other family we have met—chlorine, bromine and iodine—are in corresponding positions on the peaks of the curve, and again their regular spacing confirms the periodic nature of their properties. If the occupation of corresponding positions on the peaks of the curve is an indication of the elements being members of a family, the other families can be identified from the curve.

Evidence similar to that obtained by Lothar Meyer in his study of atomic volumes, can be obtained by plotting the melting points or the boiling points of the elements

against their relative atomic masses. Again the idea of periodicity is confirmed and there is some evidence as to which elements can be grouped into families.

Although the work of Lothar Meyer took us close to the modern Periodic Table, the contribution which is considered to have taken us closest was made in March 1869 by Dimitri Mendeleev, who taught at the University of St. Petersburg (now Leningrad). Mendeleev constructed his table using relative atomic mass as the basis of the arrangement. He put down the elements in order of increasing relative atomic mass and grouped together elements he knew to be similar in behaviour. Unlike his predecessors, however, he left gaps for elements which, he said, had not then been discovered, and he grouped together 'odd' elements which did not appear to fit into the main groups (e.g. cobalt and nickel).

Fig. 12.2 Dimitri Mendeleev (1839–1907), the Russian chemist who published the first periodic table of the elements in 1869.

12.8 Using the classification

One of the most spectacular features of Mendeleev's work was his predictions about the elements which had not been found in his day and for which he left gaps in his table. He knew from the relative atomic masses that tin could not come immediately below silicon in the group of elements containing carbon, and he suggested that this gap should be filled by an element which he called 'eka-silicon', whose properties he was able to predict by considering the properties of the elements before and after the gap in the table. When the element was discovered in 1886 and was named germanium by its discoverer, Winkler (after his homeland), its properties were shown to agree closely with those predicted by Mendeleev. The close agreement provided very strong proof that Mendeleev's ideas about classification were correct. How close his predictions were are shown in Table 12.1.

Mendeleev had similar success with his ideas about 'eka-boron', found in 1879 and named scandium, and 'eka-aluminium', found in 1875 and named gallium. The whole of his work was summarised by his 'Periodic Law' which was stated as 'the elements arranged according to the magnitude of their relative atomic masses, show a periodic change in properties'.

TABLE 12.1. *Mendeleev's predictions of the properties of germanium*

PROPERTIES OF EKA-SILICON (PREDICTED BY MENDELEEV)	PROPERTIES OF GERMANIUM
It will be a light grey metal.	It is a dark grey metal.
One atom will combine with two atoms of oxygen to form a white oxide with a high melting point.	One atom combines with two atoms of oxygen to form a white oxide with a melting point above 1000 °C.
The oxide will have a density of 4·7 g cm⁻³.	The density of the oxide is 4·703 g cm⁻³.
The chloride will have a boiling point less than 100 °C.	The chloride boils at 86·5 °C.
The density of the chloride will be 1·9 g cm⁻³.	The density of the chloride is 1·887 g cm⁻³.

Mendeleev's Periodic Table, a second form of which appeared in 1871, showed, however, a number of cases where his Periodic Law was not obeyed. Like Newlands before him and Lothar Meyer at the same time as himself, Mendeleev placed tellurium (relative atomic mass 127·6) in what appeared to be its rightful place under selenium but in front of iodine (relative atomic mass 126·9) which clearly should be grouped, as he put it, with chlorine and bromine. His justification for this was that more accurate determinations of relative atomic masses would eventually show the mass of the tellurium atom to be less than that of the iodine atom. This has not been found to be so. Why can we place the element with the greater relative atomic mass in front of the element with the smaller?

12.9
Atomic number and the Periodic Table

The answer to this question was not to appear until 1913. Lord Rutherford, working in the Physics Department of the University of Manchester, had found that an atom of an element has in it a central nucleus (13.3) which carries a positive electrical charge. Working with Rutherford in 1913, a young research student, Henry Moseley, showed by investigation of the X-ray spectra of elements that this positive charge is a definite amount, and increases regularly by an equal amount from one element to the next. If the charge on the hydrogen nucleus is given a value of $+1$, then the relative charge on the nucleus of the next element (helium) will be $+2$ and that on the nucleus of the third element (lithium) will be $+3$, and so on. The positive charge on the nucleus of the atom of an element is called the **atomic number** of the element. In the cases of iodine and tellurium, the nucleus of the atom of tellurium has a relative charge of $+52$, whereas that of the iodine atom carries a charge of $+53$. Thus, if atomic number is used as the basis of the Periodic Table, rather than relative atomic mass, tellurium and iodine occur in the order which their chemical reactions suggest.

Atomic number is the basis of the modern Periodic Table, shown on p. 382. The reasons for the relative atomic masses of elements such as tellurium and iodine being in reverse order to their atomic numbers will be discussed in 13.5.

12.10
Relationships in the Periodic Table

Before we can see how the Periodic Table helps us to understand the relationships between elements, there are two important names to be learnt. As can easily be seen, the Periodic Table contains vertical columns of elements and also horizontal rows of elements.

A vertical column is a **group**. Lithium, sodium and potassium are in the same group and this group is Group 1. Fluorine, chlorine, bromine and iodine (the halogens) are in Group 7 in the table.

A horizontal row of elements is called a **period**. The first period contains only two elements (hydrogen and helium), the second and third periods each contain eight elements, running from an alkali metal to a noble gas, and the fourth period has eighteen elements in it.

12.11
How do the elements in a period vary?

The variation within a period is best revealed by a close examination of the third period which contains the elements from sodium (atomic number 11) to argon (atomic number 18):

$$\text{Na} \quad \text{Mg} \quad \text{Al} \quad \text{Si} \quad \text{P} \quad \text{S} \quad \text{Cl} \quad \text{Ar}$$

The most obvious change in behaviour is the change from metallic to non-metallic character (Table 12.2). We cannot really include argon in either class because although it is a gas like some non-metallic elements, it forms no compounds and therefore does not react as a typical non-metal.

TABLE 12.2. *Metallic and non-metallic elements in the third period*

Na Mg Al	Si P S Cl	Ar
METALLIC	NON-METALLIC	
Oxides are basic	Oxides are acidic	
Chlorides are solids with high melting points and they are electrolytes	Chlorides have low boiling points and are non-electrolytes	
Form positive ions in most compounds, e.g. Na^+, Mg^{2+}, Al^{3+}	Form negative ions in compounds with metals, e.g. $(Na^+)_2S^{2-}$, Na^+Cl^-	

The division between metals and non-metals is not sharp. Aluminium forms an oxide which reacts with both acids and alkalis, suggesting that it is both basic and acidic and that aluminium is on the border between the metals and non-metals. The position of the division also varies from period to period in the Periodic Table. In the second period it occurs after beryllium in Group 2, while in the fourth period the Group 4 element, germanium, has metallic properties. In fact, the division forms a type of staircase which runs diagonally through the table (Fig. 12.3).

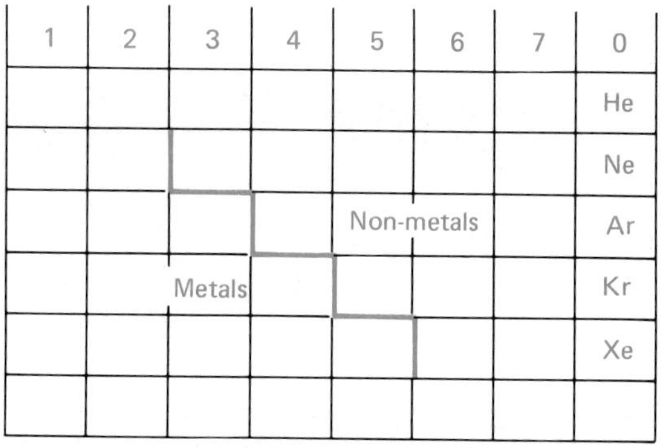

Fig. 12.3 Division of the Periodic Table (excluding the transition elements) into metals and non-metals

The elements which to the greatest extent show the character of metals are those on the extreme left of the Periodic Table—lithium, sodium and potassium—and the most non-metallic of the elements are those on the extreme right-hand side—fluorine, chlorine, bromine and iodine. The latter statement excludes the noble gases of Group 0 which show almost no reactions and therefore from the chemical point of view are difficult to classify.

There is a regular variation in the valencies or combining powers of the elements as we go across a period in the Periodic Table. The valencies which the elements in the third period show are listed in Table 12.3.

TABLE 12.3. *Variation in valency across a period*

GROUP NUMBER	1	2	3	4	5	6	7	0
ELEMENT	Na	Mg	Al	Si	P	S	Cl	Ar
HIGHEST VALENCY LOWEST VALENCY	1	2	3	4	5 3	6 2	7 1	0

If we consider the higher valencies of phosphorus, sulphur and chlorine, it is obvious that the valency increases by one as we move from group to group across the period and that the valency of the element is equal to the number of the group in which the element falls.

Considering the lower values of the valencies of these elements, the valency rises in steps of one from 1 in Group 1 to 4 in Group 4 and then falls in the same steps to 1 again in Group 7. In this case the valency is the same as the group number up to Group 4, and thereafter it is equal to 8 minus the group number.

The regular variation in valency appears within a period which has been constructed by setting down the elements in the order of increasing atomic number, and again indicates that this method of classifying the elements is sound. More fundamental reasons for the regular variation in valency will be discussed in Chapter 15.

12.12
How do the elements in a group vary?

The elements in a group in the Peroidic Table have similar properties. For example, you may have seen that lithium, sodium and potassium all react with cold water to form an alkaline solution, hydrogen being given off in the process:

$$2Li(s) + 2H_2O(l) \rightarrow 2LiOH(aq) + H_2(g)$$

This information can be used to predict that two other members of the group, rubidium and caesium, will react in a similar manner with water.

e.g.
$$2Rb(s) + 2H_2O(l) \rightarrow 2RbOH(aq) + H_2(g)$$

A closer examination of the properties of the elements within a group will show that, although they may all take part in the same type of reaction, there is often a regular change or trend in what is observed as the reactions occur. Typical of this is the reaction between alkali metals and water which becomes more vigorous as one proceeds down the group from lithium to potassium. It is reasonable to predict from this trend that rubidium will react more vigorously than potassium and that caesium will react more vigorously than rubidium.

At the other side of the Periodic Table, in Group 7, observations indicate that the elements react less vigorously with, for example, iron, as the group is descended. The iron wool glows more brightly in the reaction with chlorine than in the reaction with bromine:

$$2Fe(s) + 3Cl_2(g) \rightarrow 2FeCl_3(s)$$
$$2Fe(s) + 3Br_2(g) \rightarrow 2FeBr_3(s)$$

Both of these trends within the group may be interpreted as an increase in metallic character (or a decrease in non-metallic character) as a group is descended. This is particularly clear in the centre of the Periodic Table. Group 4 starts with carbon and silicon, both of which are clearly non-metals, these are followed by germanium which has more metallic properties, and then tin and lead which are clearly metals.

12.13
Transition elements

If you examine the fourth period of the Periodic Table (the one starting with potassium) you will see between the Group 2 element, calcium, and the Group 3 element, gallium, a series of ten elements (Table 12.4).

TABLE 12.4. *The first transition series*

Sc	scandium
Ti	titanium
V	vanadium
Cr	chromium
Mn	manganese
Fe	iron
Co	cobalt
Ni	nickel
Cu	copper
(Zn	zinc)

This is called the first **transition series**. The name was originally given to a small number of these elements which appeared to mark a transition from one set of subgroups to another in a Periodic Table which appeared soon after Mendeleev's original tables.

As you would expect from the position of these elements in the Periodic Table, they are metallic—in fact, you will recognise in the series a number of the most common metals, such as iron, nickel and copper. As well as the usual properties of metals, they have a number of properties which are peculiar to them. These can be summarised as follows:

1. They form ions which, in the presence of water, are coloured (e.g. the $Cu^{2+}(aq)$ ion is blue).

2. They can form more than one positively charged ion, which means that they can show more than one valency (e.g. iron will form Fe^{2+} and Fe^{3+} ions and is therefore said to have valencies of 2 and 3).

3. The metals and their ions often show the ability to act as catalysts in reactions (e.g. iron acts as the catalyst in the combination of nitrogen and hydrogen to give ammonia).

4. They often show magnetic properties and the metals which we tend to think of as magnetic (called the ferromagnetic metals), iron, cobalt and nickel, are found in the first transition series.

The fifth and sixth periods also contain similar series of elements.

12.14 Summary

1. In order to study what elements have in common, they have to be classified in a framework which is called the Periodic Table.

2. The first classifications of the elements used their relative atomic masses, but the basis of the modern classification is the atomic number of the element, which is the number of positive charges on the nucleus of an atom of the element. The elements are arranged in order of increasing atomic number.

3. The Periodic Table consists of vertical columns of elements, called groups, and horizontal rows of elements, called periods.

4. The elements in the same group show similarity in behaviour, but the metallic character of the elements increases as we move down the group.

5. In a period in the table, there is a transition from metallic elements, on the left, to non-metallic elements, on the right. The valency of the elements in the period either increases in steps of one, from 1 to 7, as we move across the period, or it rises in steps of one, from 1 to 4, and then falls in the same steps to 1 again. In all examinations of trends within a period, the noble gas in Group 0 has to be excluded, owing to its lack of reactivity.

6. Between Groups 2 and 3 in the fourth, fifth and sixth periods are series of elements called transition elements.

Chapter 13

What do atoms consist of?

13.1 Early ideas

In 1803 John Dalton, a teacher in Manchester, wrote down in his notebook his thoughts about the nature of gases, and so brought to life again ideas which had first been described in Ancient Greece. Between 400 and 450 B.C. Leucippus and Democritus suggested that all materials consist of very small particles, to which were given the name **atoms**, but these men were philosophers and were not able to prove their ideas in any way. The notion of the existence of atoms gradually became more popular as the years went by, and in the seventeenth century it was used by such great scientists as Robert Boyle and Isaac Newton.

Dalton's main contribution to atomic theory (4.7), which he published in 1808, was to point out the significance of the masses of atoms. In his writings he stated that atoms of one element differed in mass from atoms of another element. He coupled this with his idea that a small whole number of atoms of one element combine with a small whole number of atoms of another element (e.g. 1 atom of A with 1 atom of B or 2 atoms of A with 1 atom of B) and predicted the Law of Multiple Proportions (4.8). This law was first proved experimentally by Dalton and so provided strong evidence in support of the existence of atoms and the atomic theory on which the law was based.

Although he didn't actually state it, Dalton's mental picture of atoms must have been like a series of miniature billiard balls suspended in empty space; the differences between the atoms of one element and those of another being a difference in mass.

This picture lasted until near the end of the nineteenth century when Henri Becquerel found in 1896 that a photographic plate was fogged in the absence of light by uranium potassium sulphate. Further investigation showed that all uranium compounds exhibit this action and that they will also cause a charged electroscope to lose its charge. Becquerel had discovered the phenomenon of **radioactivity** where uranium atoms, in his experiments, were spontaneously breaking up.

A more spectacular discovery of radioactivity came two years later when Marie Curie, working in Paris, found that certain minerals containing uranium were more radioactive than was to be expected from their uranium content. Having been given a ton of the mineral pitchblende, from which uranium had been extracted, she concentrated it and extracted from it two highly radioactive elements, polonium (which she named after her native Poland) and then radium.

When radioactive elements 'decay' (i.e. when the atoms break up), an emission takes place and one of the things which can be thrown out is an α-particle, which was shown by Regener, and later by Rutherford, to be a helium atom carrying two positive charges. The decay usually occurs in several stages, each one involving the loss of either a particle (an α-particle or a β-particle) or radiation (γ-rays). When the decay of naturally occurring radioactive elements has gone as far as it can go, the product remaining is always lead.

Some elements, therefore, were found to break down, releasing helium ions and other particles, and eventually forming lead. Atoms of such elements obviously cannot be thought of as being the same all the way through like billiard balls. There are probably in an atom of radium the bits which make up a helium atom and those to make up a lead atom. So what are these bits?

Fig. 13.1 Marie Curie (1867–1934), famous for her work on radioactivity. (Courtesy Ullstein)

13.2
What cathode rays have told us

The answer to this question started to emerge in 1897 when J. J. Thomson investigated the cathode rays which were first discovered in 1859 by Plücker. At ordinary pressures gases are very poor conductors of electricity, but at low pressures their ability to conduct is considerably increased. If the gas is contained in a glass tube, fitted with an electrode at each end, and the pressure is reduced to about 0·6 kPa, a bright luminous discharge takes place when a high enough voltage is applied across the electrodes. Discharge tubes of this sort are used in strip lights and in advertising signs. If the pressure of the gas is reduced to about 0·001 kPa, the luminous discharge is replaced by very faintly visible rays, which travel from the cathode and are hence called cathode rays. These are emitted at right-angles to the cathode.

Fig. 13.2 A cathode ray tube

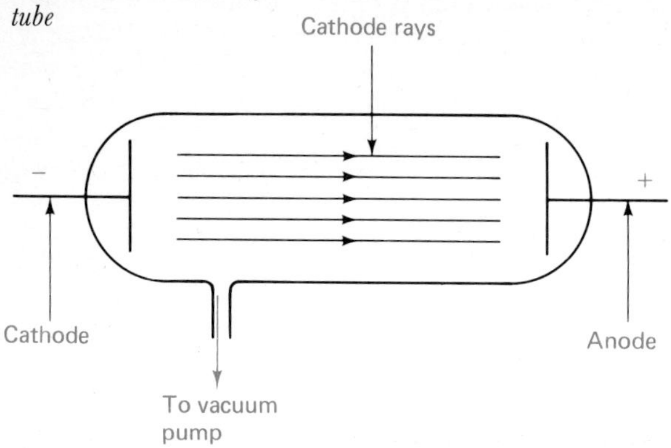

Cathode rays

−

Cathode

To vacuum pump

+

Anode

Fig. 13.3 (right) The lights for this advertising sign are obtained by passing electricity through tubes containing gases at low pressure. (Courtesy The Press Association)

The properties of these rays were investigated and they were found to be a stream of negatively charged particles of very small mass (1/1840 of the mass of a hydrogen atom).

Thomson called the particles 'corpuscles', but the name for them which was accepted, was that given in 1891 by Stoney who called them **electrons**. Thomson, in further work using his original apparatus, was able to show that, whatever gas was in the tube and whatever the material of the cathode, the value of ratio of charge to mass was always the same. This led him to suggest that electrons were to be found in all atoms. The electron is one of the bits of which an atom consists.

13.3
The nuclear atom

Fig. 13.4 Ernest Rutherford (1871–1937) who in 1911 put forward the theory of the nuclear atom. (Courtesy The Cavendish Laboratory)

Thomson's discovery that electrons are likely to be contained in all atoms posed a problem. It was known that an atom of an element, uncombined with another, is electrically neutral and therefore the negative charge, provided by the electrons, must be balanced by an equal amount of positive charge. Where the positive charge is situated in the atom was not known and Thomson was led to picture the atom, rather like a 'plum pudding', as a sphere which is positively charged and has the negatively charged electrons embedded in it. This idea was accepted until Rutherford replaced it by the **'nuclear atom'** at the beginning of the twentieth century.

Ernest Rutherford was a New Zealander, born in 1871, who started his research work at the Cavendish Laboratory in Cambridge in 1895 where his ability was soon recognised by Thomson himself. In 1898 he moved to the McGill University in Montreal and it was here that he identified α-particles, given off by the decay of radium. In 1907 he moved back to England to Manchester University where he produced his idea of the nuclear atom in 1910. In 1919 he moved back to Cambridge as Cavendish Professor and it was here that many honours, including the Nobel Prize, were deservedly given to this great physicist who gave us so many of the fundamental ideas on the structure of the atom.

When a radioactive element decays, it emits one of three sorts of radiation. One of them is easily absorbed by a sheet of paper or a few centimetres of air and this radiation was shown by Rutherford to consist of fast-moving positively charged helium atoms, called α-particles. In 1909 two of Rutherford's fellow workers, Geiger and Marsden, bombarded a very thin sheet of gold with a stream of α-particles and found that a small proportion of the particles 'bounced back' from the foil at angles of greater than 90 degrees (Fig. 13.5).

Fig. 13.5 Geiger and Marsden's experiment

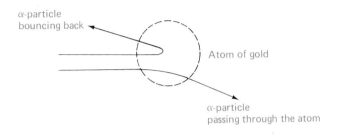

Only about one particle in 8000 bounced back in this way, but Rutherford realised the significance of these few. The fact that a large proportion of the particles passed through the foil suggests that there is a large amount of empty space in the atoms of the foil. Rutherford showed that the bouncing back could only be explained by some of the particles coming within a close distance of a large positive charge, contained in a volume at the centre of the atom, which is much smaller than the atom itself. Rutherford proposed the existence at the centre of the atom of a small, heavy, highly positively charged body, called the **nucleus** of the atom, with the electrons distributed around the nucleus and perhaps moving like the planets round the Sun. Since the mass of the electrons will be very small, very nearly all the mass of the atom will be concentrated in the nucleus. The fact that the nucleus is very small, compared with the size of the whole atom, is shown by the large majority of the α-particles managing to get through the atom, only those scoring a 'direct hit' on the nucleus bouncing back from the atom (Fig. 13.6).

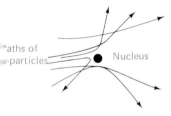

Fig. 13.6 The effect of the nucleus of an atom on bombarding α-particles

The problem which now faced Rutherford was to describe the composition of the nucleus in the light of its mass and positive charge. As has been stated in 12.9, Henry Moseley, one of Rutherford's research students, showed by examination of X-ray spectra that the positive charge on the nucleus differs from element to element. There is always the same difference in nuclear charge between elements which are next to each other in the Periodic Table. This was explained by suggesting that the charge on the nucleus was due to the presence of another type of particle in the nucleus. The number of these particles in the nucleus of an atom of an element is always one more than is present in the nucleus of the previous element in the Periodic Table. For example, if a sodium nucleus has eleven of these particles in it, a magnesium will have twelve, aluminium thirteen and so on.

Rutherford suggested that this particle was a hydrogen atom which had lost its electron, that is, a hydrogen nucleus. The charge on the particle would be equal in size to that on an electron, but opposite in sign. An electron is said to have a relative charge of −1 and a proton a relative charge of +1. At the Cardiff meeting of the British Association for the Advancement of Science in 1920, Rutherford suggested the name **proton** for this particle.

The first element in the Periodic Table, hydrogen, has 1 proton, giving its nucleus a charge of +1. Outside the hydrogen nucleus there is one electron whose charge of −1 balances that of the nucleus. The second element, helium, has 2 protons and 2 electrons, the third element 3 protons and 3 electrons, and so on. Thus, overall each

Fig. 13.7 James Chadwick, who discovered the neutron in 1932. (Courtesy Camera Press)

atom has no charge, as the positive charge of the protons is balanced by the negative charge of the electrons. The total positive charge on the nucleus of an atom (i.e. the number of protons in the nucleus) is called the **atomic number** of the element.

An important consequence of this theory is that we cannot expect to discover new elements which would be positioned between any of the known elements. For example, we cannot expect to find an element which comes between sodium and magnesium because it would have $11\frac{1}{2}$ protons in its nucleus which is not likely to be possible. This is why, when rocks are brought back from the Moon, or meteors are found, they are bound to be made up of some of the elements we know on Earth.

In the same year as he named the proton, Rutherford stated in a lecture in London that, in order to account for the mass of the atom, he predicted the existence of another particle in the nucleus. This particle, unlike the electron and the proton, would be electrically neutral and could therefore be called the **neutron**. Twelve years later Sir James Chadwick showed that the radiation produced when a beryllium foil is bombarded with α-particles consists of a stream of particles with no electrical charge and with a mass very nearly the same as the proton. Rutherford's neutron had been found and it was then possible to state more precisely what each atom consisted of and where in the atom the bits were to be found.

13.4
What do atoms consist of?

TABLE 13.1. *The fundamental atomic particles*

Table 13.1 shows the relative masses and charges of the three fundamental particles which make up atoms.

	RELATIVE MASS	RELATIVE CHARGE
proton	1	+1
neutron	1	0
electron	1/1840	−1

The relative mass of the electron is so small (about 0·0005) that it can be neglected when totalling up the relative masses of atoms.

The **atomic number** of an element is the number of positive charges on the nucleus and is therefore **the number of protons in the nucleus**.

Since the mass of the electrons in an atom can be neglected, the mass of the atom is the mass of the nucleus. The relative atomic mass must therefore be the sum of the relative masses of the protons and the relative masses of the neutrons. Since the relative mass of each is one, the relative mass of the protons would be the number of protons. Similarly, the relative mass of the neutrons would be the number of neutrons. Thus the **relative atomic mass** should be equal to the **number of protons + the number of neutrons**. (Note that after the discussion on isotopes, 13.5, it is necessary to modify this statement.)

Now let us apply these rules to determine the numbers of each type of particle in a magnesium atom. The atomic number of magnesium is 12 and there will therefore be 12 protons in the nucleus. The relative atomic mass of magnesium is 24 and the number of neutrons in the nucleus will be equal to the relative atomic mass − the number of protons (the atomic number), i.e. $24 - 12 = 12$.

In an uncombined magnesium atom the number of electrons is the same as the number of protons, i.e. 12.

The magnesium atom has, therefore, a nucleus containing 12 protons and 12 neutrons and there are 12 electrons outside the nucleus.

The compositions of other atoms can be worked out in the same way and can be represented as in Fig. 13.8.

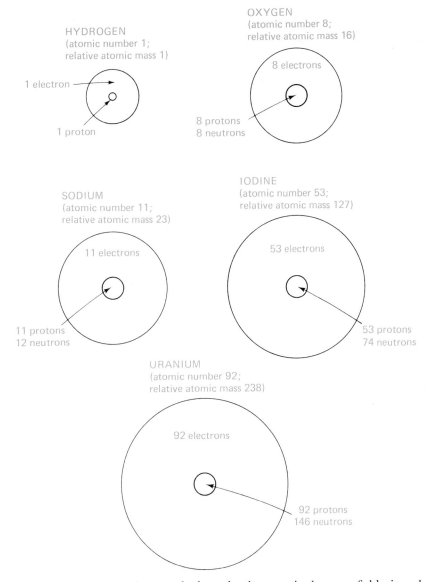

Fig. 13.8 The numbers of the fundamental particles in atoms of some elements

HYDROGEN
(atomic number 1;
relative atomic mass 1)

1 electron

1 proton

OXYGEN
(atomic number 8;
relative atomic mass 16)

8 electrons

8 protons
8 neutrons

SODIUM
(atomic number 11;
relative atomic mass 23)

11 electrons

11 protons
12 neutrons

IODINE
(atomic number 53;
relative atomic mass 127)

53 electrons

53 protons
74 neutrons

URANIUM
(atomic number 92;
relative atomic mass 238)

92 electrons

92 protons
146 neutrons

**13.5
Isotopes**

The rules above seem simple to apply, but what happens in the case of chlorine where the relative atomic mass is 35·5? The atomic number is 17, which should make the number of neutrons in the nucleus $18\frac{1}{2}$. Since we can't cut a neutron in two, this is an impossibility. So how many are there of each particle in each atom of chlorine?

In his original atomic theory Dalton stated that all the atoms of an element are alike. However, he was wrong about this. What decides the identity of an atom is the number of protons in the nucleus. An atom with 12 protons in the nucleus is an atom of magnesium, but there is nothing to say that there has to be 12 neutrons in the nucleus with the protons. In fact, it is possible to have several types of magnesium atoms, the only difference between them being the number of neutrons in the nucleus. The most ·abundant form is that which has a relative mass of 24, and nuclei containing 12 protons and 12 neutrons. This is represented as $^{24}_{12}Mg$. Also found in naturally-occurring magnesium is $^{25}_{12}Mg$, whose atoms have nuclei containing 13 neutrons, and $^{26}_{12}Mg$ with 14 neutrons in each nucleus. These three different types of magnesium atoms are called **isotopes** of magnesium.

The relative mass of an isotope is called its **mass number**. When an isotope is represented in the way used for the magnesium isotopes above, the upper number is the mass number of the isotope and the lower number the atomic number of the element.

Isotopes are different types of atoms of a single element whose nuclei contain the same number of protons but different numbers of neutrons. The word isotope means literally 'in the same place', indicating that the different forms of the element occupy the same position in the Periodic Table. Since the numbers of electrons in uncombined atoms of the different isotopes are the same, their chemical properties are the same.

Most elements consist of more than one isotope and the answer to the chlorine problem lies in the isotopes of chlorine. Chlorine prepared from naturally-occurring salt contains two isotopes, $^{35}_{17}Cl$ with 18 neutrons in the nucleus and $^{37}_{17}Cl$ with 20 neutrons in the nucleus. There are approximately three $^{35}_{17}Cl$ atoms to every $^{37}_{17}Cl$ atom and the relative atomic mass will therefore be

$$\frac{(3 \times 35) + 37}{4} = 35 \cdot 5.$$

It is now necessary to modify slightly the statement given previously (13.4) about the sum of the number of protons and the number of neutrons in an atom. The statement now becomes:

the mass number of an isotope = the number of protons + the number of neutrons.

The **relative atomic mass** of an element is the mean value of the mass numbers of the isotopes of the element which takes into account the relative abundances of the isotopes in the naturally occurring element.

13.6 Radioactive isotopes

If the only difference between isotopes of an element is a difference in the number of neutrons, why is there not an unlimited number of isotopes of a particular element? Why is it not possible to have $^{38}_{17}Cl$, $^{39}_{17}Cl$, $^{40}_{17}Cl$, $^{41}_{17}Cl$ and so on with the difference between one isotope and the next being one neutron in the nucleus? The answer to this question lies in the fact that, as the number of neutrons becomes significantly greater or less than the number of protons, the nucleus becomes more unstable and more likely to disintegrate spontaneously, thus making the isotope radioactive.

This behaviour is clearly shown with carbon. The most abundant isotope of carbon is $^{12}_{6}C$. The nucleus contains six protons and six neutrons. Here the nucleus has a proton-neutron ratio of 1 and the nucleus is stable. $^{13}_{6}C$ also has a stable nucleus with 7 neutrons to the 6 protons, but $^{14}_{6}C$, with a proton-neutron ratio of 6:8, is radioactive. Putting in more neutrons would make the nucleus so unstable that the particles could not hold together.

The radioactivity of $^{14}_{6}C$ is a useful tool for archaeologists, since every material consisting of or made from something which was once alive (e.g. wood, leather, bone) contains $^{14}_{6}C$. This radioactive isotope of carbon is present in small quantities in the atmosphere. As fast as it decays more is produced by the effect of radiation from outer space, with the result that the amount present in the atmosphere remains constant.

Plants absorb $^{14}_{6}C$ in part of the carbon dioxide which they take in during photosynthesis (9.4). Animals, in turn, absorb it by eating the plants. At the moment a plant or animal dies the intake of $^{14}_{6}C$ stops. The only change in the $^{14}_{6}C$ content after that instant is due to its decay. By measuring the level of radioactivity in an object made from once-living material and then relating it to the known rate of decay of $^{14}_{6}C$, it is possible to determine the approximate age of the object. This technique is known as radiocarbon dating and in 1960 Willard Libby, an American, was awarded the Nobel Prize for Chemistry for his work in developing the technique. He perfected the technique by checking samples of material taken from inside an Egyptian pyramid which, from other evidence, was known to be of a particular age.

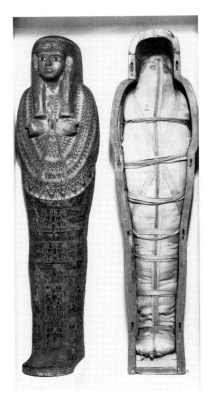

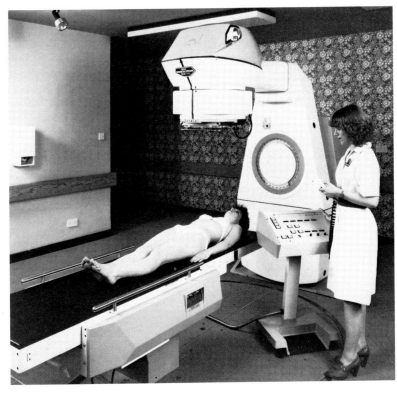

Fig. 13.9 Objects such as this coffin and the mummy of an Egyptian princess who died about 1000 BC can be dated by carbon methods. (Courtesy The Trustees of the British Museum)

Fig. 13.10 Setting up a patient for treatment on a modern cobalt-60 teletherapy unit. (Courtesy TEM Instruments Ltd)

Most applications of radioactive isotopes make use of either the effects of the radiation or the fact that the radiation can be very easily detected by means of an instrument called a Geiger counter.

People who are suffering from certain types of cancer can be treated with γ radiation from an isotope of cobalt, $_{27}^{60}$Co. The radiation can be focused very precisely on the cancer cells and so destroy them. Treatment which uses the effects of radiation in this way is called **radiotherapy**.

In other medical cases the efficiency of a particular organ in the body (e.g. the thyroid gland or a kidney) can be investigated. The patient is given either a drink or an injection containing a small quantity of a radioactive isotope of iodine, $_{53}^{131}$I. By measuring the radiation being given off from the appropriate areas of the patient's body doctors can trace the path of the iodine and so find out if the organs are working properly.

Tracer techniques are also used in many non-medical investigations. The substance which is being followed, such as a fertiliser being absorbed by a plant, is labelled by mixing with it a small quantity of a radioactive isotope. The radioactivity of various parts of the plant can then be measured to see how much of the fertiliser has been absorbed from the soil. A similar technique could be used to investigate the movement of mud in a river estuary or leaks in underground pipes.

13.7 Nuclear energy

When a radioactive substance decays (13.1) by giving off radiation, the nuclei of the atoms which are left will contain different numbers of protons and neutrons. The atoms of the radioactive element change to atoms of either another isotope of the element or to another element. This type of change is called a **nuclear reaction**.

Some chemical reactions give out quite a lot of heat (e.g. burning a fuel) but nuclear reactions often give out very much more energy. Energy given out by nuclear reactions

133

can be used for destructive or productive purposes. The atomic bomb is a **nuclear weapon** and the devastating explosion is caused by a nuclear reaction which, once started, cannot be stopped. On the other hand a **nuclear power station** controls a nuclear reaction and the energy released is used to produce electricity.

Calder Hall power station in Cumbria, which opened in 1956, was the first nuclear reactor in the world to produce electricity on an industrial scale. The nuclear fuel used in the reactor is an isotope of uranium, $^{235}_{92}U$. Atoms of this isotope, when bombarded with neutrons, split up (**nuclear fission**) into atoms of other elements which have relative masses of about half that of uranium. This nuclear reaction gives out a lot of heat which is used to convert water into steam. The steam is used to drive generators to produce electricity.

Nuclear power stations, unlike coal- and oil-powered stations, do not cause sulphur dioxide and smoke pollution of the atmosphere (9.5). However, there is concern about the possibility of leaks of radioactive material and about problems associated with the disposal of some of the radioactive waste materials from the nuclear reaction. Some of these materials decay at a slow rate and will therefore be radioactive for a very long time. There is a more general discussion on the world's energy resources in 31.6.

13.8 Summary

1. An atom of an element consists of a small, heavy, positively-charged body called the nucleus, which contains positively-charged particles called protons and uncharged particles called neutrons. The negatively charged electrons move around outside the nucleus.
2. The number of protons in the nucleus of an atom is the atomic number of the element.
3. The number of neutrons in an atom is equal to the mass number − the atomic number.
4. The number of electrons in an atom is equal to the atomic number.
5. Isotopes are forms of a particular element whose atoms contain the same number of protons but different numbers of neutrons. If the number of neutrons is significantly different to the number of protons, the isotope is likely to be radioactive.
6. The energy released during a controlled nuclear reaction can be used to produce electricity.

Fig. 13.11 Fuel elements such as these are used in the reactors built for the first phase of the U.K.'s nuclear power programme. The fuel elements consist of uranium metal rods enclosed in magnesium alloy cans. (Courtesy UKAEA and British Nuclear Fuels Ltd)

Chapter 14

How are the electrons arranged?

14.1
Bohr's theory of atomic structure

Rutherford's model of the atom included the nucleus which was small and heavy and carried a concentrated positive charge. The electrons, equal in number to the number of protons in the nucleus, revolved around the nucleus like the planets around the Sun (Fig. 14.1).

Fig. 14.1 Rutherford's model of the atom

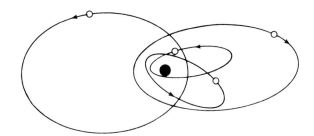

One question which Rutherford did not answer about the atom was why the electrons, being negatively charged, are not drawn into the nucleus. An answer to this question was provided in 1913 by a Danish physicist, Niels Bohr, who had spent a year working with Rutherford at Manchester. Bohr's theory used the idea, proposed in 1900 by the German, Max Planck, that when a hot body gives out radiation, the energy is emitted in the form of little 'packets', or separate fixed amounts called 'quanta'.

Niels Bohr suggested that the electrons follow fixed circular or elliptical paths which are not all the same distance from the nucleus but are situated in **'shells'** or layers. There are no electrons between the shells and the difference between one shell and the next is one quantum of energy. This last statement means that, if an electron is to be pulled from one shell to the next further from the nucleus, work has to be done and energy has to be paid into the atom to drag the negative charge away from the positive nucleus. If the electron then falls back into its original shell, the Law of Conservation of Energy applies and the energy which had to be given to the electron to move it, is given back in the form of a 'packet' or quantum of energy.

The colours obtained when portions of salts of some metals are placed in a Bunsen flame provide direct evidence of this occurring. The electrons take energy from the flame so that they are moved from an inner shell to an outer one, and when they drop back to their original energy level, a certain quantity of energy is given out in the form of light. The wavelength of the light corresponds to a certain colour, which is the characteristic flame colour of that particular metal.

An electron in an atom is therefore rather like a book in a bookcase with a number of shelves. If the book is on the bottom shelf and you want it on a higher shelf, you have to do work to lift the book against its own mass and therefore some of your energy will be transferred to the book, so that its potential energy will be increased. Now suppose the book slips off the higher shelf and falls down to the bottom shelf again. The energy which had been given to the book, will be lost by it and given to the surroundings, probably in the form of heat. The shells in an atom are similar to the shelves in the bookcase and, just as the shelves represent different levels of potential energy above the ground, whose potential energy can be considered to be zero, the shells can be thought of as energy levels for electrons outside the nucleus which, like the ground, has a potential energy of zero. Just as it would not be possible to have a book hanging, unsupported, between two shelves in the bookcase, so it is not possible to have an electron between two shells in the atom. However, it must be remembered that, unlike a shelf, an energy shell does not have any physical existence of its own.

The further out from the nucleus an electron is, the higher is its potential energy and so energy has to be given to an electron if it is to be moved from an inner shell to an outer one.

14.2 Ionization energies

The question which now has to be asked concerns the number of electrons in each shell. The answer can be obtained by studying the work which has to be done, or the energy which has to be transferred to an electron so that it is pulled completely out of the atom. Removing an electron in this way will leave one more proton than electrons in the atom and so the atom will have changed into a positively charged ion.

Obviously less energy will be needed to remove an electron from an outer shell than from an inner shell which is more tightly held by the nucleus. The energy which has to be supplied to remove a particular electron from an atom is called the **ionization energy** for that electron. The ionization energies of each electron in a sodium atom (atomic number 11) are given in Table 14.1, and in graphical form in Fig. 14.2 (the logarithm of each value for the ionization energy is used in the graph because the range of values is so great).

Fig. 14.2 The successive ionization energies of sodium

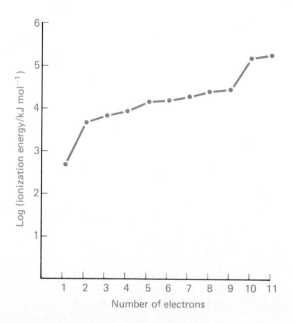

The considerable jump between the first (furthest electron from the nucleus) and second values suggests that 1 electron is much more easily removed than any of the others. This is followed by 8 electrons which require only gradually increasing energies before the next sudden jump in ionization energy. The last two electrons to be removed require much more energy than any of the others. This evidence suggests that the electrons in a sodium atom are arranged so that 2 electrons are in the innermost shell, 8 in the next shell and 1 in the outer shell. This configuration is usually represented by 2.8.1.

136

TABLE 14.1. *Ionization
energies of sodium*

ELECTRON REMOVED	ENERGY NEEDED TO REMOVE IT $/\text{kJ mol}^{-1}$
First	496
Second	4578
Third	6930
Fourth	9576
Fifth	13 440
Sixth	16 716
Seventh	20 244
Eighth	25 578
Ninth	29 022
Tenth	141 540
Eleventh	158 700

The slightly confusing thing about this argument is that the energy required to remove the outermost electron is called the first ionization energy, whereas the shell which is nearest to the nucleus is called the first shell.

14.3 Electronic configurations and the Periodic Table

The electronic configurations of all the elements are now known. The configurations for the elements with atomic number 1–20 are shown in Table 14.2.

TABLE 14.2. *The electronic configurations of the first 20 elements in the Periodic Table*

ATOMIC NUMBER	ELEMENT	ELECTRONIC CONFIGURATION
1	H	**1**
2	He	**2**
3	Li	2.**1**
4	Be	2.**2**
5	B	2.**3**
6	C	2.**4**
7	N	2.**5**
8	O	2.**6**
9	F	2.**7**
10	Ne	2.**8**
11	Na	2.8.**1**
12	Mg	2.8.**2**
13	Al	2.8.**3**
14	Si	2.8.**4**
15	P	2.8.**5**
16	S	2.8.**6**
17	Cl	2.8.**7**
18	Ar	2.8.**8**
19	K	2.8.8.**1**
20	Ca	2.8.8.**2**

If this table is examined alongside the Periodic Table (p. 381), it soon becomes obvious that the last figure in the electronic configuration of an element (the number of electrons in the outermost shell) is the number of the group in the Periodic Table in which the element is found, and therefore, if you are asked for the electronic configuration of an element, you can, at least, supply the last figure in it.

For example, sulphur (atomic number 16) is in Group 6 in the Periodic Table. There will be 6 electrons in the outermost shell and, of the remaining 10 electrons, 2 will fill up the first shell and 8 the second.

The noble gases can present a problem as in some Periodic Tables they are called Group 0 and in others Group 8. For applying the above rules it is easier if you assume they are in Group 8.

14.4 Electronic configurations and chemical properties

We have found, by examination of the electronic configurations of the elements between hydrogen and calcium in the Periodic Table, that the number of electrons in the outermost shell of electrons is the same as the number of the group in the Periodic Table in which the element is found. All the elements in a group will therefore have the same number of electrons in the outermost shell of their atoms. For example, in Group 7 (the halogens) the configurations will be the following:

fluorine	2.7
chlorine	2.8.7
bromine	2.8.18.7
iodine	2.8.18.18.7

In the previous chapter we saw that all the elements in the same group in the Periodic Table have similar properties and the reason for this must obviously be the presence of the same number of electrons in the outermost shells of the atoms. Thus, the properties of an element are decided by the number of electrons in the outermost shell.

Also in Chapter 12, we saw how the elements in Groups 1, 2 and 3 of the Periodic Table are metals and that one of the characteristics of metallic elements is that they form positive ions. Examination of the ionization energies will show why this is so. The one or two or three electrons in the outermost shell require a comparatively small amount of energy for their complete removal, but, once they have been removed, a very much larger amount of energy has to be fed into the atom to take an electron away from the next shell.

Fig. 14.3

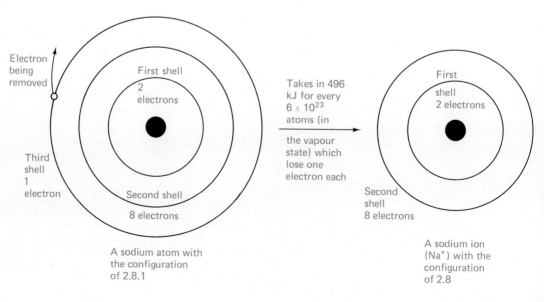

Electron being removed

First shell
2 electrons

Third shell
1 electron

Second shell
8 electrons

A sodium atom with the configuration of 2.8.1

Takes in 496 kJ for every 6×10^{23} atoms (in the vapour state) which lose one electron each

First shell
2 electrons

Second shell
8 electrons

A sodium ion (Na^+) with the configuration of 2.8

Thus, when a positive ion is formed, the outer electrons only are removed, leaving more protons in the atom than electrons and hence a net positive charge equal to the number of electrons removed. Thus sodium, for example, forms an ion with a single positive charge (Fig. 14.3). The value of the positive charge will often be the same as the number of the group in the Periodic Table in which the element occurs.*

The elements in Groups 5, 6 and 7 are non-metallic and form negative ions. This is because the amount of energy required to remove any of the 5, 6 or 7 electrons in the outermost shell is relatively large, and unlikely to be available. The atom therefore behaves differently by accepting electrons so that the incomplete outer shell gains electrons. Since there will then be more electrons in the atom than protons, it will carry a net negative charge equal to the number of electrons accepted. As a general rule, the number of electrons gained is that which brings the number in the outermost shell up to eight, so that the resulting ion has the same electron configuration as a noble gas atom. Thus chlorine, in Group 7, gains one electron (Fig. 14.4), while oxygen, in Group 6, gains two electrons.

Fig. 14.4

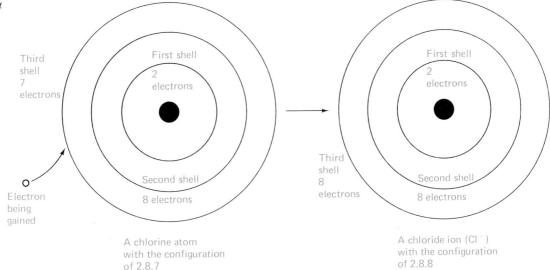

A chlorine atom with the configuration of 2.8.7

A chloride ion (Cl^-) with the configuration of 2.8.8

14.5 Summary

1. The electrons are arranged in energy levels called shells around the nucleus. Each shell can hold only a certain number of electrons and there are no electrons between the shells. The electrons in the shells nearer the nucleus have lower energies than those in shells further away and the difference between one shell and the next is one quantum or 'packet' of energy.

2. The arrangement of electrons in the shells in an atom is called the electronic configuration. This can be determined by studying the ionization energies of the atom or from the Periodic Table. If the electrons are stripped from the atom and put back, one by one, each one goes into the available shell with the lowest energy level.

3. The number of electrons in the outermost shell is the same as the number of the group in the Periodic Table in which the element occurs.

4. The properties of an element are decided by its electronic configuration and, in particular, by the number of electrons in the outermost shell.

* This simple rule cannot be reliably applied to the transition metals (12.13), most of which form more than one type of positively charged ion.

Chapter 15

How are atoms joined together?

15.1
Bonding in electrolytes

In his atomic theory in 1808, Dalton suggested that two atoms could be bound together. When Rutherford, over 100 years later, described the atom as being made up of electrically charged particles, the major problem was to understand how this bonding could take place.

The nuclei of the two atoms will tend to repel each other, rather than remain together. If, however, there is negative charge between the two nuclei, then each nucleus will be attracted to the negative charge and the two will be held together (Fig. 15.1).

Fig. 15.1

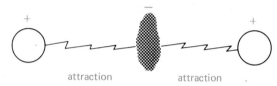

This is like sticking a photograph on to a page of a photograph album, using a piece of sticky tape which has adhesive on both sides. The photograph sticks to the tape, the page sticks to the tape and so the photograph is held on to the page. The negative charge, which holds the two nuclei together, can only be provided by the electrons in the atoms and we must therefore look at the electrons to see how the atoms can be joined.

The two most significant contributions to this enquiry were made quite separately by W. Kossel and G. N. Lewis who published their ideas in the same year, 1916, Kossel in Germany and Lewis in America. The two theories were complementary to each other. Kossel considered how the atoms are joined in electrolytes (substances which, when in solution or when melted, will conduct an electric current) while Lewis was concerned with non-electrolytes. Kossel noticed that the element immediately before a **noble gas** in the Periodic Table is a halogen which is a highly reactive non-metal, usually showing a valency of one. The element immediately following the noble gas is again highly reactive with a valency of one, but this time it is an alkali metal which is one of the most reactive metals. The element preceding the halogen is again non-metallic in character, though less so than the halogen, and has a valency of two. The element following the alkali metal is a metal with a valency of two, e.g. the elements in Table 15.1.

TABLE 15.1.
The electronic configurations of elements on either side of a noble gas

Element	S	Cl	Ar	K	Ca
Atomic number	16	17	18	19	20
Electronic configuration	2.8.6	2.8.7	2.8.8	2.8.8.1	2.8.8.2

Kossel pointed out that these facts could be connected using the simple rule that the elements close to a noble gas in the Periodic Table tend to form compounds in which they attain the configuration of that noble gas by gaining or losing the necessary number of electrons. In the series in Table 15.1, argon has the configuration of 2.8.8 and chlorine (2.8.7) could achieve this arrangement by taking up one electron and sulphur (2.8.6) by taking up two electrons. Potassium (2.8.8.1) by losing one electron can also achieve this arrangement. Each resulting ion will carry a net charge equal to the number of electrons gained or lost.

When potassium is burned in chlorine, each potassium atom with the configuration of 2.8.8.1 **transfers** its outermost electron to a chlorine atom with the configuration of 2.8.7, so that the configurations of both atoms become 2.8.8 which is the configuration of argon. Each potassium atom now has 19 protons in its nucleus and only 18 electrons outside the nucleus, so that it is an ion with a relative positive charge of 1. Each chlorine atom now has 17 protons in its nucleus and 18 electrons outside it, so that it is now a chloride ion with a relative negative charge of 1.

The transfer of an electron from potassium to chlorine occurs for all of the atoms taking part in the reaction and as the reaction mixture cools slightly, the K^+ ions and Cl^- ions are attracted to each other because of their opposite charges. The ions come together to form an orderly arrangement (Fig. 15.2). In this way crystalline potassium chloride is formed. The orderly arrangement of the ions is called a **lattice** and the electrostatic forces binding the ions together in the crystal are referred to as **electrovalent** or **ionic bonding** forces. The details of the arrangement of particles in some crystal lattices are discussed in the next chapter.

Fig. 15.2 (left)
Formation of solid
potassium chloride

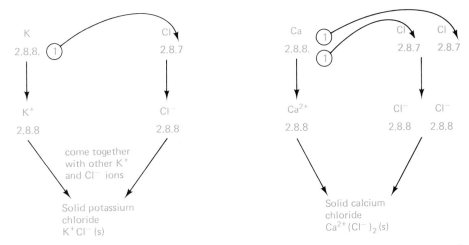

Fig. 15.3 (right)
Formation of solid
calcium chloride

A crystal of sodium chloride consists of the same type of orderly arrangement of oppositely charged ions. One electron has been transferred from each sodium atom to a chlorine atom. The resulting ions, Na^+, have an electronic configuration of 2.8 (that of neon), and the chloride ions, as above, have a configuration of 2.8.8.

Calcium chloride is also an electrolyte. A calcium atom has a configuration of 2.8.8.2 and the two shells containing 8 electrons effectively shield the two outermost electrons from the attraction of the nucleus. When calcium is burnt in chlorine, these two electrons are transferred from a calcium atom to each of the two chlorine atoms. The calcium atom then becomes a calcium ion, Ca^{2+}, with the configuration of 2.8.8 which is the configuration of argon. As was the case in the formation of potassium chloride, the chlorine atom now has in it one more electron than the number of protons and is therefore a chloride ion, Cl^-, with the configuration of 2.8.8. The calcium ions and chloride ions will now come together to form a lattice in which there are twice as many chloride ions as calcium ions, with the two types of ion being held together by the electrostatic forces (Fig. 15.3).

Electrons are transferred from one atom to another when an ionic compound is formed. The elements whose atoms lose electrons most easily are found in Groups 1 and 2 in the Periodic Table and those whose atoms have the greatest tendency to attract electrons are those in Groups 6 and 7. Ionic compounds, therefore, are usually formed when elements in Groups 1 and 2 combine with elements in Groups 6 and 7.

141

15.2
Properties of ionic compounds

In the solid state an ionic compound consists of crystals which are orderly arrangements of ions, held together by strong electrical forces. When the solid is heated, energy is absorbed and the effect of this energy is to cause greater vibrations of the ions within the arrangement. Eventually enough energy is absorbed that the vibrations become so great that the orderly arrangement breaks down. The ions, not now being rigidly held together, can move towards electrodes placed in the molten substance and so the molten compound can conduct an electric current. The electrical forces holding the ions in the crystal are so strong, however, that sufficient energy to overcome them will only be supplied when the substance is at a very high temperature and therefore the melting point will be very high. Potassium chloride, sodium chloride and calcium chloride all melt at temperatures of over 750 °C.

If an ionic compound dissolves in water, the ionic lattice is pulled apart by the water. Each ion which is separated from the lattice becomes surrounded by a 'cage' of water molecules. The 'cage' of water molecules prevents the ion getting back into the lattice. It is only when the water in the solution is evaporated so that there are not enough water molecules to form 'cages' around all the ions that the lattice can be reformed. The solute then crystallises.

When a source of electricity is connected by means of electrodes to a solution of an ionic compound, each ion is attracted to the electrode with the opposite charge to itself. This flow of ions and their discharge at the electrodes constitutes an electric current through the solution. The solution is conducting electricity.

15.3
Bonding in non-electrolytes

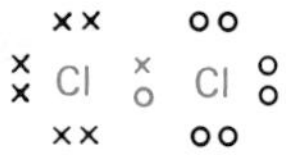

(Note that the symbol Cl in this type of diagram stands for the nucleus, together with the inner-shells of electrons.)

Fig. 15.4 The bonding between chlorine atoms in a chlorine molecule

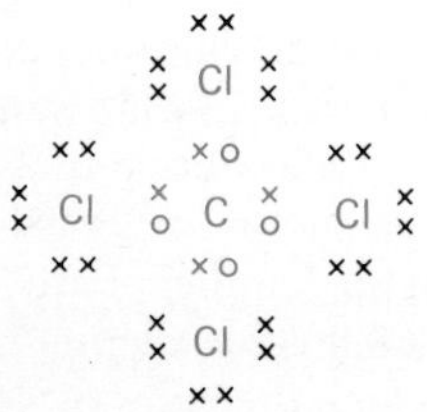

Fig. 15.5 The bonding in a molecule of tetrachloromethane

The Kossel theory has allowed us to explain how potassium combines with chlorine to form the electrolyte, potassium chloride. However, some compounds, such as tetrachloromethane (CCl_4) do not conduct electricity when molten and so do not appear to contain ions. How can we explain the bonding in compounds such as these? Also, how can we explain the bonding in the chlorine molecule (Cl_2), where two identical atoms are bonded together?

The answer to these questions was suggested by Gilbert Lewis of the University of Berkeley in California. He supposed, like Kossel, that atoms combine because the electrons rearrange themselves to form the same configurations that occur in the noble gases. What Lewis suggested which was new, was that it was possible for electrons to be shared between two atoms, so that they come under the influence of both nuclei.

A chlorine atom has an electron configuration of 2.8.7 and so requires one electron to reach the configuration of argon. When one chlorine atom combines with another, they cannot both attain the argon configuration by one electron being transferred from one atom to the other. However, if two electrons are **shared** between the atoms, each atom will have eight electrons associated with it. This can be represented as in Fig. 15.4. The crosses represent the electrons in the outermost shell of one atom and the circles those in the outermost shell of the other atom. You should realise that all the electrons are exactly the same and that the crosses and circles are only a way of showing differently the electrons from the two atoms.

The two electrons represented in blue are holding the two atoms together by attracting the two positively charged nuclei. A bond which is formed by sharing a pair of electrons between two atoms is called a **covalent bond**. In the chlorine molecule, both chlorine atoms have a share in eight electrons in their outermost shells.

In tetrachloromethane (CCl_4), each chlorine atom again requires to gain one electron, but the energy needed to remove the four outer electrons of carbon is so high that these electrons cannot be transferred completely to the chlorine atoms. The problem is solved by the carbon atom being linked to the four chlorine atoms by four covalent bonds. The carbon atom, with the configuration of 2.4, shares each of its outer electrons with a chlorine atom and, in return, receives a share in one of the outer electrons of the chlorine atom, so that a two-electron bond is formed, the resulting cloud of negative charge holding the two positive nuclei together. The molecule can be represented as in Fig. 15.5; the bonding electrons being shown in blue.

Each shared pair of electrons in a molecule makes one covalent bond. Each bond in a molecule is often represented by a single straight line —. The molecules of chlorine and tetrachloromethane can be represented by:

$$Cl-Cl \qquad\qquad Cl-\underset{\displaystyle |}{\overset{\displaystyle |}{C}}-Cl$$

with the upper and lower Cl shown above and below the carbon.

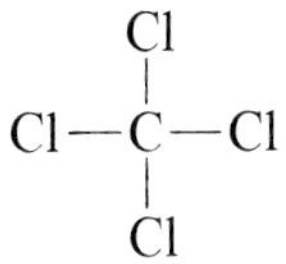

Substances which contain covalent bonds include non-metallic elements which exist as molecules made up of more than one atom of the element. For example, hydrogen exists as molecules, each of which contains two atoms bonded together by a covalent bond. In this molecule, each atom has a share of two electrons which is the configuration of the noble gas helium.

$$H \overset{\times}{_{\circ}} H \qquad\qquad \text{or} \qquad\qquad H-H$$

Molecules of some non-metallic elements contain more than one covalent bond between the atoms. Oxygen has a double covalent bond formed by two pairs of shared electrons and nitrogen has three bonds formed by three pairs of shared electrons.

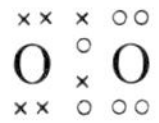

$$O \overset{\times\times\ \times\ \circ\circ}{\underset{\times\times\ \circ\ \circ\circ}{\overset{\circ}{\underset{\times}{}}} O \qquad\qquad \text{or} \qquad\qquad O{=}O$$

$$\overset{\times}{\underset{\times}{\times}} N \overset{\times}{\underset{\circ}{}} N \overset{\circ}{\underset{\circ}{}} \qquad\qquad \text{or} \qquad\qquad N{\equiv}N$$

Some others exist as molecules containing more than two atoms of the element. For example, sulphur exists as S_8 molecules (27.9).

Most of the common compounds which contain covalent bonds are formed by two or more non-metals combining together, whereas ionic compounds are formed by metallic elements combining with non-metallic elements. Some important examples of covalent compounds are water, ammonia, hydrogen chloride and carbon dioxide. The bonding in these compounds can be represented by:

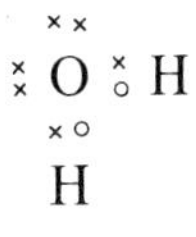

$$\overset{\times\times}{\underset{\times}{\times}} O \overset{\times}{\underset{\times\ \circ}{}} H \qquad\qquad \text{or} \qquad\qquad \underset{H}{\overset{O-H}{|}}$$

with a second H below the oxygen.

$$\underset{H}{\overset{H}{\underset{\times\ \circ}{}}} \overset{\times}{\underset{\times}{\times}} N \overset{\times}{\underset{\times\ \circ}{}} H \qquad\qquad \text{or} \qquad\qquad \underset{H}{\overset{H}{\underset{|}{\overset{|}{N-H}}}}$$

143

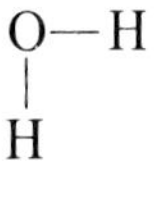

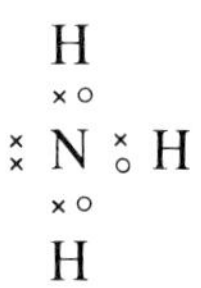

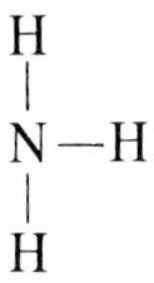

$$\text{H} \overset{\circ\circ}{\underset{\circ\circ}{\times}} \text{Cl} \overset{\circ\circ}{\underset{\circ\circ}{\circ}} \qquad \text{or} \qquad \text{H}\!-\!\text{Cl}$$

$$\text{O} \times \text{C} \times \text{O} \qquad \text{or} \qquad \text{O}\!=\!\text{C}\!=\!\text{O}$$

15.4 Properties of covalent substances

At room temperature, covalent substances are usually either gases or volatile (easily-evaporated) liquids.

The low melting points and boiling points are due to the substances existing as separate molecules. When such substances melt, the weak attractive forces between the molecules are overcome, so enabling the molecules to move around. The covalent bonds within the molecules remain intact, whereas when an ionic compound melts the ionic bonding forces do have to be overcome.

The energy, in the form of heat, which is required to overcome the forces between molecules is less than that required to overcome ionic bonding forces. This difference between ionic and covalent substances is discussed again in the next chapter when the structures of some typical examples are considered in detail.

Pure covalent liquids are poor conductors of electricity which indicates that they do not contain ions.

Many covalent substances do not dissolve in water, and the solutions of those that do usually will not conduct electricity. Two important exceptions are solutions of hydrogen chloride and ammonia. Because these solutions conduct electricity they must contain ions. The presence of the ions is due to the covalent compounds reacting with water when they dissolve. Hydrogen chloride forms an acidic solution (hydrochloric acid):

$$\text{HCl(g)} \quad + \quad \text{H}_2\text{O(l)} \quad \rightarrow \quad \text{H}_3\text{O}^+\text{(aq)} \quad + \quad \text{Cl}^-\text{(aq)}$$

Ammonia forms an alkaline solution:

$$\underset{\substack{\text{covalent} \\ \text{molecules}}}{\text{NH}_3\text{(g)} \quad + \quad \text{H}_2\text{O(l)}} \quad \rightarrow \quad \underset{\text{ions}}{\text{NH}_4^+\text{(aq)} \quad + \quad \text{OH}^-\text{(aq)}}$$

15.5 Co-ordinate (or dative) covalent bonds

In his original article, in 1916, on covalent bonding, Lewis also put forward an idea which was to provide an answer to a problem, first described by Alfred Werner in 1891. For example, when ammonia solution is added to copper(II) sulphate solution, a deep blue solution is formed. The deep blue colour is due to tetraamminocopper(II) $(\text{Cu(NH}_3)_4)^{2+}$ ions being present. In this ion complete ammonia molecules have become bonded to copper ions.

This type of ion is called a **complex ion** and such ions are very common in the chemistry of metal compounds. Indeed it is such a complex ion, the copper(II) tetrahydrate ion $(\text{Cu(H}_2\text{O})_4)^{2+}$, which is responsible for the blue colour of solutions of copper(II) salts. But how are complete molecules, like ammonia or water, joined to copper ions to form complex ions?

Lewis's answer to this question was to state that covalent bonds, as well as being formed by each atom contributing one electron to form the bond,

$$\text{A} \circ \quad \times \text{B} \rightarrow \text{A} \overset{\circ}{\underset{\times}{}} \text{B}$$

can also be made by one atom contributing both the electrons, with the other atom contributing none:

$$A \overset{\circ}{_\circ} \qquad B \rightarrow A \overset{\circ}{_\circ} B$$

This sort of covalent bond is called a **co-ordinate covalent bond** (or a dative covalent bond). For an atom to form a co-ordinate covalent bond it must have in its outermost shell a pair of electrons, not linked to another atom; such a pair is called an unshared (or lone) pair.

Lewis's co-ordinate covalent bonds provide an explanation for the existence of Werner's complex compounds. The molecules have in them at least one unshared pair of electrons which can be used to form a co-ordinate covalent bond with the metal ion. The outer shell of the metal atom will have lost some electrons when the atom ionised, so leaving a 'hole' which can receive a share in the pair of electrons donated by the molecule.

The structure of the ion formed by ammonia and copper(II) ions is:

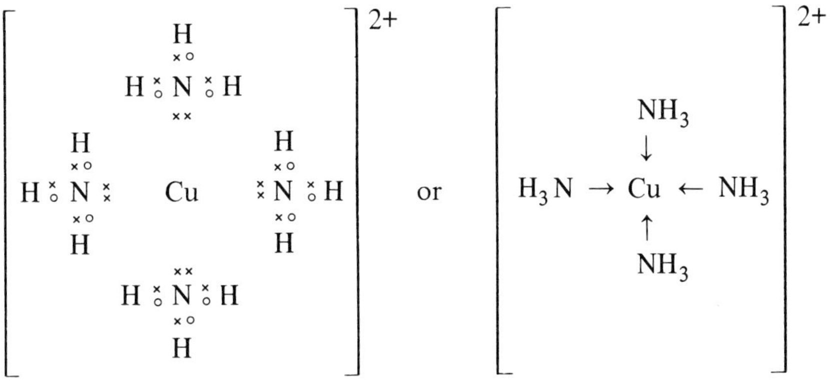

A co-ordinate covalent bond is often represented by an arrow pointing in the direction in which the pair of electrons is donated.

Probably the most important structures in which co-ordinate covalent bonds are found are those of the hydroxonium ion (sometimes called oxonium ion) and the ammonium ion. The hydroxonium ion is formed when a compound dissolves in water to give an acidic solution. Water is a covalent compound and its molecule can be represented as follows:

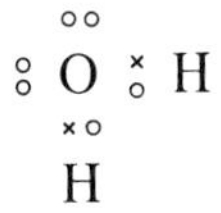

There are two lone pairs (non-bonding pairs) of electrons on the oxygen atom and one of these can form a co-ordinate covalent bond with a positively charged hydrogen ion (a proton), released from the substance forming the acidic solution. Because the hydrogen ion is positively charged, the structure formed when it is joined to an electrically neutral water molecule is also positively charged. The formation of the hydroxonium ion can be represented as follows:

$$H^+ + H_2O \rightarrow H_3O^+$$

Using lines for the covalent bonds the structure of the ion can be represented by:

$$\left[\mathrm{H} \leftarrow \mathrm{O} - \mathrm{H} \atop \mathrm{H} \right]^{+}$$

The dissolving of covalent hydrogen chloride gas in water can then be represented by the overall equation:

$$\mathrm{HCl(g)} + \mathrm{H_2O(l)} \rightarrow \mathrm{H_3O^+(aq)} + \mathrm{Cl^-(aq)}$$

It is the presence of the hydroxonium ion in the solution which is responsible for the acidity of the solution (11.11).

Ammonium ions are formed when ammonia molecules combine with a compound which is capable of releasing hydrogen ions to the ammonia molecule. The nitrogen atom of the ammonia molecule has one lone pair of electrons,

$$\mathrm{H} \overset{\times}{\underset{\circ}{\,}} \mathrm{N} \overset{\times}{\underset{\circ\times}{\,}} \mathrm{H} \atop \mathrm{H}$$

and this can be used to form a co-ordinate covalent bond with a hydrogen ion, forming an ammonium ion with the structure:

$$\left[\mathrm{H} \overset{\times}{\underset{\circ}{\,}} \mathrm{N} \overset{\times}{\underset{\times\circ}{\,}} \mathrm{H} \atop \mathrm{H} \right]^{+}$$

Again, because a positively charged hydrogen ion is combining with an electrically neutral ammonia molecule, the resulting structure carries a single positive charge. A reaction in which the ammonium ion is formed is that between ammonia gas and hydrogen chloride gas. The overall change can be represented by the equation:

$$\mathrm{NH_3(g)} + \mathrm{HCl(g)} \rightarrow \mathrm{NH_4^+Cl^-(s)}$$

The product, ammonium chloride, is formed as a white solid.

15.6 The shape of covalent molecules

While molecules of even very complicated compounds are far too small to be seen with the most powerful microscope, it is possible, by using techniques such as the diffraction of beams of electrons and X-rays, to measure the angles between bonds and the lengths of bonds in covalent molecules. When these results are put together, the shape of the molecule will emerge. A molecule of methane (CH_4), for example, has angles of 109° between the carbon-hydrogen bonds. The value of this angle shows that the molecule is **tetrahedral** in shape, with the carbon atom being at the centre of the tetrahedron and the four hydrogen atoms being at the corners (Fig. 15.6).

Fig. 15.6 The shape of the methane molecule. (Photo: Russell Edwards, BSc)

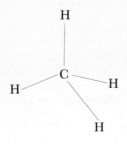

146

The fact that there is a definite angle between covalent bonds can be explained if we consider the bond to consist of a cloud of negative charge which holds together the two positive nuclei. The charge cloud will have a certain direction in space and the bond angle will be the angle between two of these directions.

The tetrahedral shape of the methane molecule is also easy to understand. The four bonding clouds of negative charge will have to touch the positively charged nucleus of the carbon atom, but, because they are all negatively charged, they will repel each other as much as they can (Fig. 15.7).

In the ammonia molecule (NH_3) there is one hydrogen atom less than in methane and therefore the molecule will have the shape which would be obtained by 'beheading' the tetrahedron (i.e. taking out one hydrogen atom), Fig. 15.8.

Fig. 15.7

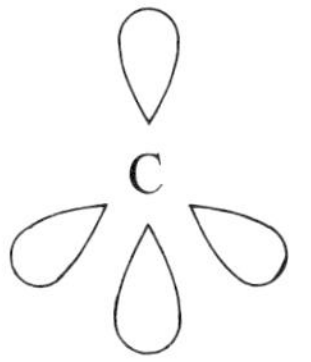

Fig. 15.8 The shape of the ammonia molecule.

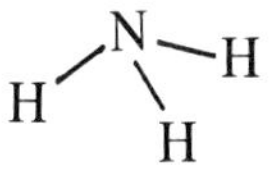

This is called a **trigonal pyramid**. At first sight it would seem that the bond angles are 109° as they are in methane, but experimental evidence shows that the angle is in this case 107°. A possible reason for this is that, although one hydrogen atom of the methane molecule is no longer there, there is still the unshared pair of electrons in the outermost shell of the nitrogen atom:

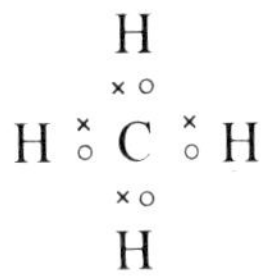

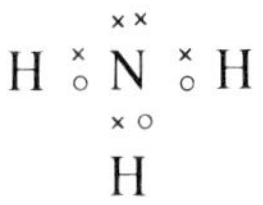

$$
\begin{array}{c}
H \\
\overset{\times\,\circ}{H \,\overset{\times}{\underset{\circ}{C}}\, H} \\
\overset{\times\,\circ}{H}
\end{array}
\qquad\qquad
\begin{array}{c}
\overset{\times\,\times}{H \,\overset{\times}{\underset{\circ}{N}}\, H} \\
\overset{\times\,\circ}{H}
\end{array}
$$

Just as you would become fat if you did no work at all, this cloud of charge, which is not doing any work by holding another atom, is fatter because it is not stretched out by being attracted to another atom. The consequence is that the three bonding clouds are pushed slightly closer together and so the bond angle becomes slightly smaller. When the ammonia molecule forms a co-ordinate covalent bond with a hydrogen ion to make an ammonium ion, the bond angle rises again to the 109° of the methane molecule because the cloud of charge has now been put to work and has become exactly the same as the other three:

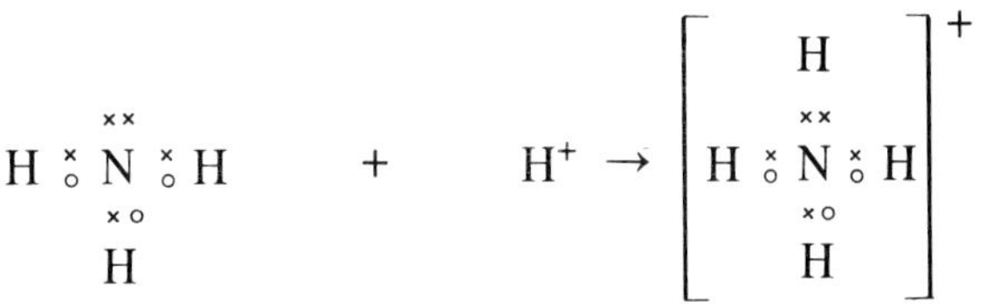

$$
\begin{array}{c}
\overset{\times\,\times}{H \,\overset{\times}{\underset{\circ}{N}}\, H} \\
\overset{\times\,\circ}{H}
\end{array}
\quad + \quad H^+ \;\rightarrow\;
\left[
\begin{array}{c}
H \\
\overset{\times\,\times}{H \,\overset{\times}{\underset{\circ}{N}}\, H} \\
\overset{\times\,\circ}{H}
\end{array}
\right]^{+}
$$

The molecule of water shows a continuation of this trend. Here there are two 'fat' non-bonding pairs of electrons in the oxygen atom and the angle between the oxygen-hydrogen bonds is pushed down to 105° (Fig. 15.8).

Fig. 15.9 The shape of the water molecule.

You should now be able to predict the approximate angle between the oxygen-hydrogen bonds in a hydroxonium ion, H_3O^+. The same cloud repulsion idea will enable us to explain the shape of a carbon dioxide molecule:

$$O \overset{\overset{\text{oo x}}{\text{o}}}{\underset{\underset{\text{oo o}}{\text{x}}}{}} C \overset{\overset{\text{x oo}}{\text{o}}}{\underset{\underset{\text{o oo}}{\text{x}}}{}} O$$

The two clouds of charge, holding each oxygen atom to the carbon atom, will merge into a big one and so we shall have two big clouds of charge, touching the carbon nucleus but trying to get as far away from each other as possible. Clearly one will be on one side of the carbon atom and the other will be in exactly the same position on the other side. The molecule will therefore be linear (Fig. 15.10).

Fig. 15.10 The shape of the carbon diacide molecule

$$O{=}C{=}O$$

15.7 Summary

1. The electrons in atoms are responsible for the formation of chemical bonds.
2. The bonding in electrolytes is of a different type to that in non-electrolytes.
3. Electrolytes are formed by electrons being transferred from one atom to another, which results in the formation of oppositely charged ions. Electrical forces of attraction hold the oppositely charged ions together.
4. Ionic compounds are formed between metals and non-metals. They are usually crystalline solids with high melting points and boiling points. When molten they conduct electricity and, if they dissolve in water, the solution conducts electricity.
5. In non-electrolytes electrons are shared between atoms, one electron coming from one atom and one from the other, the two electrons forming one covalent bond.
6. Covalent compounds are formed usually between atoms of non-metallic elements. They are usually gases or volatile liquids at room temperature and, if they dissolve in water, their solutions will not conduct electricity unless the compounds react with water to form ions.
7. Co-ordinate (dative) covalent bonds are formed by one atom contributing both electrons to be shared between itself and another atom. They occur in complex ions and in the hydroxonium ion and the ammonium ion.
8. Covalent bonds can be considered to be clouds of negative charge, having a definite direction in space. They result in covalent molecules having particular shapes. The shape of a molecule results from the repulsions between the clouds of negative charge.

Chapter 16

The structures of solid elements and compounds

**Investigation
16.1**

What happens when a molten substance is cooled?

Add naphthalene crystals to a boiling-tube to a depth of about 4 cm. Place the boiling-tube in a beaker of warm water and warm the beaker carefully on a gauze over a Bunsen flame until the naphthalene melts. Turn out the flame and allow the beaker to cool slowly. When no further change takes place, reheat the beaker until the naphthalene is molten again and then cool the boiling-tube rapidly under the tap.

Question

What is the difference between the crystals made by cooling the naphthalene slowly and cooling it quickly?

**Investigation
16.2**

Do metals form crystals?

You will need some granulated zinc, some thick copper wire, a crucible, a pipeclay triangle, a Pyrex crystallising dish and a pair of tongs.

1. To about a 4 cm depth of lead(II) nitrate solution in a test-tube, add a piece of granulated zinc and allow the tube to stand undisturbed until the end of the lesson or until the next lesson.
2. Half fill a test-tube with silver nitrate solution. Lower a thick piece of copper wire into the silver nitrate solution and leave it undisturbed until the end of the lesson or until the next lesson.

Questions

1 What do you see on the samples of metals in the tubes?
2 What products are formed on the metals in the two tubes?
3 How are these products formed?
4 In which of the two tubes is the reaction the quicker? Why is this?

3. Add a few pieces of granulated tin to a crucible and heat the crucible on a pipe clay triangle on a tripod over a Bunsen flame until the tin melts.
   Carefully, using a pair of tongs, pour the molten metal on to the bottom of a Pyrex crystallising dish or beaker. When it is cool enough, detach the metal from the glass and examine the surface which was in contact with the glass.
   Now place the metal in a solution of iron(III) chloride with the surface you examined, uppermost. This solution will dissolve some of the metal. When an obvious change has taken place, remove the metal and wash it under a cold water tap.

Questions

5 Were you able to see any crystals in the tin before and after it has been treated with iron(III) chloride solution?
6 Does the appearance of the surface suggest the metal cooled quickly or slowly?

**16.3
Crystals**

If a hot saturated solution is cooled, crystals of the solute separate from the solution. If the solution is cooled rapidly, the crystals are small; if it is cooled more slowly, they are bigger. Crystals can also be produced by cooling a molten substance. Many metals have in them crystals which have been formed in this way and sometimes, as in the case of zinc used for galvanising iron, the crystals are clearly visible. A number of minerals, like quartz and galena, which can occur as large and beautiful crystals, originally came to the crust of the Earth from its centre in the molten state, cooling when they came nearer to the surface to form the crystals. Again the slower the cooling, the larger are the crystals produced. Crystals, both permanent ones like precious stones and more short-lived ones like snowflakes, have intrigued mankind for centuries. Simply looking at them suggests that they are highly ordered and beautiful structures, but what do they consist of and how are they so ordered?

Fig. 16.1 Crystals of copper(II) sulphate

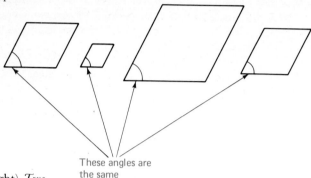

These angles are the same

Fig. 16.2 (right) *Two crystals of copper(II) sulphate. Even though they are very different in size they have the same shape. (Photo: Russell Edwards, BSc)*

These questions can be partially answered by making models of crystals of a substance. If crystals of copper(II) sulphate are made from different solutions of the compound, they will be of different sizes, but, apart from minor flaws, they will all have the same shape. This means that corresponding angles in the crystal will be the same in all the crystals (Fig. 16.1).

If we now take a series of shallow trays and a large number of marbles or plastic beads, we can make an arrangement of spheres in each tray with a line against the edge of the tray, a second line with its spheres being in the spaces between those in the first, a third line with its spheres between those in the second and so on (Fig. 16.3). If the angle shown in Fig. 16.3 is measured in each tray, we find that this angle is the same in all the trays.

Fig. 16.3

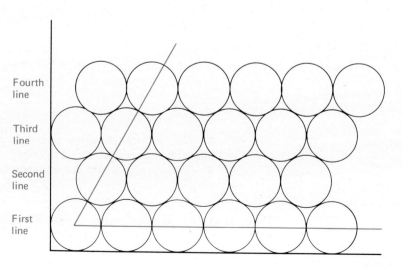

Fourth line

Third line

Second line

First line

Since a particular angle is the same in all crystals of a substance and since a particular angle is the same in all similar arrangements of spheres, it is possible that a crystal consists of an orderly arrangement of spheres, or particles, packed together in a certain way, like the marbles or beads in our trays. Different substances have different crystal shapes and a possible reason for this can be seen by putting larger spheres at regular intervals into an arrangement in one of our trays. This alters the angles in the arrangement and this would alter the shape of the crystal. It seems as if the shape could be decided by the relative numbers and sizes of the particles in the crystal.

The discovery that a crystal consists of an orderly arrangement of particles started in 1912 with the publication by von Laue (a German scientist) of a set of photographs which he had obtained by passing a narrow beam of X-rays through a crystal and placing a photographic plate on the far side. Later in the same year Lawrence Bragg, who was 22 at the time and whose father, William, was Professor of Physics at Leeds University, decided that von Laue's photographs were interference patterns, produced by the atoms in the crystals.

This pattern is made by the waves of the X-rays being scattered by the atoms in the crystals, the waves reinforcing each other in some directions and opposing in others. You can see for yourself an interference pattern with visible light by looking through a stretched handkerchief at the small bulb of a pocket torch.

The formation of these interference patterns depends on the wavelength of the visible light being similar to the sizes of the holes in the fabric. In a similar way, a crystal can only form an interference pattern when the radiation used has a wavelength which is similar to the distance between the layers of particles in the crystal. This is why X-rays must be used rather than ordinary light to investigate the structures of crystals.

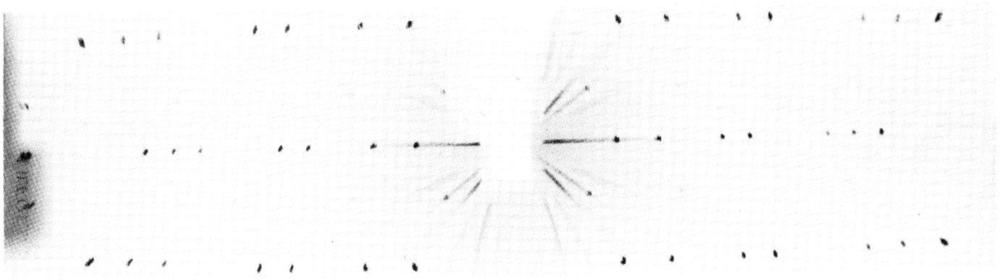

Fig. 16.4 X-ray diffraction photograph (strictly speaking a negative) from a single crystal of sodium chloride. The dark spots show where the film has been exposed to X-rays. The pattern indicates that the particles in the crystal are arranged in a regular manner. (Courtesy Bruce Gilbert)

William and Lawrence Bragg, by studying the patterns obtained in this way, were able to calculate the distances between the layers of particles in crystals. This information enabled them to work out the arrangement of the particles in the crystals. The following sections of this chapter discuss particular types of crystal structures which can be determined by X-ray analysis.

16.4 Ionic crystals

In 1913 and 1914 Lawrence Bragg was able to work out a large number of crystal structures, including that of sodium chloride. Up to that time it had been thought that common salt consists of molecules of sodium chloride with each sodium atom being joined to one chlorine atom. Bragg found that there are no molecules of sodium chloride. Each atom of sodium has six atoms of chlorine around it and each chlorine atom has six sodium atoms around it. By 1916 Kossel had suggested (15.1) that, when sodium combines with chlorine, the sodium atoms become sodium ions and the chlorine atoms become chloride ions. The crystal of sodium chloride must therefore consist of ions, not atoms. The particles are held together in the crystal by strong electrostatic attractions between the positive and negative particles. In 1916 the Braggs were jointly awarded the Nobel Prize for Physics for their work on X-ray analysis.

151

A positively charged sodium ion is capable of exerting attractive forces on chloride ions in all directions around it and the negatively charged chloride ions are able to attract sodium ions in the same way. It is the relative sizes of the two types of ions which determines the pattern of the arrangement or lattice which the ions adopt when they come together. In sodium chloride, each sodium ion has six chloride ions arranged around it as though they were at the corners of a octahedron. Each of these chloride ions has six sodium ions around it and each of these sodium ions attracts more chloride ions, and so on. In this way a giant ionic lattice is built up. As can be seen from Fig. 16.5 the lattice is cubic in shape.

Fig. 16.5 The sodium chloride lattice: (a) drawn so that the arrangement can be seen and (b) more realistically showing the ions packed together

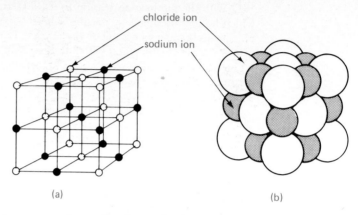

(a)　　　　　　　　　　(b)

All ionic compounds consist of crystals which are made up of orderly arrangements of ions, although not all of these lattices are cubic in shape. In each case the arrangement taken up depends on the relative sizes and charges of the ions. For example in the case of caesium chloride, the caesium ion, Cs^+, and the chloride ion, Cl^-, are very similar in size and there is room for eight chloride ions to be packed around each caesium ion. The caesium ion therefore sits at the centre of a cube, at each corner of which there is a chloride ion. Similarly each chloride ion is surrounded by eight caesium ions, Fig. 16.6.

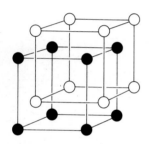

Fig. 16.6 The structure of caesium chloride

16.5 Covalent crystals

One of the crystal structures which Lawrence Bragg determined in 1913 was that of diamond and his picture of the diamond crystal was the one result which led to the considerable interest in the technique of X-ray analysis. Diamond consists of nothing but carbon, which is one of the elements showing the property of **allotropy**.

An element is said to exhibit allotropy when it can exist in more than one form in the same physical state. The different forms of the element are called **allotropes** and the allotropes of solid carbon are diamond and graphite. Since both allotropes consist of atoms of the same element and the electronic configurations of the atoms are the same in both, the chemical properties of the allotropes will be the same. There are, however, obvious differences in their physical properties.

Diamond is one of the hardest substances known, while graphite is so soft that it will mark paper (pencil 'leads' are made of graphite). Diamond is a poor conductor of electricity, while graphite is used for making electrical conductors. Both forms are crystalline and the differences in physical properties can be explained by the different ways in which the atoms are arranged in the crystals.

Both types of crystals consist of giant structures of atoms, which are held together by covalent bonds. In diamond each carbon atom has four neighbours, held by single

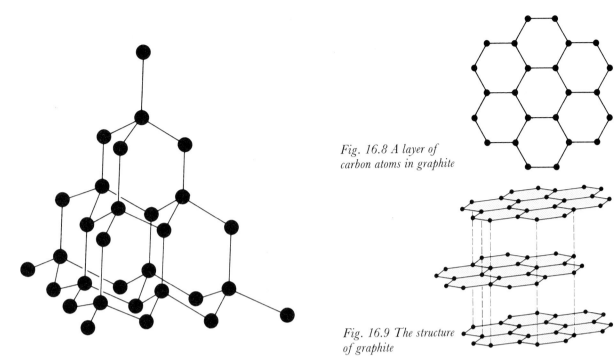

Fig. 16.8 A layer of
carbon atoms in graphite

Fig. 16.9 The structure
of graphite

Fig. 16.7 The structure
of a diamond

covalent bonds, and, if we apply the charge cloud repulsion idea which was described in 15.6, we can suggest that they will be arranged tetrahedrally. This arrangement was the one confirmed by Lawrence Bragg and is shown in Fig. 16.7. Because the repulsion of the charge clouds will create the tetrahedral arrangement, the structure of diamond will be the most stable one which the carbon atoms could take up and the carbon atoms are therefore locked together in the most efficient way. This efficient packing of a large number of covalently bonded carbon atoms to form a giant atomic lattice is responsible for the great strength of the structure which is shown in the hardness of diamond. Since all the outer electrons of the carbon atoms are being used for bonding, there are no electrons free to move through the structure and to be responsible for carrying an electric current through the substance.

The giant structure of graphite is quite different. Here the carbon atoms are in flat sheets and, within a sheet, they are arranged in hexagons, as shown in Fig. 16.8.

Graphite is a greasy substance and the greasy feeling is due to the sheets sliding easily over each other. The forces between the layers must therefore be comparatively weak and such forces are called van der Waals' forces. The structure is shown in Fig. 16.9.

Fig. 16.8 shows that each carbon atom has this time only three neighbours, to which it is joined by covalent bonds. Thus, of the four outer electrons in a carbon atom, only three are used for bonding and one is not. These 'spare' non-bonding electrons can, as in a metal, drift from atom to atom and when the drift is in one direction, the structure is conducting an electric current.

Since the van der Waals' forces are relatively weak, compared with the covalent bonds which hold the atoms together, the layers of carbon atoms will slide easily over each other and graphite can, for this reason, lubricate moving parts, slipping between them and reducing the friction. It is particularly useful when a liquid lubricant, such as an oil, cannot be used, for example when a moving part is in a vacuum or under only a low atmospheric pressure.

The fact that the layers are not so firmly held together and can be separated relatively easily makes graphite soft. The comparatively large distances between the layers are responsible for the density of graphite being significantly less than that of diamond. The different arrangements of the carbon atoms in the two allotropes of carbon have made fundamental differences in the properties of the allotropes.

16.6 Molecular crystals

The beautiful shapes of snowflakes clearly indicate that they consist of ordered arrangements of particles. The particles in snowflakes are water molecules. But how do these molecules hold together? Water is very largely a covalent compound and the atoms within each molecule are held together by electrons, or negative charges, being shared between them. Of the two elements in water, oxygen, being further over to the right in the Periodic Table and therefore having a greater tendency to accept electrons than hydrogen, has a greater pull on this negative charge and so there is a slightly greater concentration of the bonding charge near the oxygen atom than near the hydrogen atoms.

The oxygen atom will therefore be slightly negatively charged as compared with the hydrogen atoms and the hydrogen atom will be slightly positively charged as compared with the oxygen atom.

When water molecules come together to form an ice crystal, the slight positive charge on a hydrogen atom in one molecule will be attracted to the slight negative charge on the oxygen atom of a second molecule and then one of the hydrogen atoms in the second molecule is attracted by the oxygen atom of a third and so on.

Each oxygen atom has around it four hydrogen atoms, two directly attached to it by covalent bonds and two held by the attraction between its negative charge and the positive charge on the hydrogen atom. Because each hydrogen atom carries a slight positive charge, they try to place themselves as far apart from each other as possible, while still being close to the oxygen atom. The four are therefore arranged tetrahedrally and, when this arrangement is repeated through the crystal, the structure becomes that shown in Fig. 16.10.

Fig. 16.10 The structure of ice

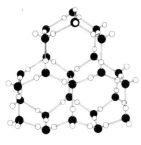

Fig. 16.11 An iceberg photographed during Scott's expedition to the South Pole in 1910. Part of the iceberg is above the water because ice is less dense than water. (Courtesy Popper foto. Photo: Herbert Ponting)

When you look at this picture of the structure of the ice crystal, the thing which might strike you about it is the considerable amount of empty space in the structure. This empty space provides the explanation for one of the remarkable properties of water. When water freezes, the ice produced has a larger volume than the liquid water from which it was made. Usually the exactly opposite behaviour is observed, the solid form having a lower volume and a higher density than the liquid. The slightly larger volume of the ice is due to the empty space in the crystal lattices and is responsible for water pipes bursting when there is a severe frost. Although the pipe actually bursts when the water freezes, it is not noticed until the ice melts as the weather becomes warmer. When the ice melts the water molecules are no longer spaced out in fixed positions, but are able to move around and come closer together and this is why water is denser than the ice. Ice crystals belong to a type of crystal called **molecular crystals** which consist of orderly arrangements of molecules. Many other substances consist of molecular lattices. Common examples include sucrose (ordinary sugar—$C_{12}H_{22}O_{11}$), naphthalene (moth balls—$C_{10}H_8$), urea (used for making plastics and as a fertilizer—$CO(NH_2)_2$) and iodine (I_2). In each case the formula represents the molecule which is used to build up the lattice.

Fig. 16.12 The structure of iodine

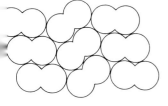

Except for iodine the arrangements of the molecules in these lattices are complicated. Iodine consists of an arrangement of diatomic molecules which are held together by weak attractive forces (similar to those between the layers of carbon atoms in graphite) called van der Waals' forces. The molecules are packed as close together as possible and this results in the arrangement shown in Fig. 16.12.

155

When iodine crystals are warmed they sublime, that is they change directly from the solid to vapour. This can be linked to the fact that the forces between the molecules are comparatively weak. When heat is supplied to the element, the vibration of the molecules increases and, at a low temperature, is sufficient to overcome the forces between the molecules so that they are no longer held together.

16.7 Metals

As with graphite and ice, the important physical characteristics of metals can be explained by their crystal structures and the ways in which the particles are held together in the crystals. It may be a surprise to you to read that metals contain crystals, but, if you keep your eyes open, even at home, you will see examples of them.

A can which has contained acid fruit such as oranges or grapefruit, shows crystals in the layer of tin on the inside of the can, the metal having been etched by the acid in the fruit juice so that the crystals are made more clearly visible. Galvanised iron consists of steel which has a coating of zinc to stop it rusting and the crystals can be clearly seen in the zinc which may have been applied by dipping the steel into molten zinc and then allowing it to cool slowly. Crystals can often be seen in a broken metal casting which has been cooled slowly.

When a sample of metal does not appear to have any crystals in it, it is probable that the crystals are there, but they are too small to be seen by the naked eye. The metal would have been formed in the molten state and must have cooled to form the solid, the crystals being formed as cooling took place.

Fig. 16.13 The spangle pattern on the surface of these zinc-coated (galvanised) columns shows that the layer of zinc is crystalline. (Courtesy Zinc Development Association)

Metals are usually hard, strong materials which are good conductors of heat and electricity. They have high melting points and boiling points, and can usually be worked by beating into sheets or by drawing or rolling into strips or wires. Just as with diamond and graphite, properties of this sort can be seen to be the consequence of the way in which the particles in the crystals are held together. A metallic element is an element whose outermost electrons are only loosely held by the nucleus of the atom and which are therefore easily removed. For example, in a piece of a metal such as copper, the electrons drift from atom to atom, their arrival from one atom causing the displacement of the similar electrons from the second atom to which they come. These displaced electrons then drift towards a third atom, from which they will displace its outer electrons, and so the movement goes on. These movements will be in all

156

directions within the metal. Only when the ends of the metal are connected to a battery, will all the drifts be in one direction, this movement then constituting an electric current flowing through the metal.

If we consider these electrons to be, not particles, but clouds of negative charge, then a crystal of a metal must consist of atoms which have lost their outer negative charge and are therefore positive ions, being submerged in a cloud of negative charge which holds the ions together. The crystal is like an orderly arrangement of billiard balls, suspended in a tank of water. The balls represent the positive ions and the water, which would fill up all the spaces between the balls, would be like the negative charge. This sort of bonding, in which closely packed positive ions are held by negative charge filling up the spaces between them, is called **metallic bonding**.

The electrical forces between the negative charge and the positive ions are strong and are responsible for the strengths of most metals. If the metal is to be worked to make something, it is usually heated, so that the ions vibrate more and they are more easily disturbed. Because the ions require a great deal of energy to separate them, the melting point and boiling point of a metal are usually high. Since the negative charge between the ions is mobile, it can be made to drift in one direction through the structure by a battery, making the metal a conductor of electricity.

The presence of crystals in the metal shows that there must be order in the arrangement of the ions and this is confirmed by X-ray diffraction techniques, as with sodium chloride crystals. There are two common types of packing which give this order and these can be conveniently seen in arrangements of beads or plastic spheres in a tray. A layer of spheres can be made by placing a line of them along the edge of the tray, then a second line in the spaces between the spheres in the first, then a third in the spaces in the second, and so on. A second layer can then be put on to the first with its spheres in the spaces between the spheres in the first (Fig. 16.14).

When we get to the stage of putting in the third layer, we find that this can be done in two ways. The spheres in this layer can be placed exactly over those in the first layer (position 1), or they can be placed in other spaces in the second layer so that they are not directly above the spheres in either of the first two layers (position 2). In the latter case the third layer covers up the holes going right through both the first and second layers of spheres.

Fig. 16.14 (a) the first layer of spheres (b) the second layer of spheres

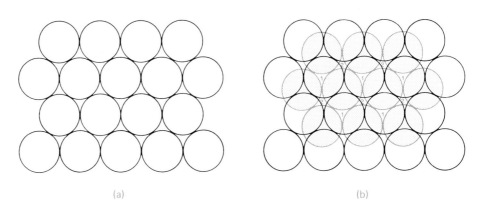

(a) (b)

The first of the arrangements for the third layer produces the type of packing which can be called the ABAB, which indicates that the third layer is a repetition of the first and the fourth a repetition of the second, and so on. This arrangement is known as **hexagonal close-packing**. The spheres are as close together as possible and when a model of the structure is viewed from a particular angle it has hexagonal symmetry (Fig. 16.15). The metals magnesium and zinc adopt this structure.

Layer *a*

Layer *b*

Layer *a*

Layer *a*

Layer *b*

Layer *a*

Layer *c*

Fig. 16.15 A model showing hexagonal close-packing of spheres (the a, b, a layers are arranged horizontally). (Model: C. Grindel. Photo: J. Olive)

Fig. 16.16 A model showing face-centred cubic packing of spheres (the a, b, c layers are arranged diagonally). (Model: C. Grindel. Photo: J. Olive)

The second arrangement for the third layer is known as ABCABC packing, as the arrangement is repeated every fourth layer. When a model of this structure is viewed from a particular angle (Fig. 16.16), it is possible to see a cube with a sphere in the centre of each face. The structure is therefore called **face-centred cubic packing**. The metals aluminium, copper and lead adopt this structure.

The fact that there are breaks or faces between individual crystals in a sample of metal can be understood by examination of a bubble raft, which can be prepared by adding a small amount of detergent to water in a Petri dish and then passing into the water small bubbles of gas until the surface of the water is covered with them. Some areas of the surface will be covered with regularly arranged bubbles and these correspond to the crystals in the metal. Between these regular arrangements will be gaps, corresponding to the absence of metal atoms, or larger bubbles, corresponding to atoms of impurities. Either gaps or atoms of impurities will define the boundaries of crystals in the metal. The working of the metal (e.g. hammering it or drawing or rolling it into wires) will alter these boundaries and so alter the forces holding certain metal atoms in the structure. This could affect the strength of an object made from the metal and this is the basis of hot and cold ways of working metals, so that they will do the jobs we want them to do.

It is a long time now since the Braggs confirmed the order which is shown in a crystal. Now we are able to understand how this order arises, we can create order in substances where previously there had been only minute ordered areas. One of the most spectacular illustrations of our ability to produce ordered arrangements of particles is in the manufacture of integrated circuits for electronics. Single crystals of silicon can be prepared large enough to be turned into complicated circuits called **silicon chips**.

It is possible to have a circuit which is equivalent to tens of thousands of transistors on a silicon chip which is 1 cm square. A chip such as this can be used to build a small computer known as a microprocessor. The chip must be made from a single crystal. Any crystal boundaries or faults would make it unusable. We may not be able to make a snow flake with as much beauty as nature, but crystals made into integrated circuits have a beauty of their own.

Potentially the microprocessor, which is a relatively cheap desk-top computer, could be put to many uses in the business world and in education, and so could be of great benefit to mankind.

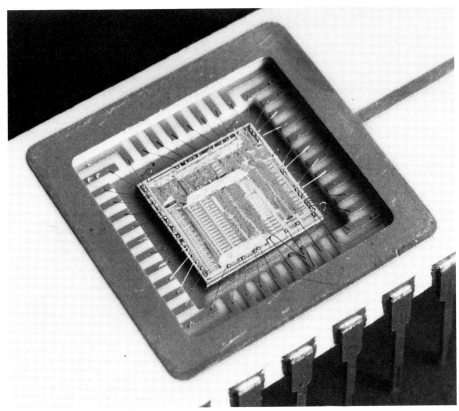

Fig. 16.17 The Ferranti F100-L microprocessor is constructed from a silicon chip 6mm square and has 100 times more capacity than their first commercial computer which used 4000 valves and 6 miles of wire.

The magnified picture shows the complexity of the circuits on the chip. (Courtesy Ferranti Ltd)

16.8 Summary

1. Crystals of substances, made by cooling a hot saturated solution or by cooling a molten substance, consist of orderly arrangements of particles, called lattices.
2. Crystals of ionic compounds consist of orderly arrangements of large numbers of ions. The shape of a lattice depends on the relative sizes and charges of the ions making up the crystal. The ions are held together by electrostatic attraction.
3. Crystals of diamond and graphite, which are allotropes of carbon, are examples of atomic lattices in which large numbers of carbon atoms are bonded together by covalent bonds. The properties of the two allotropes are decided by the ways in which the atoms are arranged in the lattices.
4. Molecular crystals are orderly arrangements of large numbers of covalent molecules, held together by weak attractive forces. The properties of a substance which consists of molecular crystals can often be explained by the arrangement of the molecules in the crystal and the way they are held together.
5. Metal crystals consist of positive ions, held in layers by negative charge between them. The strengths of metals and their electrical conductivities can be explained by these arrangements of ions surrounded by freely moving electrons.

159

Chapter 17

How do the reactivities of metals differ?

Investigation 17.1

Which metals will displace other metals from their oxides?

1. Put a mixture of zinc powder and copper(II) oxide into an ignition-tube to a depth of about 1 cm. Heat the tube with a medium Bunsen flame, and when a reaction starts remove the tube from the flame. When the tube is cool, tip out the residue and examine it.

Repeat the experiment using a mixture of iron powder and copper(II) oxide, but this time heat the tube quite strongly for about two minutes. After allowing it to cool examine the residue.

Questions

1   What kind of reaction occurs in these two experiments, and what new substances are formed?
2   Which of these two reactions is the more vigorous?
3   If you heated a mixture of copper powder and zinc oxide, what do you think would happen?

Investigation 17.2

Which metals will displace other metals from solutions of their salts?

You will need one small iron nail, and small pieces of aluminium, copper, iron, lead, magnesium and zinc. You will also need solutions of salts of each of these metals and, if available, a solution of silver nitrate.

1. Pour copper(II) sulphate solution into a test-tube to a depth of about 3 cm.

Drop a clean iron nail into the copper(II) sulphate solution and leave it undisturbed for a few minutes. Observe what happens to the surface of the nail.

Questions

1   What do you think has been formed on the surface of the iron nail?
2   What must some of the iron have been converted to?
3   Is the iron more or less reactive than the copper?

2. Take a small quantity of one of the metals and place it in a test-tube containing about 2 cm depth of a solution of a salt of one of the other metals (your teacher will tell you which combinations to try).

Look for signs of a reaction in which a new metal is deposited on the surface of the original one.

Construct a table of class results, showing in which combinations a displacement reaction occurred.

Question

4   On the basis of this table, try and put these metals into an order of reactivity, using the principle that if metal A displaces metal B from a salt, then metal A is more reactive than metal B.

17.3
How do metals differ from each other

Elements can be classified as metals rather than non-metals by noting their physical properties (e.g. shiny appearance when cut) and chemical properties (e.g. form oxides which have some basic properties). In these respects, two metals such as sodium and gold are similar, but in many other ways they are obviously different. Gold is used to make jewellery because it is not attacked by other substances and hence stays shiny. On the other hand sodium reacts so violently with water that it must be kept under oil. Also a freshly cut piece of sodium rapidly loses its shiny appearance because its surface reacts with the moisture and oxygen in the atomsphere. We can conclude that the reactivities of gold and sodium towards oxygen and water are very different and yet they are both metals.

By comparing how easily or vigorously metals react with various substances such as oxygen, water and acids, it is possible to arrange the metals into **orders of reactivity** with each of these substances and in each set of experiments the order obtained is very similar.

Some metals which you are likely to use, or see, during your chemistry course are (arranged in alphabetical order): aluminium, calcium, copper, iron, lead, magnesium, potassium, silver, sodium and zinc.

Fig. 17.1 Two metals which have very different properties (a) Sodium is so reactive that it must be kept away from air and moisture by storing it under oil. (Photo: Russell Edwards, BSc) (b) Gold is so unreactive that it never loses its shine. These statuettes of Tutankhamen were made in the 14th century BC. (Courtesy Egyptian Museum, Cairo)

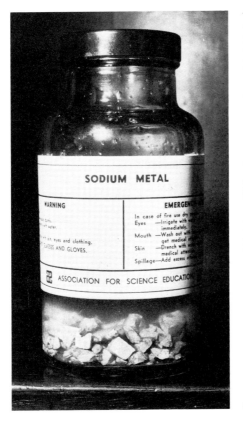

(a)

(b)

17.4
Reactions of metals with water

When each of the above metals is added to water, it is obvious that three of them—potassium, sodium and calcium—react. Potassium reacts most vigorously—it whizzes around the surface of the water and the heat produced by the reaction is sufficient to melt the metal to form a round globule and also to ignite the gas which is given off during the reaction. The reaction of sodium with water is slightly less vigorous. The heat evolved is still sufficient to melt the sodium and the molten globule does move around the surface of the water, but with a small piece of sodium no flame is observed.

161

A piece of calcium does not melt and it does not move around the surface of the water; it reacts steadily giving off bubbles of gas. If the gas is collected, Fig. 17.2, and tested with a lighted splint it ignites with a small explosion (or pop) which shows that the gas is hydrogen, as this is the only common gas which reacts in this way. In all three reactions the remaining solution turns red litmus blue showing that an alkali has been formed.

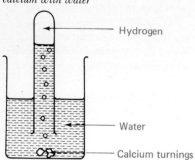

Fig. 17.2 The reaction of calcium with water

Hydrogen

Water

Calcium turnings

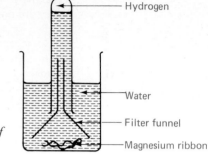

Hydrogen

Water

Filter funnel

Magnesium ribbon

Fig. 17.3 The reaction of magnesium with water.

Equations for the three reactions are:

$$2K(s) + 2H_2O(l) \rightarrow 2KOH(aq) + H_2(g)$$
$$2Na(s) + 2H_2O(l) \rightarrow 2NaOH(aq) + H_2(g)$$
$$Ca(s) + 2H_2O(l) \rightarrow Ca(OH)_2(aq) + H_2(g)$$

If a piece of magnesium is left in water with a test-tube full of water clamped above it as in Fig. 17.3, a small volume of hydrogen will be collected above the water in a few days, showing that magnesium does react very slowly with cold water.

17.5 Reactions of metals with steam

Reactions tend to occur more readily at higher temperatures and so metals which react with water would be expected to do so more vigorously with steam. Also we could expect that some metals which do not appear to react with water might react with steam.

The latter prediction can be tested on a small scale by producing steam by heating a plug of rocksil (which does not take any part in the reaction) soaked in water and then passing the steam over the heated metal either in the form of a powder or, with magnesium, as ribbon. When a reaction occurs, the gas produced can be collected over water as shown in Fig. 17.4.

Magnesium, zinc, aluminium and iron do react. The reaction with magnesium is the most vigorous, followed by zinc and aluminium, and the iron reaction is slow. The gas produced ignites with a pop indicating that it is hydrogen and the residue left in the boiling-tube is the oxide of the metal (unlike the products of the reactions of potassium, sodium and calcium with water which in each case are the hydroxides of the metals),

e.g.
$$Mg(s) + H_2O(g) \rightarrow MgO(s) + H_2(g)$$

The reaction with iron is reversible. If hydrogen is passed over heated iron oxide the reverse reaction occurs and some iron and steam are formed. The reaction is represented by:

$$3Fe(s) + 4H_2O(g) \rightleftharpoons Fe_3O_4(s) + 4H_2(g)$$

162

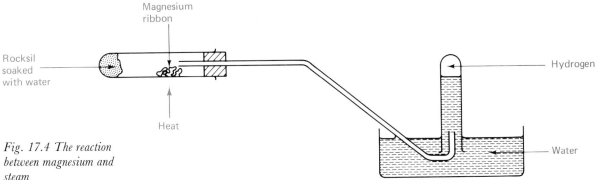

Fig. 17.4 The reaction between magnesium and steam

The sign ⇌ indicates that the reaction is reversible.

As predicted, the reactions of potassium, sodium and calcium with steam are too vigorous and dangerous to demonstrate in a school laboratory. The other metals, lead, copper and silver do not appear to react with steam.

17.6 Displacement reactions

When an iron penknife blade is dipped into copper(II) sulphate solution, it becomes pink in colour, showing that some of the copper from the solution has been deposited as copper metal on the blade. This change is accompanied by some of the iron from the blade passing into solution to form iron(II) sulphate solution. The overall reaction is:

$$Fe(s) + CuSO_4(aq) \rightarrow FeSO_4(aq) + Cu(s)$$

The change can be interpreted ionically by saying that copper(II) ions, $Cu^{2+}(aq)$, in solution have changed to copper atoms, $Cu(s)$, which are deposited on the blade:

$$Cu^{2+}(aq) + 2e^- \rightarrow Cu(s)$$

The electrons required for this change are supplied by iron atoms changing to iron(II) ions, $Fe^{2+}(aq)$, which pass into solution:

$$Fe(s) \rightarrow Fe^{2+}(aq) + 2e^-$$

The overall ionic equation for the change is obtained by adding the two ionic half-equations:

$$Fe(s) + Cu^{2+}(aq) \rightarrow Fe^{2+}(aq) + Cu(s)$$

Thus the sulphate part of the copper(II) sulphate takes no part in the reaction and it could be predicted that the same change would occur if the penknife blade was dipped into copper(II) nitrate solution or copper(II) chloride solution.

The reaction occurs because the iron is more reactive than the copper, that is, iron has a greater tendency to form ions than copper has. The general case of one metal reacting with the compound of another metal can be represented by:

metal(1) + compound of metal(2) → compound of metal(1) + metal(2)

In terms of ions:

metal(1) + ion of metal(2) → ion of metal(1) + metal(2)

This type of change can be used to sort out the metals into an order of reactivity with respect to each other. If, on adding one metal to a solution of a compound of another metal, a reaction occurs, it can be concluded that the first metal is more reactive than the second. Investigation 17.2 provides the instructions for finding such an order of reactivity.

163

The classification of metal-metal ion displacement reactions as oxidation-reduction reactions is discussed in 21.7.

17.7 Reactions of metals with hydrochloric acid

Some metals will displace hydrogen from water or steam (17.4 and 17.5). As a general rule, hydrogen is more easily displaced from dilute hydrochloric acid solution than from water itself. For this reason potassium, sodium and calcium must not be added to solutions of acids. Each of the other metals may be tested by adding dilute hydrochloric acid to a small quantity of the metal. If the reaction is slow, gentle heat may be used. A lighted splint can be used to test for hydrogen being given off.

In decreasing order of vigour, the metals which react are: magnesium, aluminium, zinc, iron and lead (the reaction with lead is only noticeable with warm acid and the reaction is complicated by the fact that the other product, lead(II) chloride, is insoluble in cold water). Copper and silver do not react, even with warm hydrochloric acid.

In each case where a metal does react, the reaction is:

$$\text{metal} + \text{hydrochloric acid} \rightarrow \text{metal chloride} + \text{hydrogen}$$

In the case of magnesium the equation is:

$$Mg(s) + 2HCl(aq) \rightarrow MgCl_2(aq) + H_2(g)$$

This reaction can be explained ionically, in that the magnesium atoms in the metal are changing into magnesium ions which pass into solution:

$$Mg(s) \rightarrow Mg^{2+}(aq) + 2e^-$$

The hydrogen ions from the acid are changing into molecules of hydrogen gas:

$$2H^+(aq) + 2e^- \rightarrow H_2(g)$$

The overall ionic equation is obtained by adding these two ionic half-equations together:

$$Mg(s) + 2H^+(aq) \rightarrow Mg^{2+}(aq) + H_2(g)$$

This reaction is rather similar to the metal-metal ion displacement reactions which were discussed in the previous section,

e.g. $$Mg(s) + Cu^{2+}(aq) \rightarrow Mg^{2+}(aq) + Cu(s)$$

Therefore the reactions of metals with dilute hydrochloric acid enable us to fix the position of hydrogen in the reactivity series (although it is not a metal it does form positive ions, $H^+(aq)$) between copper and lead.

17.8 Reactions of metals with other acids

Dilute sulphuric acid reacts in the same way as dilute hydrochloric acid and can be used to sort out the metals into a similar order of reactivity.

Concentrated sulphuric acid and both dilute and concentrated nitric acid react with metals to give a variety of products due to the oxidising action of the sulphate and the nitrate parts of the acids. The particular reactions of these acids with metals are discussed in detail in 27.26 and 34.7.

17.9 Summary

Table 17.1 gives an order of reactivity of the metals and summarises their reactions with water, steam and dilute hydrochloric acid. The positioning of aluminium is not straightforward as the metal is covered with a layer of aluminium oxide which is very resistant to attack. If the layer is removed by rubbing the metal with mercury, the results would place aluminium between magnesium and zinc, but frequently observations of reactions of aluminium would lead you to think that it ought to be placed lower in the series.

164

TABLE 17.1. *The reactivity series of the metals*

METALS	REACTIONS OF METALS			
	WITH WATER TO FORM HYDROGEN	WITH STEAM TO FORM HYDROGEN	WITH DILUTE HYDROCHLORIC ACID TO FORM HYDROGEN	WITH SOLUTIONS OF COMPOUNDS OF OTHER METALS
potassium sodium	violent reaction	violent reaction	violent reaction	displace hydrogen from water
calcium	steady reaction			
magnesium	very slow reaction	the metal burns	steady reaction	displace a metal which is lower in the series
aluminium zinc	no reaction			
iron		reversible reaction		
lead		no reaction	slight reaction	
(hydrogen) copper silver			no reaction	

Chapter 18

Hydrogen

The preparation and properties of hydrogen

Hydrogen can be produced by the action of a metal which is above hydrogen in the reactivity series (17.9) on dilute sulphuric acid or dilute hydrochloric acid, provided the metal does not form an insoluble salt from the acid. To prepare the gas in the laboratory, it is necessary to choose the metal and the acid to give a steady flow of gas. The reaction of dilute hydrochloric acid on granulated zinc is convenient for this purpose.

You will need an apparatus such as that shown in Fig. 18.1. ON NO ACCOUNT SHOULD YOU BRING A FLAME CLOSE TO THE EXIT TUBE OF THE APPARATUS, AS THIS COULD CAUSE AN EXPLOSION INSIDE THE FLASK.

Fig. 18.1

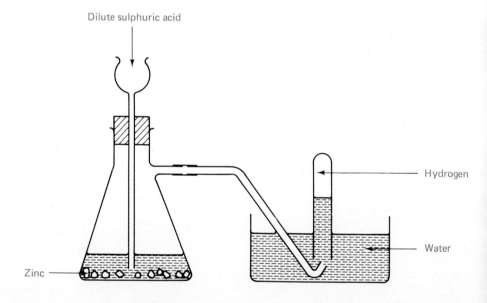

1. Place about 1 cm depth of granulated zinc in the flask, insert the bung firmly, add about 3 cm³ of copper(II) sulphate solution (which acts as a catalyst) and then add enough dilute sulphuric acid so that the bottom end of the funnel tube is covered.

Let the reaction proceed for about 2 minutes to give time for air to be displaced from the flask, then start the investigation. (If at any time the reaction becomes too slow, more acid may be added via the funnel.)

Fill a test-tube with water, put your finger over the end and invert it over the delivery tube as shown in Fig. 18.1.

When the tube is full of gas close the end with your finger before removing the tube from the water. Keeping the tube closed, take it several feet away from the apparatus (to avoid accidents), open the tube and hold the mouth

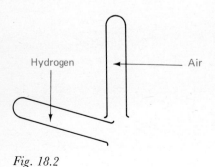

Fig. 18.2

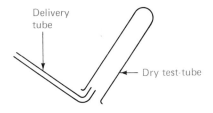

Delivery tube

Dry test-tube

Fig. 18.3

of the tube near a Bunsen flame. Note what you see and hear. This is the characteristic test for hydrogen.

2. Collect another test-tube full of the gas and try to pour the hydrogen upwards into a test-tube of air, as shown in Fig. 18.2. Find whether hydrogen has been transferred to the second test-tube by opening it close to a Bunsen flame as in part 1 of this investigation.

Question

1 What can you conclude about the density of hydrogen compared to that of air?

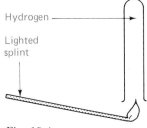

Hydrogen

Lighted splint

Fig. 18.4

3. Take away the dish of water and let the hydrogen flow from the delivery tube into a dry test-tube held at an angle of about 45° as shown in Fig. 18.3.

Let the gas flow for about twice as long as it took to collect a tube-fill over water in the previous experiment. Then lift the test-tube off the delivery tube, closing it with your finger as you do so. Take it away from the apparatus. Invert the tube and then, keeping your fingers away from the open end of the tube, hold a lighted splint to the mouth of the tube (Fig. 18.4).

Questions

2 With which part of air is the hydrogen likely to be combining when it burns?

3 What is the liquid left in the test-tube after burning hydrogen?

4 How could you prove what the liquid is?

18.2 Occurrence and importance of hydrogen

Hydrogen is the gas which is displaced from water by calcium (17.4), from steam by magnesium (17.5) and from dilute hydrochloric or sulphuric acid by zinc. Hydrogen gas is present on the Sun and on many other stars but there is very little free hydrogen on the Earth and there is almost none at all in the Earth's atmosphere. Most of the hydrogen on the Earth is combined with other elements such as with oxygen in water. It is also found in a combined form in all living things and in substances such as oil and natural gas which have been formed from living things.

If a test-tube of a colourless gas ignites with a pop then it is likely to be hydrogen. The other distinctive property of hydrogen is that it has a very low density. It is about fifteen times less dense than air and for this reason it was used in early balloon travel. It was a balloon filled with hydrogen which first carried men by air across the English Channel. A Frenchman, J. P. Blanchard, and an American, John Jeffries, made the crossing in 1785 and to prevent the balloon from falling into the sea during the flight they had to reduce their load by throwing most of their belongings (including Blanchard's trousers) overboard.

If you think back to the test for hydrogen which was mentioned above, you will realise that the use of hydrogen in balloons or airships has a serious disadvantage. This point was brought home to the world by disasters such as that which struck the giant airship *Hindenburg*. This airship, which was 240 m long and contained about 140 000 m³ of hydrogen, burst into flames at Lakehurst, U.S.A. in 1937. When you see how readily a test-tube full of hydrogen ignites, you can begin to appreciate the danger of having an airship filled with thousands of cubic metres of the gas. Many people lost their lives in the *Hindenburg* disaster and since about that time balloon travel has been confined to the use of helium-filled balloons or airships, or hot air balloons. Helium is

(a)

(b)   (c)

Fig. 18.5 The use of different gases for balloon travel (a) The Hindenburg, which was filled with hydrogen (Courtesy The Science Museum) (b) This balloon uses hot air. The heat is obtained by burning a 'bottled' gas such as butane. (Courtesy J. Walter Thompson Co for Rank Hovis Ltd). (c) The first successful balloon crossing of the Atlantic was by the helium-filled Double Eagle in 1978. (Courtesy of the Press Association Ltd)

only about seven times less dense than air, but it does not burn. Hot air balloons fly because hot air is less dense than air at normal atmospheric temperature.

The fact that hydrogen is no longer used for balloons does not mean that the gas is no longer important. It is used in enormous quantities in the chemical industry, in particular for making ammonia which is itself very important for making fertilisers (33.10). Hydrogen is also used in converting vegetable oils into margarine (31.9). Hydrogen has been brought to the public notice because of the hydrogen bomb. The mechanism of the bomb involves the combination of different types of hydrogen atom (isotopes, 13.5) to form helium. This change gives off enormous quantities of energy and is similar to the processes by which the energy in the forms of light and heat are produced on the Sun. The hydrogen bomb has nothing to do with the burning of hydrogen.

18.3 Laboratory preparation of hydrogen

In Chapter 17 several reactions which produce hydrogen were mentioned. The reaction which is most suitable for preparing and collecting a sample of hydrogen in the laboratory is the one which produces the gas at a reasonable rate without the application of heat. Zinc with dilute sulphuric acid is usually used. A trace of copper(II) sulphate solution is added to the zinc at first as this speeds up the reaction. (Some copper is displaced from the solution and deposited on the zinc. This results in the zinc being attacked more rapidly as it is the more reactive of the pair of metals present, 20.4.)

Test-tubes full of the gas can be obtained by collecting it over water as shown in Fig. 18.1. The equation for the reaction occurring is,

$$Zn(s) + H_2SO_4(aq) \rightarrow ZnSO_4(aq) + H_2(g)$$

or
$$Zn(s) + 2H^+(aq) \rightarrow Zn^{2+}(aq) + H_2(g)$$

Owing to the explosive nature of mixtures of the gas with air, all flames must be kept well away from the apparatus. If you wish to carry out a test for the presence of hydrogen, a test-tube containing the gas should be either stoppered or covered with your finger and taken well away from the apparatus before testing with a lighted splint.

The reaction of a metal with an acid is too expensive to be a suitable method for preparing the gas on a large scale. As the hydrogen is to be used in the preparation of ammonia for fertilisers and in the conversion of oils into margarine which are to be sold, it is important that the cost of producing the hydrogen is kept as low as possible. It is for this reason that the methods of producing hydrogen have changed over the years as different processes have been developed and different raw materials have become available.

From coal

In Germany in 1913, just before the First World War, the demand for hydrogen increased dramatically in order to produce ammonia by the Haber Process, 33.9. The ammonia was used to make nitric acid which was then used for making explosives. Prior to this nitric acid was made from sodium nitrate from South America. The hydrogen for the Haber Process was produced by passing steam over white hot coke (previously made from coal). The resulting mixture of carbon monoxide and hydrogen is known as **water gas**:

$$H_2O(g) + C(s) \rightarrow \underbrace{CO(g) + H_2(g)}_{\text{water gas}}$$

Water gas was then mixed with more steam and passed over a heated iron catalyst which resulted in the carbon monoxide being converted into carbon dioxide, and the formation of more hydrogen:

$$CO(g) + H_2O(g) \rightarrow CO_2(g) + H_2(g)$$

This reaction is known as the **shift reaction**. Carbon dioxide is removed by dissolving it in water under pressure. The reaction which forms water gas absorbs heat (endothermic, 22.5) and would eventually stop when the temperature of the coke becomes too low. For this reason it is necessary to pass air over the coke at intervals so that the partial combustion of the coke reheats it. The mixture of gases produced by this change is called **producer gas**:

$$\underbrace{O_2(g) + 4N_2(g)}_{\text{air}} + 2C(s) \rightarrow \underbrace{2CO(g) + 4N_2(g)}_{\text{producer gas}}$$

From petroleum

Owing to the relatively high operating costs of the water gas process and the high cost of coal, it was gradually replaced by a process which is based on petroleum.

Around the 1950s a method of converting the part of petroleum oil known as naphtha (31.5) into a mixture of carbon monoxide and hydrogen was developed by I.C.I. in the United Kingdom. Naphtha is a mixture of hydrocarbons such as C_6H_{14} and C_7H_{16}. The process which is known as steam reforming involves reacting naphtha with steam at about 900 °C. The reaction for one of the hydrocarbons is:

$$C_6H_{14}(g) + 6H_2O(g) \rightarrow 6CO(g) + 13H_2(g)$$

This mixture of products, which is called **synthesis gas** as it can be used for synthesising other important chemicals, is converted into carbon dioxide and more hydrogen by the same shift reaction as was used in the water gas process. Carbon dioxide is removed by dissolving in water as above.

From natural gas

From the 1930s in the U.S.A. natural gas (mainly methane, CH_4), which was plentiful there, was used to prepare a mixture of carbon monoxide and hydrogen. Since the discovery of natural gas under the North Sea it has become more economical in the U.K. to use methane as the main source of the mixture of carbon monoxide and hydrogen:

$$CH_4(g) + H_2O(g) \rightarrow CO(g) + 3H_2(g)$$

The mixture is then reacted with more steam in the shift reaction, and the resulting carbon dioxide is removed as in the other methods of making hydrogen.

In a particular country, the method selected for producing hydrogen will depend on which of the possible raw materials is most readily available.

18.5 Properties of hydrogen

Hydrogen is insoluble in water, colourless, and has no smell. At any particular temperature and pressure the density of hydrogen is less than that of any other gas measured at the same temperature and pressure.

It combines directly with a number of non-metallic elements.

Reactions with non-metals

When a mixture of hydrogen and oxygen is kept at room temperature there is no reaction, but the heat from a spark or a burning splint is sufficient to start the reaction, which then becomes so rapid that it is explosive.

A stream of pure hydrogen will burn quietly in air with a pale blue flame. If the flame is kept near a cold surface (10.8), the product of the burning condenses as a colourless liquid. This liquid turns white anhydrous copper(II) sulphate blue, showing that it contains water, and if sufficient of the liquid is collected, it will be found to boil at $100\,^\circ$C and freeze at $0\,^\circ$C showing that it is water. The equation for the burning is:

$$2H_2(g) + O_2(g) \rightarrow 2H_2O(l)$$

Hydrogen will also combine with oxygen which is already combined with other elements. For example, hydrogen will remove the oxygen from oxides of lead and copper forming the metal in each case. (This process is called reduction, 21.4.)

e.g.
$$CuO(s) + H_2(g) \rightarrow Cu(s) + H_2O(g)$$
$$PbO(s) + H_2(g) \rightarrow Pb(s) + H_2O(g)$$

Investigation 4.4 gives instructions and shows a diagram of the apparatus required to demonstrate these reactions. More will be said about the production of metals from their oxides in Chapter 24.

A mixture of hydrogen and chlorine kept at room temperature also appears not to react, but in this case the reaction requires even less energy to cause it to explode. The energy of the light (23.14) from burning magnesium or even direct sunlight is sufficient to start the reaction and cause the mixture to explode. For this reason mixtures of hydrogen and chlorine should not be prepared in school laboratories. The product of the reaction is hydrogen chloride:

$$H_2(g) + Cl_2(g) \rightarrow 2HCl(g)$$

Under other conditions the reaction can be controlled and is used to manufacture hydrogen chloride which is used to make products such as polyvinyl chloride (PVC).

With nitrogen, even though it is unreactive compared to chlorine and oxygen, hydrogen reacts to form ammonia. This reaction is the basis of the Haber Process for the production of ammonia which is discussed in detail in 33.9.

18.6 Summary

1. Hydrogen is very important in the chemical industry as it is used in the production of important substances such as margarine, ammonia for fertilisers, and hydrogen chloride for polyvinyl chloride.
2. It is prepared in the laboratory by the reaction of a dilute acid (sulphuric) with a metal (zinc).
3. The main industrial source of hydrogen in the U.K. is now natural gas.
4. Hydrogen reacts explosively with some non-metals, in particular oxygen and chlorine.

Chapter 19

The products formed by passing electricity through substances

**Investigation
19.1**

What patterns are there in the products formed during the electrolysis of aqueous solutions of electrolytes?

In all of the following investigations an electrolysis cell similar to that in Fig. 19.1 should be used. This cell is suitable for passing electricity through aqueous solutions of electrolytes and collecting any gaseous products which are formed.

Fig. 19.1

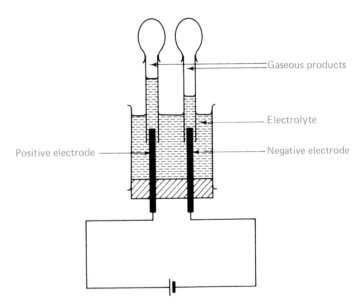

Gaseous products

Electrolyte

Positive electrode

Negative electrode

When trying to identify a gas given off during electrolysis, first make a list of the gases it could possibly be and then, if necessary, look up the tests for the gases by consulting the appropriate chapter in this book. Remember there is no point in testing for a gas which could not possibly be formed. For example, chlorine could not be formed from sodium sulphate solution as there is no chlorine in the solution.

You will need, in addition to the cell, a low voltage supply of direct current such as a 6 V dry battery, connecting wires and small glass tubes fitted with rubber teats for collecting the gases.

1. Pour sufficient dilute sulphuric acid into the cell to cover the electrodes.

By means of a rubber teat completely fill a piece of glass tubing with water and invert it over one of the electrodes as shown in Fig. 19.1. Put another tube full of water over the other electrode. Support each tube so that it covers about the top 1 cm of the electrode.

Connect each electrode to the power supply and when sufficient gas has been produced to test, try to identify the gases. Make a note of which gas is produced at the positive electrode and which at the negative.

After washing the cell out, repeat the procedure using dilute sodium hydroxide solution as the electrolyte.

Question

The ions present in each solution are as follows:

dilute sulphuric acid solution,

H^+(aq) and OH^-(aq) from H_2O

H^+(aq) and SO_4^{2-}(aq) from H_2SO_4

dilute sodium hydroxide solution,

H^+(aq) and OH^-(aq) from H_2O

Na^+(aq) and OH^-(aq) from NaOH

The same product is formed at the positive electrode in both experiments. When sodium hydroxide solution is the electrolyte, the gaseous product must have been formed by discharging hydroxide ions as these are the only negative ions present. When dilute sulphuric acid is the electrolyte it is also the hydroxide ions which are discharged, in this case in preference to the sulphate ions.

When H^+(aq) and Na^+(aq) ions are present in the same solution, which ion is discharged in preference to the other?

2. Write down the formulae of the ions present in each of the following solutions and then, using the knowledge gained in the first part of this investigation, predict what the products are likely to be when each solution is electrolysed:

sodium sulphate, Na_2SO_4
potassium hydroxide, KOH
potassium sulphate, K_2SO_4

Test your predictions for as many of the electrolytes as time permits.

Investigation 19.2

What are the products formed when solutions of chlorides are electrolysed?

You will need the same apparatus as used in Investigation 19.1 (Fig. 19.1).

Pour concentrated sodium chloride solution into the cell until the electrodes are covered. Fill the glass tubes with the sodium chloride solution (rather than water) and support them over the electrodes.

Connect each electrode to the power supply and try to identify the gases which are produced. Make a note of which gas is produced at the positive electrode and which at the negative.

Repeat the procedure using concentrated potassium chloride solution instead of sodium chloride solution.

Questions

1   There are two negative ions present in each solution, OH^-(aq) and Cl^-(aq). When a concentrated solution of a chloride is electrolysed, which of these two ions is discharged in preference to the other?
2   When very dilute sodium chloride solution is electrolysed, oxygen is the main product at the anode. Using this information and that which you obtained from Investigation 19.1, write down the ions, Cl^-(aq), OH^-(aq) and SO_4^{2-}(aq) in order of decreasing ease of discharge from dilute solutions of the ions.

What are the products formed when copper(II) sulphate solution is electrolysed using carbon electrodes?

Using the electrolysis cell shown in Fig. 19.1 pass a direct electric current through a solution of copper(II) sulphate solution. In addition to trying to identify any gases which are formed, carefully observe what happens to the electrodes.

Questions

1   The positive ions present in the solution are $H^+(aq)$ and $Cu^{2+}(aq)$. Which ion is discharged in preference to the other?
2   What must be happening to the concentration of the copper ions in the solution? What will eventually happen to the colour of the solution?
3   Using the information gained from this investigation and Investigation 19.1, write down the ions, $Cu^{2+}(aq)$, $Na^+(aq)$ and $H^+(aq)$ in order of decreasing ease of discharge from dilute solutions of the ions.

What happens when copper(II) sulphate solution is electrolysed using copper electrodes?

You will need two copper electrodes and a source of direct current similar to that used in Investigation 19.1.

Weigh two clean copper foil electrodes. Dip the electrodes into a solution of copper(II) sulphate solution in a small beaker (Fig. 19.2) and pass a small direct current through the solution. Note which electrode is connected to the positive terminal and which to the negative.

Fig. 19.2

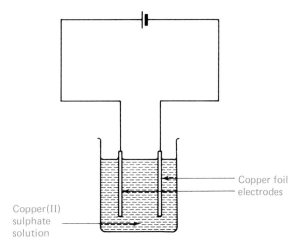

Copper foil electrodes

Copper(II) sulphate solution

After about 10 minutes, disconnect the electrodes, carefully remove them from the solution and wash them by gently dipping them firstly into a beaker of pure water and then into a beaker of propanone.
When the electrodes are dry, weigh each one again.

Questions

1   Bearing in mind that conduction by an electrolyte involves taking electrons from the cathode and giving them up to the anode (19.6), suggest equations for the changes which have occurred at each electrode.
2   During electrolysis the number of electrons transferred at each electrode must be identical. What will happen to the concentration of the copper ions in the solution during electrolysis?

19.5 Electrolysis

This chapter is concerned with the identity of the products and how they are formed when a direct current is passed through substances. Chapter 6 concentrated on what can be learnt about substances by passing electricity through them. It was noted that when an electrolyte conducts electricity, the positively charged ions (cations) which are present are attracted to the negative electrode (cathode) and the negatively charged ions (anions) to the positive electrode (anode).

Some of the ions which are attracted to the electrodes are converted into neutral atoms or molecules. For example, when electricity is passed through molten lead(II) bromide, lead is produced at the cathode and bromine at the anode. Thus the compound lead(II) bromide is separated into its two components, which means that a chemical change has occurred.

Fig. 19.3 The conduction of electricity by molten lead(II) bromide

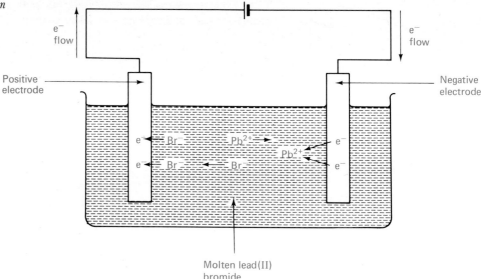

The process of producing chemical changes by passing a direct electric current through electrolytes is known as **electrolysis**.

Electrolysis is of considerable economic importance as many essential materials such as aluminium and chlorine are obtained by electrolytic methods. The plating of objects with metals such as chromium and silver is also carried out by electrolysis.

19.6 Electrode reactions

As the reactions which occur at the electrodes stop as soon as the current is switched off, it seems likely that the reactions play an essential part in the conduction of electricity by the electrolyte. A current of electricity through a metal is a flow of electrons, so the electrode reactions must be helping to maintain the flow of electrons through the wires connecting the electrodes to the battery or other source of electricity.

Lead(II) bromide contains lead ions each with a relative charge of $2+$ (i.e. Pb^{2+}) and bromide ions each with a relative charge of $1-$ (i.e. Br^-). The Pb^{2+} ions are changed to neutral atoms of lead (Pb) at the cathode and the Br^- ions are changed to neutral molecules of bromine (Br_2) at the anode.

These changes can be explained by the theory that electrons (which have a relative charge of $1-$ and are represented by e^-) are lost from the negative electrode:

$$Pb^{2+} + 2e^- \rightarrow Pb \tag{1}$$
from the cathode

The two negative charges balance out the two positive charges on the lead ion which is thus neutralised (or discharged).

At the positive electrode each bromide ion loses one electron to the electrode:

$$Br^- - e^- \rightarrow Br \qquad (2)$$
to the
anode

As Br^- loses one negative charge it becomes neutralised or discharged. Two of these neutral atoms then combine to form a bromine molecule, Br_2:

$$2Br \rightarrow Br_2$$

The battery is trying to push electrons towards the negative electrode and pull them away from the positive electrode. Electrode reaction (1) helps the former process and electrode reaction (2) helps the latter process. Fig. 19.3 shows how electrode reactions (1) and (2) are helping to maintain the flow of elctrons around the circuit.

Whenever an electrolyte conducts electricity there must be an electrode reaction using up electrons at the negative electrode and another electrode reaction giving up electrons at the positive electrode.

19.7
Moles of
electrons

The number of electrons transferred at the cathode must be equal to the number transferred at the anode. The actual numbers involved are very large so, as in the case of atoms, ions or molecules, it is more convenient to refer to moles of electrons rather than actual numbers.

In the example above, 2 moles of electrons are required to neutralise 1 mole of Pb^{2+} ions, but the transfer of 2 moles of electrons at the anode will result in the neutralisation of 2 moles of bromide ions. That is, for every mole of lead atoms produced there will be 2 moles of bromine atoms produced. This ratio is consistent with the formula of lead(II) bromide, $PbBr_2$. 2 moles of bromine atoms formed will combine together to form 1 mole of bromine molecules:

$$2Br \rightarrow Br_2$$

As the minimum relative charge on any ion is $1+$ or $1-$, and the relative charge on an electron is $1-$, the minimum number of electrons required to neutralise 1 mole of ions is 1 mole of electrons.

The transfer of 1 mole of electrons through a circuit is equivalent to the passage of the minimum quantity of electricity required to neutralise 1 mole of ions. This quantity of electricity is approximately 96 500 coulombs and is known as a Faraday (6.11). In the above example:
1 mole of Pb^{2+} ions requires 2 moles of electrons which is equivalent to 2 Faradays or
  $2 \times 96\,500$ coulombs of electricity, and
1 mole of Br^- ions requires 1 mole of electrons which is equivalent to 1 Faraday or
  96 500 coulombs of electricity.

19.8
Products from
molten
electrolytes

The products which are observed to be formed during the electrolysis of molten electrolytes are easily explained. If the electrolyte is a pure compound, there is only one type of positive ion present and one type of negative ion present. Therefore there is only one possible product at each electrode and if the relative charges on the ions are known, the electrode reactions can be worked out.

The explanations of some examples are given below.

Electrolyte: molten sodium chloride
  Electrodes: carbon
  Ions present: Na^+ and Cl^-

175

At the cathode: Na^+ ions are attracted and converted into neutral atoms of sodium by removing electrons from the electrode:

$$Na^+ + e^- \rightarrow Na$$

At the anode: Cl^- ions are attracted and converted into neutral atoms of chlorine by giving up electrons to the electrode:

$$Cl^- - e^- \rightarrow Cl$$

Chlorine atoms then combine to form chlorine molecules:

$$2Cl \rightarrow Cl_2$$

The commercial application of this process for the production of sodium and chlorine is described in 24.7.

19.9 Products from aqueous solutions of electrolytes

Although pure water is not a good conductor of electricity it does conduct slightly as it does contain some ions. The formation of hydrogen ions and hydroxide ions from water is represented by the equation:

$$H_2O(l) \rightleftharpoons H^+(aq) + OH^-(aq)$$

(The $\rightleftharpoons$ sign indicates that the change is reversible)

The concentration of ions in pure water is very small. At 25 °C there are 10^{-7} moles of H^+ ions and 10^{-7} moles of OH^- ions per dm^3 of pure water. Thus, for example, 1 mole of potassium chloride dissolved in water to form 1 dm^3 of solution will provide about 10 million times more ions than is provided by the water in which it is dissolved.

When you try to predict the products of the electrolysis of aqueous solutions, the ions from the water create problems as the solution may contain two types of positive ions (e.g. H^+ and Na^+) and two types of negative ions (e.g. OH^- and Cl^-).

When two types of positive ions are attracted to the negative electrode, one type of ion tends to be neutralised in preference to the other. This also happens when two types of negative ions are attracted to the positive electrode. Which ion is preferentially discharged can depend on three factors:

1. The nature of the ions themselves; that is, some ions gain or lose electrons more readily than others.
2. The relative concentrations of the ions. This is an important factor when Cl^- ions and OH^- ions are present together.
3. The substances from which the electrodes are made.

The products which are observed to be formed during the electrolysis of some common aqueous solutions are explained below.

Electrolyte: dilute sulphuric acid

Electrodes: platinum or carbon
Ions present: $H^+(aq)$ and $OH^-(aq)$ from the water
                  $H^+(aq)$ and $SO_4^{2-}(aq)$ from the sulphuric acid

At the cathode: $H^+(aq)$ ions, which are the only positively charged ions present, are attracted to the electrode and neutralised to form hydrogen molecules:

$$H^+(aq) + e^- \rightarrow H$$

$$2H \rightarrow H_2(g)$$

The two equations can be combined:

$$2H^+(aq) + 2e^- \rightarrow H_2(g)$$

At the anode: $OH^-(aq)$ ions and $SO_4^{2-}(aq)$ ions are attracted. $OH^-(aq)$ ions are more easily discharged forming oxygen gas:

$$OH^-(aq) \quad - \quad e^- \quad \rightarrow \quad OH$$

$$4OH \rightarrow 2H_2O(l) + O_2(g)$$

The two equations can be combined:

$$4OH^-(aq) - 4e^- \rightarrow 2H_2O(l) + O_2(g)$$

The equations for the electrode reactions show that 1 mole of hydrogen molecules are produced by the transfer of 2 moles of electrons, whereas 1 mole of oxygen molecules requires the transfer of 4 moles of electrons. The number of moles of electrons transferred at each electrode must always be equal, thus the number of molecules, and hence the volume, of hydrogen produced will be twice that of oxygen.

Electrolyte: dilute sodium hydroxide solution

Electrodes: platinum
Ions present: $H^+(aq)$ and $OH^-(aq)$ from the water
               $Na^+(aq)$ and $OH^-(aq)$ from the sodium hydroxide

At the cathode: $H^+(aq)$ ions and $Na^+(aq)$ ions are attracted; $H^+(aq)$ ions are the more easily discharged and hydrogen is formed by the same electrode reactions as in the previous example.

At the anode: $OH^-(aq)$ ions are attracted and are discharged by the same electrode reactions as in the previous example and oxygen is given off.

The products formed by the electrolysis of sodium sulphate solution can be predicted to be hydrogen and oxygen by comparing the ions present in this solution with the ions present in the solutions used in the previous two examples.

Electrolyte: sodium chloride solution

Electrodes: carbon
Ions present: $H^+(aq)$ and $OH^-(aq)$ from the water
               $Na^+(aq)$ and $Cl^-(aq)$ from the sodium chloride

At the cathode: $H^+(aq)$ and $Na^+(aq)$ are attracted and as in the example of sodium hydroxide solution, hydrogen is given off.

At the anode: $OH^-(aq)$ and $Cl^-(aq)$ are attracted. If the solution is very dilute, $OH^-(aq)$ ions are discharged in preference to $Cl^-(aq)$ ions and oxygen is formed as in the previous two examples.

When the solution is more concentrated the main electrode reaction is the discharge of the chloride ions:

$$Cl^-(aq) \quad - \quad e^- \quad \rightarrow \quad Cl$$

$$2Cl \rightarrow Cl_2(g)$$

These two equations can be combined:

$$2Cl^-(aq) - 2e^- \rightarrow Cl_2(g)$$

The electrolysis of concentrated sodium chloride solution is an extremely important industrial process (29.7). Not only does it produce hydrogen and chlorine, but also in the region of the negative electrode the removal of $H^+(aq)$ ions results in a build up of an excess of $OH^-(aq)$ ions. These ions, together with $Na^+(aq)$ ions which are attracted to the electrode, constitute the ingredients of a solution of sodium hydroxide. This is an important substance used in other branches of the chemical industry such as paper manufacture.

Electrolyte: copper(II) sulphate solution

Electrodes: platinum or carbon

Ions present: $H^+(aq)$ and $OH^-(aq)$ from water

$Cu^{2+}(aq)$ and $SO_4^{2-}(aq)$ from copper(II) sulphate

At the cathode: $H^+(aq)$ and $Cu^{2+}(aq)$ are attracted. $Cu^{2+}(aq)$ ions are more easily discharged and copper metal is deposited on the electrode:

$$Cu^{2+}(aq) + 2e^- \rightarrow Cu(s)$$

At the anode: $OH^-(aq)$ and $SO_4^{2-}(aq)$ are attracted, which is exactly the same as in the case of the electrolysis of dilute sulphuric acid with platinum or carbon electrodes, and therefore oxygen is produced by the same electrode reactions.

As the electrolysis proceeds, the concentration of $Cu^{2+}(aq)$ ions in the solution will decrease and hence the blue colour will fade. Also as $OH^-(aq)$ ions are removed from the solution, there will be a build up of $H^+(aq)$ ions and the final solution when all the copper ions have been removed will be dilute sulphuric acid.

If the positive electrode (anode) is made of copper rather than platinum or carbon, the electrode reaction which occurs most readily is that in which the copper atoms of the copper electrode itself change into copper ions which pass into solution. The ions are formed by electrons being removed from the atoms. The electrons are left on the copper electrode and hence the reaction satisfies the condition that electrons must be given up to the anode during electrolysis:

$$Cu(s) \quad - \quad 2e^- \quad \rightarrow \quad Cu^{2+}(aq)$$
$$\text{to the}$$
$$\text{anode}$$

In this case the concentration of copper ions in the solution will remain constant. For every copper ion which is deposited as a copper atom at the cathode, a copper atom will dissolve as a copper ion at the anode.

Fig. 19.4 Electrolytic purification of copper

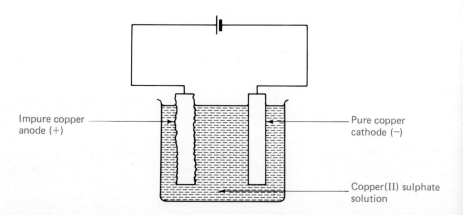

Impure copper anode (+)

Pure copper cathode (−)

Copper(II) sulphate solution

This example of electrolysis is particularly important as it is used for purifying copper during the final stage of the production of the metal from its ores. Impure copper is made the anode and a piece of pure copper the cathode (Fig. 19.4). Copper(II) sulphate solution is the electrolyte and during the electrolysis, copper is dissolved at the anode and deposited as pure copper at the cathode. Impurities in the anode which are soluble pass into solution and those which are insoluble collect as a sludge at the bottom of the container.

178

Fig. 19.5 Copper
purification in progress.
The photograph shows
rows of impure copper
anodes and pure copper
cathodes. (Courtesy RTZ
Services Ltd)

19.10
Patterns in the
products

It has already been pointed out that in certain cases the products of electrolysis can be changed by changing the concentration of the electrolyte or by changing the substances from which the electrode is made. Despite these complications, there are some patterns in the types of products formed which are worth noting.

1. At the cathode either a metal or hydrogen is produced.

2. At the anode either a non-metal other than hydrogen is produced or the metal electrode dissolves.

3. When a mixture of ions is present, there is an order of preferential discharge which, in the case of the positive ions, resembles the reverse order of the reactivity series (17.9). For positive ions, starting with the most easily discharged, the order is $Cu^{2+}(aq)$, $H^+(aq)$, $Na^+(aq)$.

For negative ions the order is $OH^-(aq)$, $Cl^-(aq)$, $SO_4^{2-}(aq)$. Remember that the order of $OH^-(aq)$ and $Cl^-(aq)$ is reversed for higher concentrations of $Cl^-(aq)$.

19.11
Uses of
electrolysis

1. **Production of useful substances**

The process of electrolysis is used in the production of the more reactive metals such as sodium and aluminium (24.7 and 24.8) for which straightforward chemical methods of extraction are either not suitable or not possible.

The electrolysis of concentrated sodium chloride solution is an essential stage in the production of chlorine, hydrogen, hydrochloric acid, sodium hydroxide and sodium hypochlorite from rock salt (29.7).

2. **Electroplating**

The chromium plating on parts of motor cars is produced by electrolysis. Chromium is more resistant to corrosion than steel and it can be polished to produce an attractive shiny surface. However, chromium is much more expensive than steel and so it is usual to cover the steel with a very thin layer of chromium. Electrolysis is particularly effective at depositing a thin, even layer of metal on an object. In this case the steel is first electroplated with a layer of copper or nickel on which the chromium sticks better than it does on steel. The object such as a car bumper or headlight trim is made the cathode and pure chromium is used as the anode. The electrolyte is a solution of chromium(VI) oxide in dilute sulphuric acid. The process which occurs is similar to that used for the purification of copper. The metal is dissolved at the anode and deposited on the cathode.

The layer of tin on the 'tin' cans used as food containers is usually electroplated on to the steel by making the steel the cathode, pure tin the anode and using an electrolyte containing $Sn^{2+}(aq)$ ions.

Tin is more resistant to corrosion than steel but it would be too expensive to make the containers entirely of tin. By similar processes, cutlery which is marked EPNS has been electroplated with nickel and then with silver.

3. **Anodising**

Aluminium which has been exposed to the atmosphere is covered with a layer of

Fig. 19.6 These motor car wheel rims have been nickel and chrome plated by being automatically dipped into a series of electrolytic cells and washing liquids. (Courtesy W. Canning Engineering Ltd)

Fig. 19.7 These aluminium wheel trims have been anodised. (Courtesy RTZ Services Ltd)

aluminium oxide, Al_2O_3, which protects the metal from further corrosion. This layer of oxide can be thickened and its protective powers increased by making the aluminium the anode with dilute sulphuric acid as the electrolyte. $OH^-(aq)$ ions are neutralised at the anode and oxygen is produced. Some of this oxygen is used up in oxidising the surface of the aluminium anode.

The oxide layer at this stage readily absorbs dyes which can then be permanently sealed into the layer by placing the object in boiling water. This process is used to produce brightly coloured, corrosion resistant articles such as saucepan lids.

19.12 Summary

1. Electrode reactions enable an electrolyte to conduct electricity by removing electrons from the negative electrode (cathode) and giving up electrons to the positive electrode (anode).
2. The transfer of 1 mole of electrons at each electrode is equivalent to the passage of 1 Faraday (96 500 coulombs) of electricity through the circuit.
3. The products of some of the more important examples of electrolysis are given in Table 19.1.

TABLE 19.1. Products of some examples of electrolysis

ELECTROLYTE		ELECTRODES		PRODUCTS	
		CATHODE	ANODE	CATHODE	ANODE
molten	lead(II) bromide	carbon	carbon	lead	bromine
	sodium chloride	carbon	carbon	sodium	chlorine
	potassium iodide	carbon	carbon	potassium	iodine
aqueous solutions	sulphuric acid (dilute)	platinum or carbon	platinum or carbon	hydrogen	oxygen
	sodium hydroxide (dilute)	platinum	platinum	hydrogen	oxygen
	sodium chloride (very dilute)	carbon	carbon	hydrogen	oxygen
	sodium chloride (concentrated)	carbon	carbon	hydrogen	chlorine
	copper(II) sulphate	platinum or carbon	platinum or carbon	copper	oxygen
	copper(II) sulphate	copper	copper	copper deposited	copper dissolved

Chapter 20

Using substances to produce electricity

Investigation 20.1

How can we obtain electricity from chemical reactions?

You will need a voltmeter reading 0–3 V, two wires fitted with crocodile clips, and clean pieces of magnesium ribbon, copper foil and zinc foil. You will also need a small beaker containing dilute (0·5 M) sulphuric acid.

1. Using the wires with clips, connect a piece of magnesium ribbon to one terminal of the voltmeter and a piece of copper foil to the other, as shown in Fig. 20.1. Now dip both the metals into the dilute sulphuric acid, without letting them touch each other, and note whether you get a reading on the voltmeter.

Fig. 20.1

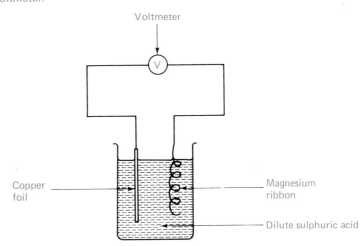

Voltmeter

Copper foil

Magnesium ribbon

Dilute sulphuric acid

2. Replace the copper foil with another piece of magnesium ribbon, dip both pieces of magnesium into the acid, and note whether you get a reading on the meter.

Now replace both pieces of magnesium ribbon with pieces of copper foil, and dip the two pieces of copper into the acid. Do you get a reading this time?

3. Replace one of the pieces of copper foil with a piece of zinc foil, but have the copper connected to the same terminal of the meter as in part 1 of the investigation. Dip both metals into the acid and see whether you get a reading on the meter.

4. Keep the metals the same as in part 3, but now try different solutions in the beaker. Possible ones to try are sodium sulphate solution, sodium chloride solution, and sugar solution. Make a note of those cases where you get a reading on the meter, and those cases (if any) where you do not.

Questions

1 In order to produce electricity by dipping two pieces of metal into a solution, do you need (a) two similar metals or two different metals, and (b) should the solution contain an electrolyte or a non-electrolyte?

2 When you have a cell producing electricity, does the less reactive metal form the positive pole or the negative pole?

3 Did the copper/zinc cell give the same reading on the voltmeter as the copper/magnesium cell, or was it higher or lower?

Investigation 20.2

Can we put metals in an order of reactivity by making cells?

You will need a voltmeter (0–3 V), wires, a beaker containing sodium sulphate solution, and pieces of copper, iron, lead, magnesium and zinc metals.

Connect the copper to the positive terminal of the voltmeter and one of the other metals to the negative terminal. Dip both metals into the sodium sulphate solution and quickly take the highest reading of the voltage.

Keeping copper as the positive pole, try the various other metals in turn as the negative pole. In each case record the highest reading on the voltmeter.

Write down the metals in an order of decreasing voltage.

Questions

1 What connection is there between the order that you have written down, and the order of reactivity of the various metals?

2 Silver is a less reactive metal than copper. In a copper/silver cell, which metal would form the positive pole?

20.3 Sources of electricity

Electricity which is supplied to your home is produced in power stations where heat from burning fuel, such as coal or oil, or from a nuclear reactor (13.7) converts water into steam. The steam is used to turn a generator to produce electricity. In the same way the movement of the back wheel of a bicycle can be used to make a dynamo rotate and produce electricity which can be used to operate the lights of the bicycle.

Not all electricity which is used in everyday life is obtained from mechanical motion. For example, a torch and a transistor radio obtain the electricity they require from dry batteries. Also the lights of motor cars and the spark to ignite the petrol-air mixture in the engine are both dependent on the car's battery which, unlike a dry battery, can be recharged.

The operation of the dry battery and the car battery are both examples of electricity being obtained from chemical reactions rather than mechanical motion.

The first recorded observation of the production of electricity by chemical reactions was by Luigi Galvani, an Italian scientist, who in 1768 was investigating the effect of atmospheric electricity, occurring during thunderstorms, on a frog's legs which were suspended by a brass hook and connected to the ground by iron wires. He noticed that if the brass and iron wires were allowed to come into contact, the frog's legs twitched in the same way as they did during a thunderstorm. Galvani thought this was due to the electricity stored in the dead animal. It was Alessandro Volta, another Italian, who first realised that the electricity was caused by chemical reactions involving the two metals.

Volta repeated the experiment of the Swiss scientist, Suzler, who reported in 1780 that if a piece of lead and a piece of silver are placed on either side of the tongue and then the free ends of the metal brought into contact an unpleasant pungent taste is produced. Volta experimented with different metals and came to the conclusion that to produce electricity two different metals had to be in contact with moisture and also with each other. He observed that some pairs of metals produced bigger effects than other pairs and that if several sets of the same pair of metals are connected in series either in a pile (Voltaic pile) or in a set of containers the effects can be increased.

Volta's pile was the first ever battery to produce a steady electric current. Its use led to the discovery of many chemical effects of electricity. For example, it enabled Sir

Humphry Davy to discover the metals sodium and potassium by electrolysing molten sodium hydroxide and molten potassium hydroxide. The first pile to be constructed in the U.K. consisted of 36 silver coins separated from 36 zinc discs by paper soaked in sodium chloride solution.

20.4 Primary cells

If two different metals are dipped into an electrolyte and the two metals brought into contact, a current of electricity (i.e. a stream of electrons) will flow through the external circuit. Such an arrangement is called a **primary cell**. The presence of the current can be detected by putting a suitable bulb in the external circuit, or a measure of the tendency of the current to flow can be obtained by placing a suitable voltmeter in the circuit, Fig. 20.1.

Volta's experiments with different metals enabled him to arrange the metals in an order which is in fact similar to the reactivity series of the metals (17.9). He found that the further apart the metals were in his series, the greater the electrical effect. Thus the production of electricity by this method seems to depend on one metal being more reactive than the other.

A comparison of the reactivities of the metals is a comparison of the tendencies of the metals to form ions. That is, in the example of zinc and copper, the tendencies for the following two reactions to occur:

$$Zn(s) \rightarrow Zn^{2+}(aq) + 2e^- \quad (1)$$
$$Cu(s) \rightarrow Cu^{2+}(aq) + 2e^- \quad (2)$$

Zinc is more reactive than copper and so it would be expected that reaction (1) would occur more readily than (2). In the zinc-copper cell, with copper(II) sulphate as electrolyte, the zinc pole or terminal is negative with respect to the copper showing that it is supplying electrons to the external circuit. The flow of electrons around the external circuit can be explained by some of the zinc atoms from the zinc pole passing into solution as zinc ions and giving up electrons to the external circuit:

$$Zn(s) \rightarrow Zn^{2+}(aq) + \underset{\substack{\text{to the external}\\\text{circuit}}}{2e^-}$$

At the copper pole, electrons are removed from the external circuit by some copper ions in the solution being converted into copper atoms which are deposited on the copper pole:

$$Cu^{2+}(aq) + \underset{\substack{\text{from the}\\\text{external}\\\text{circuit}}}{2e^-} \rightarrow Cu(s)$$

The overall reaction, adding together the two separate electrode reactions is,

$$Zn(s) + Cu^{2+}(aq) \rightarrow Zn^{2+}(aq) + Cu(s)$$

and this is just the reaction which occurs when zinc metal is placed in a solution of a copper(II) salt. Zinc, being the more reactive metal, displaces copper from the solution.

When a cell produces electricity, electrons pass into the external circuit from the more reactive metal and are removed from the external circuit by the less reactive metal.

It is important to realise that the reactions occurring at the two poles or terminals of a cell are **producing** electricity, unlike the electrode reactions occurring during electrolysis which are **using** electricity.

183

The voltage obtained from a simple cell tends to fall quite quickly owing to changes at the surfaces of the poles. For example, a zinc pole dipping into copper(II) sulphate solution will become coated with copper due to the displacement reaction which occurs even when the cell is not producing electricity:

$$Zn(s) + Cu^{2+}(aq) \rightarrow Cu(s) + Zn^{2+}(aq)$$

Also whenever hydrogen is produced at a pole the pole becomes coated with small bubbles of the gas, and this has the effect of changing the nature of the pole and hence the voltage obtained from the cell. These changes in the poles of a cell are called **polarisation** and it is necessary to mimimise polarisation if the efficiency of the cell is to be maintained.

In the case of the zinc-copper cell polarisation of the zinc pole can be stopped by separating the cell into two halves. This can be done by using two beakers connected by a 'salt bridge' made of a strip of filter paper soaked in potassium nitrate solution, Fig. 20.2(a). Alternatively, the two half-cells can be separated by a porous pot. This second method is used in the Daniell cell, Fig. 20.2(b), which will produce a fairly steady maximum voltage of about 1·1 volts for quite a long time.

Fig. 20.2 Two forms of the zinc-copper cell

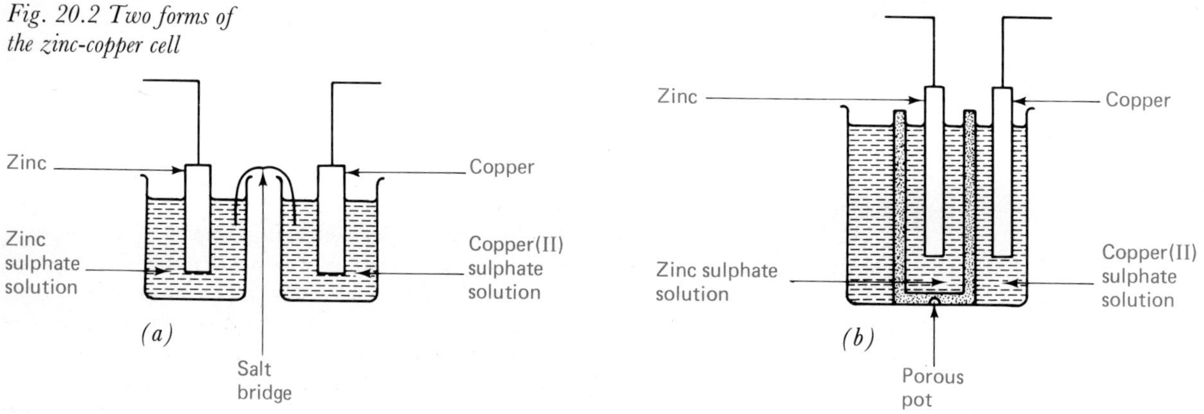

20.6
Dry batteries

The disadvantage of the Daniell cell is that it contains liquids. It is easy to visualise the problems associated with running a transistor radio from Daniell cells. The dry battery, although not entirely dry, does overcome the problems. Despite its very different appearance, the dry battery works on similar principles to the cells described so far. Fig. 20.3 shows the main components of a normal cylindrical dry battery. The zinc outer casing acts as one pole and is separated from the other pole by the electrolyte, which is a paste containing ammonium chloride. The second pole consists of a mixture of manganese(IV) oxide and carbon powder surrounding a carbon rod.

The zinc is the negative terminal of the cell by acting as a supplier of electrons:

$$Zn(s) \rightarrow Zn^{2+}(aq) + 2e^-$$

Contact is made with this terminal by pressing the bottom of the battery against a spring contact.

The other pole of the cell must be responsible for using up electrons and it is the manganese(IV) oxide which does this. Carbon powder is there to improve the conductance of the pole, and the carbon rod is there as a means of connecting the pole to the external circuit. The reactions by which the manganese(IV) oxide uses up electrons are complex and not fully understood. The products seem to vary according to how much current is drawn from the cell.

This cell produces about 1·5 or 1·6 volts. Dry batteries which produce larger voltages are made from similar materials, but with several cells arranged on top of one another rather like a Voltaic pile.

Fig. 20.3 A cross section of a dry battery

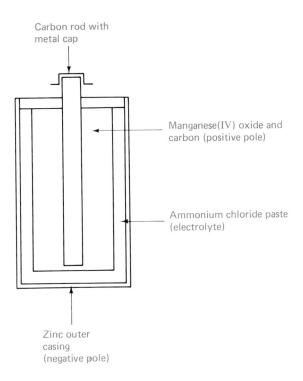

Carbon rod with metal cap

Manganese(IV) oxide and carbon (positive pole)

Ammonium chloride paste (electrolyte)

Zinc outer casing (negative pole)

20.7 Rechargeable batteries

The batteries used in motor cars are sometimes called accumulators or storage cells because, after being used to produce electricity, they can be recharged by passing electricity into them from another source. In a car the electricity for recharging the battery is obtained from a dynamo or alternator. Whenever a car's engine is running, the fan belt which passes round a wheel on the end of the dynamo causes the dynamo to rotate and produce electricity. If the fan belt is too slack or the dynamo is faulty the battery will not be recharged and its voltage will decrease, and the battery will become 'flat'. One pole of the cell consists of a lead plate and the other is a lead grid packed with lead(IV) oxide. The electrolyte is dilute sulphuric acid.

Again the chemical reactions which occur are rather complex. The lead plate is the negative pole so this can be explained by:

$$Pb(s) \rightarrow Pb^{2+}(aq) + 2e^-$$

The lead(IV) oxide plate is the positive pole which uses up electrons and so could be explained by:

$$PbO_2(s) + 4H^+(aq) + 2e^- \rightarrow Pb^{2+}(aq) + 2H_2O(l)$$

Thus both plates produce $Pb^{2+}(aq)$ ions and become coated with lead(II) sulphate, a white solid:

$$Pb^{2+}(aq) + SO_4^{2-}(aq) \rightarrow PbSO_4(s)$$

When the current is passed into the cell during charging the positive pole becomes coated with brown lead(IV) oxide, indicating that the reverse reactions are occurring.

Thus passing electricity into the cell, i.e. electrolysis, converts the cell into a form which is able to produce electricity. This can be demonstrated in the laboratory by electrolysing a dilute solution of sulphuric acid with lead plates as electrodes. When the positive plate has become coated with lead(IV) oxide, connect the plates to a voltmeter; the result shows that the cell is now able to produce electricity.

185

Fig. 20.4 A motor car
storage battery with part
of the case cut away to
show the lead plates.
(Courtesy Chloride
Automotive Batteries Ltd)

Fig. 20.5 Recharging the batteries of milk delivery vehicles at Chesterfield and
District Co-op. (Courtesy Westinghouse Brake and Signal Co Ltd)

Larger storage cells are used to supply all the power for vehicles such as milk floats
and fork-lift trucks. The batteries are recharged by connecting them to a suitable
supply of direct current while the vehicle is not in use.

20.8 Corrosion

Most of the examples of corrosion which you may have noticed are likely to be due to
the rusting of iron, but the word corrosion can be used to describe the processes by
which materials other than iron are eaten away. It can be used for the effect of air
which has been polluted by acidic gases (such as sulphur dioxide) on stone buildings, or
the conversion of the surface of copper, which is used for roofing on some buildings,
into a greenish powder. Sometimes corrosion is useful. For example, a newly cut piece
of aluminium corrodes to form a layer of aluminium oxide on its surface which then
protects the aluminium from further attack. However, the rusting of iron is
undoubtedly the most troublesome and expensive form of corrosion.

In Chapter 9 it was noted that both air and moisture are required for rusting to
occur, but to obtain a deeper understanding of what is happening when an iron object
rapidly corrodes it is necessary to apply some of the electrochemical ideas which have
been discussed earlier in this chapter.

Fig. 20.6 The corrosion
of iron being speeded up
by the presence of copper

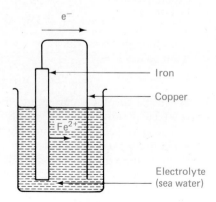

Iron corrodes more rapidly than normal when it is in contact with copper metal. This is why it is unwise to use copper rivets when fixing sheets of iron together in ship-building, or to use iron nails when fixing copper sheets on roofs. Iron corrodes more rapidly because the conditions which are required for setting up a cell have been created. That is, there are two different metals, iron and copper, in contact with each other and also in contact with an electrolyte, i.e. moisture or sea water. The arrangement can be represented in the laboratory as shown in Fig. 20.6.

The iron is the more reactive metal and therefore some of the iron atoms pass into solution as ions, leaving electrons on the piece of iron:

$$Fe(s) \rightarrow Fe^{2+}(aq) + 2e^-$$

The iron is the negative pole of the cell and electrons pass through the iron to the copper where they are used to neutralise $H^+(aq)$ ions which are present in the electrolyte:

$$2H^+(aq) + 2e^- \rightarrow H_2(g)$$

Whenever a cell is set up in which the iron is the more reactive metal, the iron will corrode more rapidly than usual.

Cells also occur during the corrosion of a piece of iron which is not in contact with another metal. One part of the iron becomes a negative pole and another part a positive pole owing to slightly different conditions existing at the two parts. In the case of an iron nail the head or the point, because they have suffered more stress, tend to be the negative regions and the stem of the nail the positive region.

The $Fe^{2+}(aq)$ ions which pass into solution when iron corrodes react with $OH^-(aq)$ ions from the water to form a precipitate of iron(II) hydroxide:

$$Fe^{2+}(aq) + 2OH^-(aq) \rightarrow Fe(OH)_2(s)$$

This precipitate is rapidly oxidised by oxygen dissolved in the water to hydrated forms of iron(III) oxide, Fe_2O_3, which is the red-brown substance known as rust.

20.9 Protection of iron

Fig. 20.7 (left) *The corrosion of iron being speeded up by the presence of tin*
Fig. 20.8 (right) *The corrosion of iron being slowed down by the presence of zinc*

The most obvious way of protecting iron is to cover it with paint, or another metal which is more resistant to corrosion. The metals used are tin and zinc, both of which form protective layers on their surfaces and are more resistant to corrosion than iron. The food containers known as tin cans are made from steel which has been plated with tin (19.11). The tin provides a protective outer covering but the disadvantage of using it is that if it is scratched, so that the steel is exposed to moisture, a cell begins to operate in which the iron is the more reactive metal and hence begins to corrode more rapidly than if the tin was not present.

Fig. 20.7 is a diagram of a magnified scratch in the tin plate, showing that the iron acts as the negative pole in the cell by dissolving as $Fe^{2+}(aq)$ ions and electrons pass through the iron to the tin.

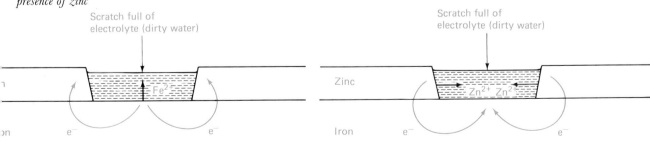

Iron coated with zinc is called galvanised iron and this method of protection has the advantage that if the zinc is scratched and a cell begins to operate, the zinc is the more reactive metal and it is corroded in preference to the iron, Fig. 20.8. Galvanised iron is used in the form of corrugated iron for the construction of storage buildings for farms and factories. It is not used for food containers as solutions of zinc compounds are poisonous.

The idea of the zinc-iron cell, in which a metal higher in the reactivity series than iron is placed in contact with the iron and hence is corroded in preference, is used to protect buried pipes and the steel plates of ships. Pieces of zinc, or better still magnesium which is even more reactive, are fixed to the pipe or ship. This results in cells being created in which the zinc or the magnesium acts as the negative pole and is corroded in preference to the iron:

$$Zn(s) \rightarrow Zn^{2+}(aq) + 2e^-$$

$$Mg(s) \rightarrow Mg^{2+}(aq) + 2e^-$$

$H^+(aq)$ ions are neutralised at the iron pole of the cell. The zinc and the magnesium are sometimes referred to as **sacrificial anodes** and the technique as anodic protection.

Fig. 20.9 Zinc anodes fixed to the hull of a ship to protect it from corrosion below the waterline. (Courtesy AM & S Europe Ltd)

20.10 Summary

1. Two different metals dipping into an electrolyte can act as a source of electricity.
2. The more reactive metal is the negative pole of the cell.
3. The further apart the metals are in the reactivity series, the greater the voltage produced by the cell.
4. Changes which occur on the surfaces of the poles of a cell (i.e. polarisation) can reduce the voltage of the cell.
5. Accumulators or storage cells, unlike dry batteries, can be recharged by passing electricity into them.
6. The corrosion of iron (rusting) is an electrochemical process.
7. A more reactive metal placed in contact with a piece of iron will corrode in preference to the iron and hence protect it from rusting.

Chapter 21

Oxidation and reduction reactions

Investigation 21.1

How can we detect the presence of an oxidising agent?

For a substance to be suitable for a test reagent for an oxidising agent, it must give an observable reaction (such as a colour change), in a very short time, with oxidising agents.

Your teacher will supply you with several solutions, each of which may or may not be an oxidising agent, and a solution which is known to be an oxidising agent.

1. Take about 1 cm³ of a solution of potassium iodide, and add an equal volume of dilute sulphuric acid.
   Now add a few drops of the oxidising agent and shake to mix. The formation of a yellow-brown solution (or perhaps even a black precipitate) indicates that iodine has been liberated from the potassium iodide by the oxidising agent.

Questions

1 Write down the formula of an iodide ion and of an iodine molecule. Are electrons added or removed when iodide ions are converted to iodine molecules?
2 Why is the liberation of iodine from potassium iodide oxidation?

2. Now use this test to find out which of the other solutions are oxidising agents.

Investigation 21.2

How can we detect the presence of a reducing agent?

Your teacher will supply you with several solutions, each of which may or may not be a reducing agent, and a solution which is known to be a reducing agent.

Take about 1 cm³ of the reducing agent in a test-tube. To it, add, drop by drop, about 4 drops of dilute potassium manganate(VII) solution which has already been acidified with dilute sulphuric acid, and shake to mix.
   Decolouration of the potassium manganate(VII) solution—i.e. loss of its purple colour—indicates the presence of a reducing agent.
   Now use this test to find out which of the other solutions are reducing agents.

21.3 Classification of reactions

You will have already met a large number of chemical reactions, and clearly there are many others. Chemists look for similarities between reactions and hope to find patterns which are then often expressed as **general reactions**. An example of a general reaction or pattern which you have met is:

$$acid \ + \ alkali \ \rightarrow \ salt \ + \ water$$

This statement tells you that you could expect a solution of any acid to react with a solution of any alkali to form a salt and water,

e.g. $$HCl(aq) \; + \; NaOH(aq) \; \rightarrow \; NaCl(aq) \; + \; H_2O(l)$$

Thus the classification of reactions in this way helps you to understand, remember and predict chemical reactions.

Another group of reactions can be classified as oxidation-reduction. Many important processes which occur in everyday life involve oxidation-reduction reactions. Burning is an oxidation-reduction process, whether it is the controlled burning of a fuel such as occurs in a motor-car engine, in a rocket engine heading for the Moon, or in a domestic fireplace; or the uncontrolled burning which occurs when a building or a forest is on fire.

(a)

(b)

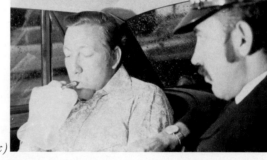

(c)

The rusting of iron, the corrosion of other metals, the production of electricity in a dry battery, and the use of electricity to produce important elements such as aluminium, chlorine and copper all involve chemical processes. Although at first glance these processes appear to be very different, they can all be classified as oxidation-reduction reactions.

21.4 Oxidation and reduction in terms of oxygen and hydrogen

As the name suggests, oxidation is used for reactions which involve the element oxygen. Whenever oxygen combines with another element, oxidation occurs and the element is said to have been oxidised,

e.g. $$2Mg(s) \; + \; O_2(g) \; \rightarrow \; 2MgO(s)$$

190

Fig. 21.1 Some important
oxidation-reduction
processes (a) The rusting
of iron such as has
happened to the body of
this motor car. (Photo:
Russell Edwards, BSc)
(b) Burning—illustrated
in this photograph by a
large fire in an empty
department store.
(Courtesy London Fire
Brigade Photographic
Service) (c) The reaction
between alcohol and
crystals in a breathalyser.
The photograph shows a
reconstruction of the scene
when a driver is asked to
blow into a breathalyser in
order to provide an
indication of the amount
of alcohol he has
consumed. (Courtesy of
the Commissioner of
Police of the Metropolis)

21.5
Oxidation and reduction together

The element which has been oxidised is sometimes indicated by an arrow as shown:

$$2Mg(s) + O_2(g) \rightarrow 2MgO(s)$$

The definition has been extended to include reactions in which elements combine with other non-metals,

e.g.

$$Mg(s) + Cl_2(g) \rightarrow MgCl_2(s)$$

Oxidation also occurs when oxygen reacts with a compound,

e.g.

$$2H_2S(g) + O_2(g) \rightarrow 2H_2O(l) + 2S(s)$$

A similar reaction occurs when chlorine reacts with hydrogen sulphide:

$$H_2S(g) + Cl_2(g) \rightarrow 2HCl(g) + S(s)$$

In these last two reactions hydrogen is removed from the compound and so it is usual to extend the definition of oxidation to include the removal of hydrogen from a compound.

The opposite of oxidation, that is, the removal of oxygen or another non-metal and the addition of hydrogen, is called reduction and indicated by an arrow as shown,

e.g.

$$2HgO(s) \rightarrow 2Hg(l) + O_2(g)$$

$$N_2(g) + 3H_2(g) \rightarrow 2NH_3(g)$$

When carbon monoxide is used to reduce copper(II) oxide to copper, the carbon monoxide is itself oxidised to carbon dioxide:

$$CuO(s) + CO(g) \rightarrow Cu(s) + CO_2(g)$$

Oxidation and reduction always occur in the same reaction. It is only possible to oxidise one element if another element is being reduced. Because the two changes occur together the overall reaction is called a **redox reaction**.

The substance which causes the oxidation (copper(II) oxide in the above example) is called the **oxidising agent** and that which causes the reduction (carbon monoxide) is called the **reducing agent**.

21.6
Redox in terms of electron transfer

The fundamental explanation as to why a lot of apparently different reactions can be classified as redox reactions is that the reactions, when they involve ionic substances, can all be explained in terms of transfer of electrons.

For example, during the combination of magnesium with oxygen,

$$2Mg(s) + O_2(g) \rightarrow 2MgO(s)$$

neutral magnesium atoms are converted to positively charged magnesium ions by the removal of electrons:

$$Mg \rightarrow Mg^{2+} + 2e^-$$

and the neutral oxygen molecules are converted to negatively charged oxide ions by gaining electrons:

$$O_2 + 4e^- \rightarrow 2O^{2-}$$

These two equations are called **ionic half-equations** and the overall equation for the reaction can be obtained by multiplying the first half-equation by 2 and adding it to the second so that all of the electrons which are removed from magnesium are gained by oxygen:

$$2Mg \rightarrow \quad 2Mg^{2+} \quad + \quad 4e^-$$
$$\underline{O_2 \quad + \quad 4e^- \rightarrow \quad 2O^{2-}}$$
$$2Mg \quad + \quad O_2 \quad \rightarrow 2Mg^{2+}O^{2-} \qquad \text{(or 2MgO)}$$

When magnesium reacts with another non-metal such as chlorine, it undergoes an identical change:

$$Mg \rightarrow Mg^{2+} + 2e^-$$

In this reaction the electrons are gained by the chlorine molecule to form chloride ions:

$$Cl_2 + 2e^- \rightarrow 2Cl^-$$

In both reactions electrons are removed from the metal (which according to the earlier definitions is being oxidised) and both non-metals gain electrons and are reduced. This leads to the more widely applicable definitions:

Oxidation occurs when electrons are removed from an atom or ion.
Reduction occurs when an atom or an ion gains electrons.

Many of the reactions which you carry out or see during chemistry lessons are reactions between substances which have previously been dissolved in water. If these reactions involve transfer of electrons they can be classified as redox reactions. The following examples illustrate the wide applicability of this method of interpreting redox reactions.

21.7 Oxidation of metals

Any reaction which converts neutral metal atoms into positively charged ions is a redox reaction as electrons have been removed from the metal atoms and gained by the other reactant, which is thus reduced. Examples of this type of redox reaction occur when the metals react with water, steam, acids, alkalis or solutions or other metal salts.

Sodium reacts with water forming sodium hydroxide solution which contains $Na^+(aq)$ and $OH^-(aq)$ ions and the water is reduced to hydrogen,

$$2Na(s) + 2H_2O(l) \rightarrow 2NaOH(aq) + H_2(g)$$

or $\qquad 2Na(s) + 2H_2O(l) \rightarrow 2Na^+(aq) + 2OH^-(aq) + H_2(g)$

When a metal is added to a solution of a salt of a less reactive metal (17.6) the more reactive metal is oxidised to its ions,

$$Fe(s) \rightarrow Fe^{2+}(aq) + 2e^-$$

and the ions of the less reactive metal are reduced to form neutral atoms of the metal:

$$Cu^{2+}(aq) + 2e^- \rightarrow Cu(s)$$

The overall reaction is,

$$Fe(s) + Cu^{2+}(aq) \rightarrow Fe^{2+}(aq) + Cu(s)$$

or $\qquad Fe(s) + CuSO_4(aq) \rightarrow FeSO_4(aq) + Cu(s)$

The metals which are highest in the reactivity series (17.9) are most easily oxidised and are therefore the most powerful reducing agents, Fig. 21.2. The ions of the metals which are lowest in the series are most easily reduced and are therefore the most powerful oxidising agents.

Fig. 21.2 Oxidation and reduction in relation to the reactivity series

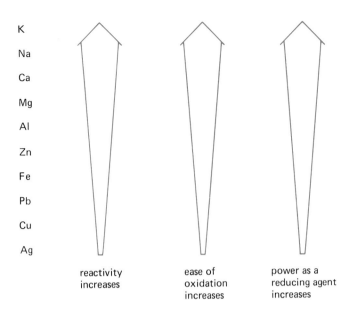

K
Na
Ca
Mg
Al
Zn
Fe
Pb
Cu
Ag

reactivity
increases

ease of
oxidation
increases

power as a
reducing agent
increases

21.8 Oxidation of metal ions

Some metals can form more than one type of ion, e.g. iron(II), Fe^{2+}, and iron(III), Fe^{3+}. When ions of the lower oxidation state (i.e. smaller positive charge) are changed to ions of the higher oxidation state (i.e. higher positive charge), oxidation has occurred as the change involves the removal of electrons. For example, the oxidation of iron(II) ions to iron(III) ions:

$$Fe^{2+}(aq) \rightarrow Fe^{3+}(aq) + e^-$$

The substance which causes the oxidation is itself reduced by taking up the electrons:

e.g. $$Cl_2(g) + 2e^- \rightarrow 2Cl^-(aq)$$

The overall equation for the reaction is,

$$2Fe^{2+}(aq) + Cl_2(g) \rightarrow 2Fe^{3+}(aq) + 2Cl^-(aq)$$

or $$2FeCl_2(aq) + Cl_2(g) \rightarrow 2FeCl_3(aq)$$

When iron(III) ions are changed to iron(II) ions reduction has taken place.

21.9 Reduction of non-metals

When a non-metal is converted to its ions,

e.g. $$Cl_2 + 2e^- \rightarrow 2Cl^-$$

reduction has occurred because electrons have been taken up by the chlorine. This change occurs when a non-metal reacts with a metal,

e.g. $$2Na(s) + Cl_2(g) \rightarrow 2NaCl(s)$$

Sodium atoms are oxidised to Na^+ ions during this reaction.

A non-metal can also be reduced when it reacts with a solution containing the ions of a less reactive non-metal. For example, chlorine will displace bromine from a solution containing bromide ions:

$$Cl_2(g) + 2NaBr(aq) \rightarrow 2NaCl(aq) + Br_2(aq)$$

In this reaction chlorine is reduced to chloride ions,

$$Cl_2(g) + 2e^- \rightarrow 2Cl^-(aq)$$

and bromide ions are oxidised to bromine:

$$2Br^-(aq) \rightarrow Br_2(aq) + 2e^-$$

The overall ionic equation for the reaction is:

$$Cl_2(g) + 2Br^-(aq) \rightarrow 2Cl^-(aq) + Br_2(aq)$$

Chlorine is more easily reduced than bromine and hence it is a more powerful oxidising agent.

21.10 Electrolysis

The production of a metal during electrolysis involves the conversion of positively charged metal ions to neutral atoms of the metal. This change is brought about by the positive ions capturing electrons from the negative electrode (cathode), 19.8,

e.g. $$Na^+ + e^- \rightarrow Na$$

This change is a reduction process as electrons are gained.

At the positive electrode (anode) the usual change involves negatively charged ions giving up electrons to the electrode,

e.g. $$2Cl^- \rightarrow Cl_2 + 2e^-$$

This change is an oxidation process as electrons are being removed from the ions.

In all examples of electrolysis, oxidation occurs at the anode and reduction at the cathode and the overall change which occurs is a redox process.

21.11 Oxidation numbers

In the previous parts of this chapter you have seen how redox reactions may be discussed in terms of electron transfer, and how oxidation and reduction may be defined in terms of loss or gain of electrons. Following on from this we have a system by which we can assign oxidation numbers to elements when they are combined with one another. This system was first invented by Johnson in 1880, and it has two advantages. Firstly, it provides a useful way of classifying compounds of an element which forms more than one type of ion, or different numbers of bonds in different compounds. Secondly, it provides a method of balancing redox equations which would be very difficult or tedious to balance by inspection. It was, in fact, for this second purpose that Johnson devised the system.

The modern system differs from Johnson's original one, in that we can define oxidation numbers in terms of ionic charges. Thus, in a simple ionic compound the oxidation numbers of the elements involved are just their ionic charges,

e.g. sodium chloride (NaCl) is Na^+Cl^-, so sodium has oxidation number $+1$ and chlorine has oxidation number -1;

iron(II) chloride ($FeCl_2$) is $Fe^{2+}(Cl^-)_2$, so iron has oxidation number $+2$ and chlorine has oxidation number -1;

iron(III) oxide (Fe_2O_3) is $(Fe^{3+})_2(O^{2-})_3$, so iron has oxidation number $+3$ and oxygen has oxidation number -2.

For the metals in the examples above, the oxidation number is just the number of electrons lost when the atom becomes an ion. The non-metals in these examples gain electrons in forming ions. They are therefore in a reduced state and so have negative

194

oxidation numbers. Note that the sum of the oxidation numbers in a compound is always zero, thus:

sodium chloride	$+1 -1 = 0$
iron(II) chloride	$+2 -1 -1 = 0$
iron(III) oxide	$+3 +3 -2 -2 -2 = 0$

It also follows that the oxidation number of a free (uncombined) element must be zero; thus the symbols Na, Fe, Cl_2, O_2, P_4, S_8 etc., represent free elements with oxidation number 0.

Now, suppose a compound is not ionic? Can we still assign oxidation numbers? The answer is that we can, and we do it by inventing a hypothetical (imaginary) ionic formula using certain simple rules. The rules mainly concern oxygen and hydrogen. When oxygen forms an ion this is normally O^{2-} as in the example of iron(III) oxide above. So, if we have a covalent (non-ionic) oxide then, for the purpose of assigning oxidation numbers, we suppose it to be ionic with oxygen present as O^{2-}, and we then work out what the ionic charge on the other element's ion would have to be in order to give a neutral compound.

When hydrogen forms an ion this is normally H^+ (ignoring hydration), so if we have a covalent hydride we assume, for the purpose of assigning oxidation numbers, that hydrogen is present as H^+ and work out the ionic charge on the other element's ion accordingly.

e.g. sulphuric acid (H_2SO_4)—assume O^{2-} and H^+ so then we need S^{6+} in order to give $(H^+)_2S^{6+}(O^{2-})_4$, so oxidation number of sulphur is $+6$, as in sulphur trioxide (SO_3) which is the anhydride of sulphuric acid;

nitric acid (HNO_3)—assume O^{2-} and H^+ so then we need N^{5+} in order to give $H^+N^{5+}(O^{2-})_3$, so oxidation number of nitrogen is $+5$.

In the case of a polyatomic ion exactly the same rules are followed, but now the sum of the oxidation numbers is equal to the net charge on the ion,

e.g. sulphate ion (SO_4^{2-})—assume O^{2-} so then we need S^{6+} in order to give $S^{6+}(O^{2-})_4$ with a net charge of $2-$; the oxidation number of sulphur is then $+6$ as in sulphuric acid;

nitrate ion (NO_3^-)—assume O^{2-} so then we need N^{5+} in order to give $N^{5+}(O^{2-})_3$ with a net charge of $1-$; the oxidation number of nitrogen is then $+5$ as in nitric acid.

21.12 Classification and nomenclature

Iron, in its compounds, shows valencies of 2 and 3; that is, it forms a series of compounds in which its oxidation number is $+2$, e.g. FeS, $FeCl_2$, $FeSO_4$, which may be referred to as iron(II) compounds, and it forms a series of compounds in which its oxidation number is $+3$, e.g. Fe_2O_3, $FeCl_3$, $Fe_2(SO_4)_3$, which may be referred to as iron(III) compounds. Thus the names iron(II) chloride and iron(III) chloride replace the old names, which were ferrous chloride and ferric chloride respectively.

Similarly, for the compounds $SnCl_2$ and $SnCl_4$ the modern names are tin(II) chloride and tin(IV) chloride, replacing the old names stannous chloride and stannic chloride.

21.13 Using oxidation numbers to balance redox equations

For example, a solution of sodium sulphite (sulphate(IV)) will decolourise an acidified solution of potassium manganate(VII). In this reaction the sulphate(IV) (SO_3^{2-}) is oxidised to sulphate(VI) (SO_4^{2-}) and so the oxidation number of sulphur changes from $+4$ to $+6$ and the half-equation must involve a two electron change:

$$SO_3^{2-}(aq) \rightarrow SO_4^{2-}(aq) + 2e^- \quad \text{unbalanced}$$

To balance the oxygen and the charges, water must be included as a reactant:

$$SO_3^{2-}(aq) + H_2O(l) \rightarrow SO_4^{2-}(aq) + 2H^+(aq) + 2e^-$$

Manganate(VII) ions (MnO_4^-) are converted to manganese(II) ions (Mn^{2+}) and so the oxidation number of manganese changes from $+7$ to $+2$. This requires the addition of five electrons:

$$MnO_4^-(aq) + 5e^- \rightarrow Mn^{2+}(aq) \quad \text{unbalanced}$$

To balance the oxygen atoms and the charges it is necessary to include hydrogen ions from the acid:

$$MnO_4^-(aq) + 8H^+(aq) + 5e^- \rightarrow Mn^{2+}(aq) + 4H_2O(l)$$

The equation for the complete reaction can be obtained by multiplying the first half-equation by 5 and the second by 2 and then adding them:

$$5SO_3^{2-}(aq) + 2MnO_4^-(aq) + 6H^+(aq) \rightarrow 2Mn^{2+}(aq) + 5SO_4^{2-}(aq) + 3H_2O(l)$$

(Note that some H_2O molecules and H^+ ions have 'cancelled out' in adding the half-equations together.)

Further examples illustrating these principles will be found in the remaining parts of this chapter.

21.14
Tests for reducing agents

For a substance to be suitable for use in a test for a reducing agent, it must be easily and rapidly reduced and the reduction must be accompanied by an obvious colour change. Two particularly suitable reagents are an acidified solution of potassium manganate(VII), $KMnO_4$, and an acidified solution of potassium dichromate(VI), $K_2Cr_2O_7$. A solution containing manganate(VII) ions, $MnO_4^-(aq)$, is purple in colour and, when reduced, manganese(II) ions, $Mn^{2+}(aq)$, are produced which form a colourless solution.

A solution of an iron(II) salt—e.g. iron(II) sulphate—will decolourise an acidified solution of potassium manganate(VII) and is therefore a reducing agent. The iron(II) salt is oxidised to an iron(III) salt.

A solution containing dichromate(VI) ions, $Cr_2O_7^{2-}(aq)$, is orange in colour and, when reduced, chromium(III) ions, $Cr^{3+}(aq)$, are produced which form a green solution.

A solution of a sulphite—e.g. sodium sulphite—will turn an acidified solution of potassium dichromate(VI) from orange to green and is therefore a reducing agent. The sulphite ions are oxidised to sulphate ions.

21.15
Test for oxidising agents

An acidified solution of potassium iodide, which is colourless, is very easily oxidised to iodine which forms a yellow-brown solution (and perhaps a black precipitate). For example hydrogen peroxide, in the presence of acid, turns potassium iodide solution to a yellow-brown colour and is therefore an oxidising agent. The half-equation for the reduction of hydrogen peroxide is:

$$2H^+(aq) + H_2O_2(aq) + 2e^- \rightarrow 2H_2O(l)$$

The oxidation number of iodine changes from -1 to 0 and the half-equation for the oxidation of iodide ions is:

$$2I^-(aq) \rightarrow I_2(aq/s) + 2e^-$$

Adding these together directly gives:

$$2H^+(aq) + H_2O_2(aq) + 2I^-(aq) \rightarrow 2H_2O(l) + I_2(aq/s)$$

21.16
Substances which are both oxidising and reducing agents

For a substance to act as an oxidising agent it must be possible for it to be reduced to something. Similarly, for a substance to act as a reducing agent it must be possible for it to be oxidised to something. Some substances, under appropriate circumstances, are capable of taking part in both types of reaction. For example, an iron(II) salt in solution can be oxidised to an iron(III) salt—e.g. by manganate(VII) ions—and so act as a reducing agent. On another occasion an iron(II) salt can be reduced to iron metal—e.g. by zinc powder—and so act as an oxidising agent.

$$\underset{\substack{\text{oxidation}\\\text{number}}}{} \quad \overset{\text{reduction}}{\underset{0}{Fe(s)} \xleftarrow{\hspace{2cm}}} \quad \underset{+2}{Fe^{2+}(aq)} \quad \overset{\text{oxidation}}{\xrightarrow{\hspace{2cm}}} \quad \underset{+3}{Fe^{3+}(aq)}$$

Hydrogen peroxide is best known as an oxidising agent, when it is reduced to water (see previous section). However, if hydrogen peroxide solution is mixed with acidified potassium manganate(VII) solution the latter is decolourised, showing that hydrogen peroxide can also act as a reducing agent. During the reaction effervescence occurs, the gas liberated being oxygen. In this reaction hydrogen peroxide is oxidised, by removal of hydrogen; in terms of a half-equation we have:

$$H_2O_2(aq) \rightarrow O_2(g) + 2H^+(aq) + 2e^-$$

(Compare this with the half-equation for its reduction—previous section.) The half-equation for reduction of manganate(VII) ions is:

$$8H^+(aq) + MnO_4^-(aq) + 5e^- \rightarrow Mn^{2+}(aq) + 4H_2O(l)$$

If we double this equation, and multiply the first half-equation by five, then on adding them together the electrons cancel out and we get:

$$6H^+(aq) + 2MnO_4^-(aq) + 5H_2O_2(aq) \rightarrow 2Mn^{2+}(aq) + 8H_2O(l) + 5O_2(g)$$

An acidified solution of potassium dichromate(VI) will also oxidise hydrogen peroxide to oxygen, the dichromate(VI) ions being reduced to chromium(III) ions.

21.17
Summary

1. Oxidation can be defined as:
   (a) combination with oxygen or another non-metal
   (b) removal of hydrogen
   (c) removal of electrons
   Reduction is the reverse of oxidation.
2. Oxidation and reduction always occur together in the same reaction and so it is called a redox reaction.
3. The substance which causes the oxidation (and is itself reduced) is called the oxidising agent. The substance which causes the reduction (and is itself oxidised) is called the reducing agent.
4. Oxidation numbers may be assigned to elements in compounds. These are useful for classifying compounds, and for helping to balance redox equations.
5. An acidified solution of potassium manganate(VII) and an acidified solution of potassium dichromate(VI) can both be used to test for a reducing agent.
6. An acidified solution of potassium iodide can be used to test for an oxidising agent.
7. Some substances can act both as an oxidising agent and a reducing agent.

Chapter 22

What influences the stability of compounds?

Investigation 22.1

A comparison of the energy changes which occur when different metals displace copper from copper(II) sulphate solution.

You will need an insulated container such as a polystyrene beaker, a 0–100 °C thermometer, 0·5M copper(II) sulphate solution, and zinc, iron and magnesium powders.

Using a measuring cylinder, transfer 20 cm³ of copper(II) sulphate solution into the polystyrene beaker. Weigh out about 1 g of zinc powder and keep it ready for use.

Holding the thermometer in the solution, take temperature measurements at 20 second intervals for about 2 minutes, then add the zinc powder all at once and keep recording the temperature at 20 second intervals, stirring with the thermometer between readings. Continue taking readings until the temperature has stopped rising, and is either constant or falling at a slow rate.

Plot a graph of temperature versus time. It should have the approximate appearance of Fig. 22.1.

Fig. 22.1 Correcting the temperature rise

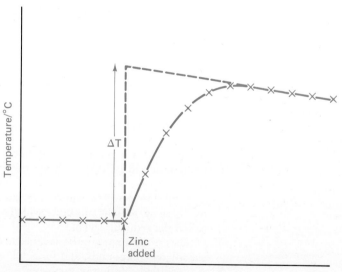

In order to obtain a reasonable measure of what the temperature rise (ΔT) would have been, had the reaction occurred instantaneously and there had been no heat losses, extrapolate back from the final part of the curve, as shown in Fig. 22.1, and obtain a value for ΔT.

Using a fresh portion of copper(II) sulphate solution repeat the experiment using the same mass of iron powder instead of zinc powder.

Finally take another portion of copper(II) sulphate solution and dilute it with its own volume (20 cm³) of water, then repeat the experiment using the same mass of magnesium powder.

1 What, roughly, would the temperature rise have been for the magnesium reaction had we not diluted the reaction mixture with an extra portion of water? (Bear in mind that doubling the volume of water will roughly double the heat capacity of the liquid.)

2 Compare the temperature rises (using the corrected value for magnesium) by placing them in order of decreasing size. What connection is there between these results and the metal reactivity series?

Investigation 22.2

A comparison of the energy changes which occur when acids and alkalis neutralise each other.

You will need an insulated container such as a polystyrene beaker, a 0–100 °C thermometer, 2M solutions of hydrochloric acid, nitric acid, sodium hydroxide and potassium hydroxide, and 1M solution of sulphuric acid.

Measure out 25 cm³ of one of the acid solutions and transfer it to the polystyrene beaker. Then measure out 25 cm³ of one of the alkali solutions and keep it in the cylinder for the present.

Take the temperature of each solution separately. If they differ, take the average and use this as the starting temperature.

Now pour the alkali all at once into the acid, stir well, and record the highest reading reached on the thermometer. (The reaction is almost instantaneous so heat losses are negligible.)

Repeat the experiment using other combinations of acid and alkali and compare the temperature rises obtained for the various combinations.

Question

Write an equation for the reaction which has occurred in each experiment and attempt to convert each equation to the simplest ionic equation for the reaction. In what way do these ionic equations help to explain the similarities in the temperature rises?

Investigation 22.3

Making reactions go either way.

Some reactions are reversible in the sense that if the conditions are changed in a particular way they can be made to go in the reverse direction.

1. Look at bottles of potassium chromate(VI) solution and potassium dichromate(VI) solution. The yellow colour of the former is due to the presence of the chromate(VI) ion (CrO_4^{2-}(aq)), while the orange colour of the latter is due to the presence of the dichromate(VI) ion ($Cr_2O_7^{2-}$(aq)).

Take 2 cm³ of potassium chromate(VI) solution in a test-tube and add an equal volume of dilute sulphuric acid. What is the colour now? See if you can get the colour to go back to the original one by adding sodium hydroxide solution.

Now pour 2 cm³ of potassium dichromate(VI) solution into a test-tube and add an equal volume of sodium hydroxide solution. What is the colour now? See if you can get the original colour back by adding dilute sulphuric acid.

1 What type of reagent causes conversion of chromate(VI) ions to dichromate(VI) ions?

2 What type of reagent causes conversion of dichromate(VI) ions to chromate(VI) ions?

3 Try and construct an ionic equation for the conversion of chromate(VI) ions to dichromate(VI) ions. Try and explain, in terms of this equation, how addition of alkali reverses the reaction.

2. Sodium sulphite solution contains sulphite ions ($SO_3{}^{2-}$(aq)). Addition of acid causes the smell of sulphur dioxide (SO_2) to appear, because of the formation of aqueous solution of sulphur dioxide which is volatile.

Take about 2 cm³ of sodium sulphite solution in a shallow dish; it should have no smell. Add a drop of litmus solution.

Now add about 2 cm³ of dilute hydrochloric acid, so that the litmus turns red. Now smell the solution cautiously; does it smell of sulphur dioxide?

Now add sufficient sodium hydroxide solution that the litmus turns blue, and smell the solution again. Does it smell of sulphur dioxide now?

4 Is the reaction reversible?

5 Try and construct an ionic equation for the conversion of sulphite ions to aqueous sulphur dioxide. Try and explain, in terms of this equation, how addition of alkali reverses the reaction.

Investigation 22.4

How can a reversible reaction be made to go to completion?

Investigation 22.3 shows that the reaction between sulphite ions and hydrogen ions is a reversible one, in the sense that the aqueous solution of sulphur dioxide produced in the reaction may be converted back into sulphite ions by addition of alkali. The purpose of this experiment is to see whether we can make the reaction go to completion, so that it cannot be reversed again.

First, pour about 2 cm³ of sodium sulphite solution into a test-tube and add enough sodium hydroxide solution to make it alkaline (check with red litmus paper). Now add two drops of potassium manganate(VII) solution. The effect which you observe is due to the presence of sulphite ions in solution, the sulphite ion being a reducing agent.

Now take a fresh portion of sodium sulphite solution in a shallow dish, and add about twice the volume of dilute sulphuric acid.

Heat the dish on a tripod and gauze until the liquid boils and then let it boil gently. Hold a piece of moist blue litmus paper in the vapour above the hot liquid and note what happens. Continue boiling for one minute and then hold another piece of moist blue litmus paper in the vapour.

Repeat this test, at one-minute intervals, until the vapour no longer turns litmus red.

Now let the liquid cool a bit, then transfer about 2 cm³ to a test-tube. Add enough sodium hydroxide solution to the test-tube to make the mixture alkaline (check with red litmus paper), then add a couple of drops of potassium manganate(VII) solution and note what happens.

1 Describe the effect which the original alkaline sodium sulphite solution had on potassium manganate(VII) solution.

2 Did the alkaline solution left at the end of the experiment have this effect on potassium manganate(VII)?

3 So did the final solution contain sulphite ions?

4 What substance, present in the vapour above the hot liquid, caused the moist litmus paper to turn red?

5 Explain clearly what effect boiling had on the reaction mixture. Did the reaction between sulphite ions and hydrogen ions go to completion under these conditions?

6 What do you think would have happened in this experiment if sulphur dioxide were not a volatile substance?

22.5 Energy changes and reactivities of elements

Chemical reactions are usually accompanied by a temperature change. Most often it is a rise in temperature, which means that as the reaction proceeds heat (i.e. energy) is given out. A process in which energy is released as heat is called an **exothermic process**. There is a small number of reactions which are accompanied by a **drop** in temperature, showing that heat is being absorbed; such a reaction is said to be **endothermic**.

In an exothermic reaction between two elements, the amount of heat given out gives an indication of their readiness to combine. This can be linked to the reactivity of the elements concerned. For example, if a mixture of powdered iron and sulphur is heated gently, a reaction starts in which the mixture glows red-hot as the iron and sulphur combine exothermically to form iron(II) sulphide (1.5). Once started, this reaction proceeds without further heating. On the other hand, it is extremely dangerous to heat a mixture of powdered magnesium and sulphur because the reaction, once started, is violently explosive. These observations indicate that magnesium has a much greater affinity for sulphur than iron does, and that magnesium is more reactive than iron.

Another example is that magnesium, once ignited, burns vigorously in air, giving out a lot of heat and light as it turns into its oxide. Copper, on the other hand, being a much less reactive metal, reacts gently with air on heating, forming a black coating of copper(II) oxide but without bursting into flame, 8.5.

22.6 Stability of compounds

Since magnesium oxide is formed more readily than copper(II) oxide, and more energy is released in the process, we might expect it to be harder to get magnesium back from magnesium oxide than it is to get copper back from copper(II) oxide. This is found to be so. Magnesium oxide does not change when it is heated in hydrogen gas, whereas copper(II) oxide is easily reduced to copper by heating it in hydrogen; indeed the reaction, once started, is exothermic:

$$CuO(s) + H_2(g) \rightarrow Cu(s) + H_2O(g) + energy$$

This observation indicates that water is more stable than the copper(II) oxide. The reaction does not go in the reverse direction, that is, copper is not oxidised when heated in steam.

What about magnesium then? If hydrogen will not reduce magnesium oxide, perhaps magnesium will reduce steam. It can easily be shown that it does (17.5). Magnesium ribbon, once ignited, will continue to burn vigorously in steam, giving out a lot of heat and light and liberating hydrogen:

$$Mg(s) + H_2O(g) \rightarrow MgO(s) + H_2(g) + energy$$

This indicates that magnesium oxide is more stable than steam.

We can now say that the order of decreasing reactivity of the elements with oxygen is magnesium, hydrogen, copper, this being indicated by:

$$2Mg(s) + O_2(g) \rightarrow 2MgO(s) + \text{a great deal of energy}$$

$$2H_2(g) + O_2(g) \rightarrow 2H_2O(g) + \text{quite a lot of energy}$$

$$2Cu(s) + O_2(g) \rightarrow 2CuO(s) + \text{a bit of energy}$$

The stability of the resulting oxides decreases from magnesium oxide to water (hydrogen oxide) to copper(II) oxide. The relationship between reactivities of the elements and the stability of the compounds is summarised in Fig. 22.2.

Observations such as those described above can be linked to the reactivity series of metals as described in 17.9 and to the suitability of various methods of extracting metals from their compounds (Chapter 24).

Fig. 22.2 Reactivities of elements and stabilities of compounds

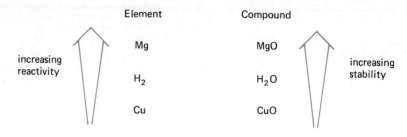

22.7 Thermal stability of compounds

Metal oxides

In principle, all compounds will decompose to their elements if heated to a high enough temperature, but in practice the number of compounds which can be decomposed into their elements at the temperature of a Bunsen flame (up to 1000 °C) is very small. Thus, although copper and lead are rather unreactive metals and their oxides are regarded as relatively unstable, it is not possible to obtain copper metal or lead metal by heating their oxides with a Bunsen flame. However, the oxide of mercury, which is an even less reactive metal, does decompose when heated in a Bunsen flame:

$$2HgO(s) + \text{energy} \rightarrow 2Hg(l) + O_2(g)$$

The decompositions of all of the metal oxides mentioned so far in this chapter are endothermic reactions, that is heat (energy) is absorbed as they decompose.

With a very unreactive metal such as gold the situation is reversed. Thus the formation of gold oxide from its elements is endothermic and the oxide decomposes exothermically into gold and oxygen on heating.

Metal hydroxides

The decomposition of a compound into its elements is rarely met in the laboratory. What is more common is the decomposition of a ternary (three element) compound into binary (two element) compounds, but here again the same reactivity and stability principles apply. The metal hydroxides provide a clear example of this. In general they decompose, on heating, into the metal oxide and water, but the ease of decomposition depends upon the position of the metal in the reactivity series. Thus, for example, the hydroxide of the very reactive metal sodium is stable at the temperature of a Bunsen flame, even though it is easily melted. The hydroxide of the slightly less reactive metal, calcium, is decomposed by very strong heat using a roaring Bunsen flame:

$$Ca(OH)_2(s) \underset{\substack{\text{about} \\ 800\,°C}}{\rightarrow} CaO(s) + H_2O(g)$$

The hydroxide of the unreactive metal copper is very easily decomposed. Copper hydroxide can be produced as a blue gelatinous precipitate by adding sodium hydroxide solution to copper(II) sulphate solution. If the test-tube containing the suspension of copper(II) hydroxide is allowed to stand in boiling water for a few minutes, the blue suspension turns to a black suspension of copper(II) oxide:

$$Cu(OH)_2(s) \quad \xrightarrow[\substack{below \\ 100°C}]{} \quad CuO(s) \; + \; H_2O(l)$$

In terms of energy changes we observe that,

$$Cu(OH)_2(s) \; + \quad \text{a little energy} \quad \rightarrow \; CuO(s) \; + \; H_2O(l)$$

but $\quad Ca(OH)_2(s) \; + \quad \text{a lot of energy} \quad \rightarrow \; CaO(s) \; + \; H_2O(g)$

We can try to do these reactions in reverse. If water is dropped on to calcium oxide, a vigorous exothermic reaction occurs in which calcium hydroxide is formed:

$$CaO(s) \; + \; H_2O(l) \rightarrow Ca(OH)_2(s) \; + \; \text{energy}$$

On the other hand, if copper(II) oxide is put into water nothing happens; the copper(II) oxide is insoluble.

22.8 Reversible reactions

Reactions in which one element displaces another are usually observed to go in the exothermic direction, so that the more reactive element displaces the less reactive one and the more stable compound is formed. For example, we have noted that magnesium burns exothermically in steam, whereas copper(II) oxide is reduced exothermically to copper by hydrogen gas and so the decreasing order of stability is MgO, H_2O, CuO. Let us consider a metal which is intermediate in reactivity between magnesium and copper, namely iron. When iron is heated to redness in steam, it is oxidised to black magnetic iron oxide, and hydrogen is liberated:

$$3Fe(s) \; + \; 4H_2O(g) \rightarrow Fe_3O_4(s) \; + \; 4H_2(g)$$

However, if iron oxide is heated in a stream of hydrogen gas it is reduced to metallic iron:

$$Fe_3O_4(s) \; + \; 4H_2(g) \rightarrow 3Fe(s) \; + \; 4H_2O(g)$$

The reaction is therefore **reversible**, and can be made to go in either direction. The symbol $\rightleftharpoons$ is used to indicate that the reaction is reversible:

$$3Fe(s) \; + \; 4H_2O(g) \rightleftharpoons Fe_3O_4(s) \; + \; 4H_2(g)$$

It could equally well be written as,

$$Fe_3O_4(s) \; + \; 4H_2(g) \rightleftharpoons 3Fe(s) \; + \; 4H_2O(g)$$

and this equation has the same meaning as the previous one.

22.9 Dynamic equilibrium

If some iron is heated in a closed flask containing steam, some of the iron is oxidised while some of the steam is reduced to hydrogen gas. After a while a state of balance is reached, in which all four substances—iron, steam, iron oxide and hydrogen—are present, and the quantities of these four substances in the flask no longer change. This is called a state of equilibrium in which there is no net reaction either forwards or backwards. The same state could have been arrived at by starting with a quantity of iron oxide heated in a closed flask containing hydrogen gas (see Fig. 22.3).

The state of balance exists because the rate of the forward reaction (say, iron with steam) is exactly equal to the rate of the backward reaction (iron oxide with hydrogen), so that there is no longer any net reaction either way. This means that,

203

whatever the quantities of each substance present at equilibrium, for every mole of iron which is oxidised by the forward reaction, another mole of iron is formed by the backward reaction. The reaction is said to be in a state of **dynamic equilibrium** in the sense that both the forward and backward reactions are still proceeding, but at equal rates. The alternative phrase 'static equilibrium' would imply a state in which nothing was happening at all.

Fig. 22.3 Approaching an equilibrium from both sides

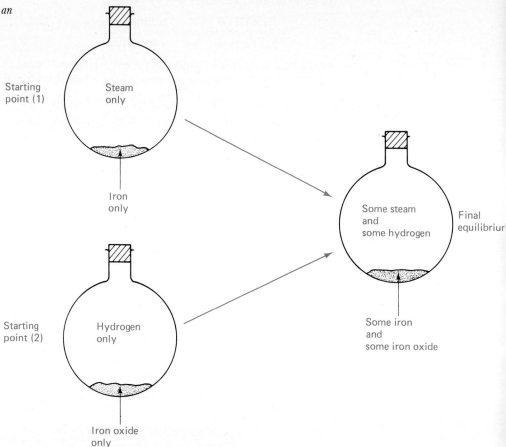

22.10
Making reversible reactions go to completion

The important point about reversible reactions is that they will, given the chance, come to a state of equilibrium in which all the substances involved are present in constant amounts. Many reactions which give commercially important products are reversible, and clearly it would be useful if these reactions could be made to go to completion rather than to the state of equilibrium. In many cases this is done by making it impossible for the equilibrium to be reached by removing one of the products as it is formed, so that the backward reaction cannot go to completion.

This can be illustrated by heating iron in a flow of steam, Fig. 22.4. As the steam, flowing through the tube from the left, meets the hot iron the forward reaction starts to go and some steam is reduced to hydrogen. If this hydrogen remained in the reaction region, the backward reaction would begin to take effect until equilibrium was reached. But hydrogen does not remain in the reaction region, since, as soon as it is formed, it is swept away in the steam flow and out of the tube at the right. Therefore the backward reaction is prevented from occurring, and in time the oxidation of the iron proceeds to completion. The same kind of argument can be used to explain how iron oxide can be completely reduced by heating in a stream of hydrogen gas. The steam which is produced is swept away by the stream of hydrogen.

Fig. 22.4 Making a reversible reaction go to completion

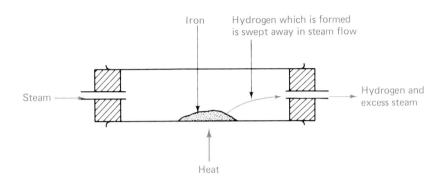

Iron   Hydrogen which is formed is swept away in steam flow

Steam   Hydrogen and excess steam

Heat

22.11 Production of quicklime

An example of the way in which these principles are put to practical use on a large scale is the manufacture of quicklime (calcium oxide) by the action of strong heat on limestone (calcium carbonate). The reaction in which calcium carbonate decomposes is reversible (this type of change is sometimes called dissociation):

$$CaCO_3(s) \rightleftharpoons CaO(s) + CO_2(g)$$

Thus, if calcium carbonate were heated in a closed container, the dissociation would proceed only until a certain steady pressure of carbon dioxide built up. The reaction would then be in a balanced state and no further dissociation would occur. In order to achieve complete conversion of limestone to quicklime in the shortest time and with the minimum expenditure of energy (i.e. fuel for the heating), it is necessary to suppress the backward reaction by removing the carbon dioxide as it is formed. The way in which the limekiln is designed to achieve this is explained in 29.3.

22.12 Preparation of volatile acids

Another example which illustrates these principles is the laboratory preparation of pure nitric acid. If potassium nitrate is mixed with concentrated sulphuric acid at room temperature, an equilibrium is set up in which both sulphuric and nitric acids are present:

$$KNO_3(s) + H_2SO_4(l) \rightleftharpoons KHSO_4(s) + HNO_3(l)$$

Of the four substances involved in this equilibrium, nitric acid (HNO_3) is by far the most volatile, having a boiling point of only 85 °C. Therefore, if the mixture is warmed, nitric acid vaporizes. This has the effect of suppressing the backward reaction by removing the nitric acid from the reaction region, and consequently the conversion of potassium nitrate to nitric acid can go to completion.

A more extreme example of this is observed if concentrated sulphuric acid is dropped on to potassium chloride, when immediate effervescence occurs as hydrogen chloride (HCl) gas is given off. The reaction is really a reversible one,

$$KCl(s) + H_2SO_4(l) \rightleftharpoons KHSO_4(s) + HCl(g)$$

but, since hydrogen chloride is a gas at room temperature, it immediately escapes from the reaction region as soon as it is formed and the reaction goes to completion without any warming being necessary. The reaction may be controlled by using moderately concentrated sulphuric acid, 25.6. The small amount of water present keeps most of the hydrogen chloride in solution until the mixture is warmed.

22.13
Position of equilibrium

In the previous section we have noted that some reactions are reversible, and that these reactions will, if given the chance, come to a state of equilibrium in which the quantities of reactants and products remain constant. We could then define the **position of equilibrium** by stating what actual quantities of the various substances are present.

We can then ask the question: If the conditions, temperature, pressure or concentrations are changed, will the position of equilibrium alter?

On many occasions, if there is a reaction which is in a state of equilibrium and we change the conditions, we are likely to find that the reaction is no longer in equilibrium and that some net reaction—either forward or backward—will occur until a new position of equilibrium is reached. The quantities of the various substances present at this new equilibrium will be different from those present when the reaction was first at equilibrium. Particularly when a reversible reaction is to be used in an industrial process, it would be very useful to be able to predict the likely effect of changing the conditions.

22.14
Effect of changes in concentrations

The way in which changing the concentrations of substances affects the position of equilibrium is most easily investigated by considering reversible reactions which occur in aqueous solution. Bismuth(III) chloride ($BiCl_3$), for example, forms a clear solution in concentrated hydrochloric acid. On adding water to the solution a white precipitate of bismuth oxychloride (BiOCl) is formed. If some more concentrated hydrochloric acid is added to the suspension, the precipitate dissolves and a clear solution is formed. On adding more water to the solution the white precipitate is formed once again. The precipitate can be dissolved and reprecipitated several times in this way:

$$
\begin{array}{ccc}
& \text{addition of water} & \\
BiCl_3 & \xrightleftharpoons{} & BiOCl \\
\text{clear} & \text{addition of conc.} & \text{white} \\
\text{solution} & \text{HCl} & \text{precipitate}
\end{array}
$$

This reversible reaction can be represented by the equation:

$$BiCl_3(aq) + H_2O(l) \rightleftharpoons BiOCl(s) + 2HCl(aq)$$

The observations indicate that adding water drives the position of equilibrium to the right, so forming more precipitate, whereas adding concentrated hydrochloric acid drives the equilibrium to the left. It thus appears that if, when a reaction is at equilibrium, the concentration of one substance is increased, the position of equilibrium moves so as to decrease the concentration of that substance.

An additional observation which can be made during the investigation is that the precipitate forms relatively slowly, showing that it takes a measurable time for the reaction to move to the new position of equilibrium. Clearly this is another important factor to be considered when designing an industrial process. An example is discussed later in this chapter.

The effect of decreasing the concentration of one of the substances involved in an equilibrium can be illustrated by the reaction involving chromate(VI) and dichromate(VI) ions:

$$
\begin{array}{ccccccc}
2CrO_4{}^{2-}(aq) & + & 2H^+(aq) & \rightleftharpoons & Cr_2O_7{}^{2-}(aq) & + & H_2O(l) \\
\text{chromate(VI)} & & & & \text{dichromate(VI)} & & \\
\text{yellow} & & & & \text{orange} & &
\end{array}
$$

The concentration of $H^+(aq)$ ions can be decreased by adding dilute alkali to the reaction mixture. $OH^-(aq)$ ions remove $H^+(aq)$ ions by combining with them:

$$H^+(aq) + OH^-(aq) \rightarrow H_2O(l)$$

The colour of the solution is seen to change from orange to yellow, showing that the equilibrium has moved to the left. Thus when the concentration of $H^+(aq)$ ions is decreased the equilibrium moves so as to produce more $H^+(aq)$ ions.

Overall it appears that a chemical equilibrium will tend to move in the direction which opposes any changes in the concentrations of the substances present in the equilibrium mixture.

This agrees with the observation, which was discussed earlier, that, when one of the products of a reversible reaction is removed as it is formed, the reaction will go to completion.

22.15 Effect of changes in temperature and pressure

Before considering the effect which changing the temperature or pressure has on a chemical equilibrium, it is easier to consider a familiar physical process, namely the boiling of water.

Water, at any temperature, has a tendency to evaporate, and it is only necessary to leave a saucer of water in a room for a day or two to observe this. If water is kept in a closed container such as a stoppered bottle, it does not evaporate as the vapour cannot escape. An equilibrium state is reached in which the space above the water in the bottle is saturated with water vapour. The water is said to be exerting its saturated (or equilibrium) vapour pressure. As with chemical equilibria, this is a dynamic equilibrium since the whole time molecules of water are moving from liquid to vapour and from vapour to liquid. The two rates are equal, so the movement in one direction is equal to the movement in the other.

The saturated vapour pressure depends strongly on temperature, as shown in Fig. 22.5. As the temperature rises, the vapour pressure rises, until at $100\,^\circ$C it has the value of 760 mmHg (1×10^5 Pa) which is the same as standard atmospheric pressure. If the atmospheric pressure is at this normal value, the water will boil at $100\,^\circ$C. Two variations on this usual behaviour are worth mentioning.

First, as any cook knows, food can be cooked more rapidly in a pressure cooker. This is a closed pan with a valve on the top which allows the steam pressure from the water inside to build up to about twice normal atmospheric pressure. It can be seen by referring to Fig. 22.5 that at this higher pressure the boiling point of water will be higher than $100\,^\circ$C; in fact about $120\,^\circ$C. This increase in the temperature causes a great increase in the rate of the reactions involved in the cooking process and so the food cooks in a shorter time.

Secondly, mountaineers know that at high altitude it is difficult to make a good cup of tea or to boil an egg. The reason for this is that at high altitude the atmospheric pressure is less than 760 mmHg (1×10^5 Pa), and consequently the boiling point of water (Fig. 22.5) is below $100\,^\circ$C. At this lower temperature the chemical processes involved in the brewing of tea or the boiling of an egg occur so slowly that the end result is not very satisfactory. Hence mountaineers prefer coffee and baked beans.

The above evidence indicates that the boiling of water is favoured by (a) high temperature and (b) low pressure. The reversible physical process involved in boiling and condensation can be represented by the equation:

$$H_2O(l) \rightleftharpoons H_2O(g)$$

The forward process (left to right) is endothermic, that is, heat must be supplied to make water boil. The reverse process is exothermic, that is, heat is given out when steam condenses (which is why a scald from steam is much worse than one from hot water). Thus the endothermic process (boiling) is favoured by a high temperature and the exothermic process (condensation) is favoured by a low temperature.

The forward process is also accompanied by a large increase in volume, since a liquid is changing into a gas. The reverse process involves a large reduction in volume.

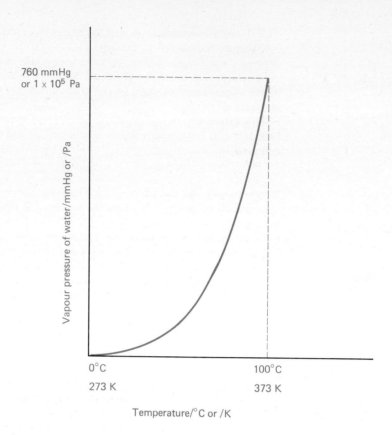

As illustrated earlier a low pressure favours boiling and a high pressure (which tends to make things contract) favours condensation. Fig. 22.6 summarises this discussion.

The principles illustrated here are found to apply to all reversible processes, irrespective of whether they involve physical or chemical changes. However, it must be stressed that the volume change is very slight when the reaction involves only solids and liquids, or when the number of molecules of gases on each side of the equation is the same. In these examples changes in pressure will have very little effect on the position of equilibrium.

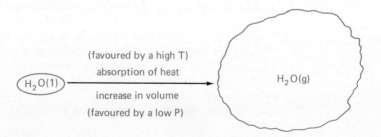

The general result was first set down in 1888 by the French scientist, Le Chatelier. He stated his principle in roughly the following words:

If a reaction is at equilibrium and one of the conditions is changed, the position of equilibrium will shift in such a way that the effect of the change tends to be opposed.

Let us see how this principle works in the case of a simple chemical equilibrium. A good example is provided by the gas which is given off when concentrated nitric acid reacts with copper. This is a brown gas which, under normal conditions, consists of a mixture of NO_2 molecules and N_2O_4 molecules in dynamic equilibrium:

$$2NO_2(g) \rightleftharpoons N_2O_4(g)$$

It is only the NO_2 molecules which cause the brown colour; the N_2O_4 molecules do not contribute to the colour.

The reaction as written above (reading from left to right) involves the combination of two molecules to form one molecule. This has two immediate consequences. Firstly, the forward reaction is exothermic, since new chemical bonds are being formed. Secondly, the forward reaction involves a decrease in volume, since the number of gaseous molecules is being halved (see Avogadro's Law, 5.6). By applying Le Chatelier's principle, what can we predict about the effect that changing the conditions of temperature and pressure will have on this equilibrium?

If we supply heat to the equilibrium mixture and so increase the temperature, then the position of equilibrium will shift in the direction in which heat is absorbed. This means that the net reaction will occur in the backward direction so that the proportion of NO_2 molecules increases (and the colour darkens) as some of the N_2O_4 molecules dissociate. On the other hand, if the reaction is at equilibrium and we decrease the volume so as to increase the pressure, the position of equilibrium will shift in the direction which tends to oppose the rise in pressure, so that it moves to the right and the proportion of N_2O_4 molecules increases (and the colour fades).

The reaction described above is not of any great practical importance but it is a very useful example for illustrating the principles involved. The important point is that the principles apply to all reactions involving gases and significant changes in volume. Two such reactions which are used in the chemical industry to produce the essential substances, ammonia and sulphuric acid, are described in the following sections.

22.16
Haber Process

The Haber Process (33.9) is used for producing ammonia and, as discussed in 33.8, is now an essential step in the nitrogen cycle which is so vital for the maintenance of the world's food supply. The chemical principles involved in using the method are dealt with in this chapter. The reaction used in the industrial process is reversible:

$$N_2(g) + 3H_2(g) \rightleftharpoons 2NH_3(g)$$

The forward reaction is exothermic and, according to Avogadro's Law, as the number of molecules is decreasing there will be a reduction in volume. The situation is therefore similar to the NO_2–N_2O_4 example discussed above. The forward reaction is favoured by a high pressure (since the position of equilibrium will move to the right in order to try to decrease the pressure) and a low temperature (since the position of equilibrium will move to the right and so tend to increase the temperature). Now these conditions are fairly easy to arrange, but the low temperature presents a problem. By using a low temperature we can get a very favourable position of equilibrium, but the trouble is that at a low temperature the rate of reaction is low.

Obviously the manufacturer wishes to produce ammonia at a reasonable rate, and it would be no good arranging the temperature to give a 99 % yield if the reaction was going to take 10 years to approach equilibrium and obtain such a yield. A compromise must be achieved, such that the temperature is high enough to give a reasonable rate of reaction, but not so high that the position of equilibrium is too unfavourable.

Faced with this sort of problem, an industrial chemist usually searches for a **catalyst** (23.15). A catalyst is a substance which speeds up a chemical reaction without being used up itself. Note that it can only speed up the attainment of equilibrium; it cannot alter the position of equilibrium. It turns out that iron, in a suitably treated form, acts as a good catalyst for the reaction between nitrogen and hydrogen.

When selecting a pressure to be used for the reaction the other factors which must be taken into consideration are the higher capital costs (for stronger reaction vessels etc.) and higher running costs (for more power to run compressors).

It is up to the chemical engineer to decide upon a set of conditions which will result in the production of ammonia at such a rate, and at such a cost, that the manufacturer can sell it at a price which is acceptable to the consumer but which still enables the manufacturer to make a profit.

Typical operating conditions for the Haber Process involve a pressure of about 200 atmospheres and a temperature of about 500 °C, with an iron catalyst to speed up the reaction. Under these conditions the proportion of the reaction mixture which, as it passes through the reaction vessel, is converted to ammonia is about 15%. The plant is

Fig. 22.7 The control panel which monitors and controls the conditions in an ammonia plant. (Courtesy ICI Ltd Agricultural Division)

therefore designed as a continuous flow system, with the ammonia being removed from the reaction mixture as it emerges from the reaction vessel, and unchanged nitrogen and hydrogen are not wasted as they are recycled. The ammonia is removed from the mixture by cooling, whereupon it condenses to a liquid and is tapped off and stored as a liquid under pressure. Liquid ammonia is transported under pressure in specially designed road and rail tankers.

22.17 Contact Process

The crucial stage in the manufacture of sulphuric acid is the oxidation of sulphur dioxide to sulphur trioxide. The reaction again involves gases and is reversible:

$$2SO_2(g) + O_2(g) \rightleftharpoons 2SO_3(g)$$

Like the Haber Process it is exothermic and, since the number of gaseous molecules decreases in the forward reaction, it involves a reduction in volume. Consequently the conditions for a favourable position of equilibrium are again low temperature and high pressure.

Again a catalyst is used, the gases being passed over vanadium(V) oxide (V_2O_5), which is why this stage is called the Contact Process. The position of equilibrium is much more favourable than that in the Haber Process. A temperature of 450 °C, at normal atmospheric pressure, is used. Under these conditions sulphur trioxide is produced in good yield (over 90%), at a satisfactory rate.

22.18 Measurement of heat changes

In the early sections of this chapter the observed energy changes associated with the reaction between two elements were used to provide an indication of the tendency of the elements to react together, and also an indication of the likely stability of the compound which had been formed. This section deals with how energy changes can be measured in a more precise manner.

210

Units

The heat change associated with a chemical reaction is the easiest form of energy change to measure. The old unit for measuring a quantity of heat was the calorie, defined as the heat required to raise 1 gram of water through 1 degree Celsius. Nowadays, however, energy in all its forms—whether it be heat, or mechanical work, or whatever—is measured in Joules (J). The Joule is defined as the work done when a force of 1 Newton moves its point of application through a distance of 1 metre. $4 \cdot 2$ J are required to raise 1 g of water through $1 \, °C$ (thus 1 calorie $= 4 \cdot 2$ J). Another way of expressing this is to say that the specific heat capacity of water is $4 \cdot 2$ J $g^{-1} \, °C^{-1}$. The heat changes during chemical reactions are often rather large, and it is convenient to use the kiloJoule (kJ), which is equal to 1000 J.

The actual quantity of heat associated with a particular reaction clearly will depend on the quantities of substances used. For example, if 100 kJ are given out when 4 g of magnesium are burned, 200 kJ will be given out when 8 g of magnesium are burned. So we need to specify the quantity of matter, and for the chemist the most fundamental measurement of quantity of a substance is the mole. Thus heat changes in chemical reactions are expressed in kJ per mole, usually written as kJ mol^{-1}.

Finally we need a convention to indicate whether the heat is being given out (exothermic reaction) or absorbed (endothermic reaction). The symbol used to mean heat change is ΔH, and the number (of kJ mol^{-1}) is then given a sign (+ or −) which indicates whether the system (i.e. the reacting mixture) is gaining or losing energy. Thus the statement,

$$C(s) + O_2(g) \rightarrow CO_2(g) \qquad \Delta H = -394 \text{ kJ mol}^{-1}$$

means that when 1 mole of carbon combines with 1 mole of oxygen molecules to give 1 mole of carbon dioxide molecules, the heat change which occurs is negative 394 kJ; that is, the reacting substances are losing 394 kJ. This means that 394 kJ are being given out to the surroundings, and the reaction is exothermic.

Fig. 22.8 (a) An exothermic reaction (b) An endothermic reaction

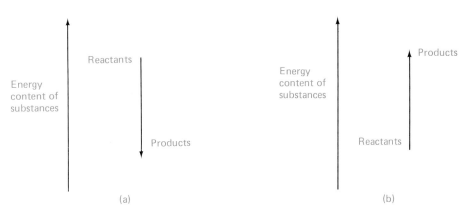

We have no means of knowing the actual energy content of substances and so all we are able to do is to measure the heat changes when the reactants change to the products. In an exothermic reaction the products contain less energy than the reactants. Therefore the change can be represented by an energy level diagram such as Fig. 22.8(a) which shows ΔH for the reaction as being negative, representing the difference between the energy content of the reactants and products.

The products of an endothermic reaction contain more energy than the reactants, therefore ΔH is given a positive sign and the change can be represented by the energy level diagram shown in Fig. 22.8(b).

211

Calorimetry

Calorimetry is the technique by which heat changes in reactions are measured. The important practical point, however, is that what we actually measure directly is not heat, but a temperature change, and so a little calculation is required to convert this to a quantity of heat.

If we consider the more common example, an exothermic reaction, there are two ways in which we can measure the heat given out. We can either allow all the heat to be conducted away from the reaction mixture and record the temperature increase of the surroundings. Alternatively, we can prevent any of the heat escaping to the surroundings and record the temperature rise of the reaction mixture itself.

In order to illustrate how a temperature rise is converted to a quantity of heat, let us consider the reaction between zinc and copper(II) sulphate solution as studied in Investigation 22.1. The reaction, which is carried out in an insulated container, is,

$$Zn(s) + CuSO_4(aq) \rightarrow ZnSO_4(aq) + Cu(s)$$

or

$$Zn(s) + Cu^{2+}(aq) \rightarrow Zn^{2+}(aq) + Cu(s)$$

The equation indicates that 1 mole of zinc displaces 1 mole of copper. In the experiment 20 cm³ of 0·5M $CuSO_4$ solution were used, and this solution contains

$$0·5 \times \frac{20}{1000} = 0·01 \text{ mol of } CuSO_4.$$

About 1 g of zinc (relative atomic mass 65) was used, and this is about

$$\frac{1}{65} = 0·015 \text{ mol of Zn.}$$

The zinc is therefore in excess, and, when all the copper(II) sulphate has reacted, there will be some zinc left. (If you did this experiment you should have observed that the blue colour of the original solution, which is due to $Cu^{2+}(aq)$ ions, disappears entirely.)

So, when all of the $Cu^{2+}(aq)$ ions have reacted, 0·01 mol of Zn(s) will have reacted with 0·01 mol of $Cu^{2+}(aq)$. Let us suppose that this reaction causes a temperature rise of 26 °C. Because the reaction was carried out in an insulated beaker, the heat which was produced by the reaction has been used to raise the temperature of the contents of the beaker through 26 °C.

At this stage, for the purpose of this experiment, we use an approximation. Since the solution is dilute we will take its specific heat capacity as being the same as that of pure water, that is, 4·2 J g^{-1} °C^{-1}. 20 cm³ of solution were used, so this corresponds to a heat capacity of 4·2 × 20 = 84 J °C^{-1}.

If the temperature rise was 26 °C then the actual quantity of heat produced was

$$84 \times 26 = 2180 \text{ J}$$

$$= 2·18 \text{ kJ}$$

This quantity of heat was produced by 0·01 mol of each reactant reacting together, so the quantity of heat produced by 1 mol would be

$$2·18 \times 100 = 218 \text{ kJ}$$

Finally, as the reaction is exothermic, ΔH is negative and the total change can be represented by:

$$Zn(s) + CuSO_4(aq) \rightarrow ZnSO_4(aq) + Cu(s) \qquad \Delta H = -218 \text{ kJ mol}^{-1}$$

The heat of combustion of a substance is defined as the heat change when 1 mole of that substance is completely burned in oxygen, and is expressed in kJ mol^{-1}. Thus the statement in a previous section regarding the combination of carbon with oxygen,

$$C(s) + O_2(g) \rightarrow CO_2(g) \qquad \Delta H = -394 \text{ kJ mol}^{-1}$$

tells us that the heat of combustion of carbon is -394 kJ mol^{-1}.

The statement that the heat of combustion of methane (CH_4) is -890 kJ mol^{-1} means that when 1 mole of methane is completely burned in oxygen, 890 kJ are produced, that is:

$$CH_4(g) + 2O_2(g) \rightarrow CO_2(g) + 2H_2O(l) \qquad \Delta H = -890 \text{ kJ mol}^{-1}$$

Similarly, the heat of combustion of octane (C_8H_{18}) is -5512 kJ mol^{-1}, meaning:

$$C_8H_{18}(l) + 12\tfrac{1}{2}O_2(g) \rightarrow 8CO_2(g) + 9H_2O(l) \qquad \Delta H = -5512 \text{ kJ mol}^{-1}$$

Clearly heats of combustion are important when studying the components of fuels such as methane and octane. The values can be used to determine the quantity of heat produced by any particular amount of the fuel. They are also important in the study of most organic compounds, as the heat of combustion for most compounds is the easiest heat change associated with that compound to measure. They can be used along with other data to determine other characteristics of compounds such as their likely stability.

Heats of neutralisation provide a good illustration of how the measurement of heat changes can be used to find out more about what is happening when reactions occur.

If you carried out Investigation 22.2 you may have found that the observed temperature rise, when equal volumes of solutions of acids and alkalis of equal concentrations were mixed, seemed to be about the same whichever acid or alkali was used. At first this might seem a remarkable coincidence. For example, the reactions,

$$NaOH(aq) + HCl(aq) \rightarrow NaCl(aq) + H_2O(l)$$

and
$$KOH(aq) + HNO_3(aq) \rightarrow KNO_3(aq) + H_2O(l)$$

would seem to have little in common except for both being acid-alkali reactions, and yet the fact that they have the same heat change associated with them suggests that they must be very similar.

If we consider the reactions from the ionic viewpoint, it is possible to suggest an explanation for these observations. Sodium hydroxide and potassium hydroxide are both strong alkalis, that is, when they are in dilute solutions they exist entirely as separate ions. Similarly, hydrochloric acid and nitric acid are both strong acids and are fully ionised in dilute solutions. Furthermore sodium chloride and potassium nitrate are both salts and therefore exist as separate ions in solution. If we rewrite the equations with these facts in mind, we obtain, respectively,

$$Na^+(aq) + OH^-(aq) + H^+(aq) + Cl^-(aq) \rightarrow Na^+(aq) + Cl^-(aq) + H_2O(l)$$

and
$$K^+(aq) + OH^-(aq) + H^+(aq) + NO_3^-(aq) \rightarrow K^+(aq) + NO_3^-(aq) + H_2O(l)$$

Ignoring ions which appear on both sides of the equation (spectator ions, which do not take part in the reaction) we obtain for both reactions the simplified equation:

$$H^+(aq) + OH^-(aq) \rightarrow H_2O(l)$$

This is all that actually happens when we mix an acid and an alkali, and so it is less surprising that the two reactions give the same temperature rise; we are in fact observing the consequences of the same process in both cases.

Heat of neutralisation is defined as the heat change which occurs when a dilute solution of an acid containing 1 mole of $H^+(aq)$ ions is neutralised by a dilute solution of an alkali. It can be calculated from typical experimental results in the following way.

When 25 cm³ of 2M hydrochloric acid and 25 cm³ of 2M sodium hydroxide solution are reacted together in an insulated container the temperature rise is about 13 °C. The total volume of solution which is raised through this temperature is 50 cm³ and we make the assumption that as the solution is dilute its specific heat capacity is that of pure water; $4 \cdot 2 \, J \, g^{-1} \, °C^{-1}$.

The heat produced in the reaction is
$$4 \cdot 2 \times 50 \times 13 = 2730 \, J$$
$$= 2 \cdot 73 \, kJ$$

25 cm³ of 2M acid contains $\quad 2 \times \dfrac{25}{1000} = 0 \cdot 05$ mol of $H^+(aq)$ ions.

Similarly the alkali solution contains $\quad 0 \cdot 05$ mol of $OH^-(aq)$ ions.

The reaction which produced $2 \cdot 73$ kJ was:

$$0 \cdot 05 \text{ mol of } H^+(aq) + 0 \cdot 05 \text{ mol of } OH^-(aq) \rightarrow 0 \cdot 05 \text{ mol of } H_2O(l)$$

If 1 mole of each reagent had been used the heat change would have been 20 times greater, which is $2 \cdot 73 \times 20 = 55$ kJ.

The accepted value for the heat of neutralisation derived by very careful experimental work is $-57 \, kJ \, mol^{-1}$.

$$H^+(aq) + OH^-(aq) \rightarrow H_2O(l) \quad \Delta H = -57 \, kJ \, mol^{-1}$$

The negative sign is used to indicate that the reaction is exothermic.

22.21 Summary

1. In chemical reactions energy is often given out as heat; such a reaction is called exothermic. When heat is absorbed during a reaction it is called endothermic. The more reactive an element is, the more energy is given out when it forms a compound with another element, and the harder it is to get the element back out of the compound.

2. When one element displaces another the reaction generally goes in the exothermic direction, so that the more reactive element displaces the less reactive one, and the more stable compound is formed at the expense of the less stable one.

3. Very few compounds can be decomposed to their elements at Bunsen temperature, but many compounds will undergo partial decomposition. The most stable hydroxides, carbonates and nitrates are those of the most reactive metals. The compounds of the less reactive metals require less energy to decompose them.

4. Some reactions are reversible. If given a chance they come to a state of dynamic equilibrium, but by arranging for one of the products to escape they may be driven to completion in that direction.

5. In a reversible reaction the position of equilibrium is generally affected by temperature, pressure and, where solutions are involved, concentrations. Le Chatelier's Principle enables us to predict the direction of the effect, which is important in many industrial processes.

6. Heat changes in chemical reactions may be measured using calorimetry. They are expressed in $kJ \, mol^{-1}$ and given the symbol ΔH. The sign of ΔH is negative for an exothermic reaction and positive for an endothermic reaction.

7. The heat of neutralisation of any strong acid by any strong alkali is virtually the same.

Chapter 23

What influences the rate of chemical reactions?

**Investigation
23.1**

Do reactions of similar kinds proceed at similar rates?

1. Pour barium chloride solution into a test-tube to a depth of about 2 cm. Then pour an equal volume of lead(II) nitrate solution into a second test-tube and an equal volume of calcium chloride solution into a third.

To each test-tube, add an equal volume of dilute sulphuric acid. Observe and compare the results, allowing a few minutes if necessary.

Questions

1 What kind of reaction is occurring in all three of the experiments?
2 Which of these reactions occurs at an untypical rate for this kind of reaction?

2. Pour dilute potassium manganate(VII) solution (previously acidified with dilute sulphuric acid) into each of three test-tubes to a depth of about 2 cm. To the first test-tube add an equal volume of sodium sulphite solution, to the second add an equal volume of sodium thiosulphate solution, and to the third add an equal volume of sodium ethanedioate (sodium oxalate) solution.

Observe and compare the results, allowing a few minutes if necessary.

Questions

3 What kind of reaction is occurring in all three experiments?
4 Which of these three reactions occurs at an untypical rate for this kind of reaction?

**Investigation
23.2**

How does the rate of the reaction of magnesium with a dilute solution of an acid vary with time?

A suitable apparatus to use for this investigation is shown in Fig. 23.1, but if gas syringes are not available an apparatus involving measuring the volume of hydrogen by displacement of water can be used.

Fig. 23.1

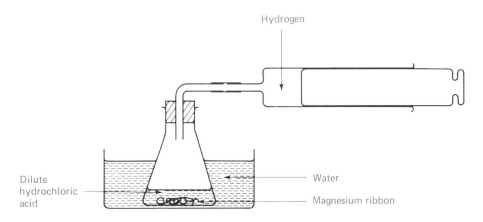

Hydrogen

Dilute hydrochloric acid

Water

Magnesium ribbon

You will need a 15 cm length of magnesium ribbon, rolled into a loose ball, and a stop-clock or a watch with a second hand.

Put 25 cm³ of 0·25M hydrochloric acid into the flask. Drop the magnesium into the acid, replace the stopper and start the clock. Record the gas volume every 30 seconds. Shake the flask very gently to keep the contents evenly mixed.

When the experiment is finished, plot a graph showing volume (vertically) against time (horizontally).

Questions

1   Why is the reaction flask surrounded by water?
2   Why does the reaction eventually come almost to a stop?
3   Which reactant is in excess in this experiment?
4   When is the rate of reaction greatest?
5   What is a possible explanation for the rate varying in the way that it does?

Investigation 23.3

What factors affect the rate of reaction of magnesium with dilute acid?

You will need the same apparatus as for Investigation 23.2. In each part of the investigation you will vary one factor only, keeping the other conditions constant.

1. Concentration of the acid

Repeat the procedure used in Investigation 23.2 three times. In each experiment add the same volume of acid solution but use three different concentrations (say 1M, 0·5M and 0·25M). Use only 5 cm of magnesium ribbon in each experiment.

In each case, after the reaction has been started, take readings of the gas volume every 30 seconds. Stop when all the magnesium has reacted, or after 10 minutes, whichever is sooner.

For each set of results plot a graph showing volume against time. Then draw, as accurately as you can, a tangent to each curve at the origin (see Fig. 23.5).

Use each tangent to work out what volume of gas would have been produced in 1 minute, if the rate of reaction had remained the same as at the start of each experiment. These values are called initial rates.

Compare the results obtained using the different acid concentrations.

Questions

1   Why is it better, in this type of experiment, to compare the initial rates rather than rates at any other stage of the reactions?
2   What happens to the rate of reaction as the concentration of acid increases?

2. Temperature

In each experiment use 5 cm of magnesium ribbon and 25 cm³ of 0·25M hydrochloric acid, and by following the procedure obtain sets of results for various temperatures.

Before starting the reaction, warm both the hydrochloric acid and the tap water surrounding the flask to about the required temperature (say 30°C or 40°C or 50°C).

After the acid flask has been standing in the warm water for about 2 minutes, measure the acid temperature and then start the reaction. Take readings of the gas volume every 30 seconds until all of the magnesium has reacted or until 10 minutes have passed.

Repeat the experiment at different temperatures and plot graphs in the usual way. Draw a tangent at the origin and obtain a value for the initial rate for each experiment.

Compare the results obtained using different temperatures.

Questions

3 What happens to the reaction rate as the temperature is increased?
4 What might you do with the results to find out whether there is a simple proportionality between rate of reaction and temperature?

3. The physical state of the magnesium
In each experiment, use 25 cm³ of 0·25M hydrochloric acid at room temperature.

Weigh 5 cm of magnesium ribbon. Carry out the experiment using this ribbon and then obtain further sets of results using firstly the same mass of small magnesium turnings and secondly the same mass of magnesium powder.

Draw a graph for each experiment and by using the tangent method compare the initial rates of reaction.

Questions

5 What happens to the rate of reaction when smaller particles of magnesium are used?
6 What explanation can you suggest for the variation in the rate, bearing in mind that the mass of the metal is about the same in each experiment?

Investigation 23.4

What factors affect the rate of the reaction between sodium thiosulphate solution and hydrochloric acid?

Rather than follow the reaction by recording a quantity of a reactant or product at regular time intervals, the most convenient method for this reaction is to record the times for the reaction to reach a certain stage when carried out under different conditions.

The equation for the reaction is:

$$Na_2S_2O_3(aq) + 2HCl(aq) \rightarrow S(s) + 2NaCl(aq) + H_2O(l) + SO_2(aq)$$

In each experiment the time is recorded for the reaction mixture to become so cloudy, owing to the precipitated sulphur, that it obscures a cross marked on a piece of paper placed under the beaker of solution.

1. Concentration of the sodium thiosulphate solution
Draw a bold cross on a piece of white paper, using a soft pencil or a dark pen. Put 20 cm³ of 0·2M sodium thiosulphate solution into a small beaker and stand it on top of the cross.

Measure exactly 2 cm³ of 2M hydrochloric acid into a test-tube; then, at a noted time (or the start of a stop-clock), add the acid all at once to the sodium thiosulphate solution and stir it well with a glass rod. Now look down at the cross through the reaction mixture and note the time when the cross is obscured by the sulphur precipitate.

Repeat the experiment using different concentrations of sodium thiosulphate solution as described below. In each case use the same piece of marked paper, the same quantity of acid and the same beaker. Make sure that the beaker is well washed between experiments.

For the repeat runs use:

   (i)   15 cm³ sodium thiosulphate solution + 5 cm³ water
  (ii)   10 cm³ sodium thiosulphate solution + 10 cm³ water
 (iii)   5 cm³ sodium thiosulphate solution + 15 cm³ water

In each case measure the time taken for the cross to become invisible.

Questions

1   Why is water added to the portions of sodium thiosulphate solution in the repeat runs?

2   If it takes a longer time for the cross to be obscured, does this mean that the rate is higher or lower?

3   What happens to the rate of reaction as the concentration of the sodium thiosulphate solution is decreased?

2. Temperature

Use the same apparatus, with 5 cm³ of sodium thiosulphate solution, 15 cm³ of water and 2 cm³ of 2M hydrochloric acid. Before adding the acid, warm the mixture of water and sodium thiosulphate solution to about the required temperature (say 30°C or 40°C or 50°C).

Measure the actual temperature, add the acid, stir well and record the time taken for the cross to disappear. Repeat the experiment at different temperatures and compare the results.

Question

4   What happens to the rate of reaction as the temperature increases?

Investigation 23.5

What factors affect the rate of reaction between marble (calcium carbonate) and dilute acids?

You will need an apparatus of the kind shown in Fig. 23.2 so that you can measure the rate of reaction by timing the production of, say, 20 bubbles of carbon dioxide gas.

Fig. 23.2

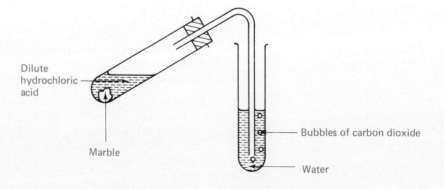

Dilute hydrochloric acid

Marble

Bubbles of carbon dioxide

Water

Put a lump of marble (about 1 cm³) in the boiling-tube and add 10 cm³ of dilute hydrochloric acid (2M). Time the reaction.

Now apply a small Bunsen flame for about 10 seconds to warm up the reaction mixture slightly; remove the flame and time the reaction again.

Now rinse out the reaction-tube and this time put some calcium carbonate powder (about 1 cm³) in the tube. Add 10 cm³ of dilute hydrochloric acid, and time the reaction.

Questions

1 What happens to the rate of reaction when the temperature is increased?
2 How does the rate of reaction using powdered marble compare with the rate of reaction using a lump of marble? Suggest an explanation for this difference in rate.

Investigation 23.6

Can the rate of a reaction be affected by the presence of other substances?

The decomposition of hydrogen peroxide to water and oxygen,

$$H_2O_2(aq) \rightarrow H_2O(l) + \tfrac{1}{2}O_2(g)$$

occurs rather slowly when a dilute aqueous solution of hydrogen peroxide is heated or exposed to sunlight.

The purpose of this experiment is to see whether the presence of small amounts of other substances (either as solids or in solution) will encourage the decomposition so that it occurs at a noticeable rate (as seen by effervescence) at room temperature.

For each test use about 2 cm³ of 20 volume hydrogen peroxide solution in a test-tube and add, without warming, a very small quantity of the substance being tested. Observe whether effervescence occurs, and make a note of how vigorous the reaction is, so that different substances may be compared. Allow a minute or so, with shaking, before you leave each test.

Here is a list of possible substances to try:

SOLIDS	SOLUTIONS
magnesium oxide	dilute hydrochloric acid
zinc oxide	dilute sodium hydroxide
manganese(IV) oxide	manganese(II) sulphate
lead(II) oxide	iron(III) chloride
lead(IV) oxide	copper(II) sulphate
copper(I) oxide	lead(II) nitrate
copper(II) oxide	silver nitrate
copper turnings	blood
iron filings	sodium chloride
sand	sugar

Classify the substances you tested, by firstly noting any which appeared to have no effect at all, and any which appear to make the decomposition vigorous. Then try to place the latter group in an order of effectiveness.

Question

What link is there between the substances which make the decomposition of hydrogen peroxide vigorous?

23.7 Rate of reaction

All around us chemical changes are going on all the time. Many natural processes occur rather slowly, for example the growth of plants and animals, the rusting of iron, the weathering of rocks, the decay of rubbish, and the formation of stalactites in caves. These processes occur on a time-scale of months, years or even centuries. There are

219

even some processes which are slower still; for example the radioactive decay of uranium-238 takes place over millions of years. In everyday life we make use of chemical processes which occur on a time-scale of minutes or hours; for example the burning of a candle or the cooking of food. In a chemistry laboratory we often meet reactions which are over in a matter of seconds, and we meet many which are instantaneous, as far as we can tell by simply looking at them. Clearly, different processes occur at different rates. By the word rate we really mean speed. To walk from your home to a shop takes longer than to cycle because walking involves a lower speed, or rate of progress. Driving gives an even higher rate of progress than cycling, so the journey time is even less.

If we take approximately equal portions of magnesium turnings and iron filings, and put them simultaneously into separate equal portions of dilute sulphuric acid, we find that the reaction between the acid and the iron takes longer to come to an end than the reaction between acid and magnesium. We can say that the rate of reaction of magnesium with acid is greater than that of iron with acid. We can link this with the metal reactivity order (17.9), magnesium being a more reactive metal than iron.

It would be a great mistake, however, to think that a discussion of reaction rates is nothing more than a discussion of reactivity. For example, hydrogen and oxygen are both reactive elements, but a mixture of these two gases is stable indefinitely at room temperature. Such a mixture can be kept in a sealed container for years without any change occurring. If, then, a flame is applied to the mixture an explosion will occur, caused by the very rapid combination of the elements to form water, liberating a large amount of energy within a millionth of a second or so. Here is a reaction which occurs at an extremely rapid rate, once it has been set off.

It should be clear, just from the few examples mentioned above, that rate of reaction depends on many things. Obviously the identity of the reactants is very important, but quite apart from that we must consider the conditions under which the reaction is occurring. By conditions we mean things like temperature, pressure, the physical state of the reactants, the presence of other substances which might affect the reaction, perhaps even the presence or absence of light. All these variables will be considered in the following sections of this chapter.

23.8 Prediction of rates

In general, it is very difficult to predict accurately the rate of any reaction. Many reactions between ions in solution are very rapid and yet exceptions are found, which you could not predict. For example, the precipitation of calcium sulphate on mixing solutions containing calcium ions and sulphate ions is slow compared to other precipitation reactions such as those forming barium sulphate or calcium hydroxide.

Similarly the oxidation of various ions by potassium manganate(VII) occurs at different rates. The reaction with sulphite ions is rapid, whereas the reaction with ethanedioate ions is slow at room temperature. We could suggest that the reason for this is that the C–C bond in the ethanedioate $(C_2O_4{}^{2-})$ ion has to be broken. This is still not a satisfactory explanation because the reaction between manganate(VII) ions and thiosulphate ions $(S_2O_3{}^{2-})$ in which S–S bonds are broken is rapid. These differences in rate are difficult to explain and certainly would have been very difficult to predict. The main purpose of this chapter is not to try to make comparisons between the rates of different reactions, but to consider how the rate of a particular reaction is likely to be affected by different conditions and how we can vary the rate. Before we can do this, we must first see what kinds of measurements we are going to need to make in order to study reaction rates.

Fig. 23.3 Apparatus used
in a research laboratory
for investigating the rates
of reactions between gases.
(Photo: J. Olive)

A study of reaction rates means a study of how fast they occur, so clearly we must make measurements involving timing—but what, exactly, should we time? In every chemical reaction certain substances are changing into other substances. The

substances referred to as reactants are being used up while the products are the substances being formed. Thus a measure of the rate of reaction will involve either measurement of how much of a reactant is used up in a certain time, or how much of a product is formed in a certain time. We would not need to measure both, since the equation for the reaction gives a direct connection between the two.

For example, suppose we are preparing copper(II) sulphate from copper(II) oxide. In this reaction the black, water-insoluble solid, copper(II) oxide, is stirred with warm aqueous sulphuric acid and gradually a blue solution of copper(II) sulphate is formed:

$$CuO(s) + H_2SO_4(aq) \rightarrow CuSO_4(aq) + H_2O(l)$$

black   colourless   blue

There are various ways in which we could follow the progress of this reaction. For example, we could take a known mass of copper(II) oxide, stir it with acid for a measured time, then quickly filter the mixture, dry the unreacted copper(II) oxide, and find its mass again. Alternatively we could let the reaction proceed for a measured time, filter the mixture, then analyse the solution to see how much copper(II) sulphate is present. Both these methods would be rather time consuming.

Perhaps a quicker way would be to use an indirect method. Instead of measuring the quantity of any substance directly, we measure some property of the reaction mixture which depends upon the quantity of a certain substance present.

Colour is a property which can often be used. For example, the reaction between potassium manganate(VII) and ethanedioic acid (oxalic acid) can be followed continuously without needing to stop the reaction to analyse the mixture. The potassium manganate(VII) is purple in colour and the intensity of the colour depends on the concentration of the manganate(VII) ions. As the ions are used up in the reaction the intensity of the colour decreases. Instruments called colorimeters are available which can measure the intensity of colours and so it would be possible to record how the intensity of the purple colour changes with time.

In an indirect method such as this, it is necessary to calibrate the instrument. This is done by making up solutions of potassium manganate(VII) of known concentrations and measuring the intensities of the colours of these solutions. In this way an intensity measured during the reaction can be converted to a concentration of potassium manganate(VII).

Other indirect methods are more suitable for other reactions and can be used to avoid stopping the reactions. For example, in reactions involving an acid:

$$MgO(s) + H_2SO_4(aq) \rightarrow MgSO_4(aq) + H_2O(l)$$

the change in conductance may be followed as magnesium sulphate has a lower conductance than sulphuric acid. The conductance method will need calibration by taking measurements with solutions of known concentrations.

Where one of the products is a gas, it is very easy to follow the reaction. Some of the investigations at the beginning of this chapter concern the reaction of magnesium metal with dilute hydrochloric acid, in which hydrogen is given off:

$$Mg(s) + 2HCl(aq) \rightarrow MgCl_2(aq) + H_2(g)$$

We can easily follow the rate of formation of hydrogen by collecting it, as it is formed, in a suitable graduated container such as a gas syringe. Using the gas laws we can, if required, convert the volume of hydrogen into the number of moles of hydrogen. We do not actually do such a conversion in the investigations in this chapter because we are concerned only in comparing rates for the reaction under different conditions, not in measuring absolute rates.

We can see, then, that many methods are available for following the rate of a reaction. In any particular case we must choose a suitable property and measure how it changes with time. Indirect methods, which do not involve stopping the reaction, are generally preferred because they are less time consuming.

23.10
Why measure rates?

Measurements of rates of reactions are important in trying to establish theories about reactions, but they are even more important in providing chemical engineers with the necessary information for designing industrial processes. When designing a particular chemical plant, it is essential to have information on how fast the reactions go, and how the rate depends upon conditions such as temperature and pressure. The following sections of this chapter are concerned with how reaction rates are affected by conditions.

23.11
Do rates depend upon concentrations?

In the study of the reaction between magnesium metal and dilute hydrochloric acid, the hydrogen gas given off can be collected in a gas syringe (Investigation 23.2) and its volume recorded at regular time intervals. The results, when plotted on a graph, look similar to Fig. 23.4. In this experiment the magnesium is in excess, so the reaction comes to an end when the acid has all been used up. One thing which is immediately noticeable is that the reaction does not stop suddenly. The dashed line on Fig. 23.4 indicates the final volume of hydrogen, and it can be seen that this maximum volume is approached slowly and gradually. As time proceeds, the reaction becomes slower and slower, as indicated by the fact that the slope of the curve becomes progressively less steep. The slope of the line at any point represents the rate of production of hydrogen in terms of volume per unit time, and is therefore a measure of the rate of reaction at that time. Thus, the most obvious conclusion from the experiment is that the reaction rate steadily decreases as the reaction proceeds, finally approaching zero as the acid supply runs out.

Fig. 23.4

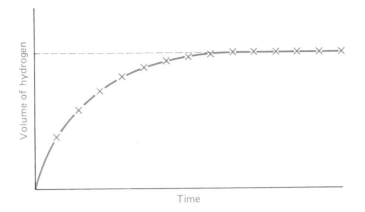

The reaction rate steadily decreases because the concentration of the acid steadily decreases as the reaction converts it gradually into magnesium chloride:

$$Mg(s) + 2HCl(aq) \rightarrow MgCl_2(aq) + H_2(g)$$

The quantity of magnesium metal is also decreasing, but since, in this experiment, the metal is in excess, it is more likely that the acid concentration is the important factor. In any case, this idea can be checked by repeating the experiment with the same quantity of magnesium ribbon, put into different solutions of hydrochloric acid. The volume of each acid solution is kept the same but the concentration of the acid is changed from one sample to another. When we plot the graphs of volume of hydrogen against time we obtain curves similar to those in Fig. 23.5. It is very clear that the more concentrated the acid solution is, the steeper is the curve and the greater is the rate of reaction. Thus our idea is correct: the rate of reaction does depend upon the acid concentration, and the reason why the reaction rate decreases as the reaction proceeds is that the acid concentration is decreasing.

Fig. 23.5

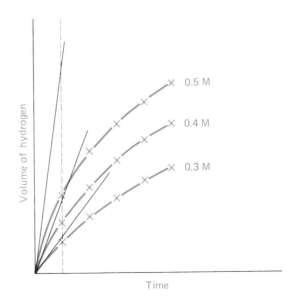

If we wish to make a numerical comparison between the reaction rates at various concentrations, a convenient way to do this is to use the method of initial rates. In order to obtain a measure of the reaction rate at the start, when the acid concentration has not fallen, we draw a tangent to the curve at the origin of the graph, as shown in Fig. 23.5, and then we measure the slope of this tangent, which gives a measure of the **initial reaction rate**. We could express this as, for example, the volume of hydrogen which would have been produced in 1 minute had the reaction rate remained at its initial value. The advantage of using initial rates is that it is only at the beginning of each reaction that we know the concentration of the acid.

If you compared the results for the various graphs you would find that the relation between reaction rate and concentration of the acid is not a simple one. You might, perhaps, expect that doubling the acid concentration would double the reaction rate but it appears that things are not quite as simple as this. In general, it has been found that in chemical reactions there may or may not be a simple relationship between rate and concentration; sometimes there is, but the results are quite impossible to predict.

The experiments discussed above involve a solid reacting with a solution. When hydrochloric acid is added to sodium thiosulphate solution, a fairly slow reaction takes place. In this reaction sulphur is precipitated, as a result of which the reaction mixture becomes cloudy:

$$Na_2S_2O_3(aq) + 2HCl(aq) \rightarrow S(s) + 2NaCl(aq) + SO_2(aq) + H_2O(l)$$

A convenient way of comparing the rate of reaction for different conditions is to measure how long it takes for the cloudiness to obscure a cross marked on a piece of paper under the reaction flask. In this case, rather than following the rate of reaction, we are comparing the times for it to reach a certain point. The faster the rate, the shorter will be the time taken. Thus the rate is inversely proportional to the time, and rates for different conditions can be compared by comparing the reciprocals of the reaction time

(i.e. $\dfrac{1}{time}$) for each set of conditions.

If a series of experiments is done in which only the concentration of the sodium thiosulphate solution is varied, it will be found that the rate increases as the concentration of sodium thiosulphate increases, but you will not necessarily find that it is a simple relationship.

The increase of reaction rate with concentration is found to apply to most reactions, although the exact nature of the relationship varies widely, and is certainly impossible to predict from the equation for the reaction. The effect applies to gaseous reagents as well as to reagents in solution, although we usually talk about gas pressures rather than concentrations. It is very important for the chemical engineer to know how the rate of a particular reaction he is using varies with concentration so that he can design a plant in such a way as to achieve concentrations which will give a suitable rate of reaction. Hence an essential preliminary stage is to investigate how the concentration of each reagent influences the rate of reaction.

23.12
Do rates depend upon temperature?

We know from everyday life that quite a lot of processes go more quickly, or more readily, if the temperature is raised. One obvious example is that sugar dissolves much more rapidly in hot water (or tea and coffee) than in cold water. Controlled experiments can be carried out to see if the rate of a particular reaction varies with temperature.

For example, equal volumes of the same hydrochloric acid solution can be put into different flasks and warmed to different temperatures, before adding equal quantities of magnesium ribbon. Temperature is the only condition which is being varied in this series of experiments.

If the rate of production of hydrogen is recorded (Investigation 23.3), the results will be similar to Fig. 23.6. The obvious conclusion is that as temperature is increased, keeping other conditions constant, the reaction rate increases. To make a numerical comparison between results it is convenient to use the method of initial rates as before, and compare the slopes of the tangents drawn as in Fig. 23.6. We can see from this comparison just how much effect the temperature has on the rate. In this particular experiment you may find that the rate is approximately doubled by raising the temperature from 20 °C to 40 °C.

Fig. 23.6

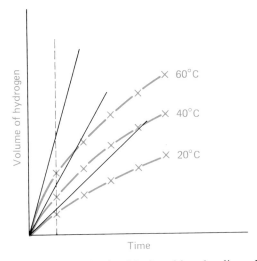

Similarly the reaction between hydrochloric acid and sodium thiosulphate solution can be carried out with equal volumes of the two solutions at a series of different temperatures. All conditions are kept constant, apart from temperature, and it is found that as temperature is increased the time taken for the cross to be obscured by precipitated sulphur decreases. Thus, again, the rate of reaction increases with temperature.

The increase of reaction rate with temperature is found to apply for virtually all reactions, but the extent to which this occurs varies from reaction to reaction. The effect is cumulative, that is, if the rate of a particular reaction is roughly doubled by a rise of 10 °C, then a further 10 °C rise doubles it again, so that a 20 °C rise quadruples the rate.

There is an important practical consequence of the fact that reaction rates generally rise markedly as temperature increases. Many chemical reactions are exothermic, that is, heat is given out as the reaction proceeds. If this heat is not allowed to escape then the reaction mixture becomes hotter. When the mixture becomes hotter the reaction rate increases, and so more heat is given out per unit time. Hence an exothermic reaction could run completely out of control as the rate increases to the point of an explosion. This is what we want to happen when petrol is ignited inside the engine of a motor car, but clearly on other occasions it is not desirable.

Generally, in reactions which we carry out in school laboratories we do not often need to worry about such extreme results because the quantities we use are so small, but on the industrial scale these considerations are very important. The chemical engineer, when designing an industrial plant, must know how the rates of the reactions concerned vary with temperature and also know how much heat, if any, is given out as the reactions proceed. It must be possible to control the temperature so that it can be kept almost constant. This is particularly important when a continuous flow process is being used. This is a process in which reactants are being continually fed into one end of the reaction container and products are continually being removed from the other end. If the temperature is not kept constant the process will either grind to a halt or go out of control.

A large proportion of the complicated network of pipes which can be seen in industrial plants is concerned simply with **temperature control**. Heat exchangers are used in which a circulating fluid removes heat where it is being generated by the reactions and transfers it to wherever it may be needed—often to pre-heat the reactants before they enter the reaction vessels.

23.13
How do solids behave in reactions?

Many of the reactions that we meet in the laboratory involve a solid reacting with an aqueous solution of a reagent. For example, most metal oxides and carbonates are insoluble in water but they will dissolve in solutions of acids because they react with the acid to form a soluble salt. As with other types of reactions, it is difficult to predict how the rates at which different solids react with acid will compare.

We can explain why copper(II) oxide reacts at a similar rate with hydrochloric acid and nitric acid of equivalent concentrations, by writing the simplest ionic equation for the reactions. In each case copper(II) oxide is reacting with hydrogen ions. The sulphate ions and the nitrate ions from the acids take no part in the reactions:

$$CuO(s) + 2H^+(aq) \rightarrow Cu^{2+}(aq) + H_2O(l)$$

As long as we have equal concentrations of hydrogen ions the reaction rates will be similar.

However, we cannot easily explain why lead(II) oxide reacts more quickly with dilute nitric acid than copper(II) oxide does. It cannot be explained by lead being higher in the reactivity series than copper because the oxide of magnesium, even though magnesium is higher in the series than lead, reacts at a similar rate to copper(II) oxide.

We have already seen how the temperature and concentration of the acid affects the rate of the reaction between magnesium and dilute hydrochloric acid. Is there anything we can now do to the magnesium which will alter the rate of the reaction? We can, of course change the mass of the magnesium present, but we can also alter the sizes of the pieces of the magnesium metal which are added to the acid.

We can take three samples of the metal, of the same mass, and put them into equal volumes of hydrochloric acid of the same concentration, at the same temperature. The difference between these magnesium samples is that one is ribbon, the second is small turnings and the third is fine powder. Typical results of the experiment are shown in Fig. 23.8, and clearly the small turnings react more quickly than the ribbon, while the powder reacts very quickly indeed.

Fig. 23.8

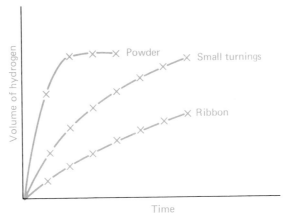

Why should magnesium particles of smaller size react more rapidly with acid than a large piece of the same total mass? To answer this question we should first note that when a solid is reacting with a liquid the only place the reactants are actually in contact is at the surface of the solid. Now let us consider what happens to the surface area of a solid if it is divided into smaller pieces. Referring to Fig. 23.9, suppose we have a 2 cm cube of a solid.

Fig. 23.9

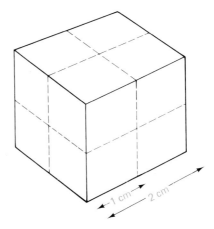

The area of each side is $2 \times 2 = 4$ cm², and since there are 6 sides the total surface area is 24 cm². If we bisect the cube in each of its three dimensions (indicated by the dashed line) we obtain 8 cubes, each of side 1 cm. Each small cube has a surface area of 6 cm², but since there are now 8 cubes the total surface area is 48 cm². This is twice the original surface area. The explanation of the experimental results would therefore seem to be that the rate of reaction of a solid with a liquid increases as the surface area of the solid increases.

Another reaction which can be examined in a similar way is that between marble (calcium carbonate) and an acid (Investigation 23.5):

$$CaCO_3(s) + 2HCl(aq) \rightarrow CaCl_2(aq) + H_2O(l) + CO_2(g)$$

The reaction involves effervescence, and merely by visual observation we can see that calcium carbonate in powder form reacts with hydrochloric acid much more rapidly than a single solid lump of marble. It is generally the case that finely divided solids undergo reactions more quickly than solids in the form of large lumps, and it is for this reason that if we want to dissolve a solid in water, we often save ourselves time by first grinding it into a powder.

We can use the surface area theory to explain some unexpected reaction rates. Firstly, when a lump of marble is dropped into dilute sulphuric acid immediate effervescence occurs but within a second or so the effervescence stops (11.9). If we write the equation for the reaction,

$$CaCO_3(s) + H_2SO_4(aq) \rightarrow CaSO_4(aq) + H_2O(l) + CO_2(g)$$

we see that one of the products is calcium sulphate, shown here as $CaSO_4(aq)$ implying that the salt is formed in aqueous solution. Calcium sulphate is only slightly soluble in water and consequently only the first little bit of calcium sulphate formed in the reaction will dissolve. The next little bit will not dissolve, and it will stay where it is formed, at the surface of the marble. The effect of this will be to break contact between the marble and the acid, and hence the reaction stops.

Many industrial processes involve a solid as one of the reactants, and since the rate of reaction depends on the surface area of the solid, and hence the size of its particles, considerable attention is paid to obtaining particles of a certain size. The suppliers of raw materials for the chemical industry state the particle size of powdered materials so that industrial users can order the most convenient materials. If powdered materials of variable grade are used, the rate of reaction will vary and this will result in extra costs as the manufacturing conditions will have to be adjusted to allow for the differing rates of reaction.

23.14 Can light affect chemical reactions?

Chemical reactions nearly always involve an energy change, and this often shows itself as heat being produced, but it may show itself in the production of light, as in the example of flame. You know that the application of energy in the form of heat will often start or accelerate a reaction, so it is natural to ask whether application of energy in the form of light will do the same. For the majority of reactions the presence or absence of light does not affect the rate at all, but there are cases where light has a considerable effect, and these reactions are very important.

One example which can be easily demonstrated involves the effect of light on silver chloride. Silver chloride is obtained as a white precipitate by mixing solutions of silver nitrate and hydrochloric acid:

$$AgNO_3(aq) + HCl(aq) \rightarrow AgCl(s) + HNO_3(aq)$$

Fig. 23.10

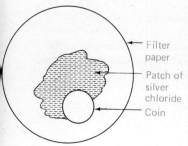

Filter paper

Patch of silver chloride

Coin

If it is then filtered off and the filter paper is opened out and placed on a bench with an object such as a coin covering part of the precipitate (Fig. 23.10), it will be observed after a few minutes that the precipitate is darkening. The effect will be more rapid if direct sunlight is falling on the paper. Alternatively, the effect can be accelerated by burning a piece of magnesium ribbon about 25 cm above the paper. When the coin is removed the contrast between the light and the dark will be obvious.

The reason for the darkening is the formation of finely-divided silver metal, caused by the decomposition of silver chloride started by light. This process is the basis of photography. The salt actually used on photographic film is silver bromide which, though more expensive than silver chloride, is more sensitive to light. The exposure time used with a modern camera is only a tiny fraction of a second, and this is why the image formed on the film must then be developed in order to make it more intense. The image must finally be fixed, by treating the film with a solution which removes unchanged silver bromide so that the film is no longer sensitive to light.

This light-sensitive property is common to silver salts in general, which is why silver nitrate solution is always kept in a dark bottle. A substance such as this which is affected by light is said to be photosensitive, and a reaction which is started or accelerated by light is referred to as a **photochemical reaction**.

Fig. 23.11 One of the first photographs ever taken. It was produced in 1826, and it is thought that an exposure time of about 8 hours was used. (Courtesy Kodak Museum)

Fig. 23.12 Polaroid film needs an exposure of a fraction of a second. The photograph is automatically ejected from the camera and the picture appears without any further treatment. (Courtesy Polaroid (UK) Ltd)

In addition to photography there are a number of photochemical reactions which are important in everyday life. One is the bleaching (or fading) of certain dyes by the action of sunlight. This problem is not as common as it used to be as a lot of research has been done into developing dyes which are not affected by sunlight. These dyes are used to produce coloured fabrics which do not fade. Actually the bleaching action of sunlight can be turned to good use. Wool, as it comes off the sheep's back, is not pure white, and woollen fabrics will have a slight yellow tinge unless something is done to prevent this. Woollen fabrics used to be left out in bleaching fields for this purpose, but this method has long since been replaced by chlorine bleaching, 25.10.

The most important photochemical reaction of all, without which life on Earth as we know it would be impossible, is photosynthesis in plants. In this reaction, water and carbon dioxide from the air combine under the influence of sunlight to form a carbohydrate called starch.

$$\text{carbon dioxide} + \text{water} \xrightarrow{\text{sunlight}} \text{starch} + \text{oxygen}$$

The reaction absorbs carbon dioxide from the air and replaces it with oxygen. In this way the composition of air is maintained, since animals and human beings when they breathe absorb some of the oxygen from the air and give out more carbon dioxide, 28.8. The reaction is also vital from the point of view of the other product, starch, which is formed in plant growth. Animals and human beings eat plants, and in their bodies the starch is oxidised to form water, carbon dioxide and energy, 9.4. Hence this reaction is an essential step in the food chain on which our existence depends. There is more to this reaction than just the action of sunlight. The reaction occurs in the leaves of plants and is associated with the presence of the green substance called chlorophyll. This is the substance which is responsible for absorbing the energy from sunlight, and making this energy available for causing the reaction to occur. Chlorophyll acts as a catalyst for the reaction. Catalysis is discussed in the following section.

23.15 Can the presence of other substances affect reaction rates?

Fig. 23.13 A collection of catalysts showing the different shapes and particle sizes which are used for different processes. (Courtesy ICI Ltd Agricultural Division)

If a solution of hydrogen peroxide is heated or exposed to light it will decompose slowly into water and oxygen:

$$2H_2O_2(aq) \rightarrow 2H_2O(l) + O_2(g)$$

We find that when various substances are added to this solution (Investigation 23.6), effervescence is observed due to oxygen being given off. The presence of these substances has accelerated the reaction so that it goes at a noticeable rate even at room temperature. A particularly effective substance for this purpose is manganese(IV) oxide (MnO_2) which, when added to hydrogen peroxide solution, causes quite vigorous effervescence.

The intriguing thing about this reaction is that, however small a quantity of manganese(IV) oxide is used, it apparently encourages the decomposition of an unlimited quantity of hydrogen peroxide solution. Furthermore, we can, at any time, recover the original quantity of manganese(IV) oxide chemically unchanged. In other words the manganese(IV) oxide is not used up in the reaction; it affects the rate of reaction, but that is all. If we use larger quantities of manganese(IV) oxide we obtain a higher rate of reaction but we do not alter the total quantity of oxygen finally obtained. This phenomenon is known as catalysis and the substance which affects the rate of a reaction by its presence is known as a catalyst.

A catalyst is a substance which speeds up a chemical reaction without itself being used up.

Catalysis is, in fact, quite common, and it is very important both in industrial processes and biological processes (e.g. photosynthesis). Many processes in the body, such as the digestion of food, are catalysed by enzymes, organic substances which enable reactions to proceed at body temperature which would otherwise proceed only at a very high temperature, if at all.

Haemoglobin, the red substance in blood, can be thought of as a catalyst for the reactions which use oxygen in the body. In the lungs oxygen combines with haemoglobin to form oxy-haemoglobin, which passes round the body in the bloodstream. The oxy-haemoglobin splits up as oxygen is used in various parts of the body, and returns as haemoglobin to the lungs to pick up more oxygen, 9.4.

This last example shows one of the ways in which a catalyst can work, namely, by the formation of an **intermediate compound** which splits up again as the products are formed and regenerates the catalyst.

This is not the only way in which a catalyst can operate. Many catalysed reactions, including many of industrial importance, operate by surface catalysis. In this case the catalyst is a solid, and provides a surface on which the reactants are adsorbed (held on the surface), and then react more quickly than they would otherwise do. For example, the metals platinum and nickel act as catalysts for many reactions involving gaseous hydrogen. Since the reaction involves the metal surface, the catalytic action is greatest if the metals are finely divided. If a mixture of hydrogen and oxygen is brought into contact with a small quantity of platinum powder, the reaction between the two gases to form water is immediately set off, just as if a spark or a flame had been applied.

What kinds of substances will act as catalysts? Much research has been carried out on this, and it has been found that catalytic activity is nearly always associated with transition metals and their compounds. The term transition metal is explained to some extent in 12.13. In particular they have variable valency (form ions with different charges) and readily form complex ions. Metals which are used as catalysts are iron, manganese, nickel, platinum, chromium, vanadium, copper, cobalt and rhodium. It is notable that the catalytic behaviour of any particular substance is usually found only to apply to a particular reaction, or type of reaction. A great deal of research continues to be done, in order both to establish general patterns of behaviour, and to develop catalysts for use in particular industrial processes.

Examples of industrial processes which use catalysts are:
(1) The Haber Process for the synthesis of ammonia (iron catalyst).
(2) The Contact Process for production of sulphur trioxide, from which sulphuric acid is made (vanadium(V) oxide catalyst).
(3) The oxidation of ammonia to nitrogen monoxide, from which nitric acid is made (platinum/rhodium catalyst).
(4) The hardening of fats and oils by hydrogenation (nickel catalyst).
(5) The reforming of petroleum hydrocarbons (platinum and other catalysts).

23.16 Summary

1. Different chemical reactions vary very widely in rate, and rates of reactions cannot, in general, be reliably predicted.
2. The rate of a reaction depends upon the conditions under which the reaction is occurring.
3. The methods available for measuring reaction rates consist of direct methods, which involve stopping the reaction and then measuring quantities of substances directly, and indirect methods, which do not involve stopping the reaction, but involve the continuous measurement of a property which is related to the concentration of one of the substances involved.
4. Rates of reactions in aqueous solution very often increase with concentrations of the reactants. Rates of reaction of gases very often increase with gas pressure. It is not possible to predict from the equation for a reaction how changing the concentration of a reactant will affect the rate.
5. Rates of virtually all reactions increase markedly with temperature. Consequently, exothermic reactions tend to become faster.
6. Reactions involving solids vary very widely in rate, and some may be extremely slow. The rate almost always increases as the surface area of the solid increases.
7. Most chemical reactions are not affected by light but some are. These are known as photochemical reactions. Photosynthesis is the most important photochemical reaction. Photography is another common example.
8. Catalysis is a common and important phenomenon, which is involved in many biological and industrial processes. Catalytic activity is associated particularly with transition metals and their compounds.

Chapter 24

The extraction and uses of some important metals

24.1
Our dependence on metals

One way of tracing the development of the way of life of mankind, from the primitive existence in Stone Age times up until the present day, is to consider what materials have been available at various times for making articles such as tools, utensils, ornaments, jewellery and weapons. Until the development of plastics during this century, the materials which have influenced our way of life most have been the metals. From the earliest times gold, silver and copper, because they can be found in the native state (that is uncombined with any other element), have been used for jewellery and ornaments. Most other metals occur combined with other elements, and each time a process was discovered for extracting a new metal it meant another material became available from which articles could be made.

In early history a new metal provided the possibility of wider political and economic power to the people who discovered it. For example, when bronze (10% tin and 90% copper) was first made about 5000 years ago in the Middle East, the nations with bronze weapons had a clear advantage over those who were still using stone axes and arrow-heads. Similarly, when iron was first made about 1500 BC, possibly by the Hittites, who lived in what is now Northern Turkey, it was found to have superior properties to bronze in that it was much harder, and easier to make into a sharper cutting edge.

Fig. 24.1 World production of aluminium

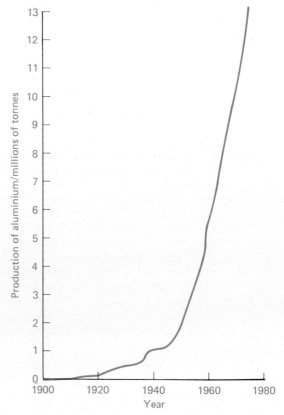

In more recent history, perhaps the most striking example of the discovery of a new extraction process is that of the electrolytic method for extracting aluminium. In the middle of the last century aluminium was still considered a rare metal, so much so that a bar of the metal was on display next to the Crown Jewels at the Paris Exhibition in 1855. In 1886 an economic method of extracting it from an abundant mineral called bauxite was developed and by 1975 (Fig. 24.1) the world production of aluminium was about 13 million tonnes per year. Aluminium, being structurally strong and yet of low density, has made the development of the modern aircraft industry possible. Without a plentiful supply of aluminium we might still be flying in aircraft made from canvas stretched over a wooden framework, or alternatively they would be made from other metals whose cost of production would be much higher.

24.2 Relative importance of different metals

The factors which influence the cost of a metal and its value to mankind are of three types; first the physical and chemical properties of the metal, second the cost of production of the metal and third the demand for the metal. Gold has always been considered a precious metal. It can be shaped easily to make different objects, it is resistant to chemical attack and therefore retains its shine even when exposed to air and moisture, and it is rare. Thus gold is a very expensive metal.

Rarity is not the only factor which contributes to the cost of a metal. For example, aluminium is the most abundant metal in the Earth's crust (Fig. 24.2) and yet, until a commercially worthwhile method of extracting it was developed, it was a very expensive substance.

Fig. 24.2 The composition of the part of the Earth's crust which is not under the sea

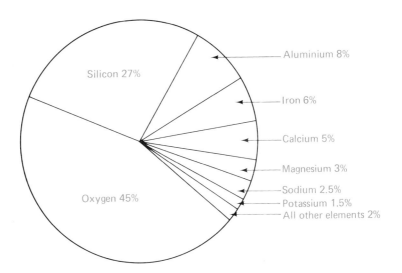

A high demand for a metal can influence the cost in two ways. If the supply cannot keep pace with the demand the cost will undoubtedly rise. If, however, an increased demand leads to an increase in production, a decrease in price can result, particularly if the chemical plant can be made to work at full capacity. This will mean that the cost of overheads (such as the cost of building the plant) relative to the turnover in cash will decrease. Economic factors are discussed further in 24.5.

The Earth's crust, the sea and the atmosphere are the only sources of elements available to man, the use of other planets or the Moon being technically and economically impracticable in the foreseeable future. When considering the metallic elements, the parts of the Earth's crust which are not under the sea are clearly the major source. However, as Fig. 24.2 shows, 72% of this part of the crust is made up of the non-metals oxygen and silicon. The only relatively abundant metals are aluminium and iron, followed by the few others mentioned in Fig. 24.2. All the elements which are not mentioned, including such important metals as copper, make up no more than 2% of the Earth's crust.

All the metals we use are extracted from **minerals** which are naturally occurring inorganic compounds. For example, an important mineral of lead is called galena which is the chemical compound lead sulphide. Table 24.1 lists some of the more important minerals.

The minerals which are used are usually found as parts of rocks or in deposits which have been formed by the weathering of rocks, though minerals which are soluble in water are sometimes extracted from the sea.

Rocks are usually made up of mixtures of minerals and when a rock is used as a source of a metal it is often called an **ore**. Haematite, which is iron(III) oxide and is used as a source of iron, is called an iron ore. Another iron ore is siderite, which is iron(II) carbonate.

In early times mineral deposits were often found by chance, but the development of the science of **geology** has reduced the chance factor. Geologists know in what type of rock formation, and near to what other types of rocks, each mineral is likely to be found. Most mineral prospecting is done now by professional geologists rather than the untrained prospector hoping to make his fortune by a lucky strike.

*Fig. 24.3 (below)
Prospecting for minerals
in Scotland, using an
electromagnetic technique.
(Courtesy RTZ Services
Ltd. Photo: Dmitri
Kasterine)*

*Fig. 24.4 (right) Iron
ore mining at
Mauretania. (Courtesy
British Steel Corporation.
Photo: Photographic
Techniques)*

Useful minerals, owing to the geological history of the Earth, tend to be concentrated in particular areas. For example, copper, which is a comparatively rare metal, is not distributed evenly throughout the Earth's crust but is found in localised mineral deposits which contain sufficient copper to make it worthwhile extracting the metal.

Rocks are classified according to how they were formed. **Igneous** rocks were formed by the cooling of molten mixtures of minerals (magmas). Granite is a common example of an igneous rock. **Sedimentary** rocks were formed when sediments, either from the weathering of igneous rocks or from the remains of very small marine creatures, were compressed to form hard rock. Thus sand, which consists of tiny grains of quartz produced by the weathering of igneous rocks, was compressed over millions of years to form sandstone. Most limestones were formed by the gradual build-up of the remains of small marine creatures, which, when compressed by overlying layers of other deposits and rocks, gradually became limestone. Sedimentary rocks often contain fossils which are the skeletal remains or the impressions of plants and animals which lived at the time the rock was being deposited. The third type of rocks are called **metamorphic** and were formed by the action of heat and pressure on either igneous or sedimentary rocks. For example, marble has been formed from limestone in this way.

TABLE 24.1. *Sources of some important metals and the main steps in their extraction*

METAL	SOURCES	METHOD OF EXTRACTION
Sodium	Rock salt, NaCl	Electrolysis of molten NaCl mixed with some $CaCl_2$
Magnesium	Dolomite, $CaCO_3.MgCO_3$ Sea water	Electrolysis of molten $MgCl_2$
Aluminium	Bauxite, $Al_2O_3.2H_2O$	Electrolysis of Al_2O_3 in molten cryolite
Zinc	Zinc blende, ZnS	Reduction of ZnO by carbon
Iron	Magnetite, Fe_3O_4 Haematite, Fe_2O_3	Reduction of the metal oxides by carbon and carbon monoxide
Lead	Galena, PbS	Reduction of PbO by carbon
Copper	Copper pyrites, $CuFeS_2$	Roasting with sand and in air

Useful mineral ores can occur in all three types of rock. The iron ore in Sweden is found in igneous rock. As the molten rock cooled the heavier iron ore (magnetite, Fe_3O_4) sank to the bottom. Due to subsequent geological movements and weathering, this high concentration of ore has become accessible and is one of the richest deposits in the world. In Britain most of the iron ore deposits are sedimentary, although there is some metamorphic magnetite in Cornwall. The main ore of aluminium, bauxite (named after Les Baux in France where it occurs), was left as a residue of hydrated aluminium oxide after the weathering of igneous rocks.

Clearly mineral resources are finite, that is, they are limited in quantity. If we were to use up all the accessible copper ore, we would not be able to produce any more copper metal. The world annual consumption of copper was about 6 million tonnes in 1975 (Fig. 24.5) and it has been predicted that it could rise to about 20 million tonnes by the end of the century. Estimations such as this are very difficult to make because, for example, the consumption of copper per head of population differs greatly from the more developed communities to those which are less developed. The annual consumption of copper in the U.S.A. in 1975 was about 10 kg per head of population whereas in India it was about 0·1 kg (Fig. 24.6). It must be anticipated that the consumption of copper (and of course most other metals) will increase considerably in developing countries over the next few decades. We need to ask how demand can be controlled at a reasonable level, and how required production rates can be maintained. These questions are very complex; for example, if the world demand for copper was drastically reduced, the economic prosperity of the countries where the ore is mined, such as some of the developing countries in Central Africa, would be threatened.

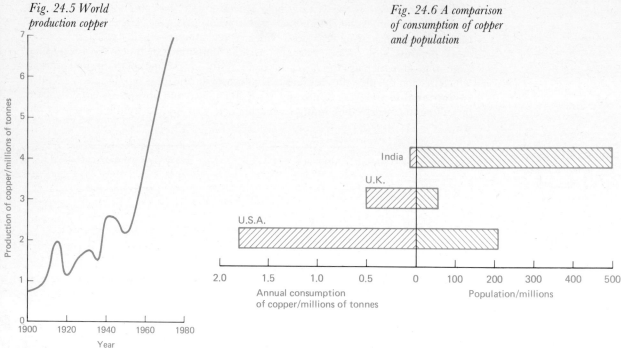

Fig. 24.5 World
production copper

Fig. 24.6 A comparison
of consumption of copper
and population

When considering whether production of a metal is likely to continue for a long time into the future, we have to take into account that it is probable that many more mineral deposits will be discovered. It is also probable that new processes will be found for extracting metals from lower-grade ores (those which contain a lower percentage of the mineral).

Another important factor is the recycling of used metals. The car-breaker's yard may be unsightly, but it is making a useful contribution to the recycling of metals.

In the future it is probable that metals will be replaced by other materials for certain uses, or more abundant metals will be used in preference to rare metals. An important example of this is the increasing use of aluminium, rather than copper, as an electrical conductor.

Nevertheless the sensible use of metals and the conservation of mineral resources are matters of concern for everybody, whether you are a technologist concerned with the production or use of metals, or simply a person who contributes to the recycling process by collecting used milk-bottle tops.

Fig. 24.7 The recycling
of metals by scrapping old
motor cars. (Photo:
Russell Edwards, BSc)

Fig. 24.8 The recycling
of aluminium after
domestic use. (Photo: by
Alcoa)

236

24.4
Factors
influencing
methods of
extracting
metals

It seems likely that the discoveries of the metals gold, silver, copper, lead and probably iron were largely accidental. The story goes that travellers banked up their fire for the night with stones which, unknown to them, contained a copper mineral, and in the morning they found beads of copper in the ashes. No doubt the methods were developed by experimentation and the knowledge acquired was passed down from generation to generation. In some less developed parts of the world the traditional methods of extracting iron by heating the ore with charcoal were used up until the middle of this century.

It was not until the science of chemistry developed during the last century that the processes involved in the extractions were understood. The essential step is the conversion of the combined form of the metal in a mineral into the uncombined element. This is a reduction process. Thus, for example, the conversion of haematite, Fe_2O_3, into iron involves the removal of oxygen. This change can also be classified as reduction by considering that the iron in the compound is in the form of Fe^{3+} ions and when converted to the metal these ions are changed to neutral atoms by gaining electrons, which is a reduction process:

$$Fe^{3+} + 3e^- \rightarrow Fe$$

Reduction can be brought about by reacting the ore with a reducing agent such as carbon (in the form of coke) which removes the oxygen from the oxide of the metal,

$$ZnO(s) + C(s) \rightarrow Zn(g) + CO(g)$$

or by electrolysis, when an electric current is passed through, for example, the molten chloride of the metal and the metal is liberated at the cathode:

$$Na^+ + e^- \rightarrow Na$$

Which of the two general methods is used depends mainly on the reactivity of the metal. A very reactive metal such as sodium has a great tendency to stay combined with other elements and is difficult to reduce by chemical means with a reducing agent, and so an electrolytic process is used. Less reactive metals such as zinc, iron, lead and copper can be obtained by chemical reduction. If the metal occurs as a sulphide (e.g. PbS and ZnS), it is roasted in air first to convert it to the oxide of the metal. Table 24.1, in which some metals are arranged in order of decreasing reactivity, shows how the method of extraction used is related to the reactivity of the metal.

When designing a process for the extraction of a metal it is necessary to consider the concentration of the mineral in the ore. Many iron ores contain as much as 70% by mass of iron whereas most gold ores contain less than 0·007% of gold. Where the ore is not rich in the metal, the first stage of the process is often the production of a more concentrated form of the mineral from the ore.

Another factor to be considered is the purity of the metal required for a particular use. Frequently the final stage involves refining the metal so that the purity is increased to the desired level. This stage sometimes involves the addition of impurities which give the metal particular required properties. When you study the extraction processes which are described later in this chapter you should be able to recognise some or all of the following stages in the processes:

1. preliminary concentration of the mineral in the ore,
2. roasting of the ore to convert sulphides to oxides,
3. reduction of the mineral to the metal, and
4. refining of the metal.

24.5
Economic factors

Overall, the process must be economically worthwhile. The factors which need to be taken into account when determining whether this is likely to be the case include:

1. The accessibility and grade of the ore. From the economic point of view a low-grade ore might be more worthwhile than a high-grade one if the latter is less accessible. The ore might be found in a remote place which would result in high transport costs, or it could be that the ore is deep underground and hence mining costs would be high.

2. The cost of the other raw materials such as the reducing agent.

3. The cost of building the chemical plant (for example the furnaces used for producing iron and the electrolytic cells used for producing sodium).

4. The cost of labour for running and maintaining the plant.

5. The cost of utilities, which is the name given to things which are required by all chemical plants, namely water and steam, and energy in the form of electricity, gas, oil or coal.

6. How the cost varies with the amount produced. Some of the costs mentioned previously are almost constant irrespective of output, and hence the greater the output from a particular chemical plant, the lower the cost per tonne of product is likely to be.

7. Finally, of course, the likely selling price of the product.

24.6
Location of the chemical industry

The site chosen for the location of a chemical plant will depend on a balance between economic and social/environmental factors. From an economic point of view it will be necessary to decide whether the plant should be situated near the source of the raw materials, so as to reduce transport costs. On the other hand, if the cost of building the plant and maintaining and running it are likely to be high, and the source of the raw materials is remote, it may be better to transport the raw materials and build the plant in a more accessible place. If the energy requirement is likely to be high, it might be necessary to build the plant near a cheap source of power.

From the environmental and social points of view it could be necessary to consider, on the one hand, the disposal of effluent and, on the other hand, the provision of work for the local community. There are two possible major sources of effluent, first any unwanted part of the ore (which for low-grade ores will be a major consideration), and secondly any unwanted by-products of the roasting, reduction or refining processes. Gaseous products could pollute the atmosphere and soluble products and liquids could pollute waterways. Where the unwanted products are solids, they can lead to the build-up of unsightly waste tips which can make the land occupied in this way unusable.

Our whole way of life depends on the products of the chemical industry and yet the problems associated with effluent and accidents can be serious. Everyone should be aware that the economic and environmental consequences of the industry can affect all our lives and it is essential that a sensible balance between them is achieved. The siting of chemical plants for the extraction of particular metals is discussed later in this chapter. Air and water pollution are discussed in greater detail in 9.5 and 10.13 respectively.

24.7
Production and uses of the more reactive metals

Of the four metals potassium, sodium, calcium and magnesium, which are some of the most reactive metals and are found at the top of the reactivity series (17.9), only sodium and magnesium are commercially important. The main use of sodium is in the production of tetraethyl lead which, as an additive for petrol, is known as antiknock. It is made by reacting chloroethane with a sodium-lead alloy:

$$4C_2H_5Cl + 4Na/Pb \rightarrow Pb(C_2H_5)_4 + 3Pb + 4NaCl$$

Antiknock prevents the petrol-air mixture in the cylinder of an engine igniting too early.

The main use of magnesium is in the production of special low density alloys with other metals such as aluminium. These are used particularly in the aircraft and aerospace industry.

Both sodium and magnesium, being very reactive, are produced by electrolysis. In each case the electrolyte contains the chloride of the metal. Sodium chloride is mined from rock salt deposits which were formed by the evaporation of seas millions of years ago. Magnesium chloride is often extracted from sea water by first precipitating it as magnesium hydroxide and then, after filtering it off, reconverting it to the chloride.

The cells in which molten sodium chloride is electrolysed are based on those developed by Down in 1921 which used electricity generated by the Niagara Falls. Fig. 24.9 is a diagram of the cell. The anode (positive) is made of carbon (graphite) and it is surrounded by a steel cathode. The electrolyte is a 40:60 mixture, by mass, of sodium chloride and calcium chloride.

The advantage of using the mixture rather than pure sodium chloride is that it melts at a lower temperature ($600\,°C$) than sodium chloride ($800\,°C$). The cost of the energy required to keep the cell at the higher temperature would obviously be greater, but also at this higher temperature the vapour pressure of the sodium becomes dangerously high and also a lot of the sodium dissolves in the electrolyte. A possible disadvantage of using the mixture is that there will be two positively charged ions (cations) present, Na^+ and Ca^{2+}. However, at the operating temperature this is not a problem, as only a small quantity of calcium is formed and the main product at the cathode is sodium which is formed by the electrode reaction:

$$Na^+ + e^- \rightarrow Na$$

The product at the anode is chlorine:

$$2Cl^- - 2e^- \rightarrow Cl_2$$

The molten sodium floats on the electrolyte and is run off. It is kept separate from the chlorine by a steel gauze diaphragm.

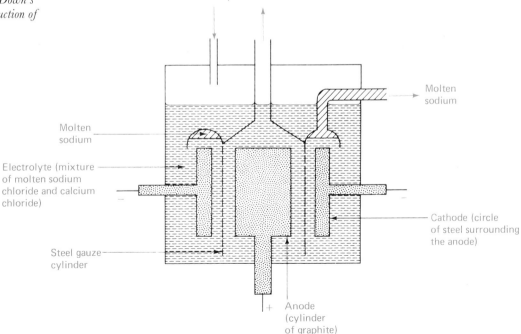

Fig. 24.9 The Down's cell for the production of sodium

239

Ninety per cent of the world's magnesium production is by electrolysis of molten magnesium chloride (mixed with other chlorides) using carbon (graphite) anodes and cathodes. This process is not used at all in the U.K. due to the high cost of electricity. In the north-east of England, magnesium is produced by extracting it from seawater and local dolomite ($CaCO_3.MgCO_3$) deposits. The dolomite is heated to produce dolime ($CaO.MgO$), which is reacted with the magnesium salts in seawater to form a precipitate of magnesium hydroxide. This precipitate is heated to form more magnesium oxide. Magnesium metal is obtained by heating the oxide to a high temperature ($1200\,^\circ C$) with an alloy of silicon and iron:

$$2MgO + Si \rightarrow SiO_2 + 2Mg$$

24.8 Production and uses of aluminium

Although aluminium is the most abundant metal in the Earth's crust (Fig. 24.2), it was regarded as a rare metal until the electrolytic extraction process was developed in 1886. Even so the world production (Fig. 24.1) did not exceed 1 million tonnes per year until the Second World War. Since the end of the war the world production has increased rapidly and is now over 13 million tonnes per year. This means that it is, except for iron, the most widely used metal in the world.

Aluminium is a very useful metal because it can be used to produce alloys which are structurally as strong as mild steel and yet its density is about one third of that of steel. Another advantage which it has over steel is that it has a high resistance to corrosion owing to the protective layer of oxide which it develops on its surface, whereas most types of steel readily rust unless protected by paint or a layer of another metal. As mentioned in 19.11 the layer of oxide can be improved by anodising, and after this process it is possible to dye the layer so that the finished product has an attractive appearance as well as being resistant to corrosion. Coloured anodised aluminium is used for making saucepans and wall panels in building construction. Other important uses of the metal include the construction of aircraft bodies and parts of engines, and in the form of aluminium foil its most common household use is for wrapping food.

In recent years it has been used to replace copper as an electrical conductor, particularly in overhead high-tension cables. Aluminium has a higher conductivity per unit mass than copper. If this trend to substitute aluminium for copper continues the demand for copper, which is a comparatively rare metal, will be reduced.

The main ore of aluminium is bauxite, which contains hydrated aluminium oxide. It is found mainly in tropical areas of the world and known deposits are likely to last into the twenty-first century, but new deposits are likely to be found in more remote tropical areas. During the last fifteen years extensive deposits have been discovered and developed in Australia, and that country has now become the major source of bauxite.

Bauxite usually occurs near the surface and is mined by open-cast methods. There are two main steps in the extraction of aluminium from bauxite. The first involves the extraction of aluminium oxide (alumina) from the bauxite and the second the electrolytic reduction of aluminium oxide to give aluminium metal. During the first stage the bauxite is heated with hot sodium hydroxide solution under pressure. Aluminium oxide, as it is amphoteric (11.20), dissolves in the sodium hydroxide solution by forming sodium aluminate:

$$Al_2O_3(s) + 2NaOH(aq) + 3H_2O(l) \rightarrow 2NaAl(OH)_4(aq)$$

The solution is filtered to remove all the insoluble impurities, which consist mainly of sand and a red mud containing iron and titanium oxides. The disposal of the red mud causes problems. It is usually pumped into the sea or into large ponds. The water which is mixed with the mud is still slightly alkaline and could be harmful to aquatic life.

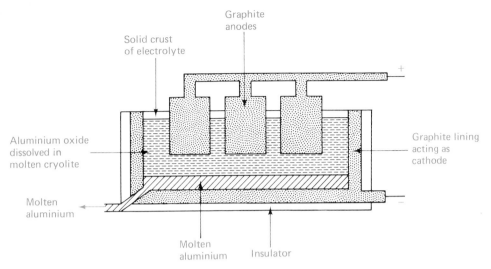

Fig. 24.10 The electrolysis cell for the production of aluminium

Graphite anodes

Solid crust of electrolyte

Aluminium oxide dissolved in molten cryolite

Graphite lining acting as cathode

Molten aluminium

Molten aluminium

Insulator

Also, if the ponds dry up the resulting red dust can be blown by the wind and cause other environmental problems. Thus, as will be discussed later, the location of the extraction plant is important.

The solution of sodium aluminate is cooled and after adding some pure hydrated aluminium oxide, which acts as a seed (it encourages the start of crystallisation), hydrated aluminium oxide is precipitated, this then being heated to drive off the water and leave pure aluminium oxide.

Aluminium oxide is very difficult to reduce and so the next stage of the process, which is called the smelting, involves electrolysis. Aluminium oxide is dissolved in molten cryolite (sodium hexafluoroaluminate, Na_3AlF_6) at 900 °C. This mixture is electrolysed using carbon (graphite) blocks as the anodes and the carbon lining of the container as the cathode, Fig. 24.10. Aluminium is produced at the cathode:

$$Al^{3+} + 3e^- \rightarrow Al$$

It collects as a liquid at the bottom of the molten electrolyte from where it can be run off and allowed to solidify.

Oxygen is produced at the anodes. The electrode reaction can be represented by:

$$2O^{2-} - 4e^- \rightarrow O_2$$

At the temperature of the cell the oxygen reacts with the carbon anode to produce carbon dioxide. The anodes are gradually burnt away and have to be replaced at regular intervals.

Both of the stages could be carried out near the bauxite mine or near the potential market for the aluminium. The third possibility, which is the most common procedure, is that the aluminium oxide could be extracted from the bauxite near the mine and the smelting process is carried out in another location. Figure 24.12 shows the extent to which the three processes, mining, alumina extraction and smelting, occur in various countries.

The factors which have been the major influence on determining the location of the chemical plants differ for different parts of the world and for different periods in the development of the industry. The cost of transporting the ore is important because for every tonne of aluminium produced, up to 5 tonnes of bauxite have to be processed.

241

Fig. 24.11 A row of cells for producing aluminium. See also Fig. 28.5. (Courtesy RTZ Services Ltd)

Thus, if the ore is to be imported, the extraction plant should be near a deep-sea port so that land transport costs are kept to a minimum. The location of the alumina extraction process must also be suitable for the disposal of unwanted residue (red mud) from the bauxite. It must be possible either to deposit it in the sea where the water is very deep, or to convert natural land formations into ponds without involving the construction of very costly dams and without using up valuable agricultural land.

Another potential pollution hazard is the possibility of fluorine being present in the waste gases from the smelting process. Fluorine is very poisonous and corrosive and hence its escape into the atmosphere must be carefully controlled. Unlike the red mud problem which involves an inevitable and bulky by-product, technological improvements in the treatment of waste gases have been able to minimise the fluorine content of the gases.

The factor which tended to outweigh other factors when determining the location of smelters was the cost of electricity. For every tonne of aluminium produced, about 15 000 kWh are required (this would run 15 000 one-bar electric fires for one hour). In the past this has resulted in aluminium smelters being located near sources of hydroelectricity which were capable of producing relatively cheap electricity. In the early part of this century two aluminium extraction plants were opened near hydroelectric schemes at Fort William and Kinlochleven in Scotland. Similar criteria operated in other parts of the world and aluminium has been produced since the early days of the industry near the St. Lawrence river system in North America, and in the French Alps and Norway in Europe.

In more recent years other sources of electricity have competed favourably with hydroelectricity; also some bauxite producing countries, such as Jamaica, have insisted on extracting aluminium oxide from the bauxite in their own countries.

Social factors which can influence the location of aluminium plants are either concerned with the effects of the red mud on the environment or the need to create jobs for the local community. Clearly these two factors can work in opposition to one another and a realistic balance between the two must be sought. It is interesting to note that in the siting of the only other three extraction plants in the United Kingdom, all of which came into operation in the early 1970s, political factors were important. The government offered the companies electricity at a cheaper rate so long as they built the plants in development areas, that is, areas where there was high unemployment and the local economy needed boosting. The extraction plants were built at Lynemouth in Northumberland, Invergordon in Scotland and in Anglesey in Wales. The establishment of these extraction plants clearly has some positive social effects for the local communities.

Fig. 24.12 The production of bauxite, alumina and aluminium in selected countries

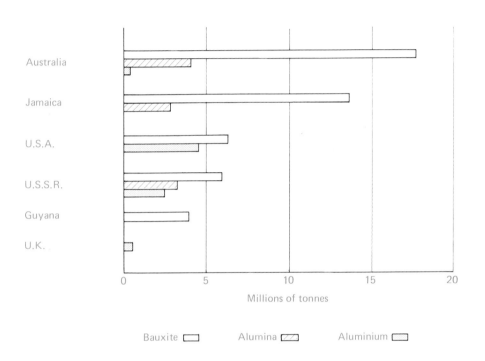

24.9 Production of iron and steel

Iron has been made for hundreds of years by the reduction of oxides of iron by carbon, the two most important starting materials being haematite, Fe_2O_3, and magnetite, Fe_3O_4. Until the fifteenth century the method of extracting the metal consisted of simply heating the ore in a charcoal fire in a shallow pit. The temperature in the fire was not sufficiently high to melt the iron and a spongy lump of the metal was recovered from the ash. This could then be hammered into tools or weapons. The process of hammering the lump caused removal of impurities, so that the final product did not corrode too quickly.

In the fifteenth century the blast furnace came into use. It is thought that the first furnace of this type was made in Belgium round about 1400. In the blast furnace a blast of air was blown into the mixture of charcoal and iron ore from bellows which were usually driven by a waterwheel. The blast produced a much higher temperature in the furnace and the iron was melted, the liquid trickling down through the solid in the furnace to collect in the hearth at the bottom. As it passed through the solid, the iron dissolved carbon from the charcoal and, when it was allowed to solidify, it was hard and strong but brittle.

The metal was usually run into a trench in a sand bed with a number of side trenches leading from the main one. Because this reminded early iron workers of a number of piglets being fed by a sow, this sort of iron was called **pig iron**, although nowadays it is more usually known as **cast iron**.

The use of charcoal as the fuel in the smelting of iron in Britain led to large areas of the country being stripped of trees. Some furnaces are recorded as having used the wood from 150 acres (about 0·6 km²) of forest every year. Since in the time of Elizabeth I this wood was also needed for shipbuilding, Parliament had to pass a law, limiting the use of wood for making iron. In 1709, however, there came the major discovery which was to change completely the making of iron and marked the beginning of the Industrial Revolution. In the Shropshire village of Coalbrookdale, Abraham Darby found that charcoal could be replaced by coke, made from certain sorts of coal, and he used coke-smelted iron for making castings from that date onwards.

By 1760 coke-smelting was being widely used for making cast iron for constructing beam engines and as a building material. Cast iron was used in 1779 to make the world's first iron bridge, which is near to Coalbrookdale. It was only after the collapse of the Tay Bridge in Scotland, while a train was going over it in a storm a hundred years later, that cast iron became unpopular as a building material.

Fig. 24.13 The first bridge to be constructed from iron still stands at Ironbridge in Shropshire. (Courtesy Ironbridge Gorge Museum Trust)

An important factor which led to Abraham Darby's success in developing coke-smelting was the supply of limestone in the region of his furnaces. Limestone had been added to the mixture in the blast furnace as early as the sixteenth century and by the eighteenth century the technique was well known. The limestone absorbs the sand and dirt in the iron ore, forming calcium silicate which, at the temperature of the furnace, is melted and can easily be run off as slag:

$$CaCO_3(s) + SiO_2(s) \rightarrow CaSiO_3(l) + CO_2(g)$$

During the nineteenth century the process was further improved. In particular Joseph Nielson of Glasgow developed a method in 1828 of using the heat from burning coal to preheat the blast of air. The efficiency of the process was improved again in 1857 by E. A. Cowper who designed stoves in which the waste gases from the blast furnace itself were burned and the heat produced was used to preheat the blast of air. Cowper stoves, first introduced at Ormesby near Middlesbrough in 1860, are still used today and their cylindrical shapes with convex tops can be seen in most modern ironworks.

Fig. 24.14 The blast furnace

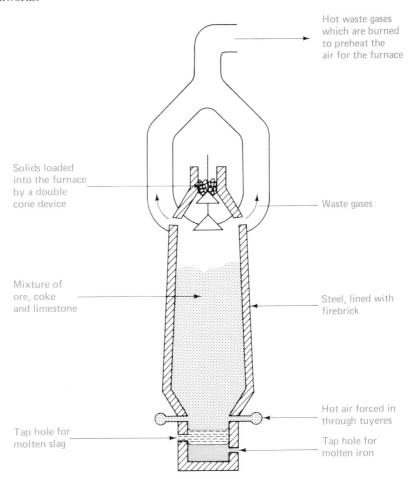

Hot waste gases which are burned to preheat the air for the furnace

Solids loaded into the furnace by a double cone device

Waste gases

Mixture of ore, coke and limestone

Steel, lined with firebrick

Hot air forced in through tuyeres

Tap hole for molten slag

Tap hole for molten iron

The discoveries of Darby, Neilson and Cowper are still included in the modern blast furnace process for making iron. The present-day furnace is shown in Fig. 24.14. It is a steel cylinder about 10 metres in diameter and up to 60 metres high, and is lined with heat-resistant firebrick. Iron ore, limestone and coke are continuously fed into the top of the furnace and the hot air blast is blown in through nozzles called **tuyeres**.

The two cones at the top of the furnace prevent waste gases escaping, so that they can be used to preheat the blast. The upper cone is lowered to let the solids into the space between the cones, and then it is raised again before the lower cone is lowered to allow the solid to fall into the furnace.

The coke is the reducing agent in the hottest part of the furnace which is opposite the tuyeres and where the temperature is about $1800\,^\circ\mathrm{C}$:

$$\mathrm{Fe_2O_3(s)\ +\ 3C(s)\ \rightarrow\ 2Fe(l)\ +\ 3CO(g)}$$

Elsewhere the oxygen is removed from the iron by carbon monoxide which is formed by the coke burning in the air blast:

$$2C(s) + O_2(g) \rightarrow 2CO(g)$$

The heat given off in this combustion causes a temperature of about $1200\,°C$ and this is sufficiently high for the iron to be produced in the molten state:

$$Fe_2O_3(s) + 3CO(g) \rightarrow 2Fe(l) + 3CO_2(g)$$

The iron ore contains some sandy and dirty impurities, which, if they were not removed, would block the furnace and prevent it operating continuously. As in Darby's process, the modern furnace uses limestone for the removal of the dirt. The limestone decomposes at the temperature of the furnace:

$$CaCO_3(s) \rightarrow CaO(s) + CO_2(g)$$

The calcium oxide then combines with the impurities which are mainly silicon dioxide (silica) to form molten calcium silicate:

$$CaO(s) + SiO_2(s) \rightarrow CaSiO_3(l)$$

This trickles down the furnace and collects above the molten iron in the hearth. The two liquid layers can be run off from separate tap holes and the calcium silicate is then allowed to solidify to form the light grey slag which, after breaking up, is mainly used for making road surfaces. The iron is tapped off from the lower tap hole and is usually transported in the molten state directly to the steel-making plant.

Because the modern blast furnace is being continuously fed with ore, coke, limestone

Fig. 24. 15 Blast furnaces at Appleby Frodingham showing the pipes which lead the waste gases away to be burned in the stoves. (Courtesy Appleby-Frodingham Steel Company)

and hot air, and because the impurities are being removed from it as slag, it will run non-stop for several years until the firebrick lining needs replacing, and it can produce up to 8000 tonnes of iron every 24 hours.

The iron which is produced by the blast furnace contains between 2·5 and 4·5 per

246

cent of carbon and this causes the iron to be strong but brittle. It has been known from the days of the charcoal blast furnaces that, if a blast of air is blown through molten iron, some impurities could be removed and the iron becomes more malleable (easy to shape) and less brittle. However, it was not until the late 1850s that methods of controlling the amounts of impurities in the iron were developed and **steel** was produced.

The major breakthrough came in 1856 with the introduction by Henry Bessemer of the **Bessemer Converter**, Fig. 24.16. The converter was charged with up to 60 tonnes of molten pig iron and then air was blown for about 15 minutes through the tuyeres in the bottom. The air oxidised the impurities (carbon, sulphur, silicon and manganese) in the iron and, as heat is given out in these reactions, the temperature of the iron actually increased. The progress of the reactions could be judged by the colour and sorts of flames appearing at the mouth of the converter. When the process was completed, the blast was turned down and then the required quantities of carbon and other elements were added to produce the type of steel required.

Fig. 24.16 The Bessemer converter

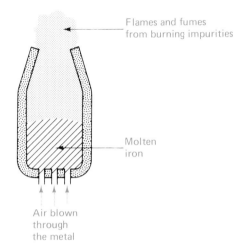

Flames and fumes from burning impurities

Molten iron

Air blown through the metal

Bessemer's original method is now obsolete, but there is a modern version called the **basic oxygen furnace**. In this furnace oxygen, obtained by distilling liquid air, is used rather than the air itself. The oxygen is blown on to the surface of the molten iron by means of a water-cooled lance, Fig. 24.17. The advantages of this method over the original Bessemer method are that the heat generated is so great that a high proportion of scrap iron (up to 30%) can be used. The possibility of nitrogen from air being absorbed in the steel is eliminated (nitrogen tends to make the steel brittle), and the process is more efficient. A modern furnace of this type can convert up to 350 tonnes of iron into steel in 40 minutes.

There are two main types of oxygen furnace in use, the **LD Process** (named after the two steel-producing towns of Linz and Donawitz in Austria) and the **Kaldo Process** (named after the person who developed the method, Kalling, and the place where he worked, Domnarvet). Although the constructions and operations of the two furnaces differ and some ore and calcium oxide are added to the iron in the Kaldo method, the processes are essentially the same. Most of the world's steel is now produced by oxygen furnaces.

As with other industries based on chemicals, steel works are usually to be found near places where the raw materials are (or were) to be found. For example in the U.K., Scunthorpe is a steel town because of the nearby deposits of iron ore and the ease with which coking coal could be supplied from the northern coalfields. Sheffield's industry grew out of the abundance of local iron ore, woodlands (for charcoal) and streams (for water power) and its development was continued by the coal from the South Yorkshire coalfield. The Welsh iron and steel industry developed for similar reasons. The survival of the industries in these areas depended on the growth of transport systems for the supply of raw materials and for the distribution of products, and also the development of special skills and processes which have taken over as the local raw materials have decreased. This is particularly true of Sheffield which has become known for its production of special steels.

Fig. 24.17 The basic oxygen furnace

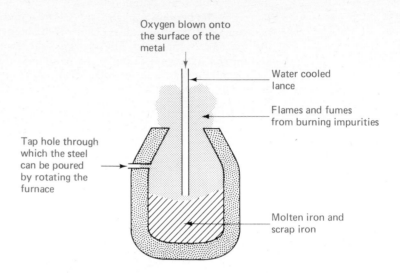

24.10 Production of zinc, lead and copper

These three metals all occur as sulphides and being less reactive metals can be extracted by chemical reduction processes rather than electrolysis. In the cases of zinc and lead the ores are first converted to the oxides by heating in air:

$$2ZnS(s) + 3O_2(g) \rightarrow 2ZnO(s) + 2SO_2(g)$$
zinc blende

$$2PbS(s) + 3O_2(g) \rightarrow 2PbO(s) + 2SO_2(g)$$
galena

The oxides are then reduced by heating with coke:

$$ZnO(s) + C(s) \rightarrow Zn(g) + CO(g)$$

$$PbO(s) + C(s) \rightarrow Pb(l) + CO(g)$$

At the temperatures reached zinc is given off as a vapour which is then cooled and condensed, whereas lead is molten and the liquid metal is run off from the furnace.

The main ore of copper is copper pyrites, $CuFeS_2$, which contains iron as well as copper. The process for extracting copper from its ore is more complex than is the case with either zinc or lead. After preliminary concentration of the ore and some roasting of it in air, it is heated with sand to convert the iron to a molten slag of iron silicate. The residue is then heated with a controlled amount of air which oxidises the sulphur present but, as copper is so unreactive, it is reduced to the metal:

$$Cu_2S(s) + O_2(g) \rightarrow 2Cu(l) + SO_2(g)$$

The impure copper is then refined by electrolysis as described in 19.11.

248

24.11 Alloys

An alloy is formed by dissolving an element (usually a metal or carbon) in a molten metal. When the solution solidifies on cooling, the solid produced is an alloy of the two metals. The compositions of some of the common alloys are given in Table 24.2. Alloys are used where they have more useful properties than the separate components. For example, one of the properties of the alloy solder which makes it more useful than the separate metals lead and tin, is that it has a lower melting point than either of the pure metals. This means that it is quicker and easier to use solder for joining pieces of other metals, such as copper wires in electrical circuits.

Pure metals such as aluminium, iron and copper, are too soft to be used for construction purposes, but the introduction of relatively small proportions of other elements increases their hardness dramatically. In the crystal lattice of a pure metal (16.7) the layers of atoms can move over each other relatively easily and so the metal is soft, ductile (can be drawn out into a wire) and malleable (can be beaten into different shapes). When atoms of another element replace a small number of atoms of the parent metal in the lattice, they have the effect of making it more difficult for the layers of atoms to move over each other and the resulting alloy is harder.

In some cases alloying produces marked changes in chemical as well as physical properties. Perhaps the best known example is the alloying of chromium with iron to produce stainless steel which is very resistant to chemical attack including atmospheric corrosion. Stainless steel is used for making cutlery and in lining containers used in the chemical industry for holding corrosive liquids.

24.12 Summary

1. Metals are extracted from minerals which are found in ores. The ores can be igneous, sedimentary or metamorphic in origin.
2. Mineral resources are finite and we ought to be concerned about the policies for their use and for their conservation.
3. The extraction of a metal from a mineral is a reduction process. Less reactive metals are extracted by a chemical reducing agent. More reactive metals are extracted by electrolytic methods.
4. Extraction processes may also involve preliminary concentration of the ore, roasting of the ore and refining of the metal.
5. The economic viability of a process is dependent on the balance between the various costs associated with acquiring and processing the mineral, and the likely selling price and market for the metal.
6. The location of a chemical industry is dependent on a balance between economic and social/environmental factors.
7. Sodium and magnesium are extracted by the electrolysis of their molten chlorides.
8. Aluminium is extracted by the electrolysis of its oxide dissolved in cryolite.
9. Iron ore is reduced by heating with coke and limestone to form cast iron, which is then refined to form steel.
10. Zinc, lead and copper are obtained by chemical reduction processes.
11. The properties of a metal can be modified by alloying it with other elements.

TABLE 24.2 The composition and uses of some alloys

ALLOY	APPROXIMATE COMPOSITION	SOME USES
Brass	70% Cn, 30% Zn	Ornaments
Bronze	95% Cn, 5% Sn	Bearings
Coinage metals	{ 97% Cn, 2.5% Zn, 0.5% Sn	Copper coins
	75% Cn, 25% Ni	Silver coins
Solder	66% Sn, 34% Pb	Joining wires and pipes
Mild steel	99.8% Fe, 0.2% C	Motor car bodies
Stainless steel	70% Fe, 20% Cr, 10% Ni	Cutlery
Duralium	96% Al, 4% Cu	Kitchen utensils, aircraft

Chapter 25

Chlorine and its compounds

Investigation
25.1

What are the properties of the gas which is given off when concentrated sulphuric acid is reacted with salt (sodium chloride)?

You will need pieces of both red and blue litmus paper, a glass rod, small volumes of ammonia solution and silver nitrate solution, and a piece of glass tubing fitted with a rubber teat.

1. Put sodium chloride into a boiling-tube to a depth of about 1 cm and clamp the tube in a retort stand.

Carefully add enough moderately concentrated sulphuric acid to the boiling-tube just to cover the solid.

Gently warm the boiling-tube with a small Bunsen flame and test the gas which is given off by observing what happens when you:
    (a) blow across the mouth of the tube,
    (b) hold pieces of moistened red and blue litmus paper in the gas,
    (c) hold a glass rod with a drop of ammonia solution on it in the gas,
    (d) hold a glass rod with a drop of silver nitrate solution on it in the gas.

Questions

1   Which of the above tests indicate the acidic nature of the moist gas?
2   Which of the above tests indicate the presence of chloride ions in the moist gas?

2. Fit a short length of dry glass tubing with a teat and, by squeezing the teat and then releasing it with the end of the tube inside the boiling-tube, draw some of the gas into the tube.

Without squeezing the teat, quickly put the open end of the glass tube under water in a small beaker and note what happens to the level of the water in the tube.

Put a drop of the solution from the glass tube on to a piece of blue litmus paper.

Question

3   Is the gas soluble or insoluble?

Investigation
25.2

What are the properties of hydrochloric acid?

Hydrochloric acid is a solution of hydrogen chloride in water.

Note what happens when each of the following are added to separate portions (about 3 cm³) of dilute hydrochloric acid in a test-tube:
    (a) a few drops of silver nitrate solution.
    (b) about 3 cm of magnesium ribbon—test the gas which is given off with a lighted splint.
    (c) a very small portion of copper(II) oxide—gently warm the mixture for a minute and allow the unreacted copper(II) oxide to settle.

Questions

1   What are the products of these reactions?

2 Which of the metals, zinc, iron and copper, would you expect not to react in a similar manner to magnesium?

Investigation
25.3

Some reactions of chlorine.

You will need a freshly prepared saturated solution of chlorine in water (chlorine water).

1. Pour about 1 cm³ of iron(II) sulphate solution into a test-tube and add about 1 cm³ of dilute sulphuric acid, followed by about 2 cm³ of chlorine water.
   Note the colour of the resulting solution and test separate portions of it with:
   (a) dilute sodium hydroxide solution,
   (b) potassium thiocyanate solution.
   Compare the results of these tests with the results obtained when the same reagents are added to separate portions of the original iron(II) sulphate solution.

Question

1 What do tests (a) and (b) tell you about the type of iron ions which are present in the solution after the reaction with chlorine?

2. Pour about 1 cm³ of a freshly prepared solution of sulphur dioxide into a test-tube and add about 2 cm³ of chlorine water. Test the resulting solution with 1 cm³ of dilute hydrochloric acid and a few drops of barium chloride solution.

Question

2 A solution of sulphur dioxide contains some sulphite ions ($SO_3{}^{2-}$(aq)). Which ions are shown to be present by the barium chloride test?

3. Pour 2 cm³ of chlorine water into a test-tube and add about 2 cm³ of hydrogen sulphide solution.

Questions

3 What is the yellowish-white product of the reaction likely to be?
4 All three of the above reactions of chlorine may be classified as the same type of reaction. What is this type of reaction called?

Investigation
25.4

How does chlorine react with alkalis?

You will need freshly prepared chlorine water.

1. Pour chlorine water into two test-tubes to a depth of about 2 cm. Add about 3 cm of magnesium ribbon to one test-tube and a few drops of silver nitrate solution to the other.

Question

Using the results of these two tests, what is the name of one substance which is present in a solution of chlorine?

2. Put about 2 cm³ of chlorine water into a small basin and test the solution with a piece of red litmus paper.
   Add about 2 cm³ of dilute sodium hydroxide solution and re-test the solution with red litmus paper.
   Now add about 4 cm³ of dilute sulphuric acid to the solution in the basin and re-test with red litmus paper. Look at 25.11 in this chapter to see how the explanation given there accounts for your results.

If you have had the misfortune to have opened your mouth too wide while swimming in the sea and to have swallowed some sea water, you will have had first-hand experience of the most common compound of chlorine to be found in our world. The taste of salt in the sea water would be obvious and this compound, **sodium chloride**, is the most abundant compound of chlorine from which the element and a number of its compounds can be obtained. Such is the concentration of salt in the sea water you swallowed that it has been calculated that, with each cubic metre containing about 100 kg of salt, evaporation of all the oceans of the Earth would yield an amount about 15 times the volume of the entire continent of Europe above sea level. This salt has been known and used since ancient times. The Greeks and Romans used to include it in their offerings to their gods as one of the fruits of the Earth and such was its value in those days that cakes of salt were used as money.

The material in the past had two main uses. It was used as a flavouring, as in the baking of bread, which, without salt, has a flat and insipid taste. It was also used in preserving meat and fish, on which the bacteria were dehydrated by the salt and were therefore unable to cause the decomposition of the flesh. It was also used to preserve animal skins and hides, where it again dehydrated bacteria and removed water from the skin to toughen it. These are still important uses of salt, although its main use now is as a starting material for the manufacture of other chemicals. A concentrated solution of salt in water, known as brine, is electrolysed to produce chlorine and sodium hydroxide. Chlorine is used for the treatment of water and for making plastics, disinfectants, bleaches, hydrochloric acid and a host of other compounds, while sodium hydroxide is used for the production of such things as soap, textile fibres and paper.

A very large and important chemical industry is associated with the products which are made directly or indirectly from salt. Some of the processes which are used are dealt with in more detail in later chapters.

This chapter is mainly concerned with the chemical and physical properties of two of the products, namely, chlorine and hydrogen chloride.

Fig. 25.1 Salt is used in the production of bacon. The photograph shows sides of pork soaking in a vat of salt solution. (Courtesy Danish Agricultural Producers).

Hydrogen chloride is a gas at room temperature and as it can be prepared from common salt (sodium chloride) it was originally called 'salt gas'.

Preparation

Concentrated sulphuric acid, which has a low volatility (i.e. a low tendency to evaporate), will produce hydrogen chloride from sodium chloride:

$$NaCl(s) + H_2SO_4(l) \rightarrow NaHSO_4(s) + HCl(g)$$

This reaction is an example of the general principle that concentrated sulphuric acid will displace more volatile acids from their salts (22.12). Thus hydrogen chloride may also be prepared by the action of concentrated sulphuric acid on other metal chlorides.

If moderately concentrated sulphuric acid is used the reaction is more easily controlled by gently heating the mixture (Fig. 25.2). The hydrogen chloride which is given off can be dried by passing it through concentrated sulphuric acid and then collected by downward delivery.

Fig. 25.2 The preparation and collection of dry hydrogen chloride

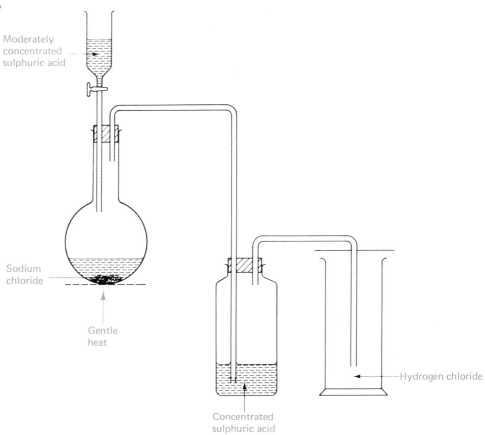

Moderately concentrated sulphuric acid

Sodium chloride

Gentle heat

Concentrated sulphuric acid

Hydrogen chloride

Physical properties

Hydrogen chloride is a colourless gas which is slightly more dense than air. The fact that it has such a low boiling point ($-85\,°C$) and that the perfectly dry gas is not acidic, indicate that hydrogen and chlorine atoms are covalently bonded.

The gas is extremely soluble in water (1 volume of water will dissolve about 500 volumes of hydrogen chloride at $0\,°C$). Such a high solubility indicates that the hydrogen chloride is reacting with the water. The resulting solution is acidic and is known as hydrochloric acid. The nature of this solution is discussed in more detail in 25.7. The misty appearance of hydrogen chloride in moist air is due to hydrogen chloride gas and water vapour in the air condensing together to form tiny droplets of hydrochloric acid.

253

The high solubility of the gas can be convincingly demonstrated by the **Fountain Experiment**. A round flask is filled with dry hydrogen chloride gas, by downward delivery, either from a cylinder of the gas or using the generating apparatus described earlier. The bung, fitted with tubes as shown in Fig. 25.3, is inserted and then the flask is positioned as shown, so that the long tube reaches almost to the bottom of the beaker. When the teat is squeezed, water is forced into the flask. This dissolves enough of the gas to create a partial vacuum within the flask, and water is then sucked up the long tube to produce a fountain at the top. As the water enters, the litmus changes from blue to red. The fountain continues until only a little air is left in the top of the flask.

Reactions of hydrogen chloride
1. Moist blue litmus paper is turned red by the gas, but perfectly dry litmus paper shows no change of colour when brought into contact with a perfectly dry sample of the gas. This shows that hydrogen chloride forms an acidic solution when it meets water.

Fig. 25.3 The Fountain Experiment

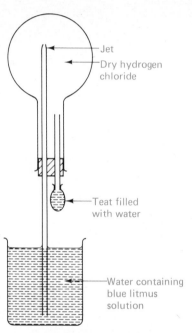

Jet
Dry hydrogen chloride

Teat filled with water

Water containing blue litmus solution

2. When hydrogen chloride gas comes into contact with ammonia, a white smoke is formed. The smoke consists of tiny solid particles of ammonium chloride:

$$NH_3(g) + HCl(g) \rightarrow NH_4Cl(s)$$

Hydrogen chloride is acting as an acid by combining with the base, ammonia, to form a salt.

3. If a drop of silver nitrate solution is held in the gas on the end of a glass rod, the liquid turns white and opaque, owing to the formation of a precipitate of insoluble silver chloride:

$$HCl(g) + AgNO_3(aq) \rightarrow AgCl(s) + HNO_3(aq)$$

When hydrogen chloride dissolves in the water in the silver nitrate solution, it forms chloride ions which combine with the silver ions from the silver nitrate to form crystals of silver chloride:

$$Ag^+(aq) + Cl^-(aq) \rightarrow AgCl(s)$$

254

25.7 Hydrochloric acid

Hydrochloric acid is an aqueous solution of hydrogen chloride.

Preparation of hydrochloric acid

There is a special technique for preparing a solution of a very soluble gas. If a narrow delivery tube were led directly into water, as in Fig. 25.4(a), the great solubility of the gas would cause a suck-back of water up the tube and into the apparatus (as in the fountain experiment). Two ways of overcoming this problem are shown in Fig. 25.4(b) and 25.4(c). In both cases, a large surface area of water is presented to the gas, so that it dissolves efficiently. When 'suck back' tries to occur, air enters to relieve the vacuum.

Fig. 25.4 Dissolving hydrogen chloride

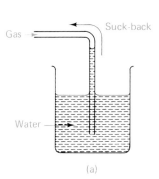

(a)

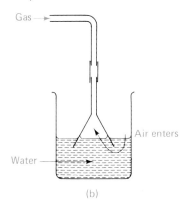

(b)

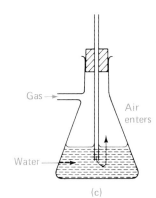

(c)

The nature of hydrochloric acid

Concentrated hydrochloric acid is a saturated solution of hydrogen chloride and smells very strongly of the gas. It contains about 35% by mass of hydrogen chloride (compared to concentrated sulphuric acid which is 98% by mass sulphuric acid). Hydrochloric acid of any concentration is a colourless liquid and if boiled, leaves no residue as the solute is volatile.

When hydrogen chloride dissolves in water, heat is evolved. This may be demonstrated by preparing a gas-jar of dry hydrogen chloride, and then dipping into the gas a thermometer whose bulb is surrounded by cotton wool moistened with water and secured with a rubber band (Fig. 25.5). As the gas dissolves in the water the thermometer reading rises, sometimes to as high as 70 °C. If the experiment is tried using solvents such as tetrachloromethane or methylbenzene, no temperature rise occurs, even though hydrogen chloride does dissolve in these solvents.

Fig. 25.5 The heat change when hydrogen chloride dissolves in water

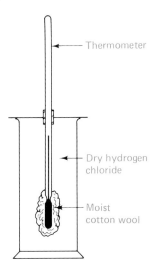

255

These observations, together with the fact that a solution of hydrogen chloride in water is a good conductor of electricity whereas solutions of the gas in the other solvents do not conduct, indicate that water reacts exothermically with the covalent hydrogen chloride to produce a solution containing ions. The dissociation of hydrogen chloride molecules into ions can be represented by the equation:

$$HCl(g) \rightarrow H^+(aq) + Cl^-(aq)$$

This equation does not reveal why the water is necessary for this ionisation to occur and a more accurate interpretation brings the water into the equation:

$$HCl(g) + H_2O(l) \rightarrow \underset{\substack{\text{hydroxonium} \\ \text{ion}}}{H_3O^+(aq)} + Cl^-(aq)$$

It is the hydration of the hydrogen ion (H^+) to form the hydroxonium ion (H_3O^+) which is so strongly exothermic. For convenience it is usual to write $H^+(aq)$ in equations rather than H_3O^+.

Hydrochloric acid is called a strong acid because in dilute solution almost all the hydrogen chloride molecules are dissociated into ions (11.14). This is why dilute hydrochloric acid does not smell of hydrogen chloride.

Reactions of hydrochloric acid

As explained above, hydrochloric acid is a strong acid and as such behaves in a typical manner with metals, bases and carbonates.

1. Metals above hydrogen in the reactivity series liberate hydrogen from the acid,

e.g. $$Zn(s) + 2HCl(aq) \rightarrow ZnCl_2(aq) + H_2(g)$$

or $$Zn(s) + 2H^+(aq) \rightarrow Zn^{2+}(aq) + H_2(g)$$

Copper will not liberate hydrogen from the acid.

An important point arises with some metals such as iron and tin which each form two chlorides. When these metals react with hydrochloric acid (or if they are heated in hydrogen chloride gas) the **lower** chloride is formed. Thus iron forms iron(II) chloride, and tin forms tin(II) chloride,

e.g. $$Fe(s) + 2HCl(aq) \rightarrow FeCl_2(aq) + H_2(g)$$

2. Bases react with hydrochloric acid forming chlorides and water,

e.g. $$CuO(s) + 2HCl(aq) \rightarrow CuCl_2(aq) + H_2O(l)$$

$$NaOH(aq) + HCl(aq) \rightarrow NaCl(aq) + H_2O(l)$$

Carbonates react with the acid to produce chlorides, carbon dioxide and water,

e.g. $$CaCO_3(s) + 2HCl(aq) \rightarrow CaCl_2(aq) + CO_2(g) + H_2O(l)$$

It is important to note that 'dilute' and 'concentrated' hydrochloric acid are both aqueous solutions, and have the same chemical properties. The only difference in behaviour is that the more concentrated acid will react more rapidly. This is not true of the other common acids, sulphuric acid and nitric acid. In these cases the concentrated acids have different properties from the dilute acids.

**25.8
Test for
chlorides**

Most metal chlorides are soluble in water, the common exceptions being silver chloride ($AgCl$) and lead(II) chloride ($PbCl_2$). Chlorides in aqueous solution may be detected by the fact that they give an immediate white precipitate with silver nitrate solution,

e.g.
$$NaCl(aq) \ + \ AgNO_3(aq) \ \rightarrow \ AgCl(s) \ + \ NaNO_3(aq)$$

sodium   silver   silver   sodium
chloride   nitrate   chloride   nitrate
(insoluble)

This forms a test for chloride ions, and the reaction can be represented by the ionic equation:

$$Ag^+(aq) \ + \ Cl^-(aq) \ \rightarrow \ AgCl(s)$$

silver ion   chloride ion   silver chloride

The test is, of course, positive for hydrochloric acid itself since the solution contains $H^+(aq)$ ions and $Cl^-(aq)$ ions.

One experimental precaution is important, however. If an unknown solution is being tested it must first be made acidic by the addition of dilute nitric acid before the silver nitrate solution is added. This is to guard against false positive results from solutions which do not contain chloride ions. For example, addition of silver nitrate solution to a solution containing carbonate ions will result in a precipitate of silver carbonate, but this precipitate will not appear if the solution is first acidified with nitric acid. Silver chloride, however, will be precipitated in the presence of nitric acid.

25.9 Chlorine

Chlorine, which is a yellowish green gas, is the most common and most widely used member of the family of elements known as the Halogens which comprise Group 7 of the Periodic Table. The similarities and trends within this group of elements are discussed in Chapter 26.

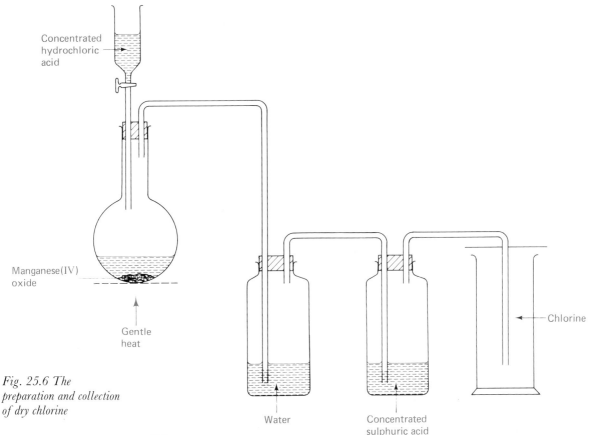

Concentrated hydrochloric acid

Manganese(IV) oxide

Gentle heat

Fig. 25.6 The preparation and collection of dry chlorine

Chlorine

Water

Concentrated sulphuric acid

Preparation of chlorine

In order to convert hydrochloric acid (or hydrogen chloride) into chlorine, it is necessary to remove the hydrogen. Removal of hydrogen is an example of oxidation and hence if hydrochloric acid is reacted with a sufficiently powerful oxidising agent, it will be converted into chlorine. Such a reaction could be represented by the simple equation:

$$[O] + 2HCl(aq) \rightarrow H_2O(l) + Cl_2(g)$$

The symbol [O] represents the oxygen provided by the oxidising agent. It is represented in this way because it is never actually free as the element.

Ionically, the oxidation can be represented by:

$$2Cl^-(aq) \rightarrow Cl_2(g) + 2e^-$$

Concentrated hydrochloric acid is oxidised to chlorine by manganese(IV) oxide, lead(IV) oxide and potassium manganate(VII) (potassium permanganate). The latter is the most convenient reagent, as the reaction, unlike the others, requires no heat and the flow of chlorine can be controlled by the rate of addition of the concentrated acid to the solid potassium manganate(VII).

The equations for the reactions are:

$$MnO_2(s) + 4HCl(aq) \rightarrow MnCl_2(aq) + 2H_2O(l) + Cl_2(g)$$

$$PbO_2(s) + 4HCl(aq) \rightarrow PbCl_2(aq) + 2H_2O(l) + Cl_2(g)$$

$$2KMnO_4(s) + 16HCl(aq) \rightarrow 2KCl(aq) + 2MnCl_2(aq) + 8H_2O(l) + 5Cl_2(g)$$

Note the similarity of the first two equations. It is unlikely that you will need to remember the potassium manganate(VII) equation.

Figure 25.6 shows the apparatus for the preparation and collection of chlorine by the manganese(IV) oxide method. The reaction should be done in a fume-cupboard since chlorine is very poisonous.

The first wash-bottle, containing water, is necessary to remove hydrogen chloride gas which has escaped oxidation and the second bottle, containing concentrated sulphuric acid, is to dry the chlorine. The gas is much denser than air and so it is collected by downward delivery. The green colour of the gas shows when the jar is full.

Physical properties of chlorine

Chlorine is a pale green gas which is denser than air. It is easily liquefied, without cooling, by simply compressing the gas. Liquid chlorine can be easily stored and transported in steel cylinders. It has a suffocating and irritating smell and it becomes intolerable at quite small concentrations. It is extremely poisonous, and should be treated with great caution.

Chlorine is moderately soluble in water and its solution is known as chlorine water.

Reactions of chlorine

The usual test for chlorine is that it will bleach moist red or blue litmus paper. This bleaching action is discussed in more detail in 25.10.

Chlorine, being a member of Group 7 of the Periodic Table, is a non-metal. A knowledge of its electron arrangement (2.8.7) enables the way in which chlorine combines with other elements to be explained by either the acceptance of a single electron to form a chloride ion (Cl^-) as when chlorine reacts with some metals, e.g. sodium,

$$Na^o \quad + \quad {}^{\times}_{\times}Cl^{\times}_{\times} \quad \rightarrow \quad [Na]^+ \left[{}^{\circ}_{\times}Cl^{\times}_{\times} \right]^-$$

or the formation of a single covalent bond from the unpaired chlorine electron and an unpaired electron from an atom of another non-metal, e.g. hydrogen,

$$H° \; + \; \overset{\times\times}{\underset{\times\times}{\times\, Cl\, \times}} \; \rightarrow \; H \overset{\circ}{\underset{\times}{\times}} \overset{\times\times}{\underset{\times\times}{Cl\, \times}}$$

All the reactions described in the following section can be explained either by the formation of a Cl⁻ ion, or a single covalent bond with each chlorine atom. The reactions of chlorine also may be interpreted as the reduction of chlorine,

$$2Cl_2 \; + \; 2e^- \rightarrow 2Cl^- \quad \text{(addition of electron)}$$

and $\qquad\qquad H_2 \; + \; Cl_2 \rightarrow 2HCl \quad \text{(addition of hydrogen)}$

The reduction of the chlorine must always be accompanied by the oxidation of the other reactant.

1. Direct combination with elements

Chlorine is a highly reactive element and combines directly with most metals and with many non-metals.

A piece of sodium, heated on a deflagrating spoon until it starts to burn, will continue to burn when lowered into a jar of chlorine, producing white clouds of sodium chloride:

$$2Na(s) \; + \; Cl_2(g) \rightarrow 2NaCl(s)$$

Similarly magnesium will continue to burn forming magnesium chloride.

Fig. 25.7 The reaction of chlorine with iron

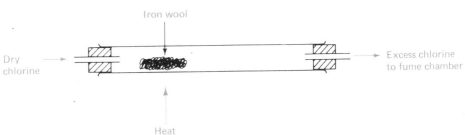

Less reactive metals are more conveniently heated in a current of dry chlorine using a combustion tube apparatus such as that in Fig. 25.7. The method can be used to prepare the anhydrous chlorides of iron, aluminium, zinc, tin and copper, the product in each case being formed as a vapour which condenses in the cooler part of the tube.

If a metal can form two chlorides, the chloride which contains the most chlorine is formed. For example, iron forms iron(III) chloride ($FeCl_3$) rather than iron(II) chloride ($FeCl_2$):

$$2Fe(s) \; + \; 3Cl_2(g) \rightarrow 2FeCl_3(s)$$

A piece of the non-metal phosphorus, in its white (yellow) form, will burn spontaneously when lowered on a deflagrating spoon into a jar of chlorine, giving whitish fumes of phosphorus(III) chloride and phosphorus(V) chloride,

$$P_4(s) \; + \; 6Cl_2(g) \rightarrow 4PCl_3(l)$$

and $\qquad\qquad P_4(s) \; + \; 10Cl_2(g) \rightarrow 4PCl_5(s)$

Other non-metals, such as sulphur and hydrogen, will burn in chlorine. For example, if hydrogen, preferably from a cylinder, is ignited at the jet of a glass tube bent upwards

259

as shown in Fig. 25.8, and then lowered into a jar of chlorine, the hydrogen continues to burn with a white flame. The green colour of chlorine disappears and the characteristic misty fumes of hydrogen chloride can be seen:

$$H_2(g) + Cl_2(g) \rightarrow 2HCl(g)$$

Figure 25.8 Burning hydrogen in chlorine

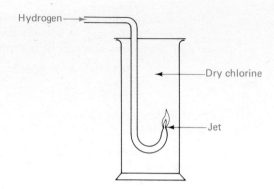

A mixture of hydrogen and chlorine will explode if subjected to an electric spark or exposed to ultra-violet light.

All of the above reactions of chlorine with other elements can be considered as involving the reduction of chlorine and the oxidation of the other reactant.

2. Reactions with compounds

Chlorine can cause a variety of chemical changes to occur in other compounds, either directly or with the aid of a third reagent, all of which can be classified as oxidation. At the same time the chlorine is always reduced to a chloride.

In all of the following reactions it is more convenient to use chlorine water, which reacts in just the same way, rather than the gas.

Iron(II) salts

If chlorine water is added to iron(II) chloride solution, iron(III) chloride is formed. The overall change is,

$$2FeCl_2(aq) + Cl_2(aq) \rightarrow 2FeCl_3(aq)$$

or

$$2Fe^{2+}(aq) + Cl_2(aq) \rightarrow 2Fe^{3+}(aq) + 2Cl^-(aq)$$

oxidation

reduction

The addition of sodium hydroxide solution to the product produces a rusty brown precipitate of iron(III) hydroxide, or the addition of potassium thiocyanate solution produces a deep red colour. Either of these two tests show that Fe^{3+} ions were formed in the reaction with chlorine. The same reaction takes place whatever iron(II) salt is used, since it is the iron(II) ions which react with the chlorine, being oxidised to iron(III) ions.

Sulphurous acid and sulphites

If a solution of sulphur dioxide, which contains sulphurous acid (H_2SO_3), or an acidified solution of sodium sulphite (Na_2SO_3) is treated with chlorine, oxidation to sulphuric acid or a sulphate occurs,

260

$$H_2SO_3(aq) \ + \ H_2O(l) \ + \ Cl_2(aq) \ \rightarrow \ H_2SO_4(aq) \ + \ 2HCl(aq)$$

and $\quad Na_2SO_3(aq) \ + \ H_2O(l) \ + \ Cl_2(aq) \ \rightarrow \ Na_2SO_4(aq) \ + \ 2HCl(aq)$,

or

$$SO_3^{2-}(aq) \ + \ H_2O(l) \ + \ Cl_2(aq) \ \rightarrow \ SO_4^{2-}(aq) \ + \ 2H^+(aq) \ + \ 2Cl^-(aq)$$

(oxidation / reduction)

This can be shown to have occurred by the addition of barium chloride solution, which reacts with the sulphate ions to form a dense white precipitate of barium sulphate.

Hydrogen sulphide

If chlorine and hydrogen sulphide (H_2S) are brought together, either directly as gases, or in aqueous solution, sulphur is seen to be deposited as a very pale yellow solid,

$$H_2S(g) \ + \ Cl_2(g) \rightarrow S(s) \ + \ 2HCl(g)$$

or $\quad H_2S(aq) \ + \ Cl_2(aq) \ \rightarrow \ S(s) \ + \ 2HCl(aq)$

(oxidation / reduction)

Bromides and iodides

Chlorine will oxidise a solution of a bromide to a yellow-orange solution of bromine. It will oxidise an iodide to a yellow-brown solution of iodine,

$$Cl_2(aq) \ + \ 2NaBr(aq) \ \rightarrow \ Br_2(aq) \ + \ 2NaCl(aq)$$

or $\quad Cl_2(aq) \ + \ 2Br^-(aq) \ \rightarrow \ Br_2(aq) \ + \ 2Cl^-(aq)$

and $\quad Cl_2(aq) \ + \ 2NaI(aq) \ \rightarrow \ I_2(aq) \ + \ 2NaCl(aq)$

or $\quad Cl_2(aq) \ + \ 2I^-(aq) \ \rightarrow \ I_2(aq) \ + \ 2Cl^-(aq)$

These reactions show that chlorine is a more powerful oxidising agent than either bromine or iodine. This trend within Group 7 will be discussed in more detail in Chapter 26.

Water (and the bleaching action of chlorine)

If a gas jar of chlorine is dried by allowing a little concentrated sulphuric acid to stand in it for a few minutes (very careful shaking will help), and then a piece of dry litmus paper is held in the gas, it will remain unbleached. If the litmus paper is then moistened and put back in the gas, it will bleach immediately. This shows that the presence of water is necessary for chlorine to act as a bleach.

Fig. 25.9 Action of light on chlorine water

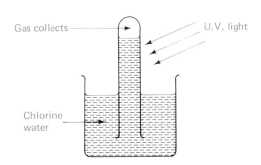

Gas collects — U.V. light

Chlorine water

If chlorine water is left in sunlight (or under an ultra-violet lamp) for a day or two in an apparatus such as that in Fig. 25.9, the green colour of the chlorine fades and a colourless gas collects in the top of the tube. The gas rekindles a glowing splint, indicating that it is oxygen. The net change is:

$$Cl_2(aq) + H_2O(l) \xrightarrow[\text{light}]{\text{u.v.}} 2HCl(aq) + \tfrac{1}{2}O_2(g)$$

25.10
The nature of chlorine water

The above equation represents the overall change when chlorine in the presence of water liberates oxygen, but the following properties indicate that the representation of chlorine water by $Cl_2(aq)$ is an over-simplification.

If a piece of magnesium ribbon is put into fresh chlorine water, effervescence is observed, the gas liberated being hydrogen. This indicates that a solution of chlorine in water is acidic, that is, it contains $H^+(aq)$ ions (this cannot be checked with litmus because the litmus is bleached). Furthermore, addition of silver nitrate solution to fresh chlorine water gives a white precipitate, showing that chloride ions, $Cl^-(aq)$, are present. So even fresh chlorine water evidently contains some hydrochloric acid. The explanation for this is that chlorine undergoes a reversible reaction with water, forming two acids:

$$Cl_2(aq) + H_2O(l) \rightleftharpoons HCl(aq) + HClO(aq)$$

The second acid is hypochlorous acid and this is the source of oxygen and is responsible for the bleaching action of chlorine water:

$$HClO(aq) + \underset{\text{(coloured)}}{Dye} \rightarrow HCl(aq) + \underset{\text{(colourless)}}{Oxidised\ Dye}$$

It is worth noting here that if chlorine (in the presence of water) is used for bleaching fabrics, then it is essential to rinse the fabric thoroughly afterwards with water in order to remove the hydrochloric acid formed during the bleaching process.

25.11
Action of alkali on chlorine water

If sodium hydroxide solution is added to chlorine water, the colour and smell of chlorine disappear, and the liquid does not bleach litmus (it stays blue). Subsequent addition of dilute acid restores the smell of chlorine and the resulting liquid bleaches litmus. The explanation of these observations is that the alkali reacts with the two acids to form salts:

$$2NaOH(aq) + HCl(aq) + HClO(aq) \rightarrow NaCl(aq) + NaClO(aq) + 2H_2O(l)$$

The second salt is sodium hypochlorite (sodium chlorate(I)). As the alkali is added and the acids are used up, the equilibrium between chlorine and the acids shifts to the right:

$$Cl_2(aq) + H_2O(l) \rightleftharpoons HCl(aq) + HClO(aq)$$

so that all the chlorine is eventually converted into the two salts. When acid is added to the mixture of salts, hypochlorous acid is reformed as it is a weak acid:

$$H^+(aq) + ClO^-(aq) \rightleftharpoons HClO(aq)$$

Thus the bleaching action returns. Sodium hypochlorite, available only in aqueous solution, is an important commercial substance being, as it were, 'packaged chlorine'. Sodium hypochlorite is the main component of commercial sterilising agents and bleaches such as Milton and Domestos.

A similar substance, less important now than it used to be, is 'bleaching powder'. This is formed by absorbing chlorine gas in solid calcium hydroxide (slaked lime). A solid substance of complex composition, sometimes written approximately as $CaOCl_2$,

262

is formed. It contains hypochlorite (ClO^-) ions, and hence chlorine is released when it is treated with acid.

Action of hot concentrated alkali on chlorine

If a mixture of chlorine and a solution of sodium hydroxide is heated, a further reaction occurs and the end products are sodium chloride and sodium chlorate(V) ($NaClO_3$):

$$3Cl_2 + 6NaCl(aq) \rightarrow 5NaCl(aq) + NaClO_3(aq) + H_2O(l)$$

Sodium chlorate(V) is used as a non-selective weed killer. Potassium chlorate(V), being less hygroscopic, is used as a constituent of fireworks, flares and explosive charges because on heating chlorates produce oxygen,

e.g. $$2KClO_3(s) \rightarrow 2KCl(s) + 3O_2(g)$$

(This reaction is catalysed by the presence of manganese(IV) oxide or, more safely, by iron(III) oxide.)

Fig. 25.10 A lot of chlorine is used to make the solvents such as tetrachloroethene (perchlorethylene, $CCl_2 CCl_2$) which are used for dry-cleaning clothes. The photographs show the front and back of a dry-cleaning machine. (Courtesy Niel & Spencer Ltd)

25.12 Manufacture and uses of hydrogen chloride and chlorine

Both the substances considered in this chapter, hydrogen chloride and chlorine, are manufactured on a large scale because of their many uses. Hydrogen chloride is produced by direct synthesis:

$$H_2(g) + Cl_2(g) \rightarrow 2HCl(g)$$

Its main use is to make hydrochloric acid by being dissolved in water, but it is also used to combine with the hydrocarbon ethyne (C_2H_2) to form monochloroethene:

$$C_2H_2(g) + HCl(g) \rightarrow H_2C = CHCl(l).$$

Polymerization of this compound gives the product commonly called polyvinyl chloride (PVC) which is an important plastic. Chlorine is manufactured by the electrolysis of brine (sodium chloride solution), in which it is liberated at the anode of a cell. This process is described more fully in 29.7.

Many of the uses of chlorine have already been mentioned but in particular it is used for making:

1. bleaches (e.g. sodium hypochlorite solution).

2. germicides (germ-killers) which can be used for sterilizing objects. The gas itself is used to kill harmful bacteria in drinking water and in swimming pool water. Disinfectants are also germ-killers and two of the most common ones are chlorine compounds. TCP is **tri**chloro**p**henol and the chemical name for Dettol is 4-chloro-3, 5-dimethylphenol.

3. sodium chlorate(V) for use as a weedkiller.

4. pesticides. DDT (**d**ichloro**d**iphenyl**t**richloroethane) was developed for use in jungle warfare during the Second World War, but it is not so widely used today because it has been found to be poisonous to wildlife. BHC (benzene hexachloride) is made by passing chlorine into benzene on which ultra-violet light is directed and it is particularly valuable in the fight against the locust.

5. solvents, such as tetrachloroethene which is used for the dry cleaning of clothes.

6. hydrogen chloride and hydrochloric acid and a large number of other chemicals.

Fig. 25.11 The water in this swimming-pool at Bletchley is purified by chlorine. (Courtesy ICI Ltd Mond Division)

25.13 Summary

1. The action of concentrated sulphuric acid on an ionic chloride produces hydrogen chloride.
2. Hydrogen chloride is very soluble in water, forming hydrochloric acid.
3. In the presence of water the covalent hydrogen chloride molecules split into ions and the solution reacts as an acid with metals, bases and carbonates, forming as one of the products in each case a chloride.
4. A solution of hydrogen chloride in water contains chloride ions. This can be shown by the action of silver nitrate solution which forms a white precipitate of silver chloride.
5. Hydrochloric acid can be oxidised to chlorine by an oxidising agent such as potassium manganate(VII) or manganese(IV) oxide.
6. Chlorine combines directly with many elements, forming chlorides.
7. Because of its ability to accept electrons, chlorine is a vigorous oxidising agent and oxidises iron(II) salts, sulphites, sulphides, bromides and iodides.
8. Chlorine combines with water to form a mixture of acids, hydrochloric acid and hypochlorous acid which is an oxidising agent. This solution, called chlorine water, bleaches certain dyes by oxidising them.
9. Chlorine dissolves rapidly in cold sodium hydroxide solution, forming in solution a mixture of sodium hypochlorite (sodium chlorate(I)) and sodium chloride. When the sodium hydroxide solution is hot, the products are sodium chlorate(V) and sodium chloride.

Chapter 26

The halogens

**Investigation
26.1**

Do bromine and iodine resemble chlorine in their reactions with dilute sodium hydroxide solution?

Instructions for investigating the reactions of chlorine with alkali are given in 25.4.

You will need a clean glass rod and a piece of red litmus paper. Pure bromine is too dangerous to use so you will need a solution of bromine in water (bromine water).

1. Put about 1 cm³ of bromine water into a test-tube. Add about an equal volume of dilute sodium hydroxide solution and observe what happens to the colour of the bromine water.
   Withdraw a drop of the liquid on a clean glass rod and smell it.

Question

1 Is free bromine still present after alkali has been added to the solution?

2. Add a piece of red litmus paper to the mixture of bromine water and sodium hydroxide solution.
   Now add about 2 cm³ of dilute sulphuric acid to the mixture and note what happens to the colour of the solution and the litmus paper.

Question

2 What has been produced by adding sulphuric acid to the alkaline solution?

3. Put a very small crystal of iodine into a test-tube and add about 2 cm³ of dilute sodium hydroxide solution. Warm the tube gently with shaking.
   To the mixture then add about 2 cm³ of dilute sulphuric acid. Now add 1–2 cm³ of trichloroethane, shake gently and hold the tube up to the light and note the colour of the trichloroethane layer.

Questions

3 Is free iodine still present in the alkaline solution?
4 What product causes the colour of the trichloroethane?
5 What has been produced by adding sulphuric acid to the alkaline solution?
6 In what ways do bromine and iodine behave similarly to chlorine (25.4) when treated with alkali and then acid?

**Investigation
26.2**

Bromine and iodine as oxidising agents.

When halogen molecules gain electrons to form halide ions, they are reduced and the halogen is acting as an oxidising agent.

1. Pour 1 cm³ of freshly prepared sulphur dioxide solution or sodium sulphite solution into a test-tube and then add 1 cm³ of dilute hydrochloric acid.
   Pour the mixture into another test-tube containing a few drops of bromine water and note what happens to the colour of the bromine water.

Now add a few drops of barium chloride solution to the mixture.

Questions

1 What do you see when barium chloride solution is added to the mixture?
2 What ions have therefore been formed in the reaction?
3 What has happened to the sulphite ions in this reaction?

2. Repeat the above experiment, using a few drops of a solution of iodine in aqueous potassium iodide, instead of bromine water.

Questions

4 In what way do bromine and iodine behave similarly when added to a solution containing sulphite ions?
5 What type of reagents are the bromine and iodine acting as in this reaction?

3. Put 1 cm³ of bromine water into a test-tube and add 1–2 cm³ of a solution of hydrogen sulphide.

Questions

6 What two changes do you see in the mixture?
7 Hydrogen sulphide solution contains sulphide ions. Into what have these been converted in the reaction with bromine water?

4. Repeat the above experiment, but this time using a solution of iodine in aqueous potassium iodide, instead of bromine water.

Question

8 In what way do bromine and iodine behave similarly when added to a solution of hydrogen sulphide?

Investigation 26.3

Bromides and iodides as reducing agents.

When a halide is oxidised to the element, it is acting as a reducing agent.

1. Put 1 cm³ of potassium bromide solution into a test-tube and add 1 cm³ of dilute sulphuric acid and then 1–2 cm³ of hydrogen peroxide solution. If there is no obvious change when cold, warm the mixture gently and note the final colour of the mixture.
   Now add a piece of red litmus paper to the mixture.

Questions

1 Which substance is likely to be causing the final colour of the solution?
2 What type of reagent is hydrogen peroxide acting as in this reaction?

2. Put 1 cm³ of potassium iodide solution into a test-tube, add 1 cm³ of dilute sulphuric acid and then about 1 cm³ of hydrogen peroxide solution. Note the colour of the reaction mixture.
   Now add 1–2 cm³ of trichloroethane. Shake the mixture and observe the colour of the trichloroethane layer.

Questions

3 What product causes the colour in the trichloroethane layer?
4 In what way does hydrogen peroxide behave similarly with bromide ions and iodide ions?

What is the order of reactivity of the halogens?

1. Put about 1 cm³ of potassium bromide solution into a test-tube. Add about 1 cm³ of chlorine water and note the colour of the solution.

Now add 1–2 cm³ of trichloroethane. Shake the mixture and observe the colour of the trichloroethane layer.

Questions

1   What product causes the colour of the reaction mixture and the colour of the trichloroethane layer?
2   Does the result indicate that chlorine is more or less reactive than bromine?
3   What type of reagent has chlorine acted as in this reaction?

2. Repeat the procedure using potassium iodide solution instead of potassium bromide solution.

Questions

4   What product causes the colour of the reaction mixture and the colour of the trichloroethane layer?
5   Does the result indicate that chlorine is more or less reactive than iodine?

3. Repeat part 2 using bromine water instead of chlorine water.

Questions

6   Is bromine more or less reactive than iodine?
7   Of the three halogens, chlorine, bromine and iodine, which oxidises the ions of the other two and is therefore more reactive than the other two?
8   Which of the three halogens has ions which are oxidised by the other two?
9   What, therefore, is the order of oxidising power of the halogens?
10   Would you expect there to be a reaction between bromine and potassium chloride solution?

A comparison of the silver halides.

1. Into three test-tubes pour separate 1 cm³ portions of solutions of potassium chloride, potassium bromide and potassium iodide respectively. Add to each tube 2 or 3 drops of silver nitrate solution. Examine the appearances of the three precipitates. Now, to each tube, add dilute ammonia solution until the tube is one third full. Stir each with a glass rod.

Questions

1   Which of the silver halides does not dissolve in the ammonia solution?
2   Which of the silver halides appears to be only partially soluble in the ammonia solution?
3   Which of the silver halides dissolves completely in the ammonia solution?
4   What sort of ion is formed when a silver halide dissolves in ammonia solution (15.5)?

2. In three more tubes repeat the precipitation of separate samples of the three silver halides. Place the tubes near a bright light, or on a window-sill, and leave them until the end of the lesson or, if possible, until the next lesson.

Questions

5   What colour changes do you see in the three samples?
6   What causes these colour changes?
7   To what use is this change in silver halides put?

26.6
The halogens

Chlorine, the subject of Chapter 25, is found in **Group 7** of the Periodic Table where its companions are fluorine, bromine and iodine. The four elements make up the family of elements known as the **halogens**, which literally means 'salt-formers'. The halogens which are the companions of chlorine in the family are worth studying, not only because they are interesting and useful elements, but also for another reason. As was mentioned in Chapter 13, the elements making up a group in the Periodic Table show at least some similar properties, and this is certainly true of the halogens. Nevertheless, as we go down a group, certain trends in the behaviour and reactions of the elements usually become clear, and it is interesting to see whether such trends appear within the halogen family and, if they do, how they can be explained.

26.7
Bromine

Bromine at room temperature is a dark red liquid which evaporates readily to give a reddish-brown vapour with a sharp choking smell. This is responsible for its name, the Greek word 'bromos' meaning 'stench'. The element is irritating to the skin and readily corrodes metals. Bromine was first isolated in France by A. J. Balard from the residue left after the evaporation of Mediterranean sea-water, and it was recognized as an element in 1826. During the second half of the nineteenth century, the majority of the world's output came from the salt deposits near Stassfurt in Germany, but at the present time most bromine is extracted from sea water, which contains about 0.075 g dm^{-3}. In the United Kingdom there is a large chemical plant on the island of Anglesey, in which sea-water is acidified with dilute acid and then reacted with chlorine to liberate the bromine:

$$2NaBr(aq) + Cl_2(g) \rightarrow 2NaCl(aq) + Br_2(aq)$$

The world's richest source of bromine is the Dead Sea which contains 4 to 4.5 g dm^{-3}, and the production from that outlet is 10 000 tonnes per year.

26.8
Uses of bromine

The main use of bromine is to prepare 1,2-dibromoethane by reacting it with ethene (31.9). 1,2-Dibromoethane is added to petrol to prevent lead (which is present in petrol, 9.5) from being deposited inside engines.

Other uses of bromine include the manufacture of soil fumigants and fire retardants, and the alkali metal bromides are required for making silver bromide for photographic films and papers.

26.9
Iodine

Iodine is the element below bromine in Group 7. It was discovered in 1811 by B. Courtois who was a manufacturer of saltpetre (potassium nitrate), used in making gunpowder. Courtois noticed that the copper vats, in which he was using an extract in water from burnt seaweed, were being corroded, and, when he treated the extract with dilute sulphuric acid, he was able to isolate a black powder. This on heating gave a violet vapour which, on cooling, gave a crop of dark shiny flakes, and it was the purple vapour which led to the name of the element, 'iodes' being the Greek word for violet.

Nowadays most of the world's supply of iodine comes from sodium iodate(V) from the solutions in the crystallisation of crude Chile saltpetre (sodium nitrate) and only a small amount of it is made from the seaweed (or kelp) in which it was first found. Sodium iodate(V) in the solutions is reduced to iodine by using sodium hydrogen-sulphite:

$$2NaIO_3(aq) + 5NaHSO_3(aq) \rightarrow I_2(s) + 3NaHSO_4(aq) + 2Na_2SO_4(aq) + H_2O(l)$$

Silver iodide, like silver bromide, is used in the emulsion layer of photographic films, and potassium iodide is used in animal feeds and for adding to table salt, producing iodized salt. The human body needs to take in about 0·07 mg of iodine per day, in the form of an iodide, so that it can be converted by the thyroid gland in the neck into a hormone called thyroxime. Normally this small daily amount will be supplied by drinking water, or by fish such as cod, salmon or herring being included in the diet, but there are some areas of the world where there is not sufficient iodine in the diet and people suffer from goitre which produces a swelling of the neck. The use of iodized salt prevents this disease, the salt being prepared by adding 1 part potassium iodide to about 40 000 parts of salt. Potassium iodide is soluble in water and is rapidly absorbed into the blood, any surplus being harmless.

Other uses for iodine include the making of drugs, the antiseptic 'iodoform' (triiodomethane), polaroid and certain dyes, some of which are used in colour photography and others in medicine for detecting gallstones and brain tumours. A solution of iodine in alcohol (tincture of iodine) has been used as an antiseptic for about 150 years.

In both industry and the laboratory, the halogens are prepared from their ionic salts, the halides. The halide ions have to be oxidised to form the halogen molecules:

$$2X^- - 2e^- \rightarrow X_2$$

Each halogen can be obtained in the laboratory by warming the appropriate halide with concentrated sulphuric acid and manganese(IV) oxide. For example, concentrated sulphuric acid will convert sodium chloride to hydrogen chloride (25.6), which is then oxidised to chlorine by manganese(IV) oxide (25.9).

As they are members of the same group (Group 7) in the Periodic Table, and have the same number of electrons in the outermost shells of their atoms, chlorine, bromine and iodine show many similar reactions. Each of the elements is immediately before a noble gas in the Periodic Table, and so the majority of their reactions are those in which the halogen atoms gain one electron to obtain the configuration of the noble gas (15.1 and 15.3). Properties and reactions which all three elements show are the following.

1. Chlorine, bromine and iodine are **diatomic**, the two atoms being held together in the molecules by covalent bonds,

e.g.

$$\overset{\text{oo}}{\underset{\text{oo}}{\text{o}}} \text{Br} \overset{\text{xx}}{\underset{\text{oo}}{\overset{\text{x}}{\underset{\text{o}}{\text{}}}}} \text{Br} \overset{\text{xx}}{\underset{\text{xx}}{\text{x}}}$$

Because of this, their solubilities in water, which is a polar solvent, are either moderate as in the cases of chlorine and bromine or low as in the case of iodine. Bromine and iodine are readily soluble in non-polar, covalent solvents, such as trichloroethane, bromine forming a deep red solution and iodine a purple solution. A saturated solution of bromine in water is red in colour and is commonly known as 'bromine water'.

Iodine readily dissolves in an aqueous solution of potassium iodide to form a brown solution, owing to the formation of a complex ion (I_3^-) in the reversible reaction:

$$\underset{\text{black}}{I_2(s)} + \underset{\text{colourless}}{I^-(aq)} \rightleftharpoons \underset{\text{brown}}{I_3^-(aq)}$$

2. **Reactions with hydrogen**
All three halogens react with hydrogen, but the vigour of the reaction and the stability of the product decreases in the order chlorine, bromine, iodine. Chlorine reacts slowly

with hydrogen in the dark, but a mixture of the two gases will explode if exposed to bright sunlight (25.9), hydrogen chloride being formed:

$$H_2(g) + Cl_2(g) \rightarrow 2HCl(g)$$

A mixture of hydrogen and bromine need to be heated to about 300 °C in the presence of a catalyst before they will react:

$$H_2(g) + Br_2(g) \rightarrow 2HBr(g)$$

A mixture of iodine vapour and hydrogen only react reversibly under the same conditions:

$$H_2(g) + I_2(g) \rightleftharpoons 2HI(g)$$

3. Reactions with alkalis

All three halogens react with alkalis to give similar compounds. With cold dilute sodium hydroxide solution, chlorine forms a mixture of sodium chloride and sodium hypochlorite (sodium chlorate(I)) (25.11). Under similar conditions, bromine forms a colourless solution of sodium bromide and sodium hypobromite,

e.g. $2NaOH(aq) + Br_2(l) \rightarrow NaBr(aq) + NaBrO(aq) + H_2O(l)$
                                                            sodium
                                                            hypobromite

With hot concentrated sodium hydroxide solution, chlorine forms a mixture of sodium chloride and sodium chlorate(V) (25.11). Under similar conditions bromine forms a mixture of sodium bromide and sodium bromate(V), and iodine forms a mixture of sodium iodide and sodium iodate(V),

e.g. $6NaOH(aq) + 3Br_2(l) \rightarrow 5NaBr(aq) + NaBrO_3(aq) + 3H_2O(l)$
                                                        sodium
                                                        bromate(V)

26.13 Differences between halogens as oxidising agents

Reactions of the halogens with other halides

When potassium bromide solution is added to a jar of chlorine and the jar is shaken, the colourless solution turns red and a red vapour is formed in the jar. If trichloroethane is then added to the jar, it forms a dark red lower layer in the liquid, showing that bromine has been liberated:

$$2KBr(aq) + Cl_2(g) \rightarrow 2KCl(aq) + Br_2(aq)$$

Chlorine is a sufficiently strong electron acceptor, or oxidising agent, to be able to take the extra electrons from bromide ions, itself being reduced to chloride ions:

oxidised

$$Cl_2(g) + 2Br^-(aq) \rightarrow Br_2(aq) + 2Cl^-(aq)$$

reduced

Addition of bromine to potassium chloride solution produces no reaction, suggesting that chlorine is a better oxidising agent than bromine and that bromine atoms are not able to take electrons away from chloride ions.

Shaking a solution of potassium iodide in a jar of chlorine causes a similar reaction to that with potassium bromide. The iodide solution turns brown and a black precipitate is rapidly formed. If trichloroethane is added, it turns to the characteristic purple colour of iodine in this solvent. Chlorine has oxidised iodide ions to free iodine and has itself been reduced to chloride ions:

$$Cl_2(g) + 2I^-(aq) \rightarrow 2Cl^-(aq) + I_2(aq)$$

Iodine solution has no effect on potassium chloride solution, showing that chlorine is a better oxidising agent than iodine.

Chlorine obviously stands at the top of the list chlorine, bromine and iodine, as far as the ability to act as an oxidising agent is concerned, but where does bromine stand in the list in relation to iodine? Is it a better oxidising agent, or a worse one, than iodine? The answer to this is supplied by adding bromine to potassium iodide solution. The colourless solution immediately turns brown and a black precipitate of iodine is formed. Bromine has oxidised iodide ions to free iodine:

$$Br_2(l) + 2I^-(aq) \rightarrow 2Br^-(aq) + I_2(aq)$$

Iodine has no effect on potassium bromide solution and bromine is obviously a better oxidising agent than iodine.

The decreasing ability of the halogens to act as oxidising agents as the group is descended is also illustrated by the observation that both chlorine and bromine will oxidise iron(II) salts to iron(III) salts, but iodine will not.

26.14
Ease of
oxidation of
halides

Reactions of halides with concentrated sulphuric acid

When cold concentrated sulphuric acid is added to solid potassium chloride, which is an ionic chloride, there is a vigorous effervescence and the mixture froths as it produces steamy white fumes of hydrogen chloride:

$$KCl(s) + H_2SO_4(l) \rightarrow KHSO_4(s) + HCl(g)$$

As the reaction proceeds, neither the reaction mixture nor the fumes show any signs of being coloured and obviously no free chlorine has been formed in the reaction.

If concentrated sulphuric acid is added to solid potassium bromide the frothing and the steamy fumes appear again, but this time the mixture rapidly turns red and a red vapour becomes mixed with the steamy fumes. The red colour is due to free bromine and this must have been formed by the oxidation of (i.e. the removal of electrons from) the bromide ions. The only substance in the reaction mixture which could have brought about the oxidation is the concentrated sulphuric acid, and this in fact acts in this way (27.26), itself being reduced to sulphur dioxide:

oxidised

$$3H_2SO_4(l) + 2NaBr(s) \rightarrow Br_2(l) + 2H_2O(l) + SO_2(g) + 2NaHSO_4(s)$$

reduced

When concentrated sulphuric acid is added to solid potassium iodide the effect is even more striking, with hardly any fumes of hydrogen iodide being given off, and a brown mess containing iodine is formed. If the mixture is warmed, purple fumes of iodine are formed and when the reaction has subsided a little, the product smells of hydrogen sulphide. Once again the concentrated sulphuric acid has acted as an oxidising agent converting the iodide ions to free iodine, but because this change occurs more easily than the corresponding change for bromide ions (that is iodide ions are better reducing agents than bromide ions) the sulphuric acid is reduced as far as hydrogen sulphide rather than sulphur dioxide:

oxidised

$$9H_2SO_4(l) + 8NaI(s) \rightarrow 4I_2(s) + H_2S(g) + 8NaHSO_4(s) + 4H_2O(l)$$

reduced

The increasing ease of oxidation of halide ions to halogens as the group is descended is also illustrated by the observation that both bromide ions and iodide ions can be oxidised by hydrogen peroxide solution, but chloride ions cannot.

26.15 The trends within the halogens

The trends within the family of the halogens are summarised in Fig. 26.1.

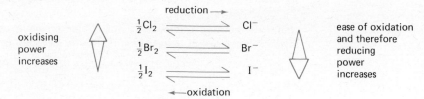

Fig. 26.1

The explanation for these trends appears in the electronic configurations of the elements and their ions:

chlorine	2.8.7		chloride ion	2.8.8
bromine	2.8.18.7		bromide ion	2.8.18.8
iodine	2.8.18.18.7		iodide ion	2.8.18.18.8

In the case of the atoms, the shell which is to gain an extra electron becomes further from the nucleus as we go down the list, and the number of 'shielding' shells between that shell and the nucleus increases, so that the outermost shell feels less attraction for the nucleus. The ability of the shell to attract an extra electron therefore decreases as we descend the list, and the power to act as an oxidising agent decreases.

The bigger the ion, the less firmly the extra electron is held as it is shielded from the nucleus by more electrons. Thus, the ability of an ion to act as a reducing agent increases as the group is descended.

26.16 Electrolysis of halides

The halides of most metals are electrolytes in aqueous solution, and electrolysis of the solutions produce the halogen at the anode (19.9),

e.g. $$2Br^-(aq) - 2e^- \rightarrow Br_2(aq)$$

As would be expected, the potential difference (or voltage) required to discharge these ions decreases as we descend the list, since it becomes easier to remove the extra electron from the ion. The molten salts behave similarly.

26.17 Fluorine

The first element in Group 7 is fluorine, which is a pale yellow, very poisonous gas. Although in reasonably plentiful supply in the form of the mineral fluorspar (calcium fluoride—CaF_2), it is only comparatively recently that compounds of this element have become of commercial importance. Fluorocarbons (compounds of carbon and fluorine) are very stable and are used as oils, sealing liquids and coolants. Polytetrafluoroethene (PTFE) is made by polymerizing (31.10) the compound,

$$F_2C = CF_2$$

PTFE is a white waxy solid, sold as Teflon, and has a low coefficient of friction. It is used for making bearings and rollers which require no lubrication (e.g. the rollers on which motorway bridges expand when they become warmer). Because it is so resistant to heat and unreactive, it is used for non-stick coatings on pans. The fluorochloro derivatives of methane (e.g. CF_2Cl_2 and $CFCl_3$) are used in refrigerators and as aerosol propellants, being sold under the name of Freons.

Metal fluorides, such as tin(II) fluoride and sodium monofluorophosphate, are now used for addition to water supplies and to toothpaste as it is believed they reduce tooth decay, particularly in children.

As would be expected from the trends in Group 7 already discussed, fluorine is the most reactive non-metallic element. It is such a powerful oxidising agent that it will oxidise chloride ions to free chlorine:

$$2Cl^-(aq) + F_2(g) \rightarrow 2F^-(aq) + Cl_2(g)$$

It will even oxidise water, liberating oxygen,

$$2F_2(g) + 2H_2O(l) \rightarrow 4HF(aq) + O_2(g)$$

Fig. 26.2 Toothpaste containing tin(II) fluoride (stannous fluoride).

and therefore it has to be extracted by the electrolysis of a molten fluoride, without any traces of water present.

Fig. 26.3 (right) *Tossing a pancake is easier when the frying-pan has been coated with PTFE. (Photo: Russell Edwards, BSc)*

26.18 Summary

1. The halogens are the family of elements making up Group 7 of the Periodic Table. Because of their similar electronic configurations, they have generally similar chemical properties.
2. With the exception of fluorine, the halogens are prepared by the oxidation of ionic halides.
3. They are diatomic elements, more soluble in non-polar solvents than in water.
4. All the halogens react in similar fashion with hydrogen, alkalis, sulphites and hydrogen sulphide. In each case the halogen atom gains an electron to form a halide ion, and therefore acts as an oxidising agent.
5. The ease of oxidation of the halide ions increases as Group 7 is descended going from fluorine to iodine, iodide ions being readily oxidised and chloride ions only being converted to chlorine molecules by powerful oxidising agents. Fluoride ions can only be oxidised in electrolysis.
6. The power of the halogens to accept electrons, and thus to act as oxidising agents, increases as we ascend Group 7, going from iodine to fluorine.
7. Fluorine is the most reactive non-metallic element. It is important for making polytetrafluoroethene (PTFE), Freons and fluorides additives for toothpaste.

Chapter 27

Sulphur and its compounds

Investigation 27.1

Fig. 27.1

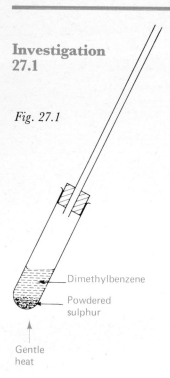

Dimethylbenzene

Powdered sulphur

Gentle heat

Questions

The allotropy of solid sulphur.

An element is said to show allotropy when it can exist in more than one form in one physical state. Allotropes of a solid element differ in physical properties, one of which could be the shape of the crystals.

1. Pour trichloroethane into a test-tube to a depth of 2–3 cm and add enough powdered sulphur to make a saturated solution. Shake the mixture thoroughly and then filter off the undissolved sulphur.

   Pour a little of the solution into a watchglass and set it aside, preferably in the fume cupboard. The trichloroethane evaporates and sulphur crystals will form on the watchglass.

   Examine the crystals closely with a hand lens.

2. To a hard glass test-tube add dimethylbenzene (xylene) to a depth of 5 cm and a 1 cm depth of powdered sulphur. Fit the tube with a bung through which passes a long glass tube and clamp the tube at a slight angle to the vertical (Fig. 27.1).

   With a small flame, gently heat the mixture, keeping the flame moving, until the liquid boils. Continue gentle boiling of the solution for a few minutes, keeping the vapour in the apparatus (you will be able to see where the vapour is by a 'ring' of liquid in the tube which acts as a condenser). Now allow the solution to cool. Crystals of sulphur should settle from the solution as it cools, but while it is still quite hot.

1   What shape are the crystals which are produced from the solution in trichloroethane?

2   What shape are the crystals which are produced from the solution in dimethylbenzene?

3   What are the names of the two allotropes of solid sulphur (27.8)?

Investigation 27.2

The action of heat on sulphur.

Put powdered sulphur into a hard glass test-tube to a depth of 3–4 cm. Gently heat the tube, taking care to keep the flame away from the open end of the tube.

The sulphur will melt and the molten sulphur goes through a series of changes of colour. As it changes colour, find how mobile the molten sulphur is by inclining the test-tube and seeing how easily the liquid will flow down the side of the tube.

When the sulphur boils, pour it quickly into a beaker of cold water. Recover the solid and examine its colour and hardness.

Finally set the solid on one side and leave it, preferably overnight, before examining it again.

1 Through what colour changes did the molten sulphur go?
2 How mobile was the molten sulphur at each colour change?
3 What changes did you observe in the solid's appearance and hardness when it was left overnight?

Investigation
27.3

What are the properties of an aqueous solution of sulphur dioxide?

1. Pour 2 cm³ of dilute sulphuric acid into one test-tube and a similar volume of sulphur dioxide solution into a second test-tube. Add a few drops of litmus solution to each test-tube.
2. Pour 2 cm³ portions of dilute sulphuric acid and sulphur dioxide solution into separate clean test-tubes. Add a small piece of magnesium ribbon to each test-tube.

Questions

1 What property of the two solutions do these changes indicate?
2 Which ion is present in both solutions?

3. Very cautiously compare the smell of sulphur dioxide solution with that of dilute sulphuric acid.
4. Boil separate portions of the two solutions, holding moist blue litmus paper in the vapour.

Question

3 In which solution is the solute volatile?

5. Pour about 2 cm³ of sulphur dioxide solution into a test-tube and add a few drops of litmus solution. Then add sodium hydroxide solution, a few drops at a time, shaking after each addition, until the litmus changes colour. Pour some of the solution into a watchglass and cautiously smell it.
   Now boil a little of this solution, holding moist blue litmus paper in the vapour.

Questions

4 Is sulphur dioxide given off from its solution to which sodium hydroxide has been added?
5 What type of reaction takes place when alkali is added to sulphur dioxide solution?
6 What type of substance is formed in this reaction?

Investigation
27.4

How can sulphate ions be detected and how can they be distinguished from sulphite ions?

1. To about 2 cm³ of dilute sulphuric acid, or sodium sulphate solution, add a few drops of barium chloride solution and shake the mixture. Then add about 2 cm³ of dilute hydrochloric acid and shake the mixture again.

Questions

1 What two products are likely to be formed when barium chloride solution is added to dilute sulphuric acid?
2 Which of the two is likely to be precipitated?
3 Does the precipitate dissolve in dilute hydrochloric acid?

2. Add about 2 cm³ of sodium sulphite solution to each of two test-tubes. To the first tube add a few drops of barium chloride solution and shake the mixture. Then add about 2 cm³ of dilute hydrochloric acid and again shake well.

To the second tube add 2 cm³ of dilute hydrochloric acid and shake thoroughly. Then add a few drops of barium chloride solution.

Questions

4 What is the precipitate in the first tube likely to be?
5 Does the precipitate in the first tube dissolve in dilute hydrochloric acid?
6 Why was there no precipitate in the second tube?
7 In testing for sulphate ions, why is dilute hydrochloric acid added before adding barium chloride solution?

3. To a test-tube add about 2 cm³ of sodium sulphite solution and to a second about 2 cm³ of sodium sulphate solution.

To each solution then add 2 cm³ of dilute hydrochloric acid and about 1 cm³ of potassium manganate(VII) solution.

Questions

8 How can you use potassium manganate(VII) solution to detect sulphite ions in solution?
9 What sort of reagent is potassium manganate(VII)?
10 Therefore as what sort of reagent is the acidified sulphite solution likely to be acting in this reaction?

Investigation 27.5

The reducing action of sulphite ions.

For these experiments, use a solution of sodium sulphite, acidified with dilute hydrochloric acid, or a freshly-prepared aqueous solution of sulphur dioxide.

1. Pour 2 to 3 cm³ of the solution containing sulphite ions into a test-tube and add about 2 cm³ of hydrogen peroxide solution.

Now acidify the mixture with dilute hydrochloric acid and then add a few drops of barium chloride solution.

Questions

1 What do you see in the mixture when the barium chloride solution is added?
2 What ion is therefore contained in the reaction mixture?
3 What type of reagent is the hydrogen peroxide acting as, when producing these ions from the sulphite ions?
4 What type of reagent are the sulphite ions acting as in this reaction?

2. Pour about 3 cm³ of your solution of sulphite ions into a test-tube, add 2 cm³ of dilute hydrochloric acid and then about 2 cm³ of potassium manganate(VII) solution. Finally add a few drops of barium chloride solution.

Questions

5 What do you see happen to the potassium manganate(VII) solution when it is added to the sulphite ions?
6 As what sort of reagent is the potassium manganate(VII) acting if you see this change in it?
7 What ions are produced from the sulphite ions when this change takes place?

3. Pour about 3 cm³ of your solution of sulphite ions into a test-tube, add 2 cm³ of dilute hydrochloric acid and then about 2 cm³ of potassium dichromate(VI) solution. Finally add a few drops of barium chloride solution.

8 What do you see happen to the potassium dichromate(VI) solution when it is added to the solution containing sulphite ions?
9 As what sort of reagent is the potassium dichromate(VI) solution acting if you see this change in it?
10 What ions are produced from the sulphate ions when this change takes place?

4. Pour about 1 cm³ of iron(III) chloride solution into a test-tube, add an equal volume of dilute hydrochloric acid and then about 2 cm³ of the solution containing sulphite ions.

Boil the mixture until no further change takes place. If it goes cloudy, add some more hydrochloric acid.

Now divide the mixture into two portions. To one add about 1 cm³ of dilute hydrochloric acid and then a few drops of barium chloride solution. To the second add a few drops of potassium hexacyanoferrate(III) solution.

11 Have the sulphite ions been oxidised in this reaction?
12 What ion is detected with potassium hexacyanoferrate(III) solution?
13 What is reduced to what in this reaction?

What are the properties of an aqueous solution of hydrogen sulphide?

HYDROGEN SULPHIDE IS A VERY POISONOUS GAS AND TEACHERS MAY PREFER TO DO THE EXPERIMENT AS A DEMONSTRATION.

In these experiments use a freshly-prepared solution of hydrogen sulphide in water.

1. To 1 cm³ of hydrogen sulphide solution in a test-tube, add a few drops of litmus solution. To the mixture then add about 2 cm³ of sodium hydroxide solution and observe what happens.

Pour a few drops of the mixture on to a watchglass and see if it now smells of hydrogen sulphide.

1 Is hydrogen sulphide solution acidic, neutral or alkaline?
2 What type of substance is formed when sodium hydroxide solution is added to the hydrogen sulphide solution?

2. Pour separate 2 cm³ portions of copper(II) sulphate solution, zinc sulphate solution and lead(II) nitrate solution into three separate test-tubes, add equal volumes of hydrogen sulphide solution to each test-tube and observe what happens. Then, preferably in a fume cupboard, add to each tube an equal volume of dilute hydrochloric acid and stir each mixture with a glass rod.

3 What are the precipitates formed in each case?
4 Which precipitate is soluble in dilute hydrochloric acid?
5 What gas is given off as this precipitate dissolves?

3. Set up six test-tubes containing 1 cm³ portions of, respectively, chlorine water, hydrogen peroxide solution, acidified potassium manganate(VII)

solution, acidified potassium dichromate(VI) solution, iron(III) chloride solution and sulphur dioxide solution. To each tube add 2 cm³ of hydrogen sulphide solution and observe what happens.

Allow the tubes to stand for about five minutes and observe what else has happened.

Finally, to the tube which contained the iron(III) chloride solution, add a few drops of potassium hexacyanoferrate(III) solution.

Questions

6 What is the milkiness which you see in all the tubes?

7 In all these reactions hydrogen sulphide acts as the same sort of reagent. What sort of reagent is this?

8 What ion is detected by potassium hexacyanoferrate(III) solution?

9 How was this ion formed in this investigation?

10 As what sort of reagent does sulphur dioxide solution act when it reacts with hydrogen sulphide?

**27.7
Sulphur for
sulphuric acid**

In 1843 a German chemist, Justus von Liebig, wrote: 'It is no exaggeration to say that we may judge the commercial prosperity of a country from the amount of sulphuric acid it consumes.' Since Liebig's time new industries not requiring sulphuric acid have been developed, and in other processes sulphuric acid has been replaced by other substances, so that this compound can no longer be regarded as the perfect indicator of industrial prosperity. Nevertheless, the annual production of sulphuric acid has continued to rise steadily over the last fifty years (from 500 000 tonnes in U.K. in 1920 to about $3\frac{1}{2}$ million tonnes in 1977). The world production in 1976 was 113 million tonnes. It is the most widely used of all manufactured chemicals. There are few materials or goods which do not require sulphuric acid at some stage of their production.

The starting material for the manufacture of this important compound is the element sulphur which makes up 0·05 % of the Earth's crust. The major source is the underground deposits of the free element. They occur in Poland, Mexico and in Louisiana in the U.S.A.

*Fig. 27.2 The Frasch
process for extracting
sulphur*

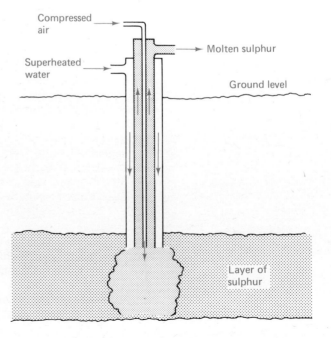

The American deposits, which are about 150 metres below the surface, were discovered in 1868 by prospectors looking for oil, but, when mining companies tried to mine the material in the usual way, they found that the beds were covered with layers of quicksand, so that deep enough mineshafts could not be sunk. For this reason the sulphur was not disturbed until 1890 when Herman Frasch suggested drilling down to the sulphur bed a 15 cm hole and then sinking into it three concentric pipes (one inside the other) (Fig. 27.2). Down the outside tube is pumped water, superheated under a pressure of 15 atmospheres. At this pressure water remains liquid up to 170 °C and this melts the sulphur at the base of the pipes. Down the central tube is pumped compressed air and this forces a mixture of molten sulphur and water up the middle tube to the surface. Here the mixture is run into vats in which the sulphur and water separate, and the sulphur is then either run, while still molten, into thermally insulated road tankers or ships (like giant thermos flasks), in which it can be transported to wherever in the world it is needed, or it is allowed to solidify to the familiar yellow solid with a purity of over 99%.

Another source of sulphur is that which can be recovered from crude oil and natural gas, but the proportion of sulphur present in these deposits does vary considerably. The oil and gas from the North Sea, for example, contains very little sulphur and makes a negligible contribution to the supply of sulphur.

Fig. 27.3 A 66-car unit of molten sulphur leaving a sulphur extraction point in Texas, USA, on its way to the port of Galveston where it will be transferred to an ocean-going tanker. (Courtesy British Sulphur Corporation)

(a)

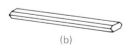

(b)

Fig. 27.4 The crystal shapes of (a) rhombic sulphur, (b) monoclinic sulphur

Alternative starting points for the manufacture of sulphuric acid are the sulphide minerals such as zinc blende (zinc sulphide) from which sulphur dioxide can be obtained (27.20). In world terms this is an important source of sulphuric acid, but its contribution to the U.K. supply is only about 5%.

27.8 Allotropy of solid sulphur

Sulphur is one of the elements which shows allotropy (16.5). An element is said to exhibit allotropy when it can exist in more than one form in the same physical state, and the allotropy of solid sulphur is interesting because one allotrope of solid sulphur is the more stable in one temperature range, while the other is stable in a different range of temperature. This sort of allotropy is called **enantiotropy**.

If powdered sulphur is dissolved in trichloroethane until the solution is saturated, and the solution is then allowed to evaporate slowly, yellow crystals separate. Examination under a microscope shows these to be octahedral in shape (Fig. 27.4(a)); they are crystals of **rhombic sulphur**.

If, however, the sulphur is dissolved by heating with methylbenzene (toluene) (boiling point 111 °C) or dimethylbenzene (xylene) (boiling point 140 °C), the crystals which separate when the solution starts to cool are needle-shaped (Fig. 27.4(b)); they are crystals of **monoclinic sulphur**.

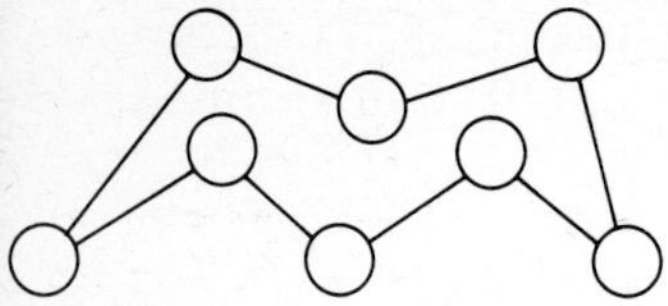

Monoclinic sulphur is formed first when molten sulphur is cooled and starts to solidify. By a simple experiment it can be shown that the temperature above which the monoclinic form is produced, and below which the rhombic form is formed, is 96 °C and this is called the **transition temperature** of sulphur. At temperatures below 96 °C the rhombic form is the more stable allotrope, whereas above 96 °C monoclinic sulphur has the greater stability. Thus, if monoclinic sulphur is stored at room temperature, the crystals very slowly turn into rhombic sulphur.

Fig. 27.5

27.9
Action of heat on sulphur

When sulphur is heated in a test-tube, the yellow solid melts at a temperature of 115 °C to form an orange-brown liquid which, when the test-tube is tilted, flows easily down the side. Then, as the temperature of the molten sulphur is raised, some interesting changes occur. The colour of the liquid darkens to deep red. At about 160 °C the liquid is very viscous and cannot be poured down the side of the test-tube. As heating is continued, the liquid turns dark brown and becomes mobile, so that it can be poured down the side of the tube again. When the temperature reaches 444 °C, sulphur boils.

If boiling molten sulphur is poured into cold water, it sets into a solid as it is cooled by the water. The solid is dark brown and is soft and pliable, rather like chewing gum. This product is called **plastic sulphur**. If it is kept for a few hours, it becomes hard and starts to turn yellow once again.

The explanations for these changes involve the ways in which sulphur atoms are joined together at the various stages of the heating. Sulphur is a molecular solid, the molecules consisting of eight atoms joined in a 'puckered' ring (Fig. 27.5). In the solid the rings are held together by van der Waals' forces (16.6) but, when the solid is heated, energy is fed into the crystals and the forces are overcome. The sulphur melts when the rings are able to move. The rings can easily pass over each other and the newly-melted sulphur is therefore runny.

As the heating continues the rings absorb more energy and some split at one point, forming a short chain of eight atoms, but, at this stage, there is not enough thermal motion to prevent these short chains joining to form longer ones. The long ones become twisted and tangled together, so that the liquid now finds difficulty in flowing. This corresponds to the dark red, extremely viscous liquid which is obtained at 160 °C.

As more energy is given to the liquid, the chains of sulphur atoms break up, forming much shorter lengths which cannot be twisted or tangled together and which can therefore easily pass over each other, the liquid now being mobile again.

When boiling molten sulphur is poured into cold water, the element is suddenly 'quenched', so that it is caught in the condition it was in just before it was poured into the water. The chains are quite short and can be pushed over each other when forces are applied to the solid. As the plastic sulphur stands, however, the short chains join up to reform the original eight-membered rings and the solid starts to become crystalline again with corresponding changes in strength and in colour.

280

Sulphur is in Group 6 of the Periodic Table and its atoms have an appreciable tendency to accept electrons to form negative ions. Sulphur will therefore combine with those elements from which electrons are comparatively easily removed—i.e. metals.

1. Reactions with metals

The ease and vigour of the reaction of a metal with sulphur, like reactions with oxygen or chlorine, depends on the position of the metal in the reactivity series (17.9).

A mixture of powdered zinc and sulphur, when warmed, burns with a green flame as zinc sulphide is formed:

$$Zn(s) + S(l) \rightarrow ZnS(s)$$

When a mixture of iron dust and powdered sulphur is heated, the reaction is again exothermic, but less so than with zinc. A red glow spreads through the mixture and a dark grey mass of iron(II) sulphide remains:

$$Fe(s) + S(l) \rightarrow FeS(s)$$

Metals lower down the series react with the evolution of less heat.

2. Reaction with oxygen

Molten sulphur burns in air or, more vigorously, in oxygen with a blue flame, forming gaseous sulphur dioxide:

$$S(l) + O_2(g) \rightarrow SO_2(g)$$

This reaction is of considerable importance, being the first stage in the manufacture of sulphuric acid from sulphur (27.20), and also a cause of atmospheric pollution (27.17).

3. Reactions with oxidising agents

*Fig. 27.6 The
preparation and collection
of hydrogen sulphide*

Sulphur is, on the whole, unreactive towards aqueous reagents. Dilute acids and alkalis have no effect on it, nor does a solution of potassium manganate(VII). It is, however, oxidised by concentrated nitric acid (34.9).

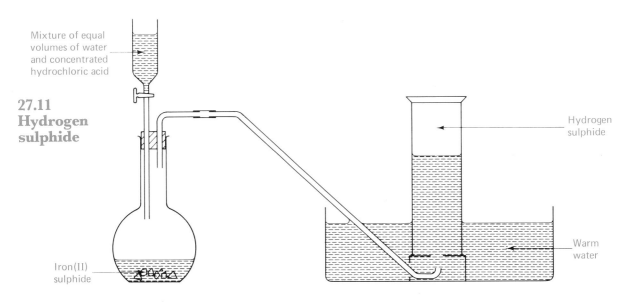

Mixture of equal
volumes of water
and concentrated
hydrochloric acid

Hydrogen
sulphide

Warm
water

Iron(II)
sulphide

If you have ever had the bad luck to crack open a rotten egg, you will have smelt the gas called hydrogen sulphide. You can detect the same smell when dilute hydrochloric acid or sulphuric acid is added to iron(II) sulphide, made by heating together powdered iron and sulphur, and this provides a convenient way of preparing the gas in the laboratory. Since the gas is moderately soluble in cold water (1 volume of cold water dissolves about 3 volumes of the gas), it is best collected over warm water. A convenient apparatus is shown in Fig. 27.6.

$$2HCl(aq) + FeS(s) \rightarrow FeCl_2(aq) + H_2S(g)$$

27.12
Properties of hydrogen sulphide

Hydrogen sulphide is a colourless gas with, as we have already described, the smell of rotten eggs. Indeed eggs, as they turn bad, give off hydrogen sulphide owing to the breakdown by bacteria of sulphur compounds in the egg-white.

It is poisonous, one volume of the gas in 600 volumes of air being a lethal mixture, but the smell of the gas announces its presence long before the concentration becomes dangerous.

Hydrogen sulphide is flammable, the products formed depending on how much oxygen is available for the combustion. In a plentiful supply of air sulphur dioxide and steam are formed:

$$2H_2S(g) + 3O_2(g) \rightarrow 2SO_2(g) + 2H_2O(g)$$

If the supply of air is limited (e.g. by holding a cold surface in the flame), the hydrogen is again converted to steam but the sulphur escapes as the element:

$$2H_2S(g) + O_2(g) \rightarrow 2S(s) + 2H_2O(g)$$

27.13
Hydrogen sulphide as an acid

Hydrogen sulphide is moderately soluble in cold water, and the resulting solution conducts electricity slightly, showing that some of the hydrogen sulphide molecules have split up into ions:

$$H_2S(aq) \rightleftharpoons 2H^+(aq) + S^{2-}(aq)$$

When litmus solution is added to the solution of the gas, it turns bluish-red, showing that the solution is weakly acidic. The solution also contains covalent hydrogen sulphide molecules as is shown by the still-present smell of bad eggs. If sodium hydroxide solution is added to an aqueous solution of the gas, the smell disappears, due to the conversion of the molecules into sulphide ions,

$$H_2S(aq) + 2NaOH(aq) \rightarrow Na_2S(aq) + 2H_2O(l)$$

or $$H_2S(aq) + 2OH^-(aq) \rightarrow S^{2-}(aq) + 2H_2O(l)$$

Most metal sulphides are insoluble in water and, if hydrogen sulphide is passed through a solution of a metal salt, a precipitate of the metal sulphide is usually formed. For example, the gas led into copper(II) sulphate solution forms a black precipitate of copper(II) sulphide,

$$CuSO_4(aq) + H_2S(g) \rightarrow CuS(s) + H_2SO_4(aq)$$

or $$Cu^{2+}(aq) + S^{2-}(aq) \rightarrow CuS(s)$$

Whether the precipitate forms or not can depend on the acidity of the solution of the metal salt. For example, copper(II) sulphide and lead(II) sulphide are precipitated from acidified solutions of the metal salts, but zinc sulphide is not.

It is the formation of one of these precipitates which gives us the characteristic test for hydrogen sulphide. If you suspect that the gas is being given off in a reaction, you should hold a piece of filter paper, dipped into lead(II) nitrate solution or lead(II) ethanoate (acetate) solution, over the reaction mixture. If the paper turns black, hydrogen sulphide is present.

As has already been described (27.12), hydrogen sulphide will burn in the oxygen of the air. Since this demonstrates the readiness of the gas to react with oxygen, it suggests that it could be a reducing agent and this is confirmed by its reactions with oxidising agents.

Further proof of the reactivity of hydrogen sulphide with air is provided by the behaviour of an aqueous solution of the gas when it is allowed to stand, open to the air, for a few hours. The solution slowly turns cloudy and a pale yellow precipitate of sulphur gradually settles, oxygen in the air oxidising the hydrogen sulphide to water and sulphur:

$$2H_2S(aq) + O_2(g) \rightarrow 2H_2O(l) + 2S(s)$$

1. Reaction with hydrogen peroxide

If hydrogen sulphide is bubbled through hydrogen peroxide solution, the solution turns cloudy as a precipitate of sulphur forms:

$$H_2O_2(aq) + H_2S(g) \rightarrow 2H_2O(l) + S(s)$$

Hydrogen sulphide has been oxidised to sulphur and, in turn, has reduced hydrogen peroxide to water.

In most of the reductions carried out by hydrogen sulphide, the white or yellow precipitate of sulphur is formed and this acts as a sign that the hydrogen sulphide has acted in that way,

$$H_2S(g) + [O] \rightarrow H_2O(l) + S(s)$$
from
the
oxidising
agent

or
$$H_2S(g) \rightarrow 2H^+(aq) + S(s) + 2e^-$$
donated
to the
oxidising
agent

2. Reaction with acidified potassium manganate(VII) solution

Purple potassium manganate(VII) solution, acidified with dilute sulphuric acid, is decolourised when hydrogen sulphide is passed through it and the colourless solution slowly turns milky as a finely divided precipitate of sulphur forms. The hydrogen sulphide has donated electrons to the manganate(VII) ions and has reduced them to manganese(II) ions, while itself it has been oxidised to sulphur.

3. Reaction with acidified potassium dichromate(VI) solution

Orange potassium dichromate(VI) solution, acidified with dilute sulphuric acid, turns dark green in the presence of hydrogen sulphide and once again the tell-tale precipitate of sulphur slowly forms, showing the hydrogen sulphide to be acting as a reducing agent. The dichromate(VI) ions have been reduced to chromium(III) ions, which are green in solution, and the hydrogen sulphide has again been oxidised to sulphur.

4. Reaction with iron(III) chloride solution

If hydrogen sulphide is passed into iron(III) chloride solution, a milkiness appears and the yellow colour of the iron(III) ions fades as they are reduced to iron(II) ions,

$$H_2S(g) + 2Fe^{3+}(aq) \rightarrow 2Fe^{2+}(aq) + 2H^+(aq) + S(s)$$

or $\qquad H_2S(g) + 2FeCl_3(aq) \rightarrow 2FeCl_2(aq) + 2HCl(aq) + S(s)$

5. Reactions with the halogens

As described in 25.9, hydrogen sulphide reacts with chlorine water, bromine water or iodine solution, forming the usual precipitate of sulphur and reducing the free halogen to halide ions,

e.g. $\qquad H_2S(g) + Br_2(aq) \rightarrow S(s) + 2H^+(aq) + 2Br^-(aq)$

6. Reaction with nitric acid

Nitric acid is a powerful oxidising agent and oxidises hydrogen sulphide, but, as explained in detail in Chapter 34, the products depend on the concentration of the acid.

7. Reaction with concentrated sulphuric acid

If hydrogen sulphide is bubbled through cold concentrated sulphuric acid, a large amount of sulphur is precipitated:

$$3H_2S(g) + H_2SO_4(l) \rightarrow 4S(s) + 4H_2O(l)$$

This reaction is unusual for two reasons. Concentrated sulphuric acid usually only acts as an oxidising agent when hot, and yet here it reacts in this way when cold. Also, concentrated sulphuric acid is usually reduced to sulphur dioxide when it oxidises something, and yet here the reduction goes all the way to sulphur. The probable reason for this is that the sulphur dioxide which would be released here, reacts with the hydrogen sulphide, liberating sulphur. Clearly concentrated sulphuric acid cannot be used to dry hydrogen sulphide; some other drying agent has to be found.

8. Reaction with sulphur dioxide

When a jar of hydrogen sulphide, to which a few drops of water have been added, is brought mouth-to-mouth with a jar of sulphur dioxide and the two jars are then opened so that the gases can mix, a yellow deposit forms on the sides of the jars. This is obviously sulphur and at least part of it must have been formed by hydrogen sulphide acting as a reducing agent. This can only have happened if sulphur dioxide (or more precisely, the sulphurous acid), which normally reacts as a reducing agent, acts as an oxidising agent and is itself converted to sulphur:

$$2H_2S(g) + SO_2(g) \rightarrow 2H_2O(l) + 3S(s)$$

The same reaction takes place when hydrogen sulphide is bubbled through an aqueous solution of sulphur dioxide, sulphur being precipitated in the solution.

The reaction between hydrogen sulphide and sulphur dioxide is of considerable importance in the extraction of the sulphur from petroleum oil and natural gas. Hydrogen sulphide is produced from the 'cracking' of fractions of the oil (31.5) or by treatment of certain fractions with hydrogen in the presence of a catalyst, and then part of it is oxidised by air to water and sulphur dioxide:

$$2H_2S(g) + 3O_2(g) \rightarrow 2H_2O(g) + 2SO_2(g)$$

The latter then combines with the unreacted hydrogen sulphide in the presence of a catalyst in the way described above, liberating the sulphur.

27.15
Sulphur dioxide

Sulphur dioxide is produced when sulphur is burnt in air or oxygen:

$$S(l) + O_2(g) \rightarrow SO_2(g)$$

Sulphur dioxide is a colourless gas, about twice as dense as air, with a pungent, choking smell and a metallic taste. It is poisonous to both plants and animals. Its boiling point is comparatively high and it is easily liquefied simply by applying pressure. The compound is in fact usually transported in the liquid form. If you have to use sulphur dioxide in the laboratory, you will probably obtain your supply from a pressurised canister which contains the liquid. As the pressure is released, the liquid evaporates, forming the gas.

Sulphur dioxide does not burn in air or oxygen and it extinguishes a burning splint. If, however, burning magnesium ribbon is lowered into a jar of the gas, the metal continues to burn and two solids, white magnesium oxide and yellow sulphur, are deposited on the sides of the jar. Magnesium, occupying a high place in the reactivity series (17.9), acts as a powerful reducing agent, and, at the temperature of the burning metal, it reduces the sulphur dioxide to sulphur:

$$2Mg(s) + SO_2(g) \rightarrow 2MgO(s) + S(s)$$

Sulphur dioxide is very soluble in water (one volume of water dissolves about 80 volumes of the gas at $0\,°C$) and the solution turns litmus bright red, showing it to be acidic. The solution is known as sulphurous acid:

$$SO_2(g) + H_2O(l) \rightarrow H_2SO_3(aq)$$

27.16
Sulphurous acid

The aqueous solution of sulphur dioxide allows an electric current to pass through it, showing that ions are present in it. Water and sulphur dioxide, from which the solution is made, are not ionic and therefore the ions must have been made in a chemical reaction between the two substances. The solution is highly acidic (e.g. magnesium ribbon readily liberates hydrogen from it) and so it must contain hydrogen ions ($H^+(aq)$). On the other hand it smells strongly of sulphur dioxide, and, on boiling, sulphur dioxide and steam are given off, leaving no residue.

Sulphurous acid must therefore be an equilibrium mixture (22.13) of water molecules, sulphur dioxide molecules, hydrogen ions and sulphite ions, represented by the equation:

$$H_2O(l) + SO_2(aq) \rightleftharpoons 2H^+(aq) + SO_3^{2-}(aq)$$

The action of heat on the mixture makes the sulphur dioxide less soluble and drives it off. The equilibrium is then disturbed and, in an attempt to restore it, some of the hydrogen ions and sulphite ions combine to form more sulphur dioxide to replace that which has gone. This will also be driven off by the heat and the process continues until the decomposition of the sulphurous acid is eventually complete, causing no residue to remain.

Similarly, when magnesium ribbon reacts with sulphurous acid, the hydrogen ions are removed from the equilibrium mixture by the reaction:

$$Mg(s) + 2H^+(aq) \rightarrow Mg^{2+}(aq) + H_2(g)$$

Again the equilibrium is disturbed and some of the sulphur dioxide molecules and water molecules in the mixture combine together in an attempt to replace the hydrogen ions which have gone. The new hydrogen ions will also eventually react with the magnesium atoms and so the reaction will also go to completion.

A similar sort of reaction takes place when a sufficient quantity of an alkali, such as sodium hydroxide solution, is added to the sulphurous acid. The smell of sulphur dioxide disappears completely and the resulting solution of sodium sulphite is a strong electrolyte, showing it to be ionic. The hydroxide ions from the alkali have combined with the hydrogen ions in the sulphurous acid and have therefore disturbed the equilibrium, so that more sulphur dioxide is turned into ions by the water to replace

285

the hydrogen ions which have been removed. Again the new hydrogen ions are absorbed by the alkali, so that the reaction goes to completion and all the sulphur dioxide (shown as H_2SO_3) is converted to sulphite ions:

$$2NaOH(aq) + H_2SO_3(aq) \rightarrow Na_2SO_3(aq) + 2H_2O(l)$$

Addition of dilute acid to the sodium sulphite immediately produces the smell of sulphur dioxide as the sulphite ions form sulphurous acid again:

$$2H^+(aq) + SO_3{}^{2-}(aq) \rightarrow H_2O(l) + SO_2(aq)$$

Sulphur dioxide, being very soluble in water, is not readily given off unless the mixture is warmed. This causes the gas to be less soluble and, as it is given off, the equilibrium is disturbed so that more will be produced. The reaction can in fact be driven to completion by boiling:

$$Na_2SO_3(aq) + 2HCl(aq) \rightarrow 2NaCl(aq) + SO_2(g) + H_2O(l)$$

The formation of sulphur dioxide on the addition of dilute acid is a general reaction of sulphites and is therefore a convenient test for the presence of a sulphite. It is also a convenient way of preparing sulphur dioxide in the laboratory.

27.17 Pollution of the air by sulphur dioxide

Whenever we burn coal, coke or fuel oil, we liberate sulphur dioxide into the atmosphere (9.5). Coal and coke contain between 1 and 2 per cent of sulphur, but the amounts in oil are frequently double these figures.   There are three main effects of this high atmospheric concentration of sulphur dioxide:
1. The gas has a strongly acidic smell and, when it is trapped with water vapour and smoke as fog, or 'smog', in an industrial region, it can participate in developing respiratory diseases, such as bronchitis.
2. Sulphur dioxide is particularly harmful to plant life. Even small amounts can destroy vegetation completely, as was shown in an extreme example in the early years of this century in Tennessee where smelting of sulphide ores was carried out at ground-level with no chimneys to carry the gases away. The sulphur dioxide collected as a concentrated low-lying cloud and killed all plant life for miles. A more recent example in Utah, also in the United States, was caused by using chimneys which were too short (about 60 m high). When the chimneys were built up to a height of 150 m and the gases were heated before they entered the chimneys, the fumes were spread over a wider area so that they were much more dilute and much less harmful.

Sulphur dioxide enters plants through small holes in their leaves, called stomata, and is then absorbed into the inner cells of the leaves where it is converted into solid sulphites. If the mass of these sulphites builds up to 0·16 per cent of the dry mass of the leaves, the cells are killed. When the leaves are below 0·16 per cent, the cells may be able to convert them by slow oxidation to sulphates which are far less poisonous. Even so, if the masses of sulphates increase, they can also become toxic and this can best be seen in pine trees. Normally a pine tree keeps its needles for about three years, but, if sulphates build up in the needles, they can fall off in one or two years. In the industrialised Ruhr region of Germany, it has proved very difficult to replace pine trees because of sulphurous fumes in the air, and the Scandinavians claim that acid rain from sulphur dioxide pollution in Britain is blown across the North Sea, settling on pine forests where it stunts the growth of the trees.
3. Sulphur dioxide in the air dissolves in rain-water, falling through the air, forming sulphurous acid. This is quite a strong acid and, if it falls on buildings made of limestone (calcium carbonate), it starts to dissolve the carbonate. It is for this reason that old buildings like York Minster or Lincoln Cathedral have suffered such damage to their outside surfaces in the period since the Industrial Revolution when the

concentration of sulphur dioxide in the air has increased. The damage, resulting in the irreplaceable loss of the centuries-old stone work, will continue until the concentration of sulphur dioxide in the air is reduced and very carefully controlled.

Fig. 27.7 A pinnacle on York Minster before it was repaired. Sulphur dioxide pollution of the atmosphere is particularly harmful to the magnesium limestone from which the Minster is built. (Courtesy York Minster Library)

27.18 Reducing action of sulphur dioxide

If barium chloride solution is added to a freshly-prepared solution of sulphur dioxide a white precipitate forms, but, if dilute hydrochloric acid is then added, the precipitate dissolves. The solution contains sulphite ions and the precipitate formed with the barium chloride solution is therefore barium sulphite which dissolves in dilute acid. If, however, the solution of sulphur dioxide is allowed to stand in contact with air for some time before barium chloride solution is added, it still gives a precipitate, but this time the precipitate does not dissolve when the dilute hydrochloric acid is added, showing it to be barium sulphate. Sulphite ions in the sulphurous acid have been oxidised to sulphate ions by oxygen in the air:

$$SO_3^{2-}(aq) + \tfrac{1}{2}O_2(g) \rightarrow SO_4^{2-}(aq)$$

Thus, this means that an aqueous solution of sulphur dioxide (sulphurous acid) is a reducing agent. This is confirmed by the reactions of the solution with oxidising agents such as hydrogen peroxide, potassium manganate(VII), potassium dichromate(VI) and the halogens, these reactions being described below.

1. Reaction of sulphurous acid with hydrogen peroxide solution
An aqueous solution of sulphur dioxide is immediately oxidised to sulphuric acid solution by hydrogen peroxide:

$$H_2SO_3(aq) + H_2O_2(aq) \rightarrow H_2SO_4(aq) + H_2O(l)$$

287

There is no visible change in the mixture as the reaction takes place. If, however, sulphurous acid solution is set up in a cell in series with an ammeter and hydrogen peroxide solution (itself not an electrolyte) is then added, the conductance of the solution is seen to rise. The sulphuric acid, formed in the reaction, is more fully ionised than sulphurous acid.

2. **Reaction with acidified potassium manganate(VII) solution**
If sulphur dioxide is passed into potassium manganate(VII) solution, acidified with dilute sulphuric acid, the purple solution is rapidly decolourised and, in contrast to the reaction with hydrogen sulphide, no precipitate is formed. The same change is observed when acidified potassium manganate(VII) solution is added to sulphurous acid or to a solution of a sulphite and makes a useful test for detecting sulphur dioxide or sulphite ions in solution. The sulphite ions are oxidised to sulphate ions.

3. **Reaction with acidified potassium dichromate(VI) solution**
Sulphur dioxide or a solution containing sulphite ions turns an orange solution of potassium dichromate(VI), acidified with dilute sulphuric acid, green as the dichromate(VI) ions ($Cr_2O_7^{2-}$) are converted to chromium(III) ions (Cr^{3+}). Again the sulphite ions release electrons and are oxidised to sulphate ions and the reaction can be used as a test for sulphur dioxide or sulphite ions in solution.

4. **Reactions with the halogens**
An aqueous solution of sulphur dioxide or of a sulphite will reduce any of the halogens to the corresponding halide ions. The halogen is decolourised as the reduction takes place:

$$SO_3^{2-}(aq) + \underset{\text{halogen}}{X_2(aq)} + H_2O(l) \rightarrow SO_4^{2-}(aq) + \underset{\substack{\text{halide}\\\text{ions}}}{2X^-(aq)} + 2H^+(aq)$$

5. **Reaction with iron(III) chloride**
An aqueous solution of sulphur dioxide or of a sulphite converts the yellow or orange solution containing iron(III) ions to a pale green one containing iron(II) ions:

$$\overset{\text{oxidised}}{\overbrace{SO_3^{2-}(aq) + H_2O(l) + \underset{\text{reduced}}{\underbrace{2Fe^{3+}(aq) \rightarrow SO_4^{2-}(aq) + 2H^+(aq) + 2Fe^{2+}(aq)}}}}$$

Again there is an obvious difference between this reaction and the corresponding one with hydrogen sulphide in which a precipitate of sulphur is formed as the reduction proceeds.

**27.19
Uses of sulphur dioxide**

Sulphur dioxide has for a long time been used as a bleach for delicate materials such as wood pulp (in paper making), wool, silk and straw, the bleaching action being due to the reduction of the coloured materials:

$$H_2O(l) + SO_2(g) + [O] \rightarrow H_2SO_4(aq)$$
$$\underset{\substack{\text{from the}\\\text{dye}}}{}$$

In some cases the bleaching is not permanent. In paper for example, the oxygen removed by the sulphur dioxide is replaced by oxygen from the air and this is the reason for old paper slowly turning yellow.

In countries which have plentiful wood supplies, sulphur dioxide is used for the production of wood pulp by the sulphite process which consists of heating chips of wood with calcium hydrogensulphite solution (made from slaked lime and sulphur dioxide) and sulphur dioxide under pressure. This dissolves all the constituents of the wood apart from the cellulose which is to be used for making the paper.

Sulphurous acid is one of the few acids which are allowed as preservatives in food, and sulphur dioxide has been used for this purpose for at least five hundred years. The gas is generally used with acid foods, such as jam, dried fruit and soft drinks, and it slows down or prevents the growth of bacteria and moulds. With fruit sulphurous acid has the advantage of being a reducing agent and it protects components in the food, like Vitamin C, which are easily destroyed by oxidation. Being a bleaching agent, it also prevents the formation of a brown colour in fruit and vegetables. A dilute solution of sulphur dioxide is used for the preservation of fruit in the manufacture of jam, the fruit simply being submerged in the solution until required. Most of the sulphur dioxide is driven off in cooking and the finished jam must not contain more than 100 parts per million of the gas.

Liquid sulphur dioxide is used as a solvent for obtaining kerosene and lubricating oils from petroleum oil fractions. The compound is also used for making a range of chemicals such as sulphites, hydrogensulphites, dyestuffs and liquid sulphur trioxide (Sulfan) which is used in the production of detergents.

27.20 Manufacture of sulphuric acid

A very small proportion of the world's sulphuric acid is made by the old Lead Chamber Process, but all of the acid produced in the United Kingdom is made by the **Contact Process**, in which sulphur dioxide is catalytically oxidised by oxygen in the air to give sulphur trioxide:

$$2SO_2(g) + O_2(g) \rightleftharpoons 2SO_3(g)$$

The product is then made to combine with water to give the sulphuric acid:

$$SO_3(g) + H_2O(l) \rightarrow H_2SO_4(aq)$$

The sulphur dioxide for this Contact Process used to be obtained from a number of sources:

1. Burning sulphur:

$$S(l) + O_2(g) \rightarrow SO_2(g)$$

2. Roasting sulphide minerals which are used for producing metals, e.g. zinc blende

$$2ZnS(s) + 3O_2(g) \rightarrow 2ZnO(s) + 2SO_2(g)$$

3. Anhydrite (anhydrous calcium sulphate) was used as a source of sulphur dioxide in the U.K. The anhydrite process was operated at Billingham from about 1930 and it was the existence of deposits of this mineral in that area which led to the development there of I.C.I.'s factory (one of the biggest chemical works in Europe). The process had the advantage of producing cement (one tonne for every tonne of sulphuric acid made) as the by-product. The method has now become too expensive because of the higher cost, in recent years, of the oil used for heating the mixture. The fuel costs associated with the sulphur process are much lower, because it is an exothermic reaction. Fig. 27.8 shows how the importance of these two processes in the U.K. has varied.

To produce sulphur dioxide from sulphur, the molten element is sprayed into excess dry air, in which it burns at a temperature of about 1000 °C, the combustion providing valuable energy for the rest of the process. Air is dried by passing through concentrated sulphuric acid before the combustion to prevent the formation of misty fumes in the plant at a later stage. The mixture of gases is cooled to 450 °C before being passed to the hot gas filter, in which particles of dust are removed by passing the gas near an electrically charged conductor. Failure to do this may result in the poisoning of the catalyst during the oxidation.

The oxidation of sulphur dioxide by oxygen, forming sulphur trioxide, is an exothermic, reversible reaction,

$$SO_2(g) + \tfrac{1}{2}O_2(g) \rightleftharpoons SO_3(g) \qquad \Delta H = -98 \text{ kJ mole}^{-1}$$

and the result of the reaction is an equilibrium mixture of the three substances (22.17).

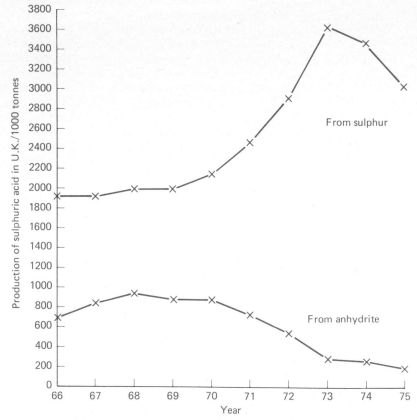

Le Chatelier's Principle (22.15) tells us that the lower the temperature, the more sulphur trioxide there is in the equilibrium mixture. Below 400 °C, however, the rate at which equilibrium is established is low (it would take far too long to obtain the product) and so it is necessary to use a temperature which is a little above that value to

guarantee a yield in a reasonably short time. Again, to ensure equilibrium is achieved as quickly as possible, a catalyst is used.

Vanadium(V) oxide (V_2O_5), which is an oxide of a transition metal, is the catalyst used. It is used on a silica support to spread it out and so give it maximum surface area.

The yield of sulphur trioxide is also theoretically improved by the use of a high pressure, but satisfactory yields can be obtained by using a pressure just slightly above that of the atmosphere.

Fig. 27.10 The preparation and collection of sulphur trioxide

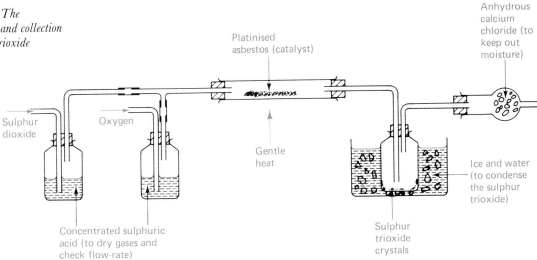

Platinised asbestos (catalyst)

Anhydrous calcium chloride (to keep out moisture)

Sulphur dioxide

Oxygen

Gentle heat

Ice and water (to condense the sulphur trioxide)

Concentrated sulphuric acid (to dry gases and check flow-rate)

Sulphur trioxide crystals

Despite the comparatively high temperature, the conversion of 98 per cent of the sulphur dioxide to sulphur trioxide can be achieved by passing a mixture of the sulphur dioxide and air through the catalyst a number of times.

Sulphur trioxide gas, formed in the converter, cannot be directly absorbed in water because it would start to dissolve in the water vapour over the water, forming a mist of droplets of sulphuric acid. This does not condense and can easily pollute the atmosphere breathed by the plant workers. Sulphur trioxide is therefore absorbed in concentrated sulphuric acid (98% sulphuric acid). The percentage of sulphuric acid in the solvent could therefore be increased to 99·5% but, as it is circulated in the absorption tower, water is added to keep the concentration at 98%. This dilution is exothermic:

$$SO_3 + H_2O \rightarrow H_2SO_4 \qquad \Delta H = -130 \text{ kJ mole}^{-1}$$

in 98% sulphuric acid

but the heat is released at low temperature and cannot be used to produce steam. The acid has therefore to be cooled by large air or water coolers.

The conversion of sulphur dioxide to sulphur trioxide can be demonstrated on a small scale in the laboratory, using the apparatus shown in Fig. 27.10. Sulphur dioxide is passed slowly, and the oxygen a little faster, while the platinised asbestos catalyst in the tube is heated gently. A white smoke, which is sulphur trioxide, forms in the tube:

$$2SO_2(g) + O_2(g) \rightleftharpoons 2SO_3(g)$$

Some of the sulphur trioxide condenses as a white solid in the cold-trap, provided the apparatus is dry (SO_3 freezes at $17\,°C$).

It is important that the catalyst is heated gently, as at high temperatures the equilibrium position lies to the left (22.17).

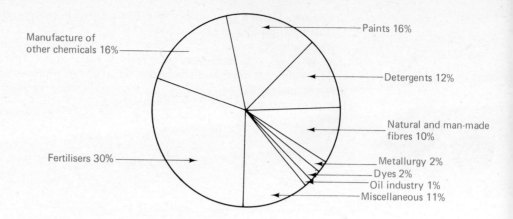

Fig. 27.11 Uses of sulphuric acid

Paints 16%

Manufacture of other chemicals 16%

Detergents 12%

Natural and man-made fibres 10%

Fertilisers 30%

Metallurgy 2%
Dyes 2%
Oil industry 1%
Miscellaneous 11%

27.21 Uses of sulphuric acid

Industry in the United Kingdom uses about 4 million tonnes of sulphuric acid which represents only about 3% of the total world production. The proportions of the U.K. produced acid which are used by the various outlets are shown in Fig. 27.11.

Superphosphate is a valuable fertiliser, made by reacting calcium phosphate, which occurs as the mineral phosphorite or rock phosphate, with 70% sulphuric acid, the product being more soluble in water and therefore a better plant food than the parent mineral:

$$Ca_3(PO_4)_2(s) \ + \ 2H_2SO_4(aq) \ \rightarrow \ \underbrace{Ca(H_2PO_4)_2(s) \ + \ 2CaSO_4(s)}_{superphosphate}$$

Sulphate of ammonia (ammonium sulphate) is a nitrogenous fertiliser (33.10) which encourages top growth of plants and is made by neutralising ammonia with sulphuric acid.

A fundamental constituent of modern paints is one of the whitest compounds known, titanium(IV) oxide, which is manufactured from a black sand called Ilmenite by treating it with sulphuric acid and then hydrolysing the titanium sulphate formed, the impurities in the sand (mainly iron) remaining in solution in the sulphuric acid.

Detergents are made by treating vegetable oils with concentrated sulphuric acid (32.12) and viscose rayon is made by reacting non-fibrous cellulose with carbon disulphide and aqueous sodium hydroxide to form a solution of a yellow substance, called cellulose xanthate, which is then squirted into dilute sulphuric acid to reform the cellulose in fibre form.

The main use of sulphuric acid in metallurgy is for the 'pickling' of steel, in which the surface of the metal is cleaned before it is galvanized, tin-plated (19.11) or enamelled.

27.22 Reactions of sulphuric acid

The reason for the large number of uses of sulphuric acid is the fact that this compound can react in a number of ways. In the following sections we are going to discuss four of the ways in which the compound will react:

1. as an acid,
2. as a sulphate,
3. as a dehydrating agent,
4. as an oxidising agent.

27.23
Sulphuric acid as an acid

When sulphuric acid is dissolved in water to give the dilute acid, the water causes the molecules to split up into ions in two ways,

$$H_2SO_4(l) + H_2O(l) \rightleftharpoons H_3O^+(aq) + HSO_4^-(aq)$$

and

$$H_2SO_4(l) + 2H_2O(l) \rightleftharpoons 2H_3O^+(aq) + SO_4^{2-}(aq)$$

Because both interactions result in the formation of hydroxonium ions (H_3O^+), dilute sulphuric acid fulfils the definition of an acid (11.9) and will show the usual properties and reactions of an acid. For simplicity, a dilute solution of sulphuric acid is usually represented by:

$$2H^+(aq) + SO_4^{2-}(aq)$$

1. **Reaction of dilute sulphuric acid with metals**
Dilute sulphuric acid reacts with the metals above hydrogen in the reactivity series (17.9), liberating hydrogen and forming the sulphate of the metal.

e.g. $$Mg(s) + H_2SO_4(aq) \rightarrow MgSO_4(aq) + H_2(g)$$

or $$Mg(s) + 2H^+(aq) \rightarrow Mg^{2+}(aq) + H_2(g)$$

The vigour of the reaction is greater, the higher in the reactivity series is the metal. Other metals do not react with the dilute acid, liberating hydrogen, although they may react with the concentrated acid, liberating sulphur dioxide.

2. **Reaction of dilute sulphuric acid with bases**
Dilute sulphuric acid reacts with a base, forming a salt (the sulphate) and water,

e.g. $$CuO(s) + H_2SO_4(aq) \rightarrow CuSO_4(aq) + H_2O(l)$$

or $$CuO(s) + 2H^+(aq) \rightarrow Cu^{2+}(aq) + H_2O(l)$$

Because the sulphuric acid can ionize in aqueous solution to produce two anions, hydrogensulphate (HSO_4^-) and sulphate (SO_4^{2-}), it is possible, by controlling the amount of base added to a sample of the acid, to prepare either a salt containing the first anion or a salt containing the other. Thus it is possible, simply by varying the volume of sodium hydroxide solution added, to prepare two sodium salts of the acid, sodium hydrogensulphate (the acid salt) and sodium sulphate (the normal salt) (11.15),

$$NaOH(aq) + H_2SO_4(aq) \rightarrow NaHSO_4(aq) + H_2O(l)$$

$$2NaOH(aq) + H_2SO_4(aq) \rightarrow Na_2SO_4(aq) + 2H_2O(l)$$

3. **Reaction of dilute sulphuric acid with carbonates**
Cold dilute sulphuric acid reacts with a carbonate, liberating carbon dioxide and forming a sulphate and water,

e.g. $$MgCO_3(s) + H_2SO_4(aq) \rightarrow MgSO_4(aq) + H_2O(l) + CO_2(g)$$

The only apparent exceptions to this rule are the carbonates of calcium, lead and barium which react only for a short time, the particles of the solid then becoming covered with a layer of the insoluble sulphate which prevents the acid coming into further contact with the carbonate (11.9).

27.24
Sulphuric acid as a sulphate

An aqueous solution of sulphuric acid contains sulphate ions and, if a solution containing lead, barium or calcium ions is added, metal ions and sulphate ions come together to form the insoluble metal sulphate:

$$Ba^{2+}(aq) + SO_4^{2-}(aq) \rightarrow BaSO_4(s)$$

In each case a white precipitate is formed.

The formation of the white precipitate of barium sulphate when barium chloride or barium nitrate solution is added is the test for sulphate ions in solution, the precipitate being insoluble in dilute hydrochloric acid, distinguishing it from barium carbonate, hydroxide or sulphite.

27.25 Sulphuric acid as a dehydrating agent

If concentrated sulphuric acid is left in a beaker open to the air for a few weeks, the level of liquid is seen to rise noticeably. This is due to the absorption of water vapour from the air and demonstrates the liking (or affinity) of concentrated sulphuric acid for water, the compound being a hygroscopic substance (10.16). This is confirmed when the concentrated acid is added to cold water, the mixture becoming very hot and perhaps even boiling with the temperature then being about 120°C. Because of this heat being given out, whenever dilute sulphuric acid is made from the concentrated acid, the acid **must** be added to the water and **not** the water to the acid (it helps to remember to follow the alphabet—a to w). If the water is poured into the concentrated acid, it floats on top of the acid, being less dense, and the heat produced at the boundary of the two liquids is sufficient to boil the water which may be violently flung out bringing some of the acid with it.

Not only will concentrated sulphuric acid absorb free water, it will also take from a substance the elements of water which are chemically combined in that substance and, when this happens, the concentrated sulphuric acid is acting as a dehydrating agent (a 'taker-away' of water). Concentrated sulphuric acid is acting in this way in the following reactions.

1. Reaction with hydrated copper(II) sulphate crystals

Cold concentrated sulphuric acid slowly turns blue hydrated copper(II) sulphate crystals white as the water of crystallization is removed to give the anhydrous salt:

$$CuSO_4.5H_2O(s) \quad - \quad 5H_2O \quad \rightarrow \quad CuSO_4(s)$$

blue   removed   white
by the
sulphuric
acid

2. Reaction with carbohydrates

Carbohydrates are compounds of carbon, hydrogen and oxygen, in molecules of which there are twice as many hydrogen atoms as oxygen atoms. A general formula for carbohydrates is therefore $C_xH_{2y}O_y$. Carbohydrates are contained in all materials obtained from plants, and examples include glucose ($C_6H_{12}O_6$) and ordinary sugar (sucrose—$C_{12}H_{22}O_{11}$).

Concentrated sulphuric acid removes the elements of water from carbohydrates, leaving carbon. The white substances therefore turn brown or black:

$$C_xH_{2y}O_y(s) \quad - \quad yH_2O \quad \rightarrow \quad xC(s)$$

removed
by the
sulphuric
acid

If the carbohydrate is glucose or sucrose, the reaction is rapid and the mixture quickly becomes hot:

$$C_{12}H_{22}O_{11}(s) - 11H_2O \rightarrow 12C(s)$$

It is the dehydration of a carbohydrate which is responsible for the destruction of paper and cotton or rayon by concentrated sulphuric acid. These materials consist largely of cellulose which is a polymer (31.10) made up of a large number of glucose molecules joined together.

The corrosive action of the acid on your skin, which can be given a very severe burn, is due to the removal of water from the skin. Thus, concentrated sulphuric acid must always be treated with great care whenever it is used in the laboratory.

3. Reaction with methanoic acid and ethanedioic acid

Cold concentrated sulphuric acid reacts with methanoic acid (formic acid), which is a liquid or sodium methanoate (which is a solid), giving an effervescence of the colourless and very poisonous gas, carbon monoxide, which can be collected over water:

$$HCOOH(l) - H_2O \rightarrow CO(g)$$

removed by
the sulphuric acid

When concentrated sulphuric acid is warmed with ethanedioic acid (oxalic acid) a mixture of carbon monoxide and carbon dioxide is formed:

$$H_2C_2O_4(s) - H_2O \rightarrow CO(g) + CO_2(g)$$

removed by
the sulphuric acid

The reaction can again be used as the basis for preparing carbon monoxide in the laboratory, but carbon dioxide has to be removed from the gaseous product by passing it through potassium hydroxide solution.

27.26 Sulphuric acid as an oxidising agent

Copper, which is below hydrogen in the reactivity series, does not react with dilute sulphuric acid, even when the mixture is warmed. However, if copper is heated with concentrated sulphuric acid, a pungent colourless gas which decolourises acidified potassium manganate(VII) solution is given off and a dark grey muddy solid is formed in the reaction mixture. If, after cooling, the reaction mixture is poured into cold water, stirred and then filtered, a clear blue filtrate separates. The filtrate is copper(II) sulphate solution. The acid is acting as an oxidising agent and the gas given off is sulphur dioxide.

If sulphuric acid is regarded as hydrated sulphur trioxide (SO_3), we can say that the copper has been oxidised to copper(II) sulphate and sulphur trioxide has been reduced to sulphur dioxide (SO_2).

Copper(II) sulphate, having been formed from the concentrated acid in the absence of water, will be in the anhydrous form and will be a white powder in the grey sludge left at the end of the reaction:

$$Cu(s) + 2H_2SO_4(l) \rightarrow CuSO_4(s) + 2H_2O(l) + SO_2(g)$$

It is only when water is added that it turns blue.

27.27 The oxidation states of sulphur

In this chapter we have discussed many sulphur compounds, but they can all be grouped under just three headings, according to the oxidation state of the sulphur. Sulphur has three oxidation states, i.e. -2, $+4$ and $+6$, which are shown in hydrogen sulphide, sulphur dioxide and sulphur trioxide respectively. All the other compounds mentioned in this chapter are related to the above three by simple acid-base reactions. Hydrogen sulphide is an acid whose salts are sulphides, sulphur dioxide is the anhydride of sulphurous acid, whose salts are sulphites and hydrogensulphites, while sulphur trioxide is the anhydride of sulphuric acid, whose salts are sulphates and hydrogensulphates. The chemistry of sulphur can then be neatly summarised in Table 27.1 Movement 'up-and-down' involves only acid-base reactions, while movement 'left-to-right' involves oxidation of the sulphur, and movement 'right-to-left' involves reduction of the sulphur.

TABLE 27.1. *The compounds and oxidation states of sulphur*

OXIDATION STATE OF SULPHUR	OXIDATION $\longrightarrow$			
	-2	0	$+4$	$+6$
Oxides		(S)	SO_2	SO_3
Acids	H_2S		H_2SO_3	H_2SO_4
Acid salts	NaHS		$NaHSO_3$	$NaHSO_4$
Normal salts	Na_2S		Na_2SO_3	Na_2SO_4
Normal ions	S^{2-}		SO_3^{2-}	SO_4^{2-}

The oxidation state (number) of sulphur itself is considered to be zero and in hydrogen sulphide and sulphide salts it is taken as -2 because the conversion of hydrogen sulphide to sulphur is a clear example of oxidation:

$$H_2S + [O] \rightarrow S + H_2O$$

A point worth noting is that for the three acids in the table, the acidic strength increases from left to right—i.e. with the oxidation state of the sulphur. Thus sulphuric acid is a stronger acid than sulphurous acid, which, in turn, is a much stronger acid than hydrogen sulphide. This behaviour is typical of an element with variable valency.

27.28 Summary

1. Sulphur is a very important non-metallic element which occurs uncombined in certain areas of the world and which can also be obtained from petroleum oil.
2. There are two allotropes of solid sulphur, rhombic sulphur and monoclinic sulphur. Rhombic sulphur is the more stable below 96°C, and monoclinic above 96°C.
3. When sulphur is heated it melts and the molten sulphur shows changes in colour and viscosity as the temperature is raised, these being explained by the arrangements of the atoms of sulphur.
4. Sulphur will combine with metals, forming sulphides, and if dilute acid is added to a sulphide, hydrogen sulphide is given off.
5. Hydrogen sulphide is a smelly poisonous gas which dissolves in water, giving a weakly acidic solution. It precipitates metal sulphides when passed into solutions of metal salts, and it is a reducing agent.
6. When sulphur is burnt in air or oxygen it forms sulphur dioxide, a colourless gas which dissolves in water to give sulphurous acid. When sulphurous acid is neutralized with alkalis, sulphites are formed. Sulphurous acid is a powerful reducing agent.
7. Sulphuric acid is a very important industrial chemical as it is used in the production of many other substances. It is manufactured from sulphur dioxide by the Contact Process.
8. Sulphuric acid, when it is dilute, will react as an acid and as a sulphate, and, when it is concentrated, as a dehydrating agent. When it is hot and concentrated, it will oxidise metals.
9. Sulphates can be tested for by adding dilute hydrochloric acid and barium chloride solution.

Chapter 28

Carbon and its compounds

What are the properties of an aqueous solution of carbon dioxide?

You will need a suitable small-scale apparatus for generating carbon dioxide, Fig. 28.1. The emerging gas should pass through water in order to remove acid spray from generator.

Fig. 28.1 The preparation of carbon dioxide

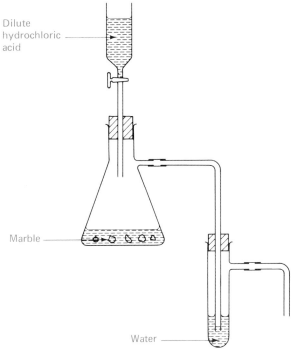

Dilute hydrochloric acid

Marble

Water

1. Take 3–4 cm³ of purified water in each of two clean test-tubes, and add a drop of litmus solution to each. Keep one tube as a 'neutral' comparison.

Bubble carbon dioxide into the other for a little while and note whether a colour change occurs.

Then boil the liquid gently for a minute or so and compare the colour once more with that of the 'neutral' tube.

Repeat the experiment using Universal Indicator instead of litmus.

Questions

1 Is carbon dioxide solution strongly acidic or weakly acidic?
2 What happens to the acidity when the solution is boiled? What is the likely explanation of this?

2. Pour about 3 cm³ of lime water into a test-tube and keep it for use in a moment or two.

Take another test-tube and allow carbon dioxide to pass into it for 10–15 seconds. Now 'pour' (Fig. 28.2) the invisible carbon dioxide gas from this tube into the tube containing the lime water. Then close the mouth of the

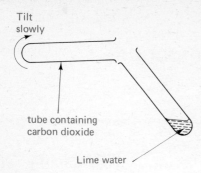

Tilt
slowly

tube containing
carbon dioxide

Lime water

Fig. 28.2 Pouring carbon dioxide

latter tube with finger or thumb and shake it vigorously up and down. The lime water should go milky (28.13).

This is the specific chemical test for carbon dioxide, and you will need to carry out this test in many of the following investigations in this chapter.

3. Pour 3–4 cm³ of sodium hydroxide solution into a test-tube and then bubble carbon dioxide through the solution for about 30 seconds. Carefully heat the liquid until it boils to see if carbon dioxide is evolved. (Look for effervescence, and test the vapour with lime water if you think necessary.)

Cool the tube under the tap for a few seconds, and cautiously add 2–3 cm³ of dilute hydrochloric acid. If effervescence occurs, test the evolved gas with lime water as in part 2 of this investigation.

Questions

3 Was any carbon dioxide absorbed by the sodium hydroxide solution? (How can you tell?)
4 If so, was it expelled on heating or was it expelled by the addition of dilute hydrochloric acid?
5 What type of chemical reaction must occur between carbon dioxide and sodium hydroxide, and what type of substance will be formed?

Investigation 28.2

What are the properties of the salt, sodium carbonate?

1. Pour separate 2 cm³ portions of sodium carbonate solution into two test-tubes and test one with litmus and the other with Universal Indicator. Then to one portion (either one) add dilute hydrochloric acid.

Observe what happens, and carry out the appropriate test to identify the gas evolved.

Questions

1 On the basis of the effect an aqueous solution of sodium carbonate has on indicators, what ion must be present in the solution?
2 What gas is evolved when sodium carbonate solution is acidified?
3 What kind of substance will be left in solution when sodium carbonate reacts with an acid?

2. In five separate test-tubes put 2 cm³ portions of solutions of barium chloride, calcium chloride, magnesium sulphate, zinc sulphate and copper(II) sulphate, respectively.

To each tube, add 1–2 cm³ of sodium carbonate solution, with stirring, and note what happens. Then to each tube add 2–3 cm³ of dilute hydrochloric acid, stir well with a glass rod, and note what happens.

Questions

4 What do you think the various precipitates were?
5 What happens to these precipitates when a strong acid is added?

3. Put a few crystals of hydrated sodium carbonate (washing soda) in a dry test-tube supported horizontally in a clamp.

Heat the crystals fairly gently until no further change occurs, making careful observations, and testing for evolution of carbon dioxide in the usual way.

Allow the residue to cool, then remove the tube from the clamp and add dilute hydrochloric acid. Observe what happens, and again test for evolution of carbon dioxide.

6 What do you think the residue was, and so what do you think happened when hydrated sodium carbonate was heated?

Investigation 28.3

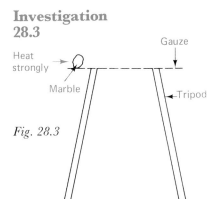

Heat strongly

Gauge

Marble

Tripod

Fig. 28.3

What is the action of heat and acids on marble?

1. Weigh a lump of marble and then place it on the edge of a gauze supported by a tripod (Fig. 28.3).

Holding the Bunsen burner in your hand, allow the hottest flame to play directly on the surface of the marble, over a large area, for 10–15 minutes. (Do not heat the marble through the gauze, as it will not get hot enough.)

Note any change in the appearance of the marble, then, when it is cool, and without touching it with your fingers, weigh again and record any change which has occurred.

Now put the lump of heated marble in a dish or basin and allow water to drip on to it, a little at a time. Observe carefully what happens. Finally add sufficient water that excess liquid is present, and test this liquid with litmus papers.

Questions

1 What is the chemical name of marble?
2 Did the marble gain something or lose something when it was heated?
3 What type of substance must have been present in the liquid at the end of the experiment?

2. Put a lump of marble in a test-tube and add some dilute hydrochloric acid. Make a mental note of the rate of reaction and do the appropriate test to confirm the identity of the gas evolved.

Now pour off the acid, rinse the marble with water, and treat it with dilute nitric acid. Compare the reaction with the previous one.

Again, pour off the acid and rinse the marble with water. Now treat it with dilute sulphuric acid, and compare the reaction with the previous ones.

Questions

4 What gas is evolved in these reactions?
5 What kind of substance will be formed in solution in each case?
6 What is a possible explanation for one of the acids being less effective in dissolving marble?

Investigation 28.4

How do metal carbonates behave when heated and when treated with acids?

Substances which you may be asked to investigate include barium carbonate, copper(II) carbonate, lead(II) carbonate, magnesium carbonate and zinc carbonate.

1. Heat a small portion of one of the substances in an ignition-tube.
Test for evolution of carbon dioxide in the usual way, and note any change of appearance which occurs in the solid. When no further change seems to occur, even on fairly strong heating (but don't melt the tube!) allow the residue to cool, watching for any change in appearance as it does so.

When the residue is cool, add dilute nitric acid and note whether or not effervescence occurs.

1 On the basis of these observations, did the substance decompose on heating or not? If it did decompose, what could the residue be?

2. Put separate small portions of the substance in three test-tubes. To these tubes add 3–4 cm³ of, respectively, dilute nitric, hydrochloric and sulphuric acids. Test for evolution of carbon dioxide in the usual way.

Compare the three results. If any reaction seems to be feeble, try warming the reaction mixture to see if this makes any difference.

Repeat the procedure with other carbonates.

Questions

2 What is the general reaction of carbonates with acids?
3 If any of your reactions failed to go readily, what could be a possible explanation?

Investigation 28.5

What are the properties of the acid salt, sodium hydrogencarbonate (sodium bicarbonate)?

1. Put a little solid sodium hydrogencarbonate in a dry test-tube supported horizontally in a clamp. Heat the solid gently, observing carefully, and test in the usual way for carbon dioxide being given off.

Allow the residue to cool, then remove the tube from the clamp and add dilute hydrochloric acid. Observe what happens, and again test for carbon dioxide.

Questions

1 Which two products were given off when the sodium hydrogencarbonate was heated?
2 In order to explain the reaction of the residue with acid, what must be present in the residue?

2. Pour about 3 cm³ of sodium hydrogencarbonate solution into a test-tube and add dilute hydrochloric acid. Observe the result, and identify the gas evolved.

Questions

3 What gas is evolved?
4 What type of substance will be left in solution?

Investigation 28.6

What happens when excess carbon dioxide is passed into lime water?

For this experiment carbon dioxide should be generated using the same apparatus as in Investigation 28.1

1. Put about 5 cm³ of lime water in a test-tube and bubble carbon dioxide through it until two changes have occurred. Save the resulting liquid for part 2 of this investigation.

Questions

1 The first change you observe is the appearance of milkiness which indicates the presence of an insoluble white substance. What is this substance?
2 The second change is due to the formation of a soluble substance. What might this substance be?

2. Warm the liquid from the first part of this investigation in a Bunsen flame until it has boiled gently for 2 or 3 minutes.

3 Boiling converts the soluble compound which was formed when excess carbon dioxide was passed through the lime water into an insoluble compound. What do you think the insoluble compound is likely to be?

28.7
Carbon itself

Carbon is one of the most widely distributed elements in nature, and one of the elements upon which all life is based. All organic (animal and vegetable) matter is based on carbon compounds, and the range of carbon compounds derived from living sources is so vast that it forms a separate branch of chemistry called organic chemistry. Some organic chemistry is covered in Chapters 30, 31 and 32 of this book; the present chapter is concerned mainly with carbon compounds of mineral origin.

In mineral sources, carbon occurs in rather small quantities as the element itself. Carbon shows allotropy (16.5), and its two allotropic forms, graphite and diamond, are both found naturally. Graphite occurs as the mineral called plumbago or 'black lead' (as in 'lead' pencils), while diamond is, as you know, a very precious stone. (For a comparison of these allotropes see 28.11.)

Carbon also occurs in the substances known as fossil fuels, namely coal, petroleum and natural gas. These substances, found in vast underground deposits, were formed by the decay and compression, over millions of years, of ancient forests and remains of marine organisms, and so are ultimately of organic origin. They are of major importance as fuels (see below).

Carbon occurs in large quantities as carbonates, one of the major examples being calcium carbonate, found as chalk, limestone and marble. It also occurs in the atmosphere, as carbon dioxide. The proportion of carbon dioxide is small, but it plays a vital part in the life-cycle of animals and plants.

28.8
The carbon cycle

The processes by which carbon is distributed around the Earth's surface are very complex, but some of the pathways in the cycle are shown in a simplified form in Fig. 28.4. A key part of the cycle is the atmosphere, which contains about 0.03% (by volume) carbon dioxide. This figure remains almost constant, because of a delicate balance between formation and removal of carbon dioxide.

The process of photosynthesis in plants is extremely important, because it not only removes carbon dioxide from the atmosphere but also regenerates the oxygen in the atmosphere which is required for animals and humans to breathe. Photosynthesis involves the building up, in the leaves of plants, of carbohydrates—in particular starch, which consists of very large molecules (relative molecular mass between 10 000 and 100 000). The molecules can be considered as being made up of units of formula $C_6H_{10}O_5$ joined together, and hence its formula is usually written as $(C_6H_{10}O_5)_n$, where n is a very large number. The overall reaction is:

$$\text{carbon dioxide} + \text{water} \rightarrow \text{starch} + \text{oxygen}$$

This is a photochemical reaction (23.14) as sunlight provides the energy necessary to make the reaction proceed. Chlorophyll in the leaves of plants acts as a catalyst for the reaction. Using formulae, the reaction can be summarised by:

$$6nCO_2 + 5nH_2O \rightarrow (C_6H_{10}O_5)_n + 6nO_2$$

When plants are eaten the starch is hydrolysed, that is, broken down by the action of water. This reaction is catalysed by enzymes in saliva and the product is glucose:

$$(C_6H_{10}O_5)_n + nH_2O \rightarrow nC_6H_{12}O_6$$

This glucose acts as the source of energy in the body, as it is slowly oxidised to carbon dioxide and water using inhaled oxygen:

$$C_6H_{12}O_6 \;+\; 6O_2 \;\rightarrow\; 6CO_2 \;+\; 6H_2O \;\text{(exothermic)}$$
$$\text{(inhaled)} \qquad \text{(exhaled)}$$

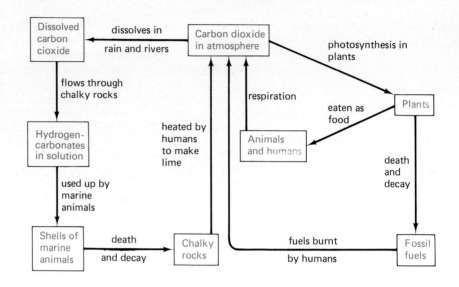

Fig. 28.4 The carbon cycle. The time scales for the various pathways differ enormously

28.9 Natural substances as fuels

The fossil fuels mentioned above consist largely of hydrocarbons (30.3), which are compounds of carbon and hydrogen only (though coal contains a high proportion of elemental carbon). Hydrocarbons, on burning, release a lot of heat and the products of combustion are fairly harmless, provided that a good oxygen (air) supply is available:

hydrocarbon + oxygen → carbon dioxide + water vapour

This is why they are so valuable as fuels.

If the oxygen supply is inadequate, some of the carbon escapes oxidation and appears as soot. This creates pollution problems if boilers and furnaces are not properly maintained. Motor vehicle engines gradually accumulate deposits of carbon inside the cylinders, and then need to be stripped down and decoked by scraping off this carbon deposit. With coal there is the further problem of pollution by sulphur compounds (9.5 and 27.17).

The advantage of the modern fuels, oil and natural gas, is that they can be purified at the refinery (31.5) so that they burn more cleanly and pollution problems are lessened.

Until the large-scale extraction of fossil fuels was developed, wood was burnt as a fuel. Again, the products of complete combustion are carbon dioxide and water vapour, but by deliberately limiting the oxygen supply carbon can be made to escape oxidation almost completely and we are then left with charcoal, which is a fairly pure form of carbon.

Coal can be subjected to carbonisation by heating it in the absence of air; in this case the solid product is known as coke (again consisting mainly of carbon). The gaseous product is coal gas (31.3). Coke is used in domestic fires in smokeless zones, since it burns quite cleanly producing little pollution. (For more discussion of fuels see 9.3 and 31.6.)

302

28.10 Properties and uses of carbon

Carbon in its usual form, graphite (of which charcoal is an impure form), is a black non-metallic element. It is one of the most involatile substances known, and it is not dissolved by any common solvent. Unlike all other non-metals, it is a good electrical conductor (this applies only to graphite, not to diamond), and so it is widely used for making electrodes for batteries and electrolytic cells, being also a rather unreactive element. Graphite is very soft (unlike diamond) and, although it is best known for its use in 'lead' pencils, its most important use is as a lubricant for metal bearings, usually being applied as an emulsion with oil or water. Its insolubility and lack of reactivity make graphite an ideal component of printers' ink and other 'indelible' inks, since it will not fade or wash away.

Fig. 28.5 Uses of the different forms of carbon (a) A diamond-tipped blade sawing a slab of stone. (Courtesy Christensen Diamond Products Ltd)

(b) Carbon blocks being prepared for use as anodes in the production of aluminium. See Fig. 24.11. (Courtest RTZ Services Ltd)

(a)

(b)

A remarkable property of charcoal (a very finely-divided form of graphite) is its ability to attract substances on to its surface (adsorption). It is used for decolourising substances containing vegetable dyes (e.g. the removal of litmus from a solution), and it is used in water-purification processes and by the food industry. Charcoal is also used in some gas-masks, and in special inner-soles for shoes to prevent foot odour.

Even when carbon is heated to a high temperature, the only common element with which it readily combines is oxygen. For example, if the charcoal is heated in air until it is red-hot, and then put into pure oxygen, it glows white-hot as it burns, the product being the colourless gas carbon dioxide:

$$C(s) + O_2(g) \rightarrow CO_2(g)$$

Alternatively, carbon will continue to burn exothermically in ordinary air provided that a good draught is available, as in the type of domestic grate which has an air-draught vent under the floor.

The readiness of hot carbon to combine with oxygen is sufficiently great to make it a useful reducing agent. Industrially, carbon (in the form of coke) is used for reducing the oxides of lead, zinc and iron to the metals (Chapter 24). In the laboratory, the reducing action of carbon can be shown by heating a mixture of lead(II) oxide with charcoal for several minutes in a crucible. After a while, a bead of molten lead forms in the bottom of the crucible:

$$PbO(s) + C(s) \rightarrow Pb(l) + CO(g)$$

The initial oxidation product is carbon monoxide but this burns, with a blue flame, to carbon dioxide at the surface of the mixture:

$$2CO(g) + O_2(g) \rightarrow 2CO_2(g)$$

Carbon monoxide is discussed later in this chapter (28.19).

28.11 Graphite and diamond

The allotropes (16.5) of carbon, graphite and diamond, have essentially the same chemical properties but differ considerably in their physical properties. These differences are summarised in Table 28.1.

TABLE 28.1. *Properties of graphite and diamond*

GRAPHITE	DIAMOND
density 2·3 g cm⁻³ opaque to light electrical conductor very soft	density 3·5 g cm⁻³ transparent to light electrical insulator extremely hard

The uses of graphite have been discussed above. The special properties of diamond make it highly prized as a precious stone for use in jewellery, but it is also important industrially in cutting tools such as circular saws. The cutting edges are impregnated with very small diamonds, and this improves both the cutting efficiency and the wearing qualities of the blade.

28.12 Carbon dioxide

Fig. 28.6 Solid carbon dioxide being used to chill shellfish. (Courtesy British Oxygen Corporation)

You are no doubt familiar with fizzy drinks such as lemonade. Have you ever wondered what causes the fizz? The answer is carbon dioxide, dissolved in the water under pressure. When you open the bottle and pour the liquid into a glass, effervescence occurs, because at normal atmospheric pressure the solution is more than saturated with the gas. Soda water, which was invented by Joseph Priestley, the discoverer of oxygen (8.7), is simply a pressurised solution of carbon dioxide. Drinks such as lemonade also contain sugar and various flavourings. An aqueous solution of carbon dioxide is weakly acidic, giving it a slightly sharp taste which is rather pleasant. In some fizzy alcoholic drinks, such as champagne, the carbon dioxide is produced by the fermentation process (32.6). At normal atmospheric pressure 1 volume of cold water dissolves about 1 volume of carbon dioxide, so it is only moderately soluble.

The reaction of hydrochloric acid with marble (calcium carbonate) is used as the standard laboratory preparation of carbon dioxide; a suitable apparatus is shown in Fig. 28.1. The mixture does not need heating and the gas is usually collected by downward delivery as it is denser than air and moderately soluble in water.

Carbon dioxide is a colourless, odourless gas. It can be liquefied by compressing it (no cooling is necessary), and can be stored under pressure in cylinders. On cooling at normal atmospheric pressure it freezes at −78 °C to a white solid known as Dry Ice. This substance is useful as a refrigerant—e.g. when foods are being transported by road or rail—because as it evaporates (i.e. sublimes) it leaves no residue.

Carbon dioxide is non-flammable and it does not support the combustion of most materials. This, coupled with the fact that it is denser than air, makes it very useful in fire-extinguishers. Some modern extinguishers contain carbon dioxide under pressure, which is released by squeezing a lever. Carbon dioxide extinguishers have an advantage over water extinguishers, since they may be used safely and effectively on electrical fires and on fires involving oil, petrol or fat.

Although most burning substances are extinguished by carbon dioxide, magnesium is not. If burning magnesium ribbon is lowered into a jar of carbon dioxide it continues to burn (with a bit of a struggle) and black specks of carbon are deposited in addition to white magnesium oxide:

$$2Mg(s) + CO_2(g) \rightarrow 2MgO(s) + C(s)$$

This reaction is very similar to the burning of magnesium in sulphur dioxide (27.15).

28.13 Carbonic acid

The solution formed when carbon dioxide dissolves in water is a very weak electrolyte (showing that a few ions are present) and it turns litmus a dull red colour, rather than bright red. This shows that hydrogen ions are present, but only in very small concentration. If carbon dioxide solution is boiled, it loses its acidity completely. These facts are interpreted by saying that carbon dioxide reacts reversibly with water to form the weak, unstable, carbonic acid (H_2CO_3) which in turn is partially ionised into hydrogen ions and carbonate (CO_3^{2-}) ions,

$$H_2O(l) + CO_2(g) \rightleftharpoons H_2CO_3(aq)$$

and
$$H_2CO_3(aq) \rightleftharpoons 2H^+(aq) + CO_3^{2-}(aq)$$

Carbonic acid, like sulphurous acid (27.16), cannot be isolated as a single substance.

Since carbon dioxide is an acidic oxide it reacts readily with alkalis to form salts (carbonates), and its reaction with sodium hydroxide is discussed in the following section. The reaction of carbon dioxide with lime water (calcium hydroxide solution) is very important because this is used as the test for carbon dioxide. If carbon dioxide is shaken with lime water (or bubbled through lime water) the liquid turns milky. This is because the salt formed (calcium carbonate) is insoluble in water and forms a precipitate,

$$CO_2(g) + Ca(OH)_2(aq) \rightarrow CaCO_3(s) + H_2O(l)$$

or as another way of looking at it:

$$H_2CO_3(aq) + Ca(OH)_2(aq) \rightarrow CaCO_3(s) + 2H_2O(l)$$

Carbon dioxide is the only gas which turns lime water milky.

28.14 Sodium carbonate

When carbon dioxide (not in excess—see 11.17) is absorbed by sodium hydroxide solution the salt formed is sodium carbonate, Na_2CO_3:

$$CO_2(g) + 2NaOH(aq) \rightarrow Na_2CO_3(aq) + H_2O(l)$$

The most common crystalline form of this salt is the decahydrate $Na_2CO_3.10H_2O$, known as washing soda because of its use in treating water hardness (10.25). If crystals of washing soda are left exposed to the air for a day or two they crumble into a white powder, undergoing efflorescence to the monohydrate:

$$Na_2CO_3.10H_2O(s) \rightarrow Na_2CO_3.H_2O(s) + 9H_2O(g)$$

If the crystals are heated they lose all their water of crystallization and the residue is anhydrous sodium carbonate, which is then stable at the temperature of a Bunsen flame. Sodium carbonate is important in glass-making (29.6).

Sodium carbonate, although it is a salt, gives a solution in water which is alkaline. This arises because carbonate ions react with water:

$$CO_3{}^{2-}(aq) + H_2O(l) \rightleftharpoons HCO_3{}^-(aq) + OH^-(aq)$$

($HCO_3{}^-$ is the hydrogencarbonate (bicarbonate) ion)

This interaction occurs because of the weakness of carbonic acid. Sodium carbonate solution reacts immediately with any strong acid at room temperature, neutralising the acid with liberation of carbon dioxide,

e.g. $$Na_2CO_3(aq) + 2HCl(aq) \rightarrow 2NaCl(aq) + CO_2(g) + H_2O(l)$$

or $$CO_3{}^{2-}(aq) + 2H^+(aq) \rightarrow CO_2(g) + H_2O(l)$$

Sodium carbonate solution may be used as a test for acids, since the resulting effervescence is easily observed.

28.15 General properties of carbonates

Most carbonates are insoluble in water (e.g. calcium carbonate formed in the lime water test), although they will usually dissolve, with vigorous effervescence, in strong acids, forming a new salt in solution,

e.g. $$CaCO_3(s) + 2HCl(aq) \rightarrow CaCl_2(aq) + CO_2(g) + H_2O(l)$$

and $$CuCO_3(s) + H_2SO_4(aq) \rightarrow CuSO_4(aq) + CO_2(g) + H_2O(l)$$

The action of strong acid is, in fact, used as a test for carbonates since the effervescence is easily observed and the gas evolved may be identified by its action on lime water.

The occasions on which carbonates do not react readily with acids arise when the salt formed in the reaction is insoluble in water. For example, calcium carbonate reacts very sluggishly with dilute sulphuric acid because the calcium sulphate formed is almost insoluble in water (11.9).

Metal carbonates in general decompose to the metal oxide and carbon dioxide on heating,

e.g. $$CaCO_3(s) \rightarrow CaO(s) + CO_2(g)$$

The ease with which this type of reaction occurs depends on the reactivity of the metal concerned. Two carbonates which are stable at the temperature of a Bunsen flame are those of the reactive metals, potassium and sodium. The carbonates of these metals are also different from most other carbonates in that they are soluble in water.

Ammonium carbonate is different from carbonates in general. It is soluble in water and when it decomposes on heating it leaves no solid residue:

$$(NH_4)_2CO_3(s) \rightarrow 2NH_3(g) + CO_2(g) + H_2O(g)$$

Ammonium carbonate smells quite strongly of ammonia even at room temperature. This accounts for its old name being sal volatile and for its being used in smelling salts.

28.16 Natural occurrence of carbonates

Many carbonates are found as minerals; for example the green mineral malachite is (basic) copper carbonate, cerussite is lead carbonate and calamine is zinc carbonate. By far the most common, however, is calcium carbonate which occurs as chalk, limestone, marble and calcite, and is a very important raw material for the alkali industry (29.2). Eggshells and the shells of marine animals consist of calcium carbonate, and so also does coral. Chalk cliffs—as in the South of England—were formed, over many thousands of years, by the gradual deposition and compression of the shells of marine animals as they died.

306

**28.17
Sodium
hydrogen-
carbonate**

Sodium hydrogencarbonate (sodium bicarbonate), also known as bicarbonate of soda and baking soda, has the formula $NaHCO_3$. It is the salt formed when only one of the hydrogen atoms of carbonic acid (H_2CO_3) had been replaced by sodium. In this sense it is an acid salt as it still contains replaceable hydrogen, but in fact, owing to carbonic acid being such a weak acid, a solution of sodium hydrogencarbonate is slightly alkaline.

It is an important intermediate in the production of sodium carbonate (29.5) which is itself used in very large quantities in the glass industry. Its two common names may remind you of two of its domestic uses. Firstly, as bicarbonate of soda (or an ingredient of health-salts or such products as Alka-Seltzer) it is used as an anti-acid in the treatment of indigestion. Excess acid in the stomach is neutralised by reacting with the sodium hydrogencarbonate:

$$NaHCO_3(aq) + HCl(aq) \rightarrow NaCl(aq) + H_2O(l) + CO_2(g)$$

Another ingredient in these products is a solid acid such as tartaric acid. When they are added to water the acid dissolves and reacts with some of the sodium hydrogen-carbonate giving off carbon dioxide which causes effervescence and makes the mixture more pleasant to drink.

The other domestic use is baking powder which also contains sodium hydrogen-carbonate and tartaric acid. The mixture is added to cakes to make them rise during baking. The heat causes the sodium hydrogen carbonate to decompose:

$$2NaHCO_3(s) \rightarrow Na_2CO_3(s) + CO_2(g) + H_2O(g)$$

The carbon dioxide which is given off causes the cake to rise. The acid is present to react with the sodium carbonate so that all the possible carbon dioxide is given off and the residue after the reaction is a tasteless salt rather than sodium carbonate which would result in an unpleasant taste.

There are very few other hydrogencarbonates known; potassium hydrogen-carbonate is the only other common one which is solid. Calcium hydrogencarbonate is important but it only exists in aqueous solution. Any attempt to evaporate the water results in it decomposing to calcium carbonate (see next section).

**28.18
Calcium
hydrogen-
carbonate**

When carbon dioxide is passed into lime water the solution becomes milky, owing to precipitation of insoluble calcium carbonate:

$$CO_2(g) + Ca(OH)_2(aq) \rightarrow CaCO_3(s) + H_2O(l)$$

If, however, passage of carbon dioxide is continued for some time the milkiness disappears again. This is due to further reaction, forming the soluble salt calcium hydrogencarbonate,

$$CaCO_3(s) + CO_2(g) + H_2O(l) \rightarrow Ca(HCO_3)_2(aq)$$

or as another way of looking at it:

$$CaCO_3(s) + H_2CO_3(aq) \rightarrow Ca(HCO_3)_2(aq)$$

The solution can be made milky again by boiling it, which results in the hydrogencarbonate decomposing back to the carbonate, the excess carbon dioxide being given off:

$$Ca(HCO_3)_2(aq) \rightarrow CaCO_3(s) + CO_2(g) + H_2O(l)$$

This is the general behaviour of hydrogencarbonates on heating and it is very important with regard to the hardness of water which is discussed in 10.23.

28.19 Carbon monoxide

It is well known that it is dangerous to run the engine of a motor car inside a closed garage, because poisonous fumes build up after a while. The main culprit here is carbon monoxide, a colourless and odourless gas which rapidly causes death if breathed at a concentration of 1 % in the air.

Carbon monoxide is a product of incomplete combustion of hydrocarbons, the product of complete combustion being of course carbon dioxide. It can be shown experimentally that carbon monoxide is produced when carbon itself is heated in carbon dioxide, using the apparatus shown in Fig. 28.7.

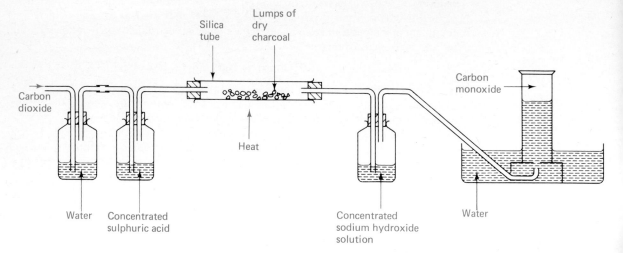

Fig. 28.7 The preparation and collection of carbon monoxide

Lumps of dry charcoal are heated in a current of dry carbon dioxide gas, and carbon monoxide, being insoluble in water, collects in the gas jar. The reaction is:

$$C(s) + CO_2(g) \rightarrow 2CO(g)$$

The purpose of passing the gas through a solution of sodium hydroxide is to ensure that unreacted carbon dioxide, being acidic, dissolves in the alkali and so does not collect in the jar.

28.20 Properties of carbon monoxide

The gas can also be prepared by reacting sodium methanoate with concentrated sulphuric acid (27.25). If, when the jar is full, a lighted taper is applied, the carbon monoxide burns with a characteristic blue flame, forming carbon dioxide:

$$2CO(g) + O_2(g) \rightarrow 2CO_2(g)$$

Carbon monoxide is a colourless odourless gas with about the same density as air. It is almost insoluble in water or in alkaline solutions, and so is classed as a neutral oxide, as opposed to carbon dioxide which is weakly acidic (28.13).

Carbon monoxide burns readily in air, forming carbon dioxide (see above), and its affinity for oxygen is such that it will reduce the oxides of the less reactive metals on heating. Its action as a reducing agent is thus similar to that of hydrogen,

e.g. $$CuO(s) + H_2(g) \rightarrow Cu(s) + H_2O(g)$$

and $$CuO(s) + CO(g) \rightarrow Cu(s) + CO_2(g)$$

Its role in the industrial extraction of iron is described in 24.9.

The reason for the extremely poisonous nature of carbon monoxide is that it combines with the haemoglobin of the blood more readily than oxygen does; the resulting compound, carboxyhaemoglobin, builds up in the bloodstream and prevents the blood from carrying a sufficient supply of oxygen from the lungs to the body tissues (9.4).

**28.21
Carbon
monoxide as a
fuel gas**

The carbonisation of coal has been mentioned earlier (28.9), the solid product of which is coke. The other products are coal tar, a complex liquid mixture, and coal gas, both of which are discussed briefly in 31.3. Coal gas contains about 8% (by volume) carbon monoxide and here carbon monoxide acts as a fuel gas, though it is only a minor constituent of the mixture. The solid residue, coke, may itself be used to produce fuel gases based wholly or partly on carbon monoxide, as described below.

If air is blown through a bed of white-hot coke, the coke (mainly carbon) is oxidised to carbon monoxide. This is actually a two-step process in which, firstly, carbon dioxide is formed,

$$C(s) + O_2(g) \rightarrow CO_2(g)$$

but then this reacts further with the hot coke forming carbon monoxide:

$$C(s) + CO_2(g) \rightarrow 2CO(g)$$

Since the original oxidising agent was air, rather than pure oxygen, the final gas is a mixture of carbon monoxide and nitrogen (1:2 by volume) and is known as **producer gas**. Because of its high nitrogen content, producer gas has a relatively low calorific (heating) value, but it is cheap and may be used, for example, in heating furnaces or boilers in industry.

A different reaction involves blowing steam through white-hot coke:

$$C(s) + H_2O(g) \rightarrow CO(g) + H_2(g)$$

**28.22
Summary**

The product, which is a mixture of carbon monoxide and hydrogen, is known as **water gas**, and is discussed in 18.4.

1. Carbon is an essential constituent of all living matter, and forms the basis of the various fossil fuels. The element itself is fairly unreactive, but combines readily with oxygen when hot. Carbon shows allotropy, existing as graphite and diamond.
2. Carbon dioxide gas dissolves in water to form carbonic acid, a weak acid, most of whose salts (carbonates) are insoluble in water. Lime water acts as a test for carbon dioxide. Sodium carbonate (washing soda) is soluble in water giving a solution which is alkaline.
3. Carbonates in general give off carbon dioxide when treated with acids. Carbonates of the less reactive metals are decomposed easily by heat alone. Many carbonates occur as minerals, e.g. marble (calcium carbonate).
4. Hydrogencarbonates (bicarbonates) are easily decomposed to carbonates on heating, giving off carbon dioxide. Only a few are known as solids (e.g. sodium hydrogencarbonate); some others are known only in solution (e.g. calcium hydrogencarbonate).
5. Carbon monoxide is produced by incomplete combustion of hydrocarbons. It is a poisonous gas, which burns to form carbon dioxide. It forms the basis of various fuel gases, and may be used as a reducing agent for some metal oxides.

Chapter 29

The alkali industry

29.1
The importance of the alkali industry

Fig. 29.1 Raw materials and products of the alkali industry

The world consumption of alkali, in various forms, is over 60 million tonnes per year. From this fact alone we can see that the alkali industry is a very important one. The major raw materials of the industry are limestone (calcium carbonate) and salt (sodium chloride). Fig. 29.1 shows the connections between the raw materials and the products. Each part of the alkali industry will be described separately, but you can refer back to Fig. 29.1 in order to remind yourself of the connections between these various parts.

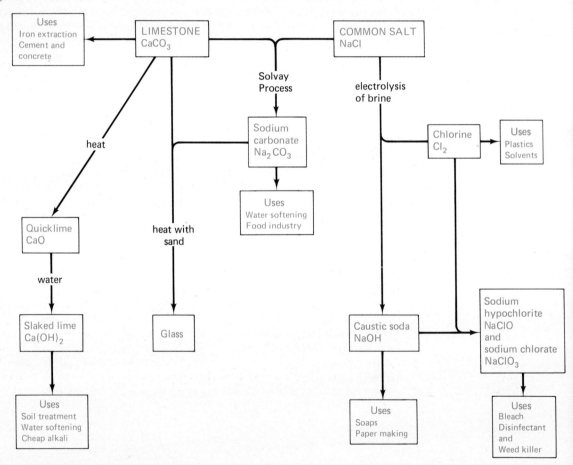

29.2
Limestone as a raw material

Limestone ($CaCO_3$) is mined directly as a mineral. For different purposes it is either left in lump form, or ground to a powder.

Direct uses of limestone

Limestone is used in the extraction of iron. It is added to the mixture of iron ore and

310

coke in the blast furnace in order to extract sandy impurities as a liquid slag, and this is explained in detail in 24.9.

Another major use is in the manufacture of cement and concrete, substances which are of vital importance to the building industry. Cement is made by heating a mixture of limestone and clay to about 1500°C in a long rotary kiln. Clay is a hydrated aluminosilicate mineral which can be thought of as containing aluminium oxide (alumina, Al_2O_3) and silicon dioxide (silica, SiO_2). Inside the kiln, limestone decomposes to quicklime (CaO), and this combines with the alumina and silica to form calcium aluminate ($Ca(AlO_2)_2$) and calcium silicate ($CaSiO_3$). The dry product is ground up and sold as a powder. When water is added to the mixture, complex chemical changes occur forming a hard interlocking mass of crystals of hydrated calcium aluminate and silicate.

Concrete is a mixture of cement and a ballast of gravel or stone chippings which gives it body. After mixing it with water it is poured into wooden moulds where it sets hard. For reinforced concrete, steel rods or mesh are incorporated into the concrete during pouring. This gives the greatly increased strength which is required when building such structures as bridges and tall buildings.

29.3 Production of lime

Lime is manufactured in very large quantities from limestone. Lumps of the rock are packed into a furnace called a limekiln and heated strongly (Investigation 28.3) for several hours. At around 900°C limestone undergoes thermal dissociation into quicklime (calcium oxide) and carbon dioxide:

$$CaCO_3(s) \rightleftharpoons CaO(s) + CO_2(g)$$

This reaction is reversible but, as discussed on 22.11, it can be made to go to completion by ensuring that the carbon dioxide formed is continually removed from the kiln. To achieve this, a good air draught is necessary and this is why vertical kilns are open at the top and have plenty of air entering at the bottom. The process is a continuous one, with lumps of quicklime being removed at the bottom and fresh limestone being added at the top. The more recently constructed kilns consist of a slightly inclined rotating tube which has the advantage that smaller lumps of limestone can be used.

Quicklime has some direct uses, but most of it is treated with water to form calcium hydroxide, in a process known as slaking:

$$CaO(s) + H_2O(l) \rightarrow Ca(OH)_2(s)$$

Water is dropped slowly on to quicklime and the lumps absorb the water, swell, break

Fig. 29.2 A modern rotary limekiln at the Swindon Quarry, Yorkshire (a) A general view showing a conveyor belt delivering the limestone to the preheaters, the horizontal kilns, the lime storage silos, and the road and rail links for distribution of the lime.

(b) A close-up of the rotary kilns. Limestone is heated in the tubes, which are rotated to ensure uniform heating. (Both courtesy Tilling Construction Services Ltd. Photos: Turners (Photography) Ltd)

apart and crumble to a white powder. A lot of heat is given out in this reaction. The product, calcium hydroxide, is known as slaked lime or hydrated lime. If an excess of water is used in the slaking process the product is a suspension of calcium hydroxide in water (since it is not very soluble) known as milk of lime.

Direct uses of lime

Firstly, since lime is the cheapest available alkali, it has many industrial uses where the precise identity of the alkali is not important. In particular, many industrial processes leave behind acidic liquid waste which must be disposed of. Obviously it cannot just be dumped in a river because it would destroy the fish and plant life. In many countries, it is the law that acidic waste must be neutralised before allowing it to run into natural waterways, and lime is used for this purpose.

Secondly, lime is used in agriculture for the treatment of 'acid soil', that is, soil in which the pH-value is too low for the essential bacteria to function properly. When you see ploughed fields covered with a sprinkling of white powder it is likely that lime-spreading has recently been carried out.

Other uses include water softening (it removes temporary hardness, 10.23), and the making of bleaching powder (25.11).

29.4 Salt as a raw material

As everyone knows, common salt (sodium chloride) is present in vast quantities in sea-water, but it also occurs in huge underground deposits as the solid mineral called halite or rock salt. These deposits were formed many thousands of years ago by the evaporation of ancient oceans, and are our main source of sodium chloride for industrial purposes. The major deposits of salt in the United Kingdom are underneath the Cheshire Plain, and salt is obtained from these deposits by drilling bore-holes, down which very hot water under pressure is injected. This water dissolves the underground salt and returns to the surface as brine (aqueous solution of salt), from which sodium chloride, if required as a solid, is obtained by evaporation. For the production of sodium carbonate, as described in the next section, the brine is used directly, after purification.

Fig 29.3 The effects of subsidence due to salt mining in Cheshire during the late 19th century. (Courtesy ICI Ltd Mond Division)

29.5 Production of sodium carbonate; the Solvay Process

Sodium carbonate is one of the major industrial chemicals as it is used in the manufacture of many other materials. The world production of sodium carbonate is now over 26 million tonnes per year.

In 1861 in Belgium, Ernest Solvay patented the principle of his ammonia-soda process for making sodium carbonate. In 1872 he made an agreement with Brunner and Mond, in the United Kingdom, to start a factory, and they decided upon

Winnington, in Cheshire, as a suitable site because it is close to the salt mines. The factory was built and, by the end of the century, 90% of the sodium carbonate in the United Kingdom was made by the Solvay Process.

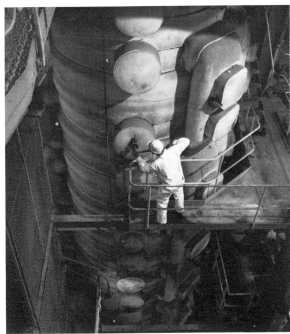

Fig. 29.4 The Solvay Process for converting sodium chloride to sodium carbonate
(a) Tall buildings are needed to house the Solvay Towers

(b) A view inside one of the buildings, showing a tower. (Both photos courtesy ICI Ltd Mond Division)

The overall change in the Solvay Process is that the ions in calcium carbonate ($CaCO_3$) and sodium chloride ($NaCl$) change places, giving sodium carbonate and calcium chloride. It is no good simply mixing limestone and brine because calcium carbonate is insoluble in water and no chemical change occurs. The Solvay Process achieves the required change by an ingenious roundabout route in which brine is treated with carbon dioxide in the presence of ammonia. The ammonia is recovered at a later stage of the process and recycled. In this sense it is not a raw material, but an intermediary. The process is a continuous one, as shown by the highly simplified flow-diagram in Fig. 29.5.

The various steps are as follows, the numbers referring to the flow-diagram.

(1) Ammonia gas is absorbed in concentrated brine to give a solution containing both sodium chloride and ammonia.

(2) Limestone is heated strongly (29.3) to give quicklime and carbon dioxide:

$$CaCO_3(s) \rightarrow CaO(s) + CO_2(g)$$

(3) This is the key stage in the process. The ammoniated brine from step (1) is passed down through the Solvay Tower, Fig. 29.6, while carbon dioxide from steps (2) and (5) is passed up it. The Solvay Tower is tall and contains a set of mushroom-shaped baffles to slow down and break up the liquid flow so that the carbon dioxide can be efficiently absorbed by the solution. Carbon dioxide, on dissolving, reacts with the dissolved ammonia to form ammonium hydrogencarbonate:

$$NH_3(aq) + H_2O(l) + CO_2(g) \rightarrow NH_4HCO_3(aq)$$

313

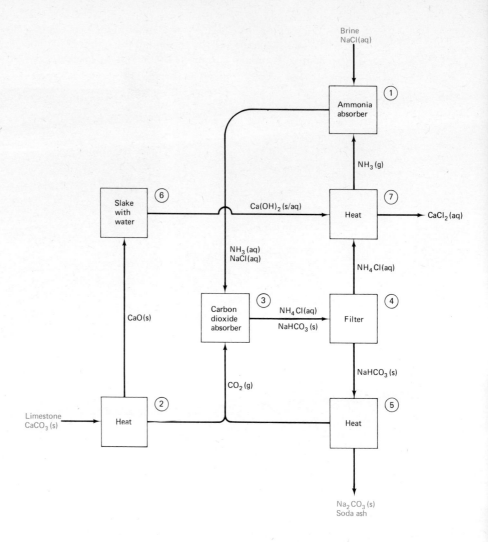

Fig. 29.5 Flow diagram of the Solvay Process

The solution now contains the ions Na^+, Cl^-, NH_4^+ and HCO_3^-. Of the four substances which could, in theory, be formed by different combinations of these ions, sodium hydrogencarbonate ($NaHCO_3$) is the least soluble and it precipitates as a solid in the lower part of the tower, which is cooled. The net process in the Solvay Tower is:

$$NaCl(aq) + NH_3(aq) + H_2O(l) + CO_2(g) \rightarrow NaHCO_3(s) + NH_4Cl(aq)$$

A suspension of solid sodium hydrogencarbonate in a solution of ammonium chloride is run out of the base of the tower.

(4) The suspension is filtered to separate the solid sodium hydrogencarbonate from the ammonium chloride solution, which is then used in stage (7).

(5) The sodium hydrogencarbonate is heated, so that it decomposes to sodium carbonate, water and carbon dioxide:

$$2NaHCO_3(s) \rightarrow Na_2CO_3(s) + H_2O(g) + CO_2(g)$$

The carbon dioxide is sent back to the Solvay Tower for use in step (3). The product of the process, anhydrous sodium carbonate, is obtained as a fine white powder known as soda ash.

The remaining two steps are concerned with recycling the ammonia.

314

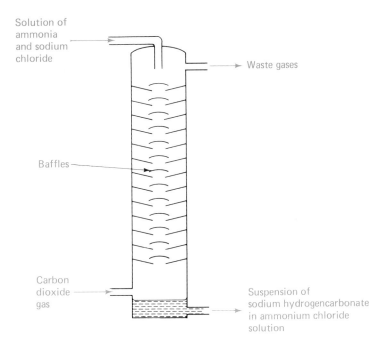

Fig. 29.6 The Solvay tower (stage 3 in the flow diagram, Fig. 29.5)

Solution of ammonia and sodium chloride

Waste gases

Baffles

Carbon dioxide gas

Suspension of sodium hydrogencarbonate in ammonium chloride solution

(6) The quicklime from step (2) is slaked with excess water giving milk of lime:

$$CaO(s) + H_2O(l) \rightarrow Ca(OH)_2(aq/s)$$

(7) This calcium hydroxide suspension is mixed with the ammonium chloride solution from step (4) and heated:

$$2NH_4Cl(aq) + Ca(OH)_2(aq/s) \rightarrow CaCl_2(aq) + 2NH_3(g) + 2H_2O(l)$$

The ammonia is thus recovered, and sent back to step (1). Calcium chloride is the only by-product of the whole process.

The overall process is an elegant one. In theory, the only raw materials are limestone and brine. Inevitably, in practice there are slight losses of ammonia, and these losses are made up for by addition of extra supplies, as required, in step (1).

The simplified flow-diagram implies that the entire output of a Solvay plant (which may be 1000 tonnes per day, or about 1 tonne per minute) is in the form of anhydrous sodium carbonate. This is not quite the case. Although the bulk of the output is sold to other industries in this form, other products of the plant are refined sodium hydrogencarbonate (very pure $NaHCO_3$) and hydrated sodium carbonate ($Na_2CO_3.10H_2O$). The latter, known as washing soda, is obtained by crystallization from an aqueous solution of sodium carbonate.

Direct uses of sodium carbonate and sodium hydrogencarbonate

The largest single use of sodium carbonate is in the manufacture of glass, which is described separately in the next section. It is also used in literally hundreds of other industries, including the manufacture of chemicals and drugs, the manufacture of soap and detergents, the dye industry, the food and drink industry, oil and gas refineries and for water softening both in the home and in laundries.

Refined sodium hydrogencarbonate (bicarbonate of soda), which is one of the purest industrial chemicals known, is of great importance in the food industry, being used in baking powder (28.17), self-raising flour, and in effervescent salts used in the treatment of indigestion.

Glass is one of the most important everyday materials of modern society, and it is manufactured on a vast scale. The raw materials for making it are sand (silica, SiO_2), ground limestone ($CaCO_3$) and sodium carbonate (soda ash, Na_2CO_3). Sodium sulphate may be used as an alternative to sodium carbonate, but the product is the same. When the mixture of substances is fused by heating to a high temperature, the following changes occur:

$$Na_2CO_3(s) + SiO_2(s) \rightarrow Na_2SiO_3(l) + CO_2(g)$$
$$CaCO_3(s) + SiO_2(s) \rightarrow CaSiO_3(l) + CO_2(g)$$

Fig. 29.7 A special type of glass: a Rover 3500 Triplex Ten Twenty laminated windscreen being inspected on the 'head impact' test rig. (Courtesy Triplex Safety Glass Co Ltd)

Fig. 29.8 Recycling glass bottles (a) The special containers used in the Glass Manufacturers Federation glass recycling scheme have separate compartments for green, brown and clear glass as mixed colours cannot be used in recycling. (b) The bottles are stored in separate bays before being transported by road to the glass works. (Both courtesy Welbeck Public Relations, London, on behalf of Glass Manufacturers Federation. Photos: Hills Harris (Oxford) Ltd)

The final product, glass, is a mixture of sodium silicate and calcium silicate with unchanged silica. This is the familiar soda glass used for bottles and ordinary windows. Nowadays many other varieties of glass are manufactured by including additional or alternative raw materials to give the glass special properties such as colour, greater physical strength, and better resistance to heat.

From the point of view of the physical scientist, glass is a peculiar substance. Although most of us would, quite naturally, regard glass as a solid, it does not have a particular crystalline structure, a constant composition, or a sharp melting point. As its temperature rises, it softens gradually before becoming liquid, a characteristic which is of vital importance in the industrial and laboratory technique of glass-blowing, that is, shaping glass while it is softened by heat. Scientists regard glass as a supercooled liquid, since there is no clear transition between the molten state and the (apparently) solid state.

Castner-Kellner Process

The original Castner-Kellner plant in U.K. was established in 1896 as a new way of making high purity sodium hydroxide. Until relatively recently sodium hydroxide was the primary product, but now chlorine is the primary product. The basic reason for the vastly increased demand for chlorine has been the development of chlorinated hydrocarbons for use as solvents and for making polymers such as PVC (31.10). British production of chlorine is around 1 million tonnes per year, mainly by the Castner-Kellner Process, and since this process produces roughly equal masses of chlorine and sodium hydroxide the figure for the latter product is similarly large.

The Castner-Kellner cell is shown diagrammatically in Fig. 29.9, and your first reaction might be to wonder why it needs to be so complicated. Consider what would happen if two electrodes were simply placed in sodium chloride solution and connected to a supply of electricity (19.9). Initially, chlorine and hydrogen would be liberated at the anode and cathode respectively, with the solute turning gradually from sodium chloride to sodium hydroxide. As the proportion of sodium hydroxide in the solution increased, two things would happen. Firstly, the chlorine liberated at the anode would become contaminated with oxygen arising from the discharge of hydroxide ions. Secondly, the chlorine would start to react with the sodium hydroxide

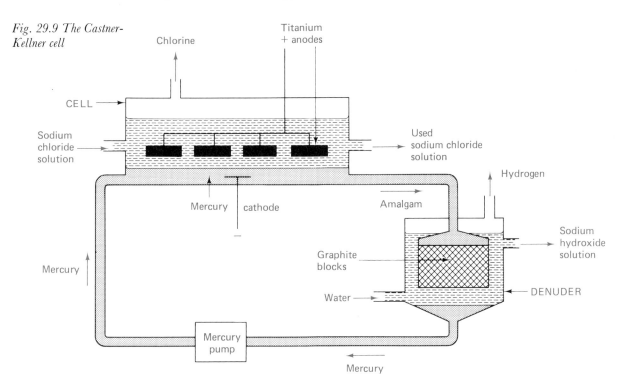

Fig. 29.9 The Castner-Kellner cell

to produce sodium hypochlorite and possibly also sodium chlorate(V) (25.11). To put it bluntly, we would end up with a real mess! In order to produce chlorine and sodium hydroxide as pure products some means of separating them must be devised. The ingenuity of the Castner-Kellner Process lies in the fact that these products are formed in separate parts of the cell, so that the problems outlined above do not arise. This is achieved by using a **flowing mercury cathode**.

Referring to the diagram, in the cell the electrolyte is saturated brine (about 25% by mass of sodium chloride). The anode is a set of titanium plates, at which chlorine is liberated:

$$2Cl^-(aq) - 2e^- \rightarrow Cl_2(g)$$

The chlorine gas is taken away through pipes, washed, dried, liquefied, and transported in cylinders or large tanks to the various industrial customers.

The cathode in the cell is a stream of flowing mercury. Normally during the electrolysis of a solution of a sodium salt, hydrogen will be liberated at the cathode since hydrogen ions are more easily discharged than sodium ions, but with a mercury cathode the situation is reversed. Hydrogen ions are not readily discharged at a mercury surface, but sodium ions are more readily discharged and the sodium metal liberated dissolves in the liquid mercury to form an amalgam (a liquid alloy):

$$Na^+(aq) + e^- \rightarrow Na(amalgam)$$

The mercury, containing the dissolved sodium, flows into the lower chamber (called the denuder), where it meets pure water flowing in the opposite direction. Hydrogen is released, and sodium goes into solution as sodium hydroxide:

$$2Na(amalgam) + 2H_2O(l) \rightarrow 2NaOH(aq) + H_2(g)$$

The denuder is packed with graphite blocks, hydrogen being readily given off at the graphite surface.

Fig. 29.10 A Castner-Kellner cell room. (Courtesy ICI Ltd Mond Division)

The hydrogen gas is taken away through a pipe, compressed, and transported in cylinders to customers. The out-flow from the denuder is sodium hydroxide solution, which is then concentrated by evaporation. Some of it is sold as solution, and some is evaporated to dryness to yield the solid which is sold as flakes, pellets or sticks.

The Castner-Kellner cell operates, typically, at about 400 000 amps at 4·3 volts, that is, about 1700 kW. Since a works may have over 100 of the cells operating at the same time, it is easy to see why it needs its own power station to supply the electricity.

Most electrolysis of sodium chloride solution has been carried out using mercury cells such as the one described above, in spite of the high capital cost of mercury. This is because the cell is very efficient and gives products of very high purity. However, a factor about which people have become more and more concerned over the last twenty years or so is the health hazard associated with the use of mercury. Although the concentration of mercury vapour in the factories is kept under strict control, it is difficult to avoid some slight loss of mercury.

In Japan in the 1950s there were a number of cases of people suffering from mercury poisoning as a result of having eaten fish which had been contaminated. It has been realised that mercury effluent can become incorporated, as organic compounds, in the tissues of marine animals and plants and so pollute the food chain. The long-term danger of this is that some of the forms of marine life which are lower in the food chain may become less abundant and hence there would be less food available for those higher in the food chain.

In the 1960s and 1970s investigations into the mercury content of fish caught in the coastal waters of Sweden, in India and in some the the Great Lakes in America showed that the mercury levels were higher than normal. As a result the laws governing the discharge of mercury by factories are now very strict. It is likely, therefore, that no new mercury cell plants will be built and in Japan, for example, their construction has been forbidden by law.

The Gibbs diaphragm cell, which is described below, offers a different approach to the problem of keeping the products of the electrolysis of sodium chloride solution separate. Although in the past the process, which does not use mercury, has taken second place to the mercury cell, it is now coming into favour.

The Gibbs diaphragm cell process

The cell is shown, in a simplified, modern form, in Fig. 29.11. There are now many variations in design, although the main principles of operation are the same. All the cathode compartments are interconnected, and each cathode consists of a steel mesh box with asbestos fibres deposited in the mesh. This is the diaphragm which, though porous, keeps the electrolysis products separate. The anodes are made of titanium, and the electrolyte, which is run into the cell via the anode compartment, is concentrated sodium chloride solution (brine).

The ions present in the electrolyte are $Na^+(aq)$, $Cl^-(aq)$, $H^+(aq)$ and $OH^-(aq)$. The $Cl^-(aq)$ ions are discharged at the anodes,

$$2Cl^-(aq) - 2e^- \rightarrow Cl_2(g)$$

and the chlorine gas which is given off is piped away from the top of the anode compartment.

$H^+(aq)$ ions are discharged at the cathodes forming hydrogen gas,

$$2H^+(aq) + 2e^- \rightarrow H_2(g)$$

which is piped away from the cathode compartments.

The removal of $H^+(aq)$ ions from the electrolyte in the cathode compartments disturbs the equilibrium (22.14),

$$H_2O(l) \rightleftharpoons H^+(aq) + OH^-(aq)$$

and causes more water to ionise and produce an excess of $OH^-(aq)$ ions. At the same time $Na^+(aq)$ ions, which are attracted by the cathode, diffuse through the asbestos diaphragm from the anode compartment to the cathode compartments. The net result is that sodium hydroxide solution ($Na^+(aq) + OH^-(aq)$) is being produced in the cathode compartments.

The liquid which is tapped off from the cathode compartments is an aqueous solution containing both sodium hydroxide and unchanged sodium chloride. The liquid is concentrated by evaporation and, since sodium chloride is less soluble than sodium hydroxide, most of the sodium chloride crystallises on cooling, leaving a solution which contains mostly sodium hydroxide. The product is not as pure as that obtained by the Castner-Kellner process but it is good enough for many of the purposes for which sodium hydroxide is used.

Current research is aimed at finding substitutes for asbestos as the diaphragm material, and at making technical improvements which will increase the purity of the product.

319

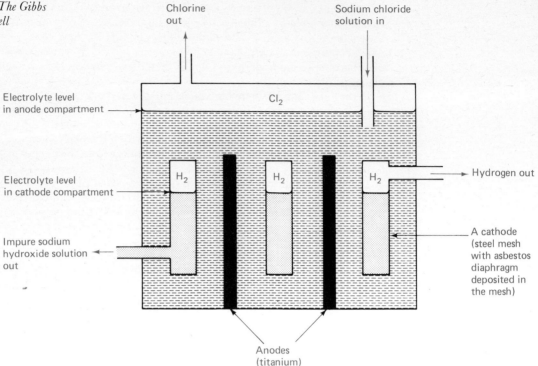

Fig. 29.11 The Gibbs diaphragm cell

Chlorine out

Sodium chloride solution in

Electrolyte level in anode compartment

Cl₂

Electrolyte level in cathode compartment

H₂ H₂ H₂

Hydrogen out

Impure sodium hydroxide solution out

A cathode (steel mesh with asbestos diaphragm deposited in the mesh)

Anodes (titanium)

Uses of sodium hydroxide

The uses of sodium hydroxide are as numerous as the uses of sodium carbonate. A few of the major ones are noted here. It is used in large quantities in the rayon (synthetic fibre) industry, the manufacture of soap and detergents (32.10), the paper industry, the purification of bauxite for the extraction of aluminium (24.8), and the dye industry. It is also used for the extraction of organic chemicals from coal-tar, and in the chemical industry generally as a strong alkali.

Production of sodium hypochlorite and sodium chlorate

If the chlorine and the sodium hydroxide from the electrolysis of sodium chloride solution are allowed to mix in a controlled manner, either sodium hypochlorite (NaClO) or sodium chlorate(V) (NaClO₃), depending on the conditions, may be produced (25.11). These are both important substances, the former being a bleaching and sterilising agent, the latter being a weed-killer.

Other uses of chlorine and those of hydrogen are discussed on 25.12 and 18.2 respectively.

29.8 Summary

1. Limestone ($CaCO_3$) is an important raw material for the alkali industry since on heating it decomposes to quicklime (CaO) which on addition to water forms calcium hydroxide ($Ca(OH)_2$). This substance is important as a cheap alkali.
2. Sodium carbonate is produced in large quantity from limestone and salt. It is a very important industrial chemical, and vital to the manufacture of glass.
3. Sodium hydroxide is produced in large quantity, along with chlorine, by the electrolysis of brine. It is also a very important industrial chemical, for example in the soap and detergent industry.

Chapter 30

Organic chemistry

30.1
What is organic chemistry?

You are probably familiar with the game 'animal, vegetable, mineral'. A living thing, whether it be of animal form or plant form, is known as an organism, and the material from which living organisms are made is known as organic matter. Historically, substances of animal or vegetable origin came to be called **organic**, while substances of mineral (non-living) origin came to be called **inorganic**. Most of the chemistry in this book deals with inorganic compounds; for example, common salt is a mineral compound and substances derived from it, such as hydrochloric acid and sodium hydroxide, are regarded as inorganic. Sugar, on the other hand, is a typical organic compound, being obtained from plants rather than mineral sources. So also is ethanol (alcohol), being obtained by the fermentation of sugar.

In the first half of the nineteenth century it was widely believed that organic compounds could only be formed by natural processes (involving a life force or vital force) and could not be made in the laboratory. This was shown to be untrue when,

Fig. 30.1 Man-made fibres and plastics are important products of the organic chemical industry. (Left) The GQ Inflated Aerofoil Parachute ('The Unit'). The canopy is made of nylon and the rigging from polyester. (Courtesy G.Q. Parachutes Ltd) (Right) Except for wooden boom being held by the surfer, the whole of this windsurfer is made from plastics and man-made fibres. (A Shell Photograph)

firstly, urea (present in urine) was made from a 'mineral' compound and, secondly, when ethanoic (acetic) acid, the acid present in vinegar (obtained by the oxidation of wine by air), was synthesised from its elements.

It was realised long ago that organic compounds have, almost without exception, something in common, namely, the presence of the element carbon. Indeed, carbon is unique in the huge number of compounds which it forms. Well over 2 million are now known, and in fact it is likely that somewhere in the world another new compound is being made or discovered during the time it takes you to read this page! In modern terminology organic chemistry means simply the **study of carbon compounds**, with the exception of compounds like carbon dioxide and metal carbonates which are traditionally considered as inorganic, since they are of mineral origin.

The importance of organic chemistry is twofold. Firstly, it is concerned with the substances of which living matter is composed—and this includes you, the reader. Organic chemistry is therefore vital to the study of agriculture, nutrition, medicine, and health in general. Secondly, organic chemistry is concerned with synthesising substances which may be useful, not only in affecting our health (drugs etc.), but also in replacing, or even improving upon, natural building materials. Thus, for example, wood is still an extremely important structural material, but supplies of wood are limited. Many of the functions served by wood can be served equally well, or better, by plastic materials made artificially. Furthermore, plastic materials are now manufactured with properties which wood and paper cannot possibly supply. Think, for example, of polythene film, or Lego bricks, or Airfix kits, or Perspex glass. The development of plastics has opened up a whole new world of materials, and there is no obvious limit to the variety of properties which may be achieved.

The scale of the modern organic chemical industry is so great that it affects almost every aspect of our lives. The products of this industry include drugs, plastics, synthetic fibres for clothes, carpets, upholstery, rope, adhesives, solvents, fuels, lubricants, detergents, cosmetics, pesticides, paints, synthetic rubber for tyres, explosives, floor coverings, ceiling tiles, furniture stuffing, perfumes, dyes, food flavourings and preservatives, sports equipment . . . the list is almost endless. Furthermore, the organic chemical industry is growing year by year, as more uses for synthetic organic compounds are devised. There can hardly be a single moment of your life when you are not holding, looking at, wearing, sitting on, eating, or sleeping in some product of this industry.

30.2 General characteristics of organic compounds

A large proportion of inorganic compounds fall into the general categories of acids, bases and salts. Such compounds are very often ionic in character, and very often contain one or more metallic elements. A large proportion of organic compounds contain no metallic element and are covalent, consisting of separate molecules. They are composed of the elements carbon and hydrogen, very often oxygen, sometimes nitrogen, and occasionally other elements such as phosphorus, sulphur, halogens and even metals, such as iron. The main common element, however, is carbon, and here two particular properties of carbon are important. Firstly, carbon in its compounds almost invariably forms covalent bonds with other elements. Secondly, carbon readily forms chains with other carbon atoms. The carbon atoms are joined to each other by covalent bonds. This is called catenation which means, literally, chain-forming.

30.3 Hydrocarbons

These characteristics of carbon are illustrated by considering, first of all, the compounds known as hydrocarbons. A hydrocarbon is a compound of carbon and hydrogen only, and the simplest type is a saturated hydrocarbon, or alkane. The simplest alkane is methane, CH_4, which is familiar to us as natural gas. It provides the flame for our Bunsen burners and the heat of our kitchen stove (and perhaps our

322

central heating as well). Its molecular structure may be shown as

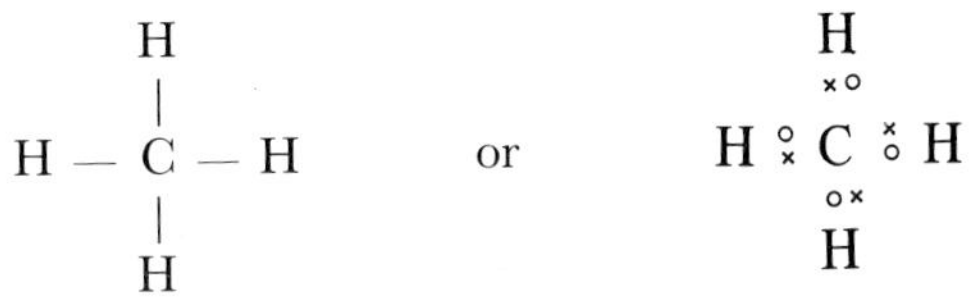

$$H - \overset{\displaystyle H}{\underset{\displaystyle H}{C}} - H \qquad \text{or} \qquad H \overset{\displaystyle H}{\underset{\displaystyle H}{\overset{\times\circ}{\underset{\circ\times}{C}}}} H$$

with carbon forming four covalent bonds to hydrogen, each bond being a pair of shared electrons. It is important to note, however, that the shape implied by a 2-dimensional picture is wrong. The molecule is not planar (flat) with bonds at 90° angles. It is 3-dimensional, with the hydrogen atoms arranged around the carbon atom in such a way that the angles between the bonds are as great as possible. In this arrangement the four bonds are directed towards the corners of a **regular tetrahedron** (15.6):

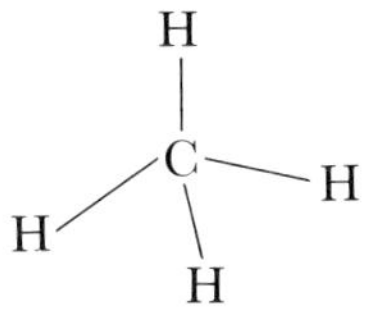

$$H - \overset{\displaystyle H}{\underset{\displaystyle H}{C}} \overset{}{\underset{}{}} H$$

Thus the angles between the bonds are about 109°, not 90°.

The next simplest alkane is the gas ethane, C_2H_6, whose molecular structure may be shown 2-dimensionally as

$$H - \overset{\displaystyle H}{\underset{\displaystyle H}{C}} - \overset{\displaystyle H}{\underset{\displaystyle H}{C}} - H$$

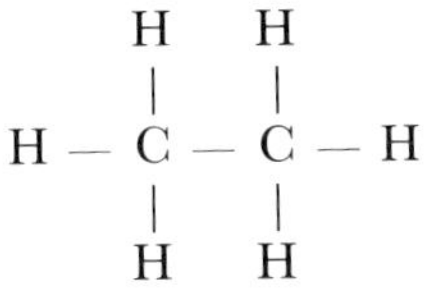

with each carbon again forming four covalent bonds, though one of these bonds is used to form the chain of two carbon atoms. Note that in the picture as drawn above it appears as if the two 'end' hydrogen atoms are different from the four 'side' atoms. This is not so, as you will see if you make a model of the molecule with tetrahedral angles between the four bonds of each carbon atom.

The next member of the alkane series is the gas propane, C_3H_8, shown 2-dimensionally as

$$H - \overset{\displaystyle H}{\underset{\displaystyle H}{C}} - \overset{\displaystyle H}{\underset{\displaystyle H}{C}} - \overset{\displaystyle H}{\underset{\displaystyle H}{C}} - H$$

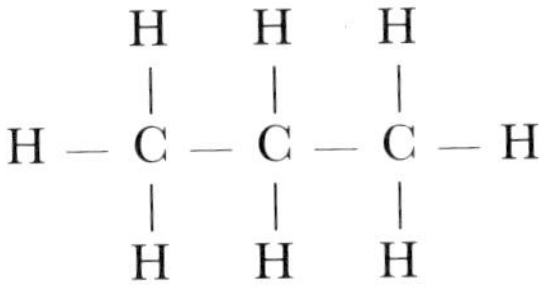

Again each carbon atom forms four bonds, but now we have a chain of three carbon atoms.

30.4 Homologous series

You may have realised by now that the molecular formulae of alkanes follow a general pattern. If the number of carbon atoms is n then the number of hydrogen atoms is 2n + 2 (Table 30.1).

This is known as a **homologous series**, that is, a series of compounds of similar properties whose molecular formulae differ by 'CH_2' between adjacent members of the series. Different members of the series are called **homologues** of one another. The general formula of the homologous series of alkanes is C_nH_{2n+2}.

TABLE 30.1. *The alkane series*

n	2n + 2	MOLECULAR FORMULA	NAME
1	4	CH_4	methane
2	6	C_2H_6	ethane
3	8	C_3H_8	propane
4	10	C_4H_{10}	butane
5	12	C_5H_{12}	pentane
6	14	C_6H_{14}	hexane
7	16	C_7H_{16}	heptane
8	18	C_8H_{18}	octane
9	20	C_9H_{20}	nonane
10	22	$C_{10}H_{22}$	decane

30.5 Isomerism

The fourth member of the alkane series is butane, C_4H_{10}. This is sold, as liquid under pressure, as Buta-gas. The interesting point is that it is possible to separate two different compounds with the molecular formula C_4H_{10}. They are both gases, with the same density, but they have different boiling points ($-0.5\,°C$ and $-11.7\,°C$). How can it be that two distinct substances with the same molecular formula may exist? The answer lies in the way in which the atoms are bonded together within the molecule. For the formula C_4H_{10} there are two ways in which the molecule can be put together, and these may be shown as,

$$
\begin{array}{cccc}
H & H & H & H \\
| & | & | & | \\
H-C-C-C-C-H \\
| & | & | & | \\
H & H & H & H
\end{array}
\quad \text{and} \quad
\begin{array}{ccc}
H & H & H \\
| & | & | \\
H-C-C-C-H \\
| & | & | \\
H & H-C-H & H \\
& | & \\
& H &
\end{array}
$$

In the first case we have a straight chain of four carbon atoms, while in the second we have a branched chain in which one carbon atom is bonded directly to the other three. These two compounds are called isomers of one another;

isomers are substances with the same molecular formula but different structural formulae.

Another example of isomerism is given by the compounds of molecular formula C_2H_6O. The two ways of putting the molecule together are shown in Table 30.2, with a comparison of the two compounds represented.

TABLE 30.2. *Isomers with the molecular formula C_2H_6O*

$$\begin{array}{cc} H & H \\ \| & \| \\ H-C-C-O-H \\ \| & \| \\ H & H \end{array}$$	$$\begin{array}{cc} H & H \\ \| & \| \\ H-C-O-C-H \\ \| & \| \\ H & H \end{array}$$
ETHANOL	METHOXYMETHANE
Liquid b.p. $78.5\,°C$	Gas b.p. $-24.8\,°C$
Flammable	Flammable
Miscible with water	Immiscible with water
Reacts with aqueous potassium manganate(VII)	
Reacts with metallic sodium	Does not react with any of the reagents mentioned on the left
Reacts with chlorides of phosphorus	

30.6 Functional groups

Ethanol and methoxymethane, although they have the same molecular formula, have different properties, and this is because their molecules contain different **functional groups**. Ethanol contains the functional group

$$-\overset{\displaystyle |}{\underset{\displaystyle |}{C}} - O - H,$$

which is characteristic of alcohols, while methoxymethane contains the functional group

$$-\overset{\displaystyle |}{\underset{\displaystyle |}{C}} - O - \overset{\displaystyle |}{\underset{\displaystyle |}{C}} -,$$

which is characteristic of ethers. Thus, for example, methanol (methyl alcohol) has the structure,

$$H - \overset{\displaystyle H}{\underset{\displaystyle H}{C}} - O - H$$

which contains the same functional group

$$-\overset{\displaystyle |}{\underset{\displaystyle |}{C}} - O - H$$

as ethanol. Because of this, methanol and ethanol have very similar chemical properties. All the statements under ethanol in the table above apply also to methanol, except that the boiling point is lower $(64.5\,°C)$. Similarly, ethoxyethane (diethyl ether),

$$H - \overset{\displaystyle H}{\underset{\displaystyle H}{C}} - \overset{\displaystyle H}{\underset{\displaystyle H}{C}} - O - \overset{\displaystyle H}{\underset{\displaystyle H}{C}} - \overset{\displaystyle H}{\underset{\displaystyle H}{C}} - H$$

has very similar properties to methoxymethane, except that the boiling point is higher $(34.5\,°C)$.

30.7 Alkyl groups

We can now distinguish between functional groups and **alkyl groups**. Table 30.3 gives some examples of alkyl groups.

TABLE 30.3. *The formulae and names of the three simplest alkyl groups*

$H - \overset{\displaystyle H}{\underset{\displaystyle H}{C}} -$ or	$CH_3 -$	methyl
$H - \overset{\displaystyle H}{\underset{\displaystyle H}{C}} - \overset{\displaystyle H}{\underset{\displaystyle H}{C}} -$ or	$C_2H_5 -$	ethyl

325

$$
\begin{array}{ccccc}
& \text{H} & \text{H} & \text{H} & \\
& | & | & | & \\
\text{H} - \text{C} & - & \text{C} & - & \text{C} & - \\
& | & | & | & \\
& \text{H} & \text{H} & \text{H} &
\end{array}
\qquad \text{or} \qquad C_3H_7 - \qquad \text{propyl}
$$

The chemical properties of a compound are determined mainly by its functional groups rather than its alkyl groups. However, the size of the alkyl group affects the relative molecular mass of the compound, and hence its physical properties.

A convention which is sometimes used when referring to the general properties of a class of compounds (e.g. alcohols) is to use the letter R to stand for an alkyl group of unspecified size. Thus alcohols are compounds of the type R – OH where R may be methyl (CH_3-) or ethyl (C_2H_5-) etc.

30.8
Variety of
compounds

It was mentioned earlier in this chapter that over 2 million carbon compounds are known, and you may now be starting to see how it is possible for there to be so many. For a start, we have the tendency of carbon to form bonds with itself, so that alkyl groups of widely varying size are formed. Then we have isomerism, so that alkyl groups of the same overall size can have different patterns of chain-branching; also, many compounds contain rings of atoms as well as chains. Then we have a wide variety of functional groups which may be incorporated into molecules. Furthermore, the same functional groups may be incorporated at different positions in the molecule, giving another source of isomerism. In addition to all this, a single organic molecule may contain any number of different functional groups. It seems reasonable to conclude that the number of possible carbon compounds is really without limit.

You may like to try to prove, as an exercise, that the relatively 'small' formula $C_4H_{10}O$ could represent any one of seven different compounds, three of which are ethers and four are alcohols. You may also like to try to show that there are no less than eighteen isomers of the hydrocarbon octane, C_8H_{18}!

30.9
Nomenclature

With such an enormous variety of organic compounds, it is essential to have an agreed system for naming them. As a result of work done by the International Union of Pure and Applied Chemistry (IUPAC) in recent years, a system has been worked out and has now been generally adopted. However, many of the older names are still around and so in this book we use the modern name but put the older name after it in brackets, if it is still in common use; e.g. ethanoic acid (acetic acid). It is neither possible nor necessary in this book to explain nomenclature in detail, but it will be helpful to point out the basic principles.

Names for several of the simple alkanes have already been given (30.4). These names form the basis of the modern system of nomenclature, and the system is best introduced by considering the naming of halogenated alkanes which are alkanes in which one or more hydrogen atoms are replaced by atoms of a halogen, such as chlorine. The name for the substance,

$$
\begin{array}{c}
\text{H} \\
| \\
\text{H} - \text{C} - \text{Cl} \\
| \\
\text{H}
\end{array}
\qquad \text{or} \qquad CH_3Cl
$$

is chloromethane. The substance,

$$
\begin{array}{c}
\quad\quad H \\
\quad\quad | \\
H - C - Cl \quad\text{or}\quad CH_2Cl_2 \\
\quad\quad | \\
\quad\quad Cl
\end{array}
$$

in which two chlorine atoms have taken the place of hydrogen atoms in the methane molecule, is called dichloromethane. Similarly, $CHCl_3$ is trichloromethane and CCl_4 is tetrachloromethane.

If two atoms of hydrogen in an ethane molecule are replaced by chlorine atoms, the resulting compound is $C_2H_4Cl_2$. The name dichloroethane does not distinguish between the two isomers whose structural formulae are

$$
\begin{array}{cc}
H \quad H & H \quad H \\
| \quad\; | & | \quad\; | \\
H - C - C - Cl & \text{and} \quad H - C - C - H \\
| \quad\; | & | \quad\; | \\
H \quad Cl & Cl \quad Cl
\end{array}
$$

In the first case both chlorine atoms are bonded to the same carbon atom, while in the second case the two chlorine atoms are bonded to different carbon atoms. To resolve the difficulty we regard the two carbon atoms as being numbered *1* and *2*. The name for the first substance then becomes 1,1-dichloroethane while the second becomes 1,2-dichloroethane. (We could also call the first one 2,2-dichloroethane, but the rule is that we number the carbon atoms in such a way that the smallest possible numbers appear in the name.)

In the case of alcohols, the older names for the substances,

$$
\begin{array}{cc}
H & H \quad H \\
| & | \quad\; | \\
H - C - O - H & \text{and} \quad H - C - C - O - H \\
| & | \quad\; | \\
H & H \quad H
\end{array}
$$

$$
\text{or } CH_3OH \quad\quad\quad \text{or } CH_3CH_2OH \text{ or } C_2H_5OH
$$

were methyl alcohol and ethyl alcohol respectively. In the modern system these names are shortened to methanol and ethanol. But the name propanol does not distinguish between the isomers,

$$
\begin{array}{cc}
H \quad H \quad H & H \quad H \quad H \\
| \quad\; | \quad\; | & | \quad\; | \quad\; | \\
H - C - C - C - OH & \text{and} \quad H - C - C - C - H \\
| \quad\; | \quad\; | & | \quad\; | \quad\; | \\
H \quad H \quad H & H \quad OH \quad H
\end{array}
$$

These compounds are called propan-1-ol and propan-2-ol respectively.

30.10 Summary

1. Organic chemistry is the chemistry of carbon compounds, with the exception of carbonate salts. Most organic compounds are molecular covalent substances. A huge number of carbon compounds exist, because of isomerism and carbon's tendency to form chains of atoms.

2. The simplest organic compounds are hydrocarbons, which contain only hydrogen and carbon. In other types of compounds the third element present is often oxygen. The chemical properties of a compound depend mainly on what functional group is present in the molecule.

Chapter 31

Sources of organic compounds: hydrocarbons

Investigation 31.1

How do the properties of an alkane and an alkene differ?

You will need a liquid alkane such as hexane and a liquid alkene such as cyclohexene.

1. Put 5 drops each of the alkane and alkene separately into dry crucibles or porcelain basins. Try to set fire to both liquids using a lighted taper.

Questions

1 Are these hydrocarbons readily flammable?
2 How do the appearances of the two flames differ?
3 What will be their products of combustion?

2. Put 2 cm³ of bromine water into each of two test-tubes. Add 2 drops of the alkane to one test-tube and 2 drops of the alkene to the other. Shake both test-tubes and note what happens in each case.

Questions

4 Which of the hydrocarbons reacts with bromine under these conditions?
5 How can you tell that a reaction has occurred?
6 Why is it necessary to shake the test-tubes?

3. Into each of two test-tubes put 1 cm³ of potassium manganate(VII) solution and 1 cm³ of dilute sulphuric acid. Now add 2 drops each of the alkane and alkene separately. Shake the test-tubes and note what happens in each case.

Questions

7 Which of the hydrocarbons reacts with manganate(VII) ions under these conditions?
8 How can you tell that a reaction has occurred?
9 What type of reagent is acidified potassium manganate(VII) solution and therefore what has happened to the hydrocarbon in this reaction?

Investigation 31.2

How does a bromoalkane differ from a metal bromide?

1. Shake a small crystal of sodium (or potassium) bromide with about 2 cm³ of water in a test-tube and note what happens. Then add a few drops of silver nitrate solution.

Questions

1 What happens when the metal bromide is shaken with water?
2 What happens when silver nitrate solution is added, and what does this indicate?

2. Shake about 2 drops of bromopropane with about 1 cm³ of water and note the result. Then add about 1 cm³ of ethanol and shake again. Note any difference which you observe. Now add about 1 cm³ of silver nitrate solution and mix well. Finally, stand the test-tube in fairly hot water for several minutes.

3 Is bromopropane miscible with water?

4 What difference does addition of ethanol make?

5 What is the immediate result of adding silver nitrate solution, and how does it compare with part 1 of this investigation?

6 What does this suggest about the way in which bromine is chemically bonded in bromopropane, as compared to a metal bromide?

7 What slowly happens as the test-tube stands in hot water?

8 How can this be explained in terms of a reaction between bromopropane and water?

31.3 Coal

Fig. 31.1 Coal being cut at Wath Colliery, Yorkshire. (Courtesy The National Coal Board)

Fig. 31.2 A coking plant for converting coal into coke. (Courtesy the National Coal Board)

Coal is an example of a fossil fuel. It was formed over the course of millions of years by the gradual decay and compression of the buried remains of ancient forests, whose trees died and fell long before man appeared on the earth. Coal has been burnt in homes for hundreds of years, but its use as domestic fuel has now declined considerably. There are various reasons for this. For one thing, coal is not the most convenient fuel to transport and store. For another, its price has risen considerably in recent years. But, more importantly, coal burnt in an ordinary domestic grate is not a very efficient fuel. It burns in a rather smoky manner, owing to unburnt carbon escaping as soot. This means, firstly, that a lot of energy is wasted and, secondly, that the atmosphere becomes polluted (9.5).

The situation now is that the use of solid fuel in any form for domestic heating has decreased enormously, owing to the rapid expansion of the petroleum and natural gas industries. Oil and gas are now very widely used both as domestic and industrial fuels, and if they have been refined before use, they burn much more cleanly than coal.

It was discovered near the end of the eighteenth century that if coal is heated strongly in the absence of air an inflammable gas (coal-gas) can be obtained from it.

TABLE 31.1. *The approximate composition by volume of coal-gas*

GASES PRESENT IN COAL GAS	
hydrogen (H_2)	50%
methane (CH_4)	30%
carbon monoxide (CO)	8%
other gases	12%

In the early part of the nineteenth century the coal-gas industry became established, the gas being used originally for lighting. Later, with the development of electric lighting, this use for coal-gas died out, and it was supplied instead for heating and cooking. In the 1960s, as coal became more expensive, the British gas industry started to use oil as an alternative source of domestic gas. Then, around 1970, the British domestic supply was changed over to natural gas (methane from the North Sea), and a major use for coal-gas disappeared almost overnight. With it disappeared the local gas-works, once a familiar sight in almost every town, with its huge gas-holders used for storing the day's supply of gas.

Of course, in the destructive distillation of coal, as it is called, gas is not the only product. A complex liquid mixture known as coal-tar is also obtained, and this can be distilled to yield many important organic compounds. However, with the rapid growth of the petroleum industry the production of chemicals from coal no longer holds the central position that it once did.

Fig. 31.3 A natural gas pipeline which is almost 1 metre in diameter being laid in the North Sea (Both Esso Photographs)

The solid residue from the distillation of coal is coke, which is mainly carbon. Coke burns fairly cleanly, and is an important industrial fuel. It is also very important as a large-scale reducing agent, one of its major applications being the extraction of iron by reduction of iron oxide in the blast furnace (24.9).

**31.4
Natural gas**

Natural gas is very important both as a fuel and as raw material for the organic chemical industry. Like coal it is a fossil fuel, and it is found trapped in 'pockets' in rock deep under the ground. It is located and released by drilling shafts deep into the earth. Sometimes the pockets of gas are beneath the sea bed, and huge drilling rigs have to be built on land and then towed out to sea, where they either float in the water or stand on the sea bottom on long legs.

Natural gas consists largely of methane (CH_4), together with variable amounts—depending on the source—of other small alkanes such as ethane (C_2H_6), propane (C_3H_8) and butane (C_4H_{10}). North Sea gas is almost entirely methane. Some gas fields also yield hydrogen sulphide (H_2S), which is a source of sulphur (27.20), while some provide helium. This gaseous element, of which there is hardly any in the Earth's atmosphere, is important as a safe substitute for hydrogen in balloons. It is not quite as light as hydrogen, but it has the advantage of being completely non-flammable (18.2).

In Britain the domestic gas supply is now provided by the North Sea gas field, and it is hoped that the supply will last for several decades.

31.5 Petroleum

Fig. 31.4 An aerial view of the Esso oil refinery at Fawley. (An Esso Photograph)

Petroleum, or crude oil, is another fossil fuel. Like natural gas it is found trapped in pockets in underground rock, from which it must be obtained by drilling. Oil fields, like gas fields, are sometimes located under the sea bed; indeed, the two very often occur together in the same region.

Crude oil is a complex mixture of hydrocarbons, mainly alkanes, with numbers of carbon atoms in the molecule varying from 1 up to over 100. Because of this it is necessary to separate the mixture into different components by fractional distillation.

Deep sea terminal for receiving crude oil and despatching products

Oil refining and chemical manufacturing plant

Intermediate product storage tanks

Administration block

Crude oil storage tanks

Refinery tree belt

This is a method of separating a mixture of substances with different boiling points, to obtain a number of parts or fractions, 2.9. Each fraction is, in the case of petroleum distillation, not a single substance but a mixture of substances with similar boiling points. Consequently, further separation of each fraction has to be carried out if pure substances are required. It is not necessary to separate every fraction completely into single substances; for example, the parts to be used as fuels can be left as mixtures, provided that the components of the mixture have boiling points within a certain specified range. The whole process of separating crude oil into its different parts is called **refining**, and an oil refinery is often situated on the coast, away from towns. This enables the oil to be delivered easily to the refinery by tanker ships coming from abroad, and also the fire hazard arising from the storage of huge quantities of flammable liquids is minimised.

Fig. 31.5 shows, diagrammatically, the main fractions obtained by the distillation of crude oil, and some of the uses of these different fractions. The picture is, of course, highly simplified.

Fig. 31.5 Distillation of petroleum

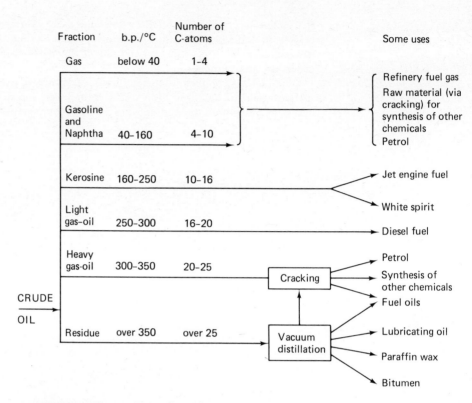

Fraction	b.p./°C	Number of C-atoms	Some uses
Gas	below 40	1–4	Refinery fuel gas / Raw material (via cracking) for synthesis of other chemicals / Petrol
Gasoline and Naphtha	40–160	4–10	
Kerosine	160–250	10–16	Jet engine fuel / White spirit
Light gas-oil	250–300	16–20	Diesel fuel
Heavy gas-oil	300–350	20–25	Cracking → Petrol / Synthesis of other chemicals / Fuel oils
Residue	over 350	over 25	Vacuum distillation → Lubricating oil / Paraffin wax / Bitumen

CRUDE OIL

Fig. 31.6 Refuelling a jet aeroplane from underground tanks with aviation fuel which is an important fraction obtained during the refining of oil. (An Esso Photograph)

The process called **cracking** is very important, and involves the breaking-up of hydrocarbon molecules into smaller ones. Originally this was done simply by heating the hydrocarbon mixture to a high temperature in the absence of air. This is called thermal cracking, and is rather difficult to control. Now lower temperatures are used, the process being aided by the use of a catalyst (23.15). This is called catalytic cracking, or cat-cracking for short. As a result of this process some of the heavy gas-oil fraction, for example, ends up as petrol. One very important substance obtained by cracking the naphtha fraction is ethene (ethylene), C_2H_4, which is a raw material for the manufacture of plastics, ethanol, and tetraethyl lead ('anti-knock' for petrol). More will be said about ethene later in this chapter.

It will be seen, even from this very brief outline, that the petroleum industry involves a lot more than just making petrol for motor cars. Over 80% of all organic chemicals are now obtained from petroleum and natural gas, and the industry is still growing. For example, almost all industrial ethanol is now produced from petroleum, whereas in the past it was obtained by the large-scale fermentation of sugars. Similarly, some of the vinegar now supplied to the catering trade contains ethanoic (acetic) acid produced from petroleum, rather than by the oxidation of wine (32.6).

31.6
Fuels and
energy—some
alternatives

During the first 75 years of this century in the United States, for example, there was a 900% increase in energy production. The world demand for energy will continue to increase, but the future rate of increase is difficult to predict. Clearly the pattern of energy production and use is of vital importance and is likely to remain so.

The relative importance of the various sources of energy is different for different parts of the world and will be influenced by such factors as the local availability of fuels, and access to imported fuels which in turn will depend on economic and political factors. However, there have been recognisable world-wide trends during the last 150 years. Up until the middle of the last century wood was the major source of energy, but as supplies became depleted coal took over as the major source. During the last few decades oil and natural gas have become more important in many parts of the world. As mentioned earlier, these two fuels tend to have lower extraction and transport costs, and they create fewer pollution problems than coal.

Coal, oil and natural gas are all fossil fuels and supplies of them in the Earth cannot last for ever. Figure 31.7 shows the number of years that known world supplies of these fuels can be expected to last at current production rates. It is significant that although oil and natural gas are very popular at the moment, reserves of coal will easily outlast them, and so we are likely to see coal revert to its earlier popularity in the future. The supplies of oil might be made to last longer if it proves possible to extract economically a higher proportion of oil from known wells. At the moment the proportion of the oil which can be extracted from a well, by the natural pressure under which the oil is trapped underground, can be as low as 40%. Nevertheless, sooner or later, the world will have to rely less on these fossil fuels, and already a great deal of thought and effort has been put into the problem. An obvious alternative is, of course, nuclear power (13.7); indeed, nuclear power stations have been running successfully for some years. There is, however, a considerable amount of opposition to the growth of the nuclear power industry because of the possible health hazards from radioactive materials, and especially the problem of disposing of radioactive waste material.

As a result of the widespread reluctance to become involved too hastily in the large-scale use of nuclear energy, various alternatives are under consideration by scientists and engineers. A common feature of these alternatives is that they make use of what one might call natural forces. Thus, three possible large-scale energy sources involve (a) harnessing tidal energy, (b) harnessing wind energy—not exactly a new idea!— and (c) making direct use of solar energy (sunlight). Hydrogen could become the fuel of the future if an economical method could be devised for using natural forces to provide the energy to extract hydrogen from water. The advantage of using hydrogen as a fuel is that the product of burning it is water, and hence the starting material would be continually replaced. At the moment the state of technology associated with

Fig. 31.7

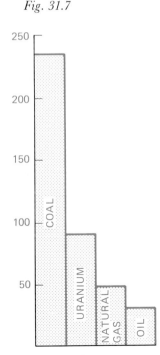

these methods is not sufficiently advanced to make them economically suitable for large-scale use.

Clearly, as our energy requirements increase year by year, the discussion of alternative sources is going to become increasingly important over the next few decades.

31.7 Properties of hydrocarbons— a comparison

There are three main series of hydrocarbons—the alkanes, the alkenes and the alkynes. As you might guess from their names, they are closely related.

The members of all three series are similar in that they burn readily in air, but the flames produced become more sooty in the order, alkanes, alkenes, alkynes. This last observation can be explained by the proportion of carbon in the compounds increasing in the same order. When the formulae of corresponding members of each series are examined this is found to be the case. For example, the compounds in each series with two carbon atoms per molecule have the formulae, C_2H_6 (alkane), C_2H_4 (alkene) and C_2H_2 (alkyne).

If the atoms in these molecules are forming their usual number of bonds, their structural formulae must be:

ethane   ethene   ethyne

The alkane with two carbon atoms per molecule is called ethane. The corresponding alkene is therefore called ethene and the alkyne is called ethyne. Previously these last two compounds have been called by the less systematic names, ethylene and acetylene.

The alkane with three carbon atoms is called propane and the corresponding alkene is propene and the alkyne is propyne:

propane   propene   propyne
(alkane)   (alkene)   (alkyne)

Ethane and propane molecules contain only carbon-carbon single bonds and are said to be **saturated**. Ethene and propene molecules each contain a carbon-carbon double bond (formed from two pairs of shared electrons). Ethyne and propyne molecules each contain a carbon-carbon triple bond (formed from three pairs of shared electrons). Compounds containing carbon-carbon double or triple bonds are described as **unsaturated** since they do not contain the maximum number of atoms per molecule that the electrons available for bonding would allow. If the electrons which are used to form the extra bonds between the carbon atoms were used to bond to other atoms, the compounds would become saturated.

Members of all three series react with bromine and chlorine. Bromine is decolourised by shaking a solution of it with an alkene or alkyne at room temperature. Alkanes will only react with bromine if the two substances are left together in light for several minutes or heated together.

The marked difference between the rate of the reaction with a saturated hydrocarbon and the rate with an unsaturated compound can be explained by the reactions being of different types. The alkane reaction is called a **substitution reaction** as hydrogen atoms in the molecule are gradually replaced by halogen atoms. It is difficult to stop the reaction at a particular stage and therefore a mixture of products is obtained.

In the case of methane, the first member of the alkane series, reacting with chlorine, the sequence of reactions is:

$$CH_4 \ + \ Cl_2 \ \rightarrow \ \underset{\text{monochloromethane}}{CH_3Cl} \ + \ HCl$$

$$CH_3Cl \ + \ Cl_2 \ \rightarrow \ \underset{\text{dichloromethane}}{CH_2Cl_2} \ + \ HCl$$

$$CH_2Cl_2 \ + \ Cl_2 \ \rightarrow \ \underset{\substack{\text{trichloromethane} \\ \text{(chloroform)}}}{CHCl_3} \ + \ HCl$$

$$CHCl_3 \ + \ Cl_2 \ \rightarrow \ \underset{\substack{\text{tetrachloromethane} \\ \text{(carbon tetrachloride)}}}{CCl_4} \ + \ HCl$$

During a substitution reaction of this type only one of the two atoms in each halogen molecule finishes in the organic product, the other forming a molecule of hydrogen halide.

The higher rates of the reactions of alkenes and alkynes with halogens are explained by these reactions being **addition reactions** rather than substitution reactions. During the addition reactions the double or triple bonds become single bonds and the unused bonding electrons form bonds with halogen atoms. The result is that the two reactants add together to form one product,

e.g.

$$\underset{H}{\overset{H}{\diagdown}}C=C\underset{H}{\overset{H}{\diagup}} \ + \ Br-Br \ \rightarrow \ H-\underset{Br}{\overset{H}{\underset{|}{\overset{|}{C}}}}-\underset{Br}{\overset{H}{\underset{|}{\overset{|}{C}}}}-H$$

All of the hydrocarbons are colourless substances which are insoluble in water. Within each series the boiling points of the compounds increase with relative molecular mass, the lower members being gases and the higher members liquids or solids. Table 31.2 lists this information, but you should realise that this information has been given so that you can see the trend; you are not expected to remember the actual values or the names of all but the first two or three members of each series.

TABLE 31.2. *Members of the three hydro-carbon series*

ALKANES C_nH_{2n+2}			ALKENES C_nH_{2n}			ALKYNES C_nH_{2n-2}		
Name	Formula	b.p./°C	Name	Formula	b.p./°C	Name	Formula	b.p./°C
methane	CH_4	−162						
ethane	C_2H_6	−89	ethene	C_2H_4	−104	ethyne	C_2H_2	−84
propane	C_3H_8	−42	propene	C_3H_6	−48	propyne	C_3H_4	−23
butane	C_4H_{10}	−0·5	butene	C_4H_8	−6	butyne	C_4H_6	8
pentane	C_5H_{12}	36	pentene	C_5H_{10}	30	pentyne	C_5H_8	40
hexane	C_6H_{14}	69	hexene	C_6H_{12}	64	hexyne	C_6H_{10}	71

31.8 Properties and uses of alkanes

The alkanes are rather unreactive towards most common reagents. For example, they do not react with acids or alkalis, nor do they react with aqueous potassium manganate(VII), a powerful oxidising agent.

In spite of their general lack of reactivity, the alkanes will burn readily in air, hence their importance as fuels. For example octane, C_8H_{18}, is a major constituent of petrol. It is from this alkane that the term octane number is derived. The grade of petrol is indicated by its octane number which is obtained by experimentally comparing its efficiency to that of an isomer of octane which is given an octane number of 100. In more recent years the system has been simplified by replacing the octane number by a star rating. Four star petrol has an octane number of between 97 and 100. High compression engines tend to need petrol with a high octane number (95–100). The use of lower octane petrol with such engines would result in pre-ignition, that is, the petrol-air mixture would ignite in the hot cylinder before the piston was in the correct position and this would lead to a reduction in efficiency.

In a good supply of air the products of combustion are the harmless substances, carbon dioxide and water (vapour),

e.g.

$$CH_4(g) \quad + \quad 2O_2(g) \quad \rightarrow \quad CO_2(g) \quad + \quad 2H_2O(g)$$
methane

$$C_8H_{18}(l) \quad + \quad 12\tfrac{1}{2}O_2(g) \quad \rightarrow \quad 8CO_2(g) \quad + \quad 9H_2O(g)$$
octane

If the air supply is restricted, soot may be produced (unburnt carbon) or, more dangerously, some carbon monoxide (CO) may be formed. This gas is extremely poisonous and this is why good ventilation must be employed when gas fires are used. This is also why it is very dangerous to run a motor-car engine inside a closed garage; the engine always produces some carbon monoxide among its exhaust products.

Methane in the form of natural gas (31.4) is a very important domestic and industrial fuel. Propane, C_3H_8, can be liquefied by compressing the gas (unlike methane, no cooling is necessary) and hence it can be stored in the liquid form in metal containers. If some of the liquid is allowed to escape from the container it immediately changes back to a gas because of the reduction of pressure. A small volume of the liquid will produce a large volume of the gas. These characteristics make propane particularly suitable for use as a portable fuel and it is the gas which is used in most bottled gas or camping gas cylinders.

The controlled oxidation of methane is important. If it is burned with its flame against a cold surface carbon black is produced, which is an important constituent of rubber tyres and printing ink:

$$CH_4(g) + O_2(g) \rightarrow C(s) + 2H_2O(g)$$

On the other hand, at high pressure in the presence of a nickel catalyst, methane may be oxidised to a mixture of carbon monoxide and hydrogen known as synthesis gas:

$$CH_4(g) + \tfrac{1}{2}O_2(g) \rightarrow CO(g) + 2H_2(g)$$

This mixture may then be converted, with the aid of a catalyst, into the important chemical, methanol:

$$CO(g) + 2H_2(g) \xrightarrow{\text{catalyst}} CH_3OH(l)$$

The only reactions, other than burning in air, which alkanes take part in are those with chlorine and bromine. As mentioned in the previous section, these reactions involve the substitution of hydrogen atoms by halogen atoms.

**31.9
Properties and
uses of alkenes**

Alkenes, unlike alkanes, contain a distinct functional group in the molecule, namely, the carbon-carbon double bond

$$\begin{array}{c} \diagdown \qquad \diagup \\ C = C \\ \diagup \qquad \diagdown \end{array}$$

The first member of the series, C_2H_4 ethene, can be conveniently prepared in the laboratory by passing ethanol vapour over hot aluminium oxide (Investigation 32.2). The ethanol is soaked up in fine dry sand, or alternatively rocksil, in a boiling-tube then, with the tube clamped horizontally, some aluminium oxide is placed in it. After connecting the delivery tube, the aluminium oxide is warmed with a small Bunsen flame. The heat conducted down the tube causes ethanol to vaporize slowly, and its vapour is dehydrated as it passes over the hot aluminium oxide:

$$C_2H_5OH(g) \rightarrow H_2O(g) + C_2H_4(g)$$

The gas, ethene, can be collected over water.

Because of the presence of a double bond, alkenes are much more reactive than alkanes. Some of the addition reactions of alkenes which are of industrial importance are described below.

Addition of hydrogen

If a mixture of ethene and hydrogen is passed over a hot nickel catalyst, ethane is formed,

$$C_2H_4(g) + H_2(g) \rightarrow C_2H_6(g)$$

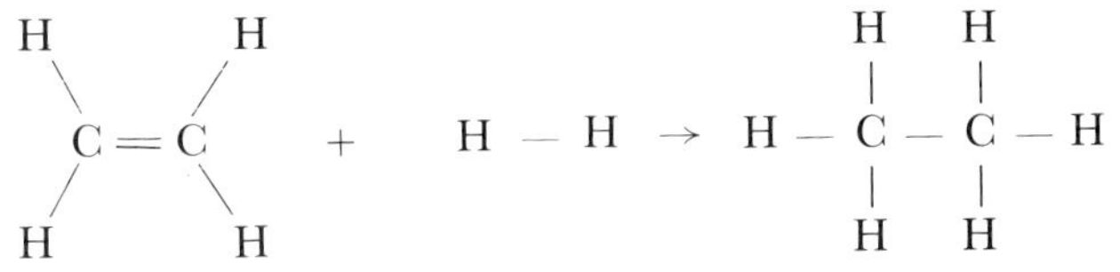

ethene   ethane

This process is called hydrogenation; it converts an unsaturated hydrocarbon into a saturated one.

Hydrogenation is industrially important in the hardening of edible oils for making margarine. Oils such as those extracted from ground nuts and soya beans contain unsaturated compounds. Hydrogenation converts them to saturated compounds which have higher melting points and hence are solids rather than liquids.

Addition of hydrogen chloride

Industrially a mixture of ethene with hydrogen chloride gas is passed over a hot catalyst, when chloroethane is formed,

$$C_2H_4(g) + HCl(g) \rightarrow C_2H_5Cl(g)$$

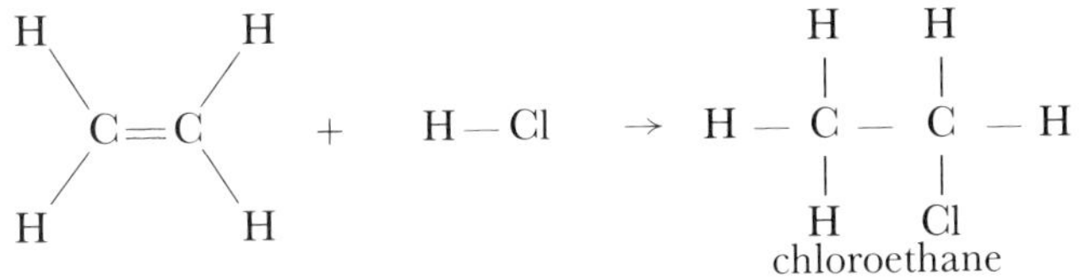

chloroethane

This product is used to make tetraethyl lead, used as anti-knock in petrol (9.5).

Addition of chlorine or bromine

Whereas alkanes react with these halogens only in sunlight (31.7) or at high temperature, alkenes react readily at room temperature even in the dark,

e.g. $$C_2H_4(g) + Cl_2(g) \rightarrow C_2H_4Cl_2(l)$$

337

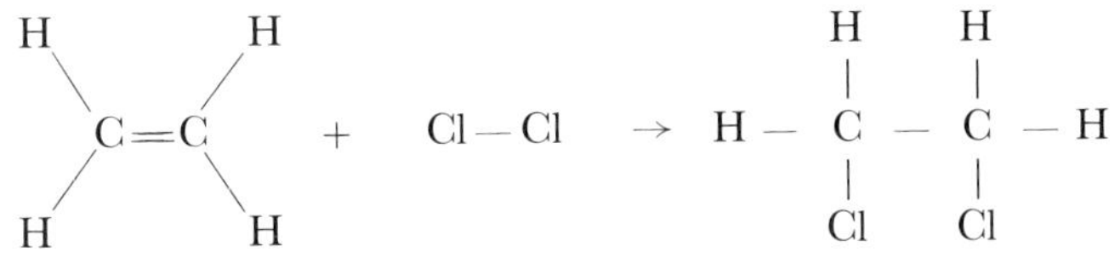

The product here is 1,2-dichloroethane; note that addition occurs 'across the double bond', so that one halogen atom becomes connected to each carbon atom. This reaction is used industrially, and the product is then heated under pressure, when it breaks down into hydrogen chloride and chloroethene (vinyl chloride), $CH_2 = CHCl$. This is used to make the important plastic, PVC (31.10).

Alternatively, chloroethene is manufactured from ethene in a single step process in which ethene, hydrogen chloride and oxygen are passed over a catalyst:

$$\begin{array}{c} H \\ \diagdown \\ \diagup \\ H \end{array} C = C \begin{array}{c} H \\ \diagdown \\ \diagup \\ H \end{array} + HCl + \tfrac{1}{2}O_2 \rightarrow \begin{array}{c} H \\ \diagdown \\ \diagup \\ H \end{array} C = C \begin{array}{c} H \\ \diagdown \\ \diagup \\ Cl \end{array} + H_2O$$

Addition of water

This process, called hydration, is very important in the manufacture of ethanol. Ethene and steam, at high temperature and pressure, are passed over a catalyst of phosphoric acid on silica,

$$C_2H_4(g) + H_2O(g) \rightarrow C_2H_5OH(g)$$

or

$$\begin{array}{c} H \\ \diagdown \\ \diagup \\ H \end{array} C = C \begin{array}{c} H \\ \diagup \\ \diagdown \\ H \end{array} + \begin{array}{c} H \quad H \\ \diagdown \diagup \\ O \end{array} \rightarrow \begin{array}{cc} H & H \\ | & | \\ H - C - C - H \\ | & | \\ H & OH \end{array}$$

Most ethanol is now made in this way.

Oxidation

Whereas an alkane does not react with aqueous potassium manganate(VII), an alkene reacts readily in the cold. If an alkene is shaken with the acidified aqueous reagent the purple colour rapidly disappears, leaving a colourless solution.

The reaction for ethene may be shown simply as,

$$C_2H_4(g) + H_2O(l) + [O] \rightarrow C_2H_4(OH)_2(aq)$$

or

$$\begin{array}{c} H \\ \diagdown \\ \diagup \\ H \end{array} C = C \begin{array}{c} H \\ \diagup \\ \diagdown \\ H \end{array} + \begin{array}{c} H \\ \diagdown \\ \diagup \\ H \end{array} O + [O] \rightarrow \begin{array}{cc} H & H \\ | & | \\ H - C - C - H \\ | & | \\ OH & OH \end{array}$$

The product, a liquid soluble in water, is ethane-1,2-diol (ethylene glycol). It is a type of alcohol, but with two hydroxy-groups in the molecule.

Ethane-1,2-diol is manufactured from ethene to be used as antifreeze to prevent the water in the cooling system of car engines freezing during cold weather, and as a raw material for the production of polyester fibres such as Terylene (32.9).

The industrial oxidation does not use potassium manganate(VII), but goes in two stages. Ethene is catalytically oxidised by air to epoxyethane (ethylene oxide),

$$\begin{array}{c} CH_2 - CH_2 \, , \\ \diagdown \quad \diagup \\ O \end{array}$$

and this is then hydrated using steam at high temperature and pressure to give ethane-1,2-diol.

31.10 Polymerisation: plastics

Polymerisation is the vital stage in the manufacture of plastics, and in the case of polyalkenes (e.g. polythene, PVC and polystyrene) this involves the building up of giant molecules by successive addition reactions between thousands of alkene molecules. Polythene—more properly called poly(ethene)—is made by subjecting ethene to very high pressures (perhaps 2000 atmospheres) in the presence of a catalyst. The overall change when a lot of ethene molecules (monomers) combine to form a very long molecule of polyethene (polymer) can be represented by,

$$\cdots \; + \quad \underset{H}{\overset{H}{>}}C{=}C\underset{H}{\overset{H}{<}} \quad + \quad \underset{H}{\overset{H}{>}}C{=}C\underset{H}{\overset{H}{<}} \quad + \quad \underset{H}{\overset{H}{>}}C{=}C\underset{H}{\overset{H}{<}} \quad + \quad \cdots$$

monomer units

$$\rightarrow \qquad \cdots - \overset{\displaystyle H}{\underset{\displaystyle H}{\overset{\mid}{\underset{\mid}{C}}}} - \overset{\displaystyle H}{\underset{\displaystyle H}{\overset{\mid}{\underset{\mid}{C}}}} - \overset{\displaystyle H}{\underset{\displaystyle H}{\overset{\mid}{\underset{\mid}{C}}}} - \overset{\displaystyle H}{\underset{\displaystyle H}{\overset{\mid}{\underset{\mid}{C}}}} - \overset{\displaystyle H}{\underset{\displaystyle H}{\overset{\mid}{\underset{\mid}{C}}}} - \overset{\displaystyle H}{\underset{\displaystyle H}{\overset{\mid}{\underset{\mid}{C}}}} - \cdots \qquad \text{polymer}$$

or more briefly $\qquad n\ CH_2 = CH_2 \rightarrow (- CH_2 - CH_2 -)_n$

The resulting molecule contains something like 10 000–100 000 carbon atoms, and you can see that it is simply an alkane with a very long chain. The fact that the product is a saturated hydrocarbon accounts for its lack of chemical reactivity, which makes it useful for a wide variety of purposes. For example, polythene containers may be used for storing acids, alkalis and other corrosive substances. Glass can, of course, serve this purpose but it has the disadvantages of being much heavier and much more fragile.

Fig. 31.8 Uses of two common plastics
(a) Pipes which are made from PVC are very light

(b) Refuse sacks made of polythene. (Both Shell Photographs)

(a)

(b)

339

Another important plastic is polyvinyl chloride (PVC)—more properly called poly(chloroethene). Chloroethene (vinyl chloride monomer) is made from ethene (31.9) and then heated in the presence of an initiator to set the reaction going,

$$
\cdots \quad + \quad
\begin{array}{c} H \\[-2pt] \diagdown \\ C = C \\ \diagup \\ H \end{array}
\begin{array}{c} H \\ \diagup \\ \\ \diagdown \\ Cl \end{array}
\quad + \quad
\begin{array}{c} H \\[-2pt] \diagdown \\ C = C \\ \diagup \\ H \end{array}
\begin{array}{c} H \\ \diagup \\ \\ \diagdown \\ Cl \end{array}
\quad + \quad
\begin{array}{c} H \\[-2pt] \diagdown \\ C = C \\ \diagup \\ H \end{array}
\begin{array}{c} H \\ \diagup \\ \\ \diagdown \\ Cl \end{array}
\quad + \quad \cdots
$$

monomer units

$$
\rightarrow \quad \cdots -
\begin{array}{c} H \\ | \\ C \\ | \\ H \end{array} -
\begin{array}{c} H \\ | \\ C \\ | \\ Cl \end{array} -
\begin{array}{c} H \\ | \\ C \\ | \\ H \end{array} -
\begin{array}{c} H \\ | \\ C \\ | \\ Cl \end{array} -
\begin{array}{c} H \\ | \\ C \\ | \\ H \end{array} -
\begin{array}{c} H \\ | \\ C \\ | \\ Cl \end{array} - \cdots \qquad \text{polymer}
$$

or more briefly

$$
n\ CH_2 = CHCl \rightarrow (-CH_2 - \underset{\underset{Cl}{|}}{CH} -)_n
$$

PVC is, like polythene, chemically unreactive and may be used for many purposes. Among other things, it is used for upholstery, piping, electric cable sheathing and gramophone records.

It should be mentioned before leaving this section that the chemical inertness (unreactivity) of these plastics is not entirely a blessing. The problem comes when objects made from them are thrown away, having outlived their usefulness. If a paper cup is thrown away it eventually rots and disappears, whereas a plastic cup does not. It is in some ways unfortunate that the development of cheap plastics has resulted in the proliferation of disposable containers—disposable they may be, but they end up as piles of rubbish which do not go away. A considerable amount of research is now going on into biodegradable plastics, that is, plastics which will gradually decay when left in the open air. Some progress has been made in producing plastics which are degraded by ultra-violet light and some department stores supply plastic carriers of this type.

31.11
Properties and uses of alkynes

Alkynes are hydrocarbons with a greater degree of 'unsaturation' than alkenes, having a carbon-carbon triple bond in the molecule. The simplest alkyne is the gas ethyne (acetylene), C_2H_2. Its molecular structure may be shown as $H - C \equiv C - H$.

It is formed when calcium dicarbide, CaC_2 (made by reacting calcium oxide, quicklime, with coke), is treated with water at room temperature:

$$CaC_2(s) + 2H_2O(l) \rightarrow Ca(OH)_2(aq/s) + C_2H_2(g)$$

The gas burns readily in air with a luminous smoky flame (a lot of carbon escaping unburnt as soot), and in the early days of cycling and motoring the reaction of calcium dicarbide with water was used as the basis of the acetylene lamp, a temperamental device which would not compare very favourably with the modern electric light!

Nowadays the combustion of ethyne is used to generate high temperatures for welding and metal-cutting, using the oxy-acetylene torch. The reaction between ethyne and oxygen (both supplied from cylinders under pressure) is so highly exothermic that temperatures of around 3000 °C are obtained:

$$C_2H_2(g) + 2\tfrac{1}{2}O_2(g) \rightarrow 2CO_2(g) + H_2O(g)$$

Ethyne is an important intermediate in the organic chemical industry, and is made on a large scale by heating methane (from petroleum or natural gas) to a temperature of 1500 °C in the absence of air:

$$2CH_4(g) \xrightarrow{1500\,^\circ C} C_2H_2(g) + 3H_2(g)$$

Ethyne, like ethene, undergoes addition reactions, being an unsaturated hydrocarbon. Thus, treatment with hydrogen over a hot nickel catalyst yields ethane:

$$C_2H_2(g) + 2H_2(g) \rightarrow C_2H_6(g)$$

Addition of hydrogen chloride yields 1,1-dichloroethane:

$$C_2H_2(g) + 2HCl(g) \rightarrow CH_3 - CHCl_2(l)$$

(Note that both halogen atoms become connected to the same carbon atom in this reaction.) Addition of chlorine yields 1,1,2,2-tetrachloroethane (bromine reacts similarly):

$$C_2H_2(g) + 2Cl_2(g) \rightarrow CHCl_2 - CHCl_2(l)$$

This reaction is used industrially, following which the product is heated so that it breaks up into hydrogen chloride and trichloroethene (trichloroethylene), $CHCl = CCl_2$. The latter is an important solvent used, for example, in degreasing metal surfaces.

Fig. 31.9 (a)
Oxyacetylene cutting
(b) Oxyacetylene welding.
(Both Courtesy British
Oxygen Company)

(a)

(b)

**31.12
Summary**

1. The fossil fuels, coal, petroleum and natural gas not only supply most of our energy but also act as the major sources of organic chemicals. Alternative energy sources for the future need to be considered since reserves of fossil fuels are limited.

2. Alkanes, which are important fuels, are saturated hydrocarbons. Apart from being combustible they are chemically unreactive under most conditions, though they will undergo substitution reactions with halogens.

3. Alkenes and alkynes are unsaturated hydrocarbons. They are chemically reactive, undergoing a variety of addition reactions, and are important industrial intermediates. Alkenes undergo polymerisation, the reaction involved in the manufacture of plastics.

341

Chapter 32

Sources of organic compounds: alcohols, acids and esters

Investigation 32.1

The oxidation of an alcohol.

1. Put 5 drops of ethanol in a dry crucible or porcelain basin and try to set fire to it with a lighted taper.

Question

1 Since ethanol is a compound of carbon, hydrogen and oxygen what are its combustion products likely to be?

2. Put 5 drops of ethanol in a test-tube, add about 1 cm³ of dilute sulphuric acid, then 2 drops of potassium manganate(VII) solution. Warm gently and note what happens.
3. Repeat experiment (2) using potassium dichromate(VI) solution in place of potassium manganate(VII).

Questions

2 Does ethanol react with either of these oxidising agents?
3 How can you tell that reaction has occurred?
4 Describe any smell that you might have noticed in either of these reactions, other than that of ethanol itself.
5 When wine, which contains ethanol in aqueous solution, is left open to the air it turns sour. How can this be explained?

Investigation 32.2

What are the properties of the gas formed by dehydrating ethanol?

You will need a boiling-tube, fitted with a bung and delivery tube, so that the gas can be collected over water (Fig. 32.1).

Fig. 32.1

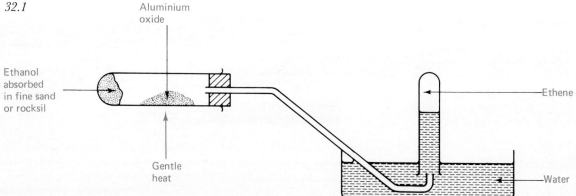

Ethanol absorbed in fine sand or rocksil — Aluminium oxide — Gentle heat — Ethene — Water

Put some loosely packed Rocksil (or fine dry sand) in the boiling-tube to a depth of about 2 cm.

Using a teat pipette, add ethanol to the tube so that the rocksil is thoroughly soaked and there is only a very thin layer of ethanol above the solid.

Clamp the tube horizontally and put a layer of dry aluminium oxide halfway down the tube (Fig. 32.1).

Put the bung and delivery tube in the boiling-tube.

Heat the aluminium oxide gently. The heat which is conducted down the tube will be sufficient to vaporize the alcohol which will then be dehydrated by the aluminium oxide.

After the air has been expelled from the apparatus, collect three test-tubes full of the gas.

Test the three samples of gas in the following ways:

(a) open the first tube and place a lighted splint in the gas which escapes at the top of the tube;

(b) add a few drops of bromine water and shake the tube;

(c) add a few drops of acidified potassium manganate(VII) solution and shake the tube.

Questions

1 The gas is a hydrocarbon. From the evidence of the three tests, to which series of hydrocarbons could it belong?

2 The formula of ethanol is C_2H_5OH. If the elements which make up water are removed, what will be the formula of the remaining gas?

Investigation 32.3

What are the properties of an organic acid?

1. Put 2 cm³ of sodium hydroxide solution in a test-tube and add a drop of litmus (or Universal Indicator) solution. Now add dilute ethanoic (acetic) acid a little at a time, with shaking, until a change is observed.

2. Put a small lump of marble (calcium carbonate) in a test-tube and add about 2 cm³ of dilute ethanoic acid. Observe what happens, and identify the gas evolved.

3. Put about 2 cm³ of dilute ethanoic acid into a test-tube and add a piece of magnesium ribbon. Observe what happens, and identify the gas evolved.

Questions

1 On the evidence of these experiments, in what ways does ethanoic acid behave as a typical acid?

2 Dilute ethanoic acid is smelly, and is a weak electrolyte. What conclusions can you draw from this?

4. Put 5 drops of ethanol into a test-tube and add cautiously 5 drops of concentrated sulphuric acid, followed by 5 drops of glacial (pure) ethanoic acid. Stand the test-tube in hot water for a few minutes, then pour the contents into 20–50 cm³ of cold water contained in a basin or small beaker. Smell the resulting mixture.

Questions

3 How would you describe the final smell?

4 Have you ever come across this smell before? If so, where? (The smelly substance formed in this reaction is an ester, ethyl ethanoate (32.9).)

Investigation 32.4

Converting a plant oil into a soap.

Oils from plants contain esters of the alcohol, propane-1,2,2,-triol, and carboxylic acids with large numbers of carbon atoms in their molecules (32.10). When the oil is heated with aqueous sodium hydroxide, the esters are hydrolysed and a solution of the sodium salts of the acids is produced. These are soaps and they can be isolated by adding salt to the solution.

You will need some castor oil, some 5M sodium hydroxide solution (this is more concentrated than the usual bench solution and must be treated with great care) and some solid sodium chloride.

Add about 2 cm³ of castor oil to a small beaker, together with about 10 cm³ of 5M sodium hydroxide solution.

Warm the beaker and stir the mixture until the liquid boils. Boil the mixture gently for a few minutes, adding more water, if necessary, so that the volume does not become too small.

Now add another 10 cm³ of distilled water and about three teaspoonsful of solid sodium chloride. Boil the mixture gently for two to three minutes, stirring with a glass rod.

Allow the mixture to cool, continuing to stir to break up any large lumps of solid.

Filter off the solid and wash it with a little distilled water.

Finally add a little of the product to some distilled water in a test-tube. Shake and observe what happens.

Questions

1 Why is the hydrolysis of the ester in castor oil faster with aqueous sodium hydroxide than with water?

2 Why was the product shaken with distilled water, and not with tap water, when finding whether it was a soap?

3 How does the solubility of a soap in salt solution differ from its solubility in water? Try to explain your answer.

32.5 Alcohols

Drinks such as wine, beer and spirits have probably been known to man for over 3000 years. The substance which is present in all of them is commonly known as alcohol. The proper chemical name for this substance is ethanol and it is only one member of a series of alcohols. As it is the most common one it is often called alcohol. The first three members of the homologous series of alcohols are listed in 30.9, and inspection of their formulae shows that they can be represented by the general formula $C_nH_{2n+1}OH$, or, more simply, ROH where R is an alkyl group (30.7). The functional group present in every member of the series is the hydroxy group $-OH$, bonded covalently to carbon.

Ethanol is the only member of the series which, if suitably diluted, as in alcoholic drinks, can be consumed by human beings. The consumption of undiluted ethanol would be extremely harmful and as you probably know, excessive, regular consumption of alcoholic drinks, particularly spirits, can lead to ill-health. The other members of the homologous series of alcohols cannot be drunk safely. Methanol is added to ethanol to make methylated spirit (meths) to make it unsafe to drink.

Ethanol in the form of alcoholic drinks is produced by the fermentation of sugars, under the influence of enzymes (biological catalysts) present in yeast. In the case of beer and also some spirits such as whisky the starting material is starch in the form of barley, and so a preliminary stage is the hydrolysis of the starch to form sugars. When fruits such as grapes are used as the starting material, the sugars are already present. The main reactions, starting, for example, with sucrose (cane sugar) as the raw material, are firstly the hydrolysis of sucrose to glucose and fructose,

$$C_{12}H_{22}O_{11}(aq) + H_2O(l) \rightarrow 2C_6H_{12}O_6(aq)$$

and secondly the decomposition of glucose to ethanol and carbon dioxide (hence the formation of froth):

$$C_6H_{12}O_6(aq) \rightarrow 2C_2H_5OH(aq) + 2CO_2(g)$$

Fermentation is carried out at warm room temperature, about 25 °C; at lower temperatures the reaction is far too slow, while at higher temperatures the yeast is 'killed'. At 25 °C the production of beer from barley malt takes about a week. When the concentration of alcohol reaches about 10 % (wine strength), fermentation stops and, if a more concentrated alcohol solution is needed, such as in the production of spirits, it is necessary to remove some of the water by distillation. Complete fractional distillation of a solution of ethanol and water yields a mixture containing 96 % ethanol which is used, for example, as surgical spirit.

Fig. 32.2 The fermenting room at John Smith's Tadcaster Brewery. The froth is caused by the carbon dioxide given off during the fermentation. A sample is being taken so that its specific gravity can be measured to check how the fermentation is proceeding. (Courtesy John Smith's Tadcaster Brewery Ltd)

Until the growth of the petroleum industry, fermentation of sugars was the main source of ethanol for industrial purposes. Now, however, most industrial ethanol is manufactured by the hydration of ethene (31.9).

32.6 Properties and uses of ethanol

Ethanol undergoes a wide variety of chemical reactions, and some of the more important ones are discussed below.

Combustion in air
Ethanol, in the form of methylated spirit, is used as a fuel in small spirit burners for camping stoves etc., though it has now been superseded by 'bottled gas' such as propane.

When ethanol burns the products are carbon dioxide and water:

$$C_2H_5OH(l) + 3O_2(g) \rightarrow 2CO_2(g) + 3H_2O(g)$$

Oxidation to ethanoic (acetic) acid
Vinegar, a dilute (4 %) aqueous solution of ethanoic acid, has been known as long as wine; in fact, the old name 'acetic acid' comes from the Latin name for vinegar. If wine is left open to the air for some weeks it goes sour, because of oxidation of the ethanol by air, aided by bacteria. The product of this oxidation is ethanoic acid:

$$C_2H_5OH(aq) + O_2(g) \rightarrow CH_3COOH(aq) + H_2O(l)$$

In terms of molecular structures, the reaction involves conversion of the $-CH_2-OH$ group of the alcohol to the carboxyl group,

$$-C\underset{\textstyle O-H}{\overset{\textstyle O}{<}}$$

345

Vinegar is still made by the air-oxidation of wine, but most ethanoic acid for industrial purposes is now obtained from petroleum products.

Fig. 32.3 Conversion of ethanol to ethanoic acid

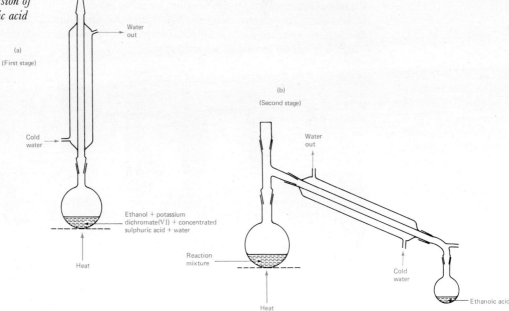

Ethanol can also be oxidised to ethanoic acid using chemical oxidising agents such as potassium dichromate(VI) solution. An aqueous solution of ethanol, potassium dichromate(VI), and sulphuric acid is refluxed for about half an hour using the apparatus shown in Fig. 32.3(a). The purpose of the vertical reflux condenser is to prevent volatile substances—such as ethanol itself—from escaping while the reaction proceeds. When the reaction is complete the condenser is turned to its 'normal' position as in Fig. 32.3(b) and the reaction mixture is distilled. The distillate, collecting in the receiver, is an aqueous solution of ethanoic acid.

Dehydration to ethene

Ethanol may be dehydrated to give ethene (ethylene), the reaction being,

$$C_2H_5OH(l) \rightarrow H_2O(l) + C_2H_4(g)$$

or

$$H - \underset{\underset{H}{|}}{\overset{\overset{H}{|}}{C}} - \underset{\underset{H}{|}}{\overset{\overset{H}{|}}{C}} - OH \quad \rightarrow \quad H_2O \quad + \quad \overset{H}{\underset{H}{>}}C = C\overset{H}{\underset{H}{<}}$$

This reaction is the reverse of the process by which ethanol is manufactured from ethene (31.9), but is useful in the laboratory as a method of preparing the gas, ethene, by passing ethanol vapour over heated aluminium oxide (31.9 and 32.2).

Reaction with sodium metal

As you know, sodium metal reacts quite violently with water, liberating hydrogen and forming sodium hydroxide solution:

$$H_2O(l) + Na(s) \rightarrow Na^+OH^-(aq) + \tfrac{1}{2}H_2(g)$$
$$(HOH)$$

Ionic charges have been included on this occasion to emphasise the fact that sodium hydroxide is an ionic substance.

In a similar fashion, sodium will react with an alcohol such as ethanol. The reaction is much less vigorous than that with water, though effervescence is quite brisk. The reaction is of the general type:

$$ROH(l) + Na(s) \rightarrow RO^-Na^+ \text{ (alc. soln.)} + \tfrac{1}{2}H_2(g)$$

The product, which, like sodium hydroxide, is ionic, is called an 'alkoxide'; thus the product from ethanol is sodium ethoxide. It is not very soluble in cold ethanol, and readily crystallises if the solution is cooled.

Formation of esters

An important reaction of alcohols is their reaction with acids (usually organic acids) to form substances known as esters. This reaction is discussed in 32.9.

32.7 Carboxylic acids

The carboxylic acids (formerly called **fatty acids** because of their presence in natural fats) have the general formula $C_nH_{2n+1}COOH$ or, more simply, $RCOOH$ where 'R' is an alkyl group or a hydrogen atom. The simplest carboxylic acids are methanoic acid (formic acid), $HCOOH$, and ethanoic acid (acetic acid), CH_3COOH. The latter is the acid present in vinegar, while the former is the acid present in the sting of an ant—hence the old name, from the Latin 'formica' = ant. The structural formulae for methanoic acid and ethanoic acid are,

Carboxylic acids are colourless substances with characteristic odours. They are either liquid or solid at room temperature depending on their relative molecular mass. The 'lower' members of the series are completely miscible with water, while the higher members become less soluble in water as the size of the alkyl group increases.

32.8 Properties of ethanoic acid

Ethanoic acid (acetic acid), CH_3COOH, the most familiar carboxylic acid, is typical of this series of compounds. It is a colourless liquid at normal room temperature, with a pungent vinegary smell. It freezes at $17\,°C$, so that in cold weather crystallisation may occur—hence the old name glacial acetic acid for pure ethanoic acid.

Ethanoic acid is a covalent compound, the pure substance being virtually a non-electrolyte and having no effect upon dry litmus paper. It is, however, very hygroscopic. Its aqueous solution is a weak electrolyte, showing that some ions are present, and it turns litmus red. Even dilute solutions (such as the household vinegar) have the characteristic smell of the pure acid, indicating that much of the solute is in the molecular form. Thus, it is a weak acid, only slightly ionised in aqueous solution; e.g. the degree of ionisation in vinegar is about 1%.

Only the hydrogen atom of the carboxyl group undergoes ionisation, the hydrogen atoms of the alkyl group being rather inert like those in an alkane. The ionisation in water is,

or $CH_3COOH(aq) \rightleftharpoons CH_3COO^-(aq) + H^+(aq)$

The anion formed is the ethanoate (acetate) ion.

Acidic behaviour of ethanoic acid

Ethanoic acid behaves as a typical acid in aqueous solution. The more reactive metals liberate hydrogen from it. Also it reacts with alkalis and bases to form salts.

e.g. $\qquad NaOH(aq) + CH_3COOH(aq) \rightarrow CH_3COONa(aq) + H_2O(l)$

The salt formed here is sodium ethanoate (sodium acetate). It is a typical colourless odourless crystalline solid, fully ionised in solution and therefore a good electrolyte. Its ionic structure is,

$$
\begin{array}{c}
H \\
| \\
H - C - C \overset{\displaystyle O}{\underset{\displaystyle O^-}{\diagup\diagdown}} \quad Na^+ \\
| \\
H
\end{array}
$$

(Don't be confused by the fact that the metal is written at the end of the formula; this is just a convention peculiar to organic chemistry, and it has no particular significance.)

Most ethanoates (acetates) are soluble in water; indeed lead(II) ethanoate is, apart from lead(II) nitrate, the only common lead salt which is soluble in water.

32.9 Properties and uses of esters

Esters are covalent substances in which the 'acidic' hydrogen atom of an acid is replaced by an alkyl group. Thus, for example, ethyl ethanoate (ethyl acetate) has the structural formula,

$$
\underbrace{H - \overset{\displaystyle H}{\underset{\displaystyle H}{\overset{\displaystyle |}{\underset{\displaystyle |}{C}}}} - C \overset{\displaystyle O}{\diagdown O} - }_{\text{ethanoate}} \underbrace{\overset{\displaystyle H}{\underset{\displaystyle H}{\overset{\displaystyle |}{\underset{\displaystyle |}{C}}}} - \overset{\displaystyle H}{\underset{\displaystyle H}{\overset{\displaystyle |}{\underset{\displaystyle |}{C}}}} - H}_{\text{ethyl}}
$$

usually written more simply as $CH_3COOC_2H_5$.

Esters are generally liquids (though may be solids), immiscible with water and having characteristic fruity smells. Many esters occur naturally in fruits and flowers, and quite large quantities of manufactured esters are used to provide the flavour and odour of products such as sweets, ice-cream and perfumes.

The formation of an ester by reaction between an alcohol and a carboxylic acid is readily shown (Investigation 32.3). Small quantities of ethanol, concentrated sulphuric acid and glacial ethanoic (acetic) acid are cautiously mixed in a test-tube and then the mixture is warmed gently for a few minutes, after which the reaction mixture (when cool) is poured into cold water in a beaker. The fruity smell of ethyl ethanoate (ethyl acetate) will now be detected, and you may recognise it as being similar to that of certain adhesives, in which this ester is used as a solvent.

$$CH_3COOH(l) + C_2H_5OH(l) \rightleftharpoons CH_3COOC_2H_5(l) + H_2O(l)$$

Note that this reaction is reversible, so it does not go to completion. The sulphuric acid in the reaction mixture acts as a catalyst.

348

Esters in industry

Apart from their uses in foods etc. mentioned above, esters have some very important large-scale uses. They are employed as solvents, for example in adhesives and emulsion paints, and they are polymerised to form various plastics and synthetic fibres. Polyvinyl ethanoate is used to make plastic film, glues and paints; polymethyl methacrylate is used for Perspex, while various polyesters such as Terylene are used to make clothing.

Terylene is made by esterification of ethane-1,2-diol (ethylene glycol) with benzene-1,4-dicarboxylic acid (terephthalic acid), to produce a long-chain molecule:

$$\cdots + HO-CH_2-CH_2-OH + HO\underset{O}{\overset{\diagup}{>}}C-\bigcirc-C\underset{O}{\overset{OH}{\diagdown}} + \cdots$$

$$\left[-O-CH_2-CH_2-O\underset{O}{\overset{\diagup}{>}}C-\bigcirc-C\underset{O}{\overset{\diagdown}{<}}\right]_n + nH_2O$$

Terylene unit

This is not, strictly, a polymerisation reaction since water is also formed, but the product is nevertheless called a polyester. A polymer formed by this type of reaction, which is not the straightforward addition of monomers, is sometimes called a condensation polymer.

(The symbol ⬡ is used to represent the compound benzene which consists of a ring of six carbon atoms, each of which is bonded to a hydrogen atom. To save time the carbon atoms and hydrogen atoms are not written in the formula, but, if any of the hydrogen atoms are replaced by other groups, such as the carboxylic acid groups in terephthalic acid, it is obviously necessary to write them out fully.)

Hydrolysis of esters

As mentioned earlier, the esterification reaction is reversible. Consequently, an ester can be hydrolysed by water to reform the alcohol and the carboxylic acid, thus:

$$CH_3COOC_2H_5(l) + H_2O(l) \rightarrow CH_3COOH(aq) + C_2H_5OH(aq)$$

ethyl ethanoate ethanoic acid ethanol

In practice, the hydrolysis must be catalysed by the presence of acid or alkali for it to occur at a reasonable rate. An ester, if refluxed with aqueous acid, will be partly hydrolysed according to the equation above.

Much more important, however, is hydrolysis using aqueous alkali; this is called saponification. The reaction goes to completion (if excess alkali is used) since the carboxylic acid ends up as a salt,

e.g. $CH_3COOC_2H_5(l) + Na^+OH^-(aq) \rightarrow CH_3COO^-Na^+(aq) + C_2H_5OH(aq)$

The alkali is not simply a catalyst in this reaction, since it is converted to sodium ethanoate. The name saponification is used because a reaction of this type is the key stage in the making of soap (next section).

32.10 Soap

The French chemist Chevreul, in the early nineteenth century, carried out research into the nature of soap. It had been known for a long time that soap could be made by boiling vegetable oils or animal fats with caustic soda (sodium hydroxide) solution, and in 1816 Chevreul established that soap is formed by combination of the alkali with an acidic constituent of the fat—or in modern terms, soap is the sodium salt of an organic acid.

Animal fats are esters formed between the alcohol propane-1,2,3-triol (also known as glycerol or glycerine) and long-chain carboxylic acids such as stearic acid (octadecanoic acid)—hence the old name 'fatty acid' for this type of acid. The alkaline hydrolysis (saponification) of a typical fat involves the reaction:

$$\begin{array}{ccc}
C_{17}H_{35}COOCH_2 & & HOCH_2 \\
| & & | \\
C_{17}H_{35}COOCH & + 3Na^+OH^- \rightarrow & 3C_{17}H_{35}COO^-Na^+ + HOCH \\
| & & | \\
C_{17}H_{35}COOCH_2 & & HOCH_2
\end{array}$$

glyceryl tristearate (fat) · sodium stearate (soap) · glycerol

The structural formula for this soap is:

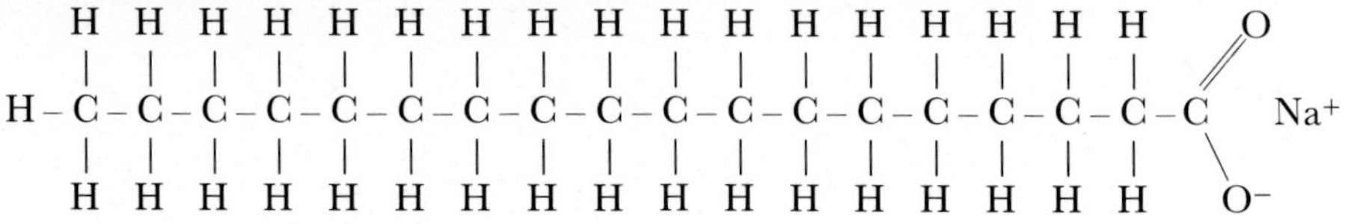

The exact length of the alkyl group is somewhat variable, depending on the identity of the original fat, but the essential feature of the structure is that there is a long-chain hydrocarbon with an ionic group at one end. This is what gives soap its remarkable and useful properties.

A simple hydrocarbon such as $C_{17}H_{36}$ is immiscible with (insoluble in) water, but miscible with (soluble in) substances of a similar nature such as oil and grease. On the other hand a simple salt like sodium carbonate, Na_2CO_3, with an ionic structure,

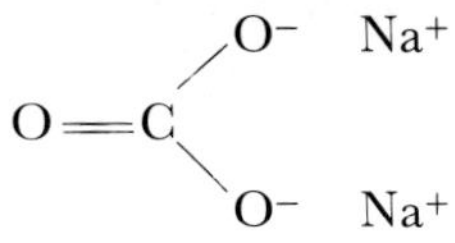

is soluble in water but insoluble in substances like oil and grease. Soap, because of its peculiar molecular/ionic structure, is both hydrocarbon and salt at the same time, and so it shares the solubility properties of both. It can therefore act as an **emulsifying agent**, bringing together as an emulsion the normally immiscible liquids oil and water. If we represent the structure of soap for simplicity as

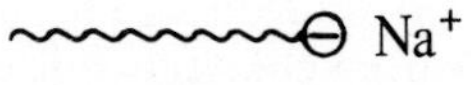

where  ∿∿∿ represents the long alkyl group and ⊖ stands for the negatively charged carboxyl group, then the emulsifying action of soap is illustrated by Fig. 32.4.

The oil droplet, with the alkyl groups of the soap embedded in it and the negatively charged carboxyl groups sticking out is, in effect, a giant anion. The oil, broken up into these tiny droplets, is now dispersed in the water as an emulsion.

The action of soap as a cleansing agent is twofold. Not only does it emulsify oil and grease, but it also lowers the surface tension of the water, as a result of which the water wets things more effectively. Normally if a small quantity of water is placed on the surface of a piece of fabric, because the molecules of water exert such strong attractive forces on each other they tend to stay as close to each other as possible (Fig. 32.5(a)) and do not soak into the fabric. If a drop of soap solution is added to this water, the hydrocarbon parts of the soap molecules try to keep away from the water and the salt parts are attracted to the water; thus a lot of the soap molecules arrange themselves

350

between the water molecules on the surface of the water (Fig. 32.5(b)). This in turn makes the attractive forces between the water molecules less effective and allows the water to spread out and wet the fabric. The reduction of surface tension in this way is more easily demonstrated by adding some soapless detergent (next section) to water.

When the fabric is agitated in water containing soap (or soapless detergent), either by hand or in a washing machine, not only is greasy material emulsified but also solid particles of dirt are loosened and removed.

Fig. 32.4 The emulsifying action of soap

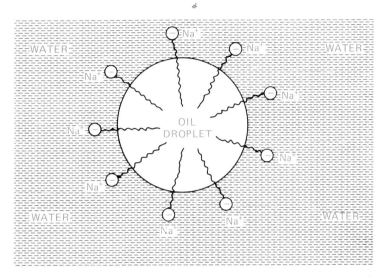

Fig. 32.5 A drop of water will soak into fabric better when detergent has been mixed with it

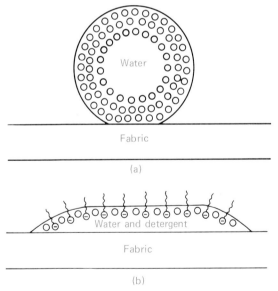

32.11 Modern detergents

The word detergent means 'cleaning agent' and so strictly speaking soap should be regarded as a detergent and the cleaning agents commonly known as detergents should be called soapless detergents.

Modern synthetic detergents are designed in such a way that problems with water hardness do not arise. Many are made by reacting hydrocarbons from petroleum with concentrated sulphuric acid (a major use for sulphuric acid), and converting the product into its sodium salt. A typical sodium alkylsulphate is $C_{15}H_{31}OSO_3^-Na^+$, that is:

351

$$\text{H–C–C–C–C–C–C–C–C–C–C–C–C–C–C–O–S–O}^{-}\ \text{Na}^{+}$$

You will see that the structure is essentially similar to that of soap, in that there is a long alkyl group with an ionic group at one end, and hence they operate in the same way as soap. They reduce the surface tension of water, so enabling the water to wet the fabric more thoroughly, and they emulsify oil by the same mechanism as soap. Their ability to emulsify oil makes them particularly useful when a shipping accident results in an oil-slick on the sea. The oil is sprayed with a solution of detergent which emulsifies the oil and hence disperses it before it is washed into shallower water or on to the coast where it would be more harmful to marine and bird life and also pollute beaches.

Fig. 32.6 Treatment of oil pollution at sea by spraying with detergent. (Crown copyright and courtesy of Warren Spring Laboratory)

One of the important advantages of synthetic detergents is that precipitation with small concentrations of Ca^{2+} or Mg^{2+} ions does not occur and so the problem of scum in hard water does not arise.

Problems have, however, occurred with pollution. Some of the early synthetic detergents were found to be non-biodegradable, i.e. they were not broken down by bacteria and they accumulated in rivers, polluting the water. This problem was overcome when the manufacturers found that if the carbon chain is straight and not branched, the detergent is biodegradable—i.e. broken down by bacteria. Another problem has arisen with certain phosphate additives, which upset the balance of river life by acting as nutrients for algae, to the extent that they proliferate as green scum on the water. Action has now been taken to solve this problem also.

32.12 Summary

1. Ethanol is produced naturally in the fermentation of sugar. It undergoes oxidation to ethanoic (acetic) acid, which is present in vinegar. Ethanol may be dehydrated to ethene.
2. Carboxylic acids are present, as esters, in natural fats. Esters are formed by reaction between a carboxylic acid and an alcohol. They may be hydrolysed using alkali in the saponification reaction employed in the manufacture of soap. Carboxylic acids are weak acids, only slightly ionised in solution. Certain synthetic fibres such as Terylene are polyesters.
3. Soap is the sodium salt of a long-chain carboxylic acid. It has the property of lowering the surface tension of water and emulsifying oil and grease. Modern synthetic detergents act in a similar manner, but are not affected by hard water.

Chapter 33

Ammonia

Investigation 33.1

Fig. 33.1

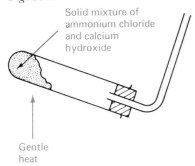

Solid mixture of ammonium chloride and calcium hydroxide

Gentle heat

Question

Questions

What are the properties of ammonia?

You will need a supply of ammonia gas. This can be conveniently prepared by heating a solid mixture of ammonium chloride and calcium hydroxide, which gives off ammonia gas on gentle heating. A suitable apparatus is shown in Fig. 33.1.

Note that the mouth of the test-tube should incline slightly downwards, as water is formed and this would otherwise run back and crack the hot glass.

Gently heat the mixture to produce some ammonia to carry out the following investigations.
1. Invert a test-tube over the delivery tube so as to collect ammonia by upward delivery (ammonia is less dense than air). Leave the tube in place for about half a minute and then place it with its mouth under water in a beaker of water. Observe what happens.

1 How would you describe the solubility of ammonia from what you see in this experiment?

2. Wet the end of a glass rod with concentrated hydrochloric acid, and hold it near the end of the delivery tube. Observe what happens.

2 Since ammonia is an alkaline gas and hydrochloric acid gives off an acidic gas (hydrogen chloride), what type of compound is likely to be formed in this reaction?
3 What is the substance formed?

3. Hold a burning splint close to the end of the delivery tube to see whether ammonia gas burns in air.

Investigation 33.2

Questions

What are the properties of an aqueous solution of ammonia (ammonium hydroxide)?

1. Put 3–4 cm³ of dilute ammonia solution in a clean evaporating basin. Cautiously smell the liquid and test the liquid with red and blue litmus papers.
   Now, with the basin on a tripod and gauze, gently boil the liquid and test the vapour with moist red litmus paper.
   Allow the liquid to boil away, but stop heating just before it goes dry, and observe whether any significant residue is left.

1 Is the liquid acidic or alkaline?
2 Is the solute given off with the steam on boiling or is it left as a residue?
3 Is the solute volatile?

2. To about 2 cm³ of dilute ammonia solution in a test-tube, add 2 or 3 drops of litmus solution.

Now add dilute hydrochloric acid, a few drops at a time, with shaking, until the indicator changes colour.

Pour about half the liquid into a clean dish and smell it to see if it smells of ammonia.

To the remaining liquid in the test-tube, add about an equal volume of sodium hydroxide solution and smell the mixture.

Finally, warm the tube, holding moist red litmus paper at the mouth, and observe what happens to the litmus.

Questions

4 What type of reaction takes place between ammonia solution and an acid?

5 What type of substance will be formed in this reaction?

6 Is ammonia given off after addition of sodium hydroxide solution?

3. Into four separate test-tubes put about 2 cm³ of solutions of iron(II) sulphate, iron(III) chloride, magnesium sulphate and copper(II) sulphate respectively.

To each tube add about 1 cm³ (no more) of dilute ammonia solution and stir with a glass rod. Observe what happens.

Now, fill each tube to about two-thirds full with ammonia solution and stir well. You should observe that one mixture behaves differently from the others; decide which one it is and note what has happened.

Questions

1 What were the four precipitates?

2 In which tube did the precipitate behave differently when excess ammonia solution was added?

3 What happened to the precipitate in this example?

Investigation 33.3

What are the properties of ammonium chloride?

Ammonium chloride is the salt formed when ammonia reacts with hydrogen chloride, or when ammonia solution reacts with hydrochloric acid.

1. Put solid ammonium chloride into a test-tube to a depth of about 0·5 cm. Fold a piece of moist red litmus paper over the rim of the tube and gently heat the ammonium chloride. Continue heating for some time.

Observe what happens (i) at the bottom of the tube, (ii) in the middle of the tube, (iii) near the top of the tube.

Question

1 What is the word used to describe the type of change in state that ammonium chloride undergoes when heated?

2. Put about 2 cm³ of ammonium chloride solution in a test-tube and note whether it smells of ammonia.

Add an equal volume of sodium hydroxide solution and smell it again.

Now heat the liquid gently, holding moist red litmus paper at the mouth of the tube.

Questions

2 Does ammonium chloride solution have a definite smell?

3 What gas is given off when ammonium chloride is warmed with sodium hydroxide solution?

3. Repeat part 3, using ammonium sulphate solution instead of ammonium chloride solution.

Questions

4 What gas, if any, is given off this time?
5 What ion in the salt reacts with sodium hydroxide solution, liberating this gas?
6 From the results of parts 3 and 4, what is a test for an ammonium salt?

33.4
The elements of life

All types of life form part of a natural cycle. A plant or animal is born, then it takes in matter from its surroundings as it grows and lives, then eventually it dies. If it dies naturally it rots away and its constituent matter returns to the surroundings. If it is eaten for food, then its constituent matter becomes part of another living organism which, in its turn, will die.

The four most important elements from which living things are made up are carbon, hydrogen, oxygen and nitrogen. However, for healthy growth, plants and animals need many other elements from the soil or from the atmosphere. The elements which are removed in the greatest quantities from soil when plants grow in it are nitrogen, phosphorus and potassium. For a soil to be fertile these elements must be present. If they are missing, or if they are removed by growing successive crops in the soil, they must be added in the form of **artificial fertilisers** (33.10) which will make the soil fertile.

Other elements such as iron and manganese are also essential, but they are removed in such small quantities by growing plants that they rarely need to be added in fertilisers.

The part which carbon plays in the life cycle has been discussed in some detail (9.4, 28.8). The carbon cycle involves the use of the sun's energy to build up starch (carbohydrate) in plants from carbon dioxide in the atmosphere. This starch eventually becomes our food, from which we obtain energy (9.4). This chapter is mainly concerned with the important part played by **nitrogen** in the life cycle.

33.5
Proteins

Quite a large proportion of animal matter is protein—about 15% for a human being. We have probably all heard of protein, but what is it? Protein is a vital constituent of living cells, and it occurs particularly in muscles, blood, cartilage and hair. It is essential for us to have a supply of protein in our diet. Foods that are particularly rich in protein are meat, fish, milk, cheese, and various kinds of nuts and beans.

Proteins are organic nitrogen compounds, and their importance to living matter is that they consist of very long-chain molecules which can be used for building up fibrous material as required for muscles, hair, cartilage, skin.

The long-chain molecules of protein are built up of amino acid units. Amino acids contain the functional group,

$$
\begin{array}{c}
H \qquad\qquad H \\
\diagdown \qquad\quad \diagup \\
N \qquad\quad O \\
| \qquad\quad \diagup\!\!\diagup \\
-C - C \\
| \qquad\quad \diagdown \\
H \qquad\quad O-H
\end{array}
$$

the amino part being the $-NH_2$ group.

There are many amino acids, the simplest being aminoethanoic acid, which is usually known as glycine:

355

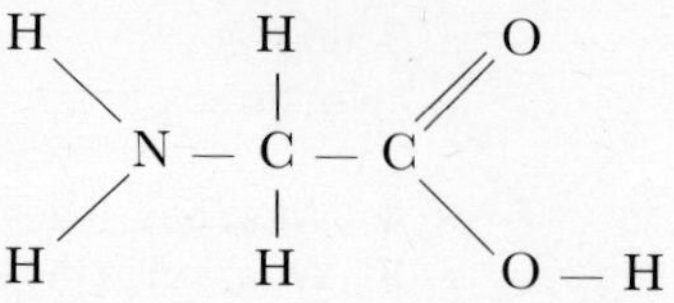

$$\text{H} \diagdown \text{N} - \underset{|}{\overset{|}{\text{C}}} - \text{C} \diagup \overset{\text{O}}{\diagdown} \text{O} - \text{H}$$

Protein molecules are built up by condensation (elimination of water) between amino acids:

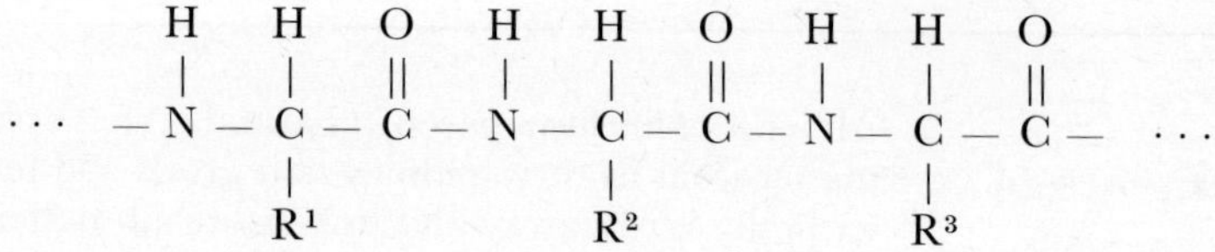

$$\cdots - \underset{\underset{R^1}{|}}{\overset{\overset{H}{|}}{N}} - \underset{\underset{|}{|}}{\overset{\overset{H}{|}}{C}} - \underset{}{\overset{\overset{O}{\|}}{C}} - \underset{}{\overset{\overset{H}{|}}{N}} - \underset{\underset{R^2}{|}}{\overset{\overset{H}{|}}{C}} - \overset{\overset{O}{\|}}{C} - \overset{\overset{H}{|}}{N} - \underset{\underset{R^3}{|}}{\overset{\overset{H}{|}}{C}} - \overset{\overset{O}{\|}}{C} - \cdots$$

The section of chain shown here contains three amino acid units ('R' stands for some unspecified organic group). An actual protein may contain thousands of such units.

When we eat proteins (as in meat, for example), the protein molecules are hydrolysed into their separate amino acids by enzymes in the digestive system. These amino acids are then used by the body for replenishing its own protein. The condensation of the amino acids in the correct order to synthesise bodily protein is controlled by the nucleic acids such as DNA (deoxyribonucleic acid).

Some of the amino acids from the protein in food are not used in building up new protein, but are simply oxidised to obtain energy, just as carbohydrate is oxidised. In this case some of the nitrogen from these amino acids is converted to carbamide (urea),

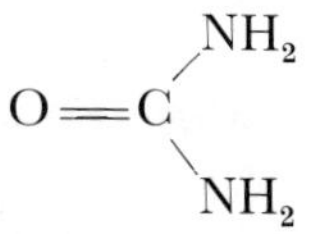

$$\text{O} = \text{C} \diagup^{\text{NH}_2}_{\diagdown \text{NH}_2}$$

which is excreted in the urine.

Where does the nitrogen in proteins come from? The answer is, of course, that ultimately it comes from the air, which is almost 80% nitrogen. The way in which this happens is considered in the next section.

33.6 The nitrogen cycle

A simplified form of the nitrogen cycle is shown in Fig. 33.2.

Some plants of the Leguminosae family, meaning certain types of pod-bearing plants like peas and beans, can absorb atmospheric nitrogen directly through their roots, but most plants must absorb nitrogen as nitrates from the soil. How do nitrates get into the soil? This happens naturally in two ways, as shown in Fig. 33.2. Firstly, death and decay of plants and animals (and also animal excretion) return nitrogen to the soil as ammonium salts. These undergo oxidation, with the aid of bacteria, to nitrates. Secondly, lightning flashes in thunderstorms cause partial combination of oxygen and nitrogen in the air to form nitrogen monoxide:

$$N_2(g) + O_2(g) \rightleftharpoons 2NO(g)$$

This undergoes further reaction with air to form nitrogen dioxide, $NO_2(g)$, which then dissolves in rain and, with further oxidation, forms nitric acid:

$$4NO_2(g) + O_2(g) + 2H_2O(l) \rightarrow 4HNO_3(aq)$$

This reacts with minerals in the soil to form nitrates.

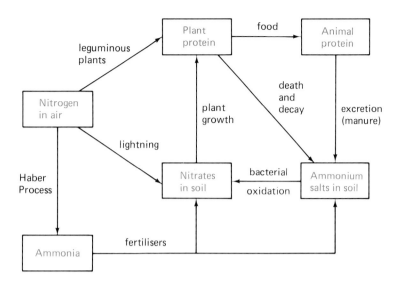

Fig. 33.2 The nitrogen cycle

These natural processes, though once sufficient to maintain the nitrogen balance, are no longer sufficient because of the greatly increased demands of modern agriculture. Consequently, the nitrogen supply in the soil must now be supplemented by synthetic fertilisers. Modern fertilisers are obtained mainly from **ammonia**, manufactured by the Haber Process (33.9) and then converted to salts such as ammonium sulphate and ammonium nitrate. In this way the conversion of atmospheric nitrogen to nitrogen compounds in the soil enables us to keep pace with the rapid removal of nitrogen from the soil by intensive farming.

33.7 Nutrition

As may be seen from the last section, man's food requirements involve more than just a source of energy. Carbohydrates provide energy, but nothing else. Proteins are essential in the diet, to replace the protein of the body. In addition to this we need various mineral compounds, such as those containing calcium and phosphorus, to build up our bones. We also need vitamins, and many trace elements such as iodine, iron, zinc and chromium. The study of nutritional requirements is very important in the maintenance of health in the population, especially to make sure that in the modern large-scale food industry the processes used to prepare food do not result in a deficiency of any essential nutritional substances. If deficiencies are found, they must be made up by addition of suitable synthetic materials.

Food additives have come under fashionable criticism in recent years, but they are not necessarily a bad thing, indeed they may be essential to our health. The idea that 'chemical' additives are in some way inferior to 'natural' substances is nonsense, and would seem to be a legacy of the 'Vital Force' theory of organic chemistry which was discredited over a century ago. Vitamin C is Vitamin C, whether it is obtained by eating oranges or whether it is made in a factory and bought as a solid in a bottle.

33.8 Ammonia and the nitrogen cycle

If you have a baby brother or sister and you have been there during the changing of the nappy, you may have noticed a rather pungent and distinctive smell which, if very strong, brings tears to your eyes. This is the smell of ammonia which has been formed by the action of water in the presence of bacteria on a substance called urea, which is present in urine. You may also have noticed the same smell in some domestic cleansers.

Ammonia has an important part to play in modern life. Fig. 33.3 shows the growth in production of ammonia compared to the growth in the world population. The increased demand for ammonia is due to the increasing need for artificial fertilisers to supplement the nitrogen which is naturally absorbed into the soil.

One of the first artificial fertilisers to be added to the soil to supply soluble nitrogen was sodium nitrate, or Chile Saltpetre. This is a mineral found, as its name suggests, in Chile, which has been exported as a fertiliser since about 1830. In time, however, the supplies of this substance became smaller and the cost of transporting it to Europe became too high, so alternatives had to be sought.

Fig. 33.3 The growth in world ammonia production compared to the growth in world population

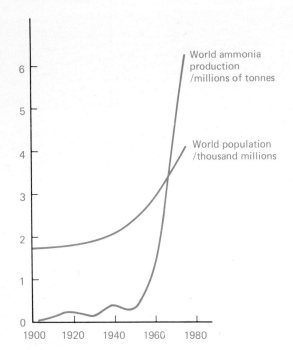

Fig. 33.4 An ammonia plant. (Courtesy ICI Ltd Agricultural Division)

About 1900 two processes for converting the nitrogen in the air into soluble nitrogen compounds were in use. The cyanamide process, which produced calcium cyanamide ($CaCN_2$), and the electric arc process, which resulted in nitric acid, required too much energy, however, and these soon faded into the background when Fritz Haber, working in Karlsruhe, managed in 1908, at the end of four years work, to find the conditions necessary for nitrogen and hydrogen to combine directly to give a reasonable yield of ammonia. Although himself not skilled in making equipment, Haber designed, and had made, an apparatus which in 1909 he was able to use to show that an acceptable yield of ammonia could be produced. The rights to the process were bought in that year by a large German chemical firm, Badische Anilin- und Soda-Fabrik (BASF), and one of their brilliant chemical engineers, Carl Bosch, undertook the huge task of developing Haber's experimental apparatus into a large manufacturing plant.

The main problem which Bosch had to overcome was that Haber had suggested the use of high pressure, which meant that strong plant was required, and the second was to find the best catalyst for the process to replace the expensive osmium which Haber had suggested and used. Bosch's team eventually found that finely-divided iron, containing oxides of potassium, aluminium and calcium, gave the best results, and this catalyst is still used today. They also found a cheap method of getting the large amounts of hydrogen needed, steam being passed over red hot coke (18.4). By 1913 Bosch had a complete works operating at Oppau on the Rhine and was making 30 tonnes of ammonia a day. By the end of 1915, with the Oppau plant having been enlarged and a second one built, the German works were fixing 180 000 tonnes of nitrogen a year (largely at that time for conversion to explosives for the German Army in the 1914–18 war). Haber was awarded the Nobel Prize for Chemistry in 1918 and Bosch shared the prize in 1931.

33.9 The Haber Process

The reaction at the heart of the **Haber Process** for the manufacture of ammonia is represented by the equation:

$$N_2(g) + 3H_2(g) \rightleftharpoons 2NH_3(g)$$

As the equation shows, the reaction is reversible and the problem which Haber had to solve was to obtain a satisfactory yield at a cost low enough for the product to be cheap. It would be no good manufacturing ammonia at the price of gold; no one would be able to buy enough. A prime consideration is that the reaction has to be reasonably fast, and this means using a catalyst and temperatures which, to the chemical engineer, are moderately high (about 500 °C). Temperatures of this order create a problem. The reaction between nitrogen and hydrogen is exothermic and, according to Le Chatelier's Principle (22.16), we could get a better yield at equilibrium by lowering the temperature. If we did this, however, the reaction would be far too slow and it would take much too long for equilibrium to be reached. So, we have to use moderately high temperatures. There still remains, however, one thing we can vary—the pressure.

The reaction involves a decrease in the number of molecules of gas from 4 to 2, and so by Avogadro's Law (5.6) it follows that if the reaction were carried out at constant temperature, the volume of gas would decrease. Le Chatelier's Principle then tells us that we can improve the equilibrium yield by carrying out the reaction at high pressure, and this also gives the bonus of making the reaction faster. In the modern process a pressure of 200 atmospheres is used and, as was mentioned earlier, this considerably increases the cost of the process, since the reaction vessels have to have thick walls to withstand this pressure (200 atmospheres is nearly $\frac{1}{4}$ tonne per square centimetre). Considerable energy also has to be used in compressing the gases, but, even despite these factors, the process is still superior to the older ones as far as cost is concerned.

A mixture of nitrogen and hydrogen, in the ratio of 1 volume to 3 volumes, is produced by the action of steam (18.4) and then air on methane from natural gas. After compression to 200 atmospheres, it is passed at 500 °C over the iron catalyst. The mixture of gases emerging from the reaction vessel contains about 15% of ammonia. As it is cooled, because it is at a high pressure, the ammonia liquefies and is run off. The unchanged mixture of nitrogen and hydrogen is recycled until it is all eventually converted to ammonia. The product is stored and transported as a liquid under pressure in specially-designed tanks.

33.10 Fertilisers

About 80% of ammonia manufactured by the Haber Process is used for making fertilisers. In the United States ammonia is sometimes injected directly into the soil, but it is more usual to react it with acid to produce a solid salt, which is easy to store and transport. The two most important ones are ammonium sulphate and ammonium nitrate (sold as Nitram), both containing high proportions of nitrogen (21% in the sulphate and 35% in the nitrate). These two fertilisers are made by reacting ammonia gas with solutions of the appropriate acid, sulphuric for ammonium sulphate and nitric for ammonium nitrate.

The two most important elements, besides nitrogen, which need to be added to the soil by means of fertilisers are phosphorus (P) and potassium (K) (33.4). Most fertilisers are used either to add nitrogen (N) or to add all three elements, N, P and K. The mixed fertiliser usually contains ammonium nitrate, ammonium phosphate and

Fig. 33.5 A sack of fertilizer showing the NPK values of the fertilizer. (Courtesy ICI Ltd Agricultural Division)

Fig. 33.6 The wheat in the centre has not been treated with the appropriate fertiliser. (Courtesy ICI Ltd Agricultural Division)

TABLE 33.1

potassium chloride in varying proportions. The N.P.K analysis of the mixture is usually quoted on the container.

A fertiliser with appropriate N.P.K values is selected for a particular soil and crop. The selection must be done carefully, because clearly it would be wasteful to add more of an element than was necessary and also there is a possibility of polluting waterways if excessive quantities of fertilisers are washed into them by rain (10.13). The composition of a soil can be found by analysis, but an indication that there is probably a shortage of a particular element can be obtained by recognising symptoms such as those in Table 33.1.

MISSING ELEMENT	SYMPTOM
Nitrogen (N)	Harsh fibrous leaves
Phosphorus (P)	Grey leaves, stunted growth
Potassium (K)	Premature death of leaves

Fig. 33.7 summarises the processes for making N.P.K fertiliser. A mixture of ammonium nitrate and ammonium phosphates is made by reacting ammonia gas with a mixture of nitric acid and phosphoric acid solutions:

$$NH_3(g) \quad + \quad HNO_3(aq) \quad \rightarrow \quad NH_4NO_3(aq)$$

$$2NH_3(g) \quad + \quad H_3PO_4(aq) \quad \rightarrow \quad (NH_4)_2HPO_4(aq)$$
$$\text{phosphoric}$$
$$\text{acid}$$

The proportion of nitrogen and phosphorus in the product which can be crystallised from the solution can be varied by varying the composition of the acid mixture.

The mixture of ammonium nitrate and ammonium phosphate is then mixed with potassium chloride which is obtained from mineral deposits.

Phosphoric acid is made from a mineral called rock phosphate (calcium phosphate, $Ca_3(PO_4)_2$) and the nitric acid is made from ammonia (34.4).

The conversion of ammonia to nitric acid is its second major use. Other uses include the Solvay Process for making sodium carbonate (29.5), the manufacture of nylon and polyurethane, and the production of dyestuffs and explosives.

360

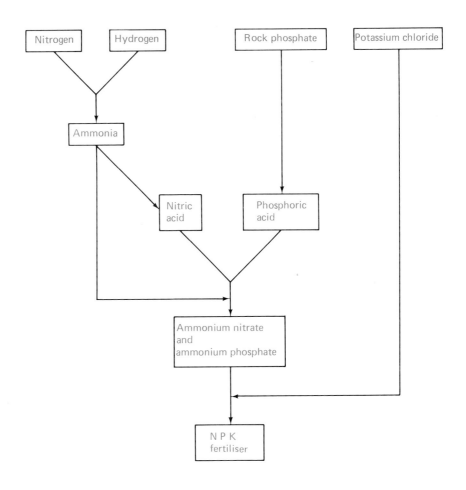

Fig. 33.7 Production of NPK fertiliser

**33.11
Properties of
ammonia**

Ammonia is a colourless gas with a distinctive pungent smell which can bring tears to the eyes if the concentration is high enough. It is less dense than air and is easily liquefied, either by applying pressure or by cooling to $-33\,°C$ at atmospheric pressure. The gas is very soluble in water, one volume of water at room temperature dissolving about 700 volumes of ammonia gas. Since its solubility is of the same order as that of hydrogen chloride, the Fountain Experiment (25.6) can be done with ammonia gas in the same way as with hydrogen chloride, the only difference being that with ammonia the beaker should contain red litmus solution rather than blue. This is because ammonia is the only common gas which is alkaline to moist litmus, and so the red litmus solution turns blue when it enters the flask of ammonia.

**33.12
Ammonia
solution
(ammonium
hydroxide)**

Amongst the bottles of solutions in your laboratory, there is probably at least one labelled 'Ammonium Hydroxide'. This is a widely-used reagent, particularly in analysis to find the identity of unknown compounds, so what is it and how is it made?

If you remove the stopper from the bottle you will notice immediately that the solution smells of ammonia, and if the solution is warmed with a piece of red litmus paper above it the paper turns blue, showing that ammonia is being given off. Evaporation to dryness leaves no residue. Thus it seems that ammonium hydroxide solution is simply a solution of ammonia in water.

361

There are, however, two bits of evidence to contradict this. Firstly, its ability to conduct electricity indicates that it is an electrolyte and must contain ions. However, it does not conduct as well as sodium hydroxide solution and must therefore contain fewer ions. Secondly, the pH of the solution is high (11.13), that is, the solution is alkaline, containing hydroxide ions (OH^-). This suggests that as the ammonia dissolves it reacts to some extent with the water, producing hydroxide ions and therefore ammonium ions as well:

$$NH_3(g) + H_2O(l) \rightleftharpoons NH_4^+(aq) + OH^-(aq)$$

The solution is probably an equilibrium mixture of the four things, free ammonia, free water, ammonium ions and hydroxide ions. It smells of ammonia because of the free ammonia, and it is alkaline because of the hydroxide ions which are present. There is much more free ammonia in the solution than ammonium ions, and this is why the electrical conductivity is small and why ammonium hydroxide is considered to be a **weak alkali**. A solution of sodium hydroxide of the same concentration would contain a greater concentration of hydroxide ions and is therefore classed as a **strong alkali**.

Although $NH_3(aq)$ is a more accurate representation of ammonia solution than $NH_4OH(aq)$, it is sometimes more convenient to use the latter. The formula NH_4OH has a useful similarity to the formulae of the common strong alkalis, sodium hydroxide ($NaOH$) and potassium hydroxide (KOH), to whose properties in aqueous solution ammonia solution shows a clear resemblance. The reactions of aqueous ammonia show that it contains both hydroxide ions and free ammonia. In its reactions with acid and with solutions of metal salts, the hydroxide ions are reacting, while in its reaction with some metal hydroxides the free ammonia is showing its presence.

1. Reaction of ammonia solution with acids

Because of the presence of hydroxide ions, ammonia solution will react with acids just as any other alkali does. Neutralisation occurs and a salt is formed,

e.g. $$NH_4OH(aq) + HCl(aq) \rightarrow NH_4Cl(aq) + H_2O(l)$$

or $$NH_3(aq) + HCl(aq) \rightarrow NH_4Cl(aq)$$

Since the ammonium ion has a charge of $+1$, ammonium salts have similar formulae to the corresponding sodium and potassium salts. Like them also, all common ammonium salts are soluble in water.

You might think that since ammonia solution is a weak alkali, containing comparatively few hydroxide ions, it would only take a tiny bit of acid to neutralise it. This is not so. The addition of acid removes hydroxide ions from the solution:

$$H^+(aq) + OH^-(aq) \rightarrow H_2O(l)$$

By Le Chatelier's Principle, more ammonia molecules will react with water to form hydroxide ions to replace those lost in the reaction,

$$NH_3(aq) + H_2O(l) \rightarrow NH_4^+(aq) + OH^-(aq)$$

and these immediately combine with hydrogen ions from the acid to form more water. The process will continue until virtually all the ammonia has combined with the water and has been converted into ammonium ions. The resulting solution has no smell because all the ammonia has been converted into ions.

2. Reactions of ammonia solution with solutions of metal salts

Ammonia solution will, like sodium hydroxide solution, precipitate an insoluble metal hydroxide when added to a solution of a metal salt,

e.g. $$3NH_4OH(aq) + FeCl_3(aq) \rightarrow Fe(OH)_3(s) + 3NH_4Cl(aq)$$

(compare this with:

$$3NaOH(aq) + FeCl_3(aq) \rightarrow Fe(OH)_3(s) + 3NaCl(aq) \quad)$$

The hydroxide ions in the ammonia solution come together with the metal ions in the metal salt solution, forming the solid metal hydroxide:

$$Fe^{3+}(aq) + 3OH^-(aq) \rightarrow Fe(OH)_3(s)$$

As in the reaction of ammonia solution with acids, all the dissolved ammonia takes part in the reaction, even though only a small part of it is initially ionised.

3. **Reactions of ammonia solution with some metal hydroxides**

If ammonia solution is added to a solution of copper(II) sulphate, a light blue precipitate of copper(II) hydroxide first forms, but this then dissolves as more of the ammonia solution is added, giving a deep blue solution. The latter contains complex ions, in which ammonia molecules from the ammonia solution have been joined by co-ordinate or dative covalent bonds to the copper(II) ions (15.5):

$$Cu^{2+}(aq) + 4NH_3(aq) \rightarrow Cu(NH_3)_4^{2-}(aq)$$

The complex ion is the tetraamminecopper(II) ion. Zinc hydroxide and silver oxide will also dissolve in ammonia solution, forming complex ions.

<table>
<tr><td>

**33.13
The action of strong alkalis on ammonium salts**

</td><td>

A solution of ammonium chloride does not smell of ammonia because the salt is completely ionic. If sodium hydroxide solution is added, however, there is a clear smell of ammonia and, if the mixture is heated, ammonia gas is given off. This behaviour can be explained by applying Le Chatelier's Principle once again to the equilibrium represented by the equation:

</td></tr>
</table>

$$NH_3(aq) + H_2O(l) \rightleftharpoons NH_4^+(aq) + OH^-(aq)$$

The sodium hydroxide solution brings along a large concentration of hydroxide ions and Le Chatelier's Principle tells us that the equilibrium is disturbed in such a way that the hydroxide ions are absorbed. This can only happen by the hydroxide ions combining with the ammonium ions in the ammonium chloride solution, forming more ammonia in the solution. When the mixture is heated, the ammonia becomes less soluble and some of it is driven out as the gas. Again Le Chatelier's Principle operates, the equilibrium being disturbed to produce more ammonia to replace that which has gone. Eventually, as more ammonia is driven off, the reaction will go to completion:

$$NH_4Cl(aq) + NaOH(aq) \rightarrow NaCl(aq) + H_2O(l) + NH_3(g)$$

The same thing will happen when any ammonium salt is warmed with any strong alkali and it follows that this can be the basis of the test for an ammonium salt (or the ammonium ion). The test simply consists of warming the compound with sodium hydroxide solution. An ammonium salt will give ammonia gas which can be identified by its smell and by its ability to turn moist red litmus paper blue.

<table>
<tr><td>

**33.14
Laboratory preparation of ammonia gas**

</td><td>

The laboratory preparation of ammonia is based on the principle discussed in the previous section, the action of a strong alkali on an ammonium salt. A convenient ammonium salt to use is ammonium chloride but, because sodium and potassium hydroxides are deliquescent and therefore difficult to work with in the solid state, it is preferable to use, as the alkali, slaked lime (calcium hydroxide). Solid ammonium chloride and solid calcium hydroxide are thoroughly mixed and when the mixture is gently heated, ammonia is given off at a convenient rate:

</td></tr>
</table>

$$Ca(OH)_2(s) + 2NH_4Cl(s) \rightarrow CaCl_2(s) + 2NH_3(g) + 2H_2O(l)$$

363

If dry ammonia gas is required, it must be dried with quicklime (calcium oxide). The dry gas is collected by upward delivery, since it is less dense than air.

A suitable apparatus for the preparation and collection of the dry gas is shown in Fig. 33.8.

The other drying agents which are normally used for drying gases will not work in this preparation. Concentrated sulphuric acid would absorb the ammonia, forming ammonium sulphate, while anhydrous calcium chloride reacts with the gas, forming complex ions.

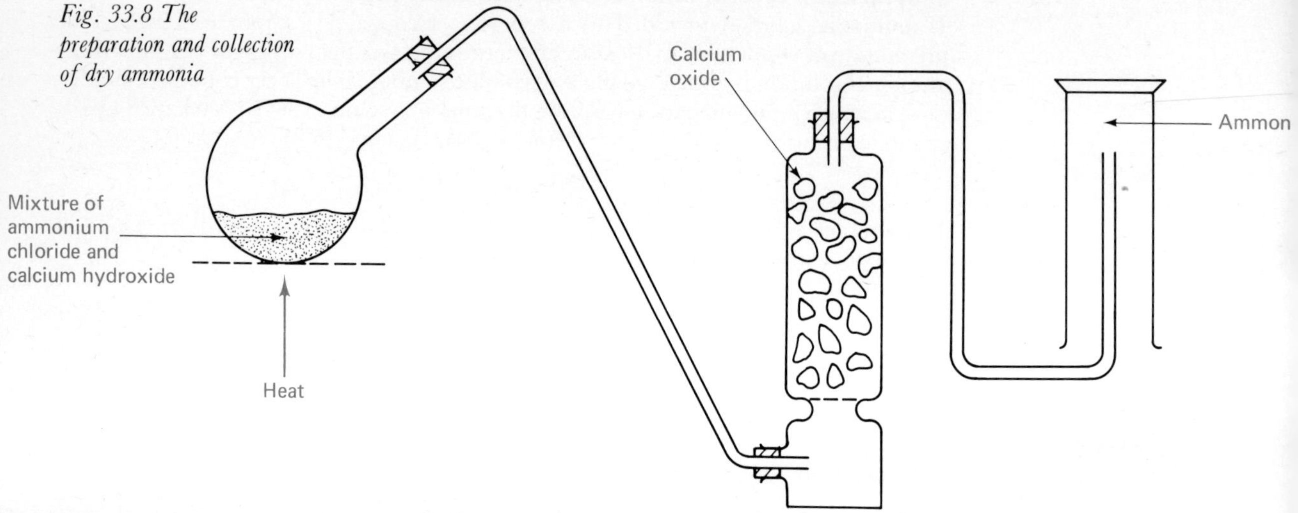

Fig. 33.8 The preparation and collection of dry ammonia

Calcium oxide

Ammon

Mixture of ammonium chloride and calcium hydroxide

Heat

33.15 Reactions of ammonia

1. **Reaction with hydrogen chloride**

The reaction of a solution of ammonia in water with acids has already been mentioned, but ammonia will also act as a base when it is alone. This is shown when a jar of dry ammonia is held mouth-to-mouth with a jar of dry hydrogen chloride. The two gases are colourless, but when they mix the jars are filled with a white 'smoke' consisting of solid particles of ammonium chloride:

$$NH_3(g) + HCl(g) \rightarrow NH_4Cl(s)$$

If the jars are allowed to stand for a few minutes, the white solid settles to the bottom of the jars.

Fig. 33.9 Action of heat on ammonium chloride

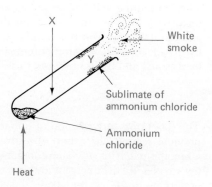

X

White smoke

Y

Sublimate of ammonium chloride

Ammonium chloride

Heat

If some solid ammonium chloride is taken from the bottle in the laboratory and gently heated in a test-tube, the white smoke appears again and the solid seems to **sublime**, i.e. it vaporizes, without melting to a liquid, and on cooling the vapour condenses to reform the solid (known as the **sublimate**). Fig. 33.9 shows what happens in the test-tube.

Salts do not usually evaporate like this, and the fact that the heating of ammonium chloride is not just simple sublimation is indicated when a piece of damp red litmus paper is held in the tube at Y. The litmus paper turns blue, showing that ammonia is being given off, and what is happening is that the ammonium chloride breaks down on heating into ammonia gas and hydrogen chloride gas, a mixture of the two being present in the region X in the test-tube:

$$NH_4Cl(s) \rightleftharpoons NH_3(g) + HCl(g)$$

This is the reverse of the reaction which takes place when the two gases are mixed at room temperature and, as the mixture rises up the tube, it becomes cooler so that the two can recombine to form the white smoke of ammonium chloride particles and a deposit of solid ammonium chloride near the top of the tube. The breakdown of ammonium chloride on heating is called **thermal dissociation**, rather than thermal decomposition, because the products of the breakdown recombine on cooling to reform the original substance. In thermal decomposition the reaction is not reversible and no recombination of the products takes place.

2. **Oxidation of ammonia**
Ammonia in aqueous solution is not easily oxidised, even by a strong oxidising agent such as potassium manganate(VII) solution, but the gas itself can be oxidised in several ways.

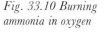

Fig. 33.10 Burning ammonia in oxygen

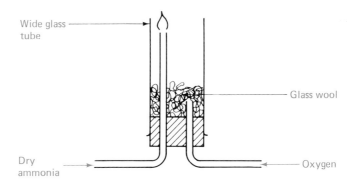

Wide glass tube

Glass wool

Dry ammonia

Oxygen

Reaction with air
Ammonia gas does not burn unaided in ordinary air but, if the air is mixed with oxygen so that the percentage of oxygen in the mixture is over 40, the ammonia burns with a yellow flame. This can be shown using the apparatus in Fig. 33.10. Under these conditions only the hydrogen ends up as an oxide, the nitrogen being set free:

$$4NH_3(g) + 3O_2(g) \rightarrow 2N_2(g) + 6H_2O(g)$$

Reaction with hot copper(II) oxide
A similar reaction to that with oxygen-enriched air takes place when ammonia is passed over hot copper(II) oxide. Once again only the hydrogen becomes an oxide, the nitrogen being set free, but this time the oxygen to combine with the hydrogen is supplied by the copper(II) oxide which is reduced to copper:

$$2NH_3(g) + 3CuO(s) \rightarrow 3Cu(s) + N_2(g) + 3H_2O(g)$$

The black solid turns brown, the reaction being very similar to the action of hydrogen on hot copper(II) oxide:

$$H_2(g) + CuO(s) \rightarrow Cu(s) + H_2O(g)$$

365

Catalytic oxidation

If a mixture of ammonia and oxygen is passed over the surface of a piece of platinum, the oxygen combines with both elements in the ammonia, forming steam and nitrogen monoxide:

$$4NH_3(g) + 5O_2(g) \rightarrow 4NO(g) + 6H_2O(g)$$

The reaction is highly exothermic, as can be shown by holding a coil of warm platinum wire above the surface of some concentrated ammonia solution with air being blown through a tube towards the coil (Fig. 33.11).

The coil glows red hot and it may be possible to see brown fumes in the flask as the nitrogen monoxide combines with oxygen in the air, forming nitrogen dioxide:

$$2NO(g) + O_2(g) \rightarrow 2NO_2(g)$$

This catalytic oxidation of ammonia is the essential step in the manufacture of nitric acid from ammonia (34.4).

Fig. 33.11 The catalytic oxidation of ammonia

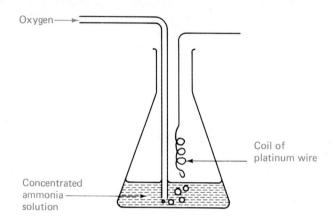

Oxygen

Coil of platinum wire

Concentrated ammonia solution

33.16 Summary

1. Nitrogen, an element present in protein, plays a vital part in the life cycle. We need proteins in our diet, and these come from foods such as meat, fish and cheese. To keep pace with modern agriculture, nitrogen must be returned to the soil by means of artificial fertilisers such as ammonium nitrate made via the Haber Process.
2. In the Haber Process for synthesising ammonia, a mixture of nitrogen (1 volume) and hydrogen (3 volumes) at about $500\,^\circ$C and a pressure of about 200 atmospheres is passed over an iron catalyst.
3. Ammonia is very soluble in water, producing ammonium hydroxide. The solution contains free ammonia, ammonium ions and hydroxide ions and is a weak alkali. It is neutralised by acids forming ammonium salts.
4. Ammonium salts are ionic compounds which are decomposed by heat. They react with strong alkalis, liberating ammonia.
5. Ammonia is prepared in the laboratory by heating ammonium chloride with calcium hydroxide. It is dried by passing through quicklime and is collected by upward delivery.
6. Ammonia can be oxidised by oxygen in the presence of a platinum catalyst forming nitrogen monoxide and steam.

Chapter 34

Nitric acid, nitrates and related compounds

Investigation 34.1

What are the properties of nitric acid?

1. Put about 2 cm³ of sodium hydroxide solution in a test-tube and add 2 or 3 drops of litmus solution. Then add dilute nitric acid, a few drops at a time, with shaking, until you see a definite change.

2. Put a very small portion of copper(II) oxide into a test-tube and add about 3 cm³ of dilute nitric acid. Warm the mixture gently and observe what happens.

3. Add about 3 cm³ of dilute nitric acid to a small portion of copper(II) carbonate (or any other carbonate provided) in a test-tube. Observe what happens and pour the gas which is given off into a test-tube containing a little lime water. Close the tube of lime water with your thumb and shake it vigorously.

Question

1 On the evidence of the first three parts of this investigation, in what ways does nitric acid behave as a typical acid?

4. For this investigation use each of the metals, magnesium, zinc, iron and copper.

Place a small portion of one of the metals in a test-tube and add about 3 cm³ of dilute nitric acid. If it appears that no reaction is taking place, warm the acid gently (do not boil). Look for effervescence occurring, and if it does, test for hydrogen by holding a second test-tube mouth-to-mouth with the reaction tube and then opening it near a Bunsen flame.

Also look for any colour in the vapour and very cautiously (by waving some of it towards you with your hand) smell it.

Repeat the procedure with each of the other metals.

Questions

For each of the metals you have used:
2 Does it appear to react with dilute nitric acid?
3 If it does react, is hydrogen given off?
4 What would happen if dilute hydrochloric acid or dilute sulphuric acid was added to this metal?

Investigation 34.2

How does nitric acid react with an iron(II) salt?

To a few crystals of iron(II) sulphate in a boiling-tube add about 4 cm depth of dilute sulphuric acid and shake the mixture until the crystals dissolve.

Pour a small portion of the solution into a test-tube and then add dilute sodium hydroxide solution until an obvious change occurs.

To the remaining solution in the boiling-tube add about 4 cm³ of dilute nitric acid. Warm the mixture until a definite change takes place. Pour some of the resulting solution into a test-tube and add dilute sodium hydroxide solution until a definite change occurs.

1 What did you see in the original iron(II) sulphate solution when sodium hydroxide solution was added?

2 Which of the ions present in iron(II) sulphate solution cause this change?

3 What did you see when sodium hydroxide solution was added to the mixture after warming with nitric acid?

4 What type of ion must therefore be present in the reaction mixture after warming with nitric acid?

5 What type of change has resulted from warming the original solution with nitric acid?

Investigation 34.3

What happens to nitrates when they are heated?

You will need a wooden splint, pieces of blue litmus paper and pieces of filter paper (or blotting paper) which can be soaked in acidified potassium manganate(VII) solution.

1. Place copper(II) nitrate crystals in an ignition-tube to a depth of about 3 mm. These crystals contain water of crystallisation and the first effect of heating them is to drive off the water. While this is happening, the tube should be held almost horizontally so that condensed steam does not run back and crack the hot glass.

Heat the ignition-tube and when coloured fumes begin to appear, hold a glowing splint in the vapour above the solid. Now hold in the vapour a piece of moist blue litmus paper and then a strip of paper soaked in acidified potassium manganate(VII) solution.

Finally examine the residue left after fumes are no longer given off.

Questions

1 What gas is shown to be given off by its action on a glowing splint?

2 What colour are the fumes which are given off?

3 What property of this coloured gas is indicated by its action on litmus paper?

4 What property of the coloured gas is indicated by its action on acidified potassium manganate(VII) solution?

5 What do you think the final solid residue was?

2. Repeat the procedure with other nitrates selected from calcium nitrate, magnesium nitrate and lead(II) nitrate.

3. Repeat the above procedure using either sodium nitrate or potassium nitrate. The solid will melt and then it will begin to effervesce. At this stage, with the tube still in the Bunsen flame, hold a glowing splint in the tube just above the molten compound, but do not let the splint touch the compound.

When no more effervescence takes place, let the tube cool and then add a little water. Warm the tube gently so that some of the residue dissolves. Then add a few drops of a very dilute acidified solution of potassium manganate(VII).

Questions

6 What gas was given off when the nitrate was heated?

7 Was there any colour in the gas which was given off?

8 What property of the solid residue is shown by its action on acidified potassium manganate(VII) solution?

9 Would sodium oxide have reacted in this way with acidified potassium manganate(VII) solution?

34.4 Manufacture of nitric acid

A simple question to begin this chapter. What property do the following have in common:

> trinitrotoluene,
> nitroglycerine,
> ammonium nitrate?

If you are told that the first one is generally known as T.N.T., the second is a constituent of dynamite and the third a constituent of amatol, the answer should be obvious—they are all explosives. Examination of the names shows that they contain the word nitro, or the word nitrate. Thus one important use of nitric acid, the compound which we are going to consider first in this chapter, is in the manufacture of explosives. However, this only accounts for about 15% of the nitric acid which is produced. By far the most important use (about 75%) is in the production of fertilisers. The main fertiliser which is made from nitric acid is ammonium nitrate.

Fig. 34.1 The use of explosives in a copper mine. (Courtesy RTZ Services Ltd)

Nitric acid is made on the large scale by the oxidation of ammonia. If the oxidation is carried out simply by burning ammonia in air which has been enriched with oxygen (33.15), only the hydrogen ends up as an oxide and the nitrogen is liberated:

$$4NH_3(g) + 3O_2(g) \rightarrow 2N_2(g) + 6H_2O(g)$$

If, however, a mixture of ammonia gas and oxygen is passed over a suitable catalyst, nitrogen monoxide is formed:

$$4NH_3(g) + 5O_2(g) \rightarrow 4NO(g) + 6H_2O(g)$$

The catalyst for the oxidation, which has been known and used for a long time, is a fine gauze made of an alloy of two transition elements, platinum and rhodium. This is expensive and, because of the considerable heat evolved in the oxidation, which causes evaporation of the metals, the gauzes have to be replaced at comparatively frequent intervals. Modern plants tend to carry out the oxidation at a pressure of about 4 atmospheres and the gaseous products of the oxidation are compressed to about 10 atmospheres for the absorption process. In this process, air, water and the nitrogen monoxide are passed into the absorption tower, in which the nitrogen monoxide is oxidised by the oxygen in the air to nitrogen dioxide, which then reacts with water and more air to produce a solution of nitric acid:

$$2NO(g) + O_2(g) \rightarrow 2NO_2(g)$$

$$4NO_2(g) + 2H_2O(l) + O_2(g) \rightarrow 4HNO_3(aq)$$

The nitric acid solution is concentrated by fractional distillation to 68% concentration, which is the usual composition of concentrated nitric acid.

34.5 Laboratory preparation of nitric acid

In the preparation of pure nitric acid in the laboratory we use the principle that a strong non-volatile acid will displace volatile acids from their salts (22.12). Nitric acid boils at 86 °C and so it can be displaced from a nitrate by warming with concentrated sulphuric acid which is much less volatile:

$$H_2SO_4(l) + KNO_3(s) \rightarrow KHSO_4(s) + HNO_3(g)$$

Nitric acid vapour is highly corrosive to cork and rubber and so an all-glass apparatus must be used, a suitable one being shown in Fig. 34.2. A readily available starting material is potassium nitrate. This is placed, with concentrated sulphuric acid, in the flask and the mixture is warmed gently. The vapour is cooled by the running water passing through the condenser and the product condenses, collecting as a pale yellow distillate in the receiver.

Fig. 34.2 The preparation of nitric acid

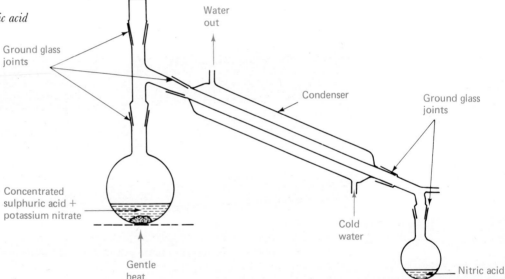

34.6 Properties of nitric acid

Pure nitric acid is a colourless liquid, although in fact the liquid is nearly always tinged yellow by decomposition products. Because it releases brown fumes when open to the air, it is commonly known as **fuming nitric acid**. Commercial concentrated nitric acid also usually has a slight yellow tinge, again caused by decomposition products, whereas dilute nitric acid is colourless. Pure or fuming nitric acid, like other pure acids, is covalent, consisting of unionized molecules, but it dissolves in water with the evolution of much heat, ionizing as it does so:

$$HNO_3(l) + H_2O(l) \rightarrow H_3O^+(aq) + NO_3^-(aq)$$

NO_3^- is the nitrate ion and, since the acid is almost fully ionized in dilute solution, it is a good electrolyte in water. The presence of a high concentration of hydroxonium ions, $H_3O^+(aq)$, (or $H^+(aq)$) makes the solution a strong acid and this is confirmed by its reactions.

1. Reactions with bases

Dilute nitric acid readily reacts with bases (the oxides and hydroxides of metals), forming the metal nitrate (a salt) and water. Since all common nitrates dissolve in water, any base will react with warm dilute nitric acid, forming a clear solution of salt,

e.g. $$CuO(s) + 2HNO_3(aq) \rightarrow Cu(NO_3)_2(aq) + H_2O(l)$$

$$NaOH(aq) + HNO_3(aq) \rightarrow NaNO_3(aq) + H_2O(l)$$

Nitric acid is **monobasic**, i.e. it contains only one mole of replaceable hydrogen per mole of acid, and therefore will form only normal salts, unlike sulphuric acid (H_2SO_4), which being dibasic, forms both normal and acid salts (11.15).

The most important reaction of nitric acid with a base is its reaction with ammonia to form ammonium nitrate:

$$NH_3(g) + HNO_3(aq) \rightarrow NH_4NO_3(aq)$$

The ammonium nitrate which is crystallised from the solution is used as a fertiliser (33.10) and as a component of some explosives.

2. Reactions with carbonates

Like almost all other acids, dilute nitric acid reacts with metal carbonates, causing carbon dioxide to be given off and a solution of the metal nitrate to be formed,

e.g. $$ZnCO_3(s) + 2HNO_3(aq) \rightarrow Zn(NO_3)_2(aq) + H_2O(l) + CO_2(g)$$

Since all common nitrates dissolve in water there are no exceptions to this rule, whereas with, for example, dilute sulphuric acid the carbonate of a metal which forms an insoluble sulphate (e.g. lead) reacts with the acid for only a few seconds as a layer of the insoluble sulphate forms around each particle of the carbonate (11.9).

3. Reactions with metals

Comparing dilute nitric acid with dilute solutions of other acids, we would expect that it would react with any metal above hydrogen in the reactivity series (17.9), liberating hydrogen and forming a solution of the metal nitrate. If, however, the reactions of this acid with these metals below calcium are investigated, it is found that with only one metal, magnesium, will it react in this way, and this only when the acid is very dilute:

$$Mg(s) + 2HNO_3(aq) \rightarrow Mg(NO_3)_2(aq) + H_2(g)$$

Other metals which are above hydrogen in the series do react with the dilute acid, but as the reaction proceeds, the space above the solution in the test-tube is often seen to contain fumes of a brown gas, nitrogen dioxide. The gas is also formed when some metals below hydrogen in the reactivity series react with the acid.

The formula of nitrogen dioxide is NO_2 and, if we try subtracting from $2HNO_3$ (i.e. two moles of nitric acid) H_2O (a mole of water) and $2NO_2$ (two moles of nitrogen dioxide), we get an indication of what is happening in this sort of reaction:

$$2HNO_3(aq) - H_2O(l) - 2NO_2(g) = [O]$$
(a mole
of oxygen
atoms)

This mole of oxygen atoms could be handed over by the nitric acid to the metal, in which case the acid is acting as an oxidising agent.

The product which is formed from the nitric acid when it acts as an oxidising agent actually depends on the concentration of the acid and this is clearly illustrated by its reactions with copper.

371

34.7
Oxidation of metals by nitric acid

If concentrated (68%) nitric acid is added to copper, a lot of brown fumes of nitrogen dioxide are given off and the colourless or light yellow liquid rapidly turns blue-green as copper(II) ions are formed in the solution. In this reaction the nitric acid is reduced to nitrogen dioxide and the copper is oxidised to copper(II) ions, forming a solution of copper(II) nitrate,

$$Cu(s) + 4HNO_3(aq) \rightarrow Cu(NO_3)_2(aq) + 2NO_2(g) + 2H_2O(l)$$
$$68\%$$

or $\quad Cu(s) + 2NO_3^-(aq) + 4H^+(aq) \rightarrow Cu^{2+}(aq) + 2NO_2(g) + 2H_2O(l)$

It is interesting to compare this reaction with that which takes place between copper and concentrated sulphuric acid (27.26).

If concentrated nitric acid is diluted by adding it to its own volume of water, and then added to copper, again brown fumes appear to be given off, but, if the gaseous product is collected over water, a colourless gas is obtained. This colourless gas is nitrogen monoxide (NO), which is the main product of the reaction,

$$3Cu(s) + 8HNO_3(aq) \rightarrow 3Cu(NO_3)_2(aq) + 2NO(g) + 4H_2O(l)$$

or $\quad 3Cu(s) + 2NO_3^-(aq) + 8H^+(aq) \rightarrow 3Cu^{2+}(aq) + 4H_2O(l) + 2NO(g)$

However, some of the nitrogen monoxide combines with the oxygen in the air above the reaction mixture, forming brown fumes of nitrogen dioxide:

$$2NO(g) \quad + \quad O_2(g) \quad \rightarrow \quad 2NO_2(g)$$
$$\text{colourless} \qquad\qquad\qquad\qquad \text{brown}$$

When the gas is collected over water, the nitrogen dioxide dissolves and only nitrogen monoxide goes through the water to be collected.

While the reactions between nitric acid and metals are therefore complicated, it is possible to produce a short summary of what happens:
1. with concentrated nitric acid the reduction product is often nitrogen dioxide, particularly with the less reactive metals;
2. with dilute nitric acid the reduction (if it takes place at all) always goes further than with the concentrated acid, yielding products such as nitrogen monoxide (NO), dinitrogen oxide (N_2O), or even ammonia, the latter being formed by more reactive metals.

34.8
Passivity

Some metals, however, do not appear to fit into this pattern. If an iron nail is dropped into concentrated nitric acid, nothing appears to happen. If the nail is then removed and dropped into dilute sulphuric acid, there is again no sign of a reaction. Normally iron would react steadily with the dilute sulphuric acid, liberating hydrogen, and so the iron in the nail is said to have been made **passive** by the concentrated nitric acid. The surface of the metal has been oxidised to a thin, tough layer of iron oxide, which then protects the nail from attack by the sulphuric acid. Scraping the metal dislodges the layer of oxide and the metal is no longer passive.

The nail only becomes passive when it is put into concentrated nitric acid; dilute nitric acid reacts with it as expected, giving oxides of nitrogen and a solution of iron(III) nitrate. Aluminium also behaves in the same way with the concentrated acid.

34.9
Other oxidations by nitric acid

The reactions of both concentrated and dilute nitric acid with metals indicate that nitric acid is a powerful oxidising agent and is therefore likely to oxidise the substances which usually react with such a reagent.

1. Reactions with iron(II) compounds

If a few drops of concentrated nitric acid are added to a pale green solution of iron(II) sulphate in dilute sulphuric acid, the mixture becomes very dark brown in colour. On warming, brown fumes are evolved, showing that the nitric acid is being reduced and is therefore acting as an oxidising agent, and the colour of the solution suddenly changes to the yellow or orange colour of a solution of an iron (III) salt:

$$6FeSO_4(aq) + 3H_2SO_4(aq) + 2HNO_3(aq) \rightarrow 3Fe_2(SO_4)_3(aq) + 2NO(g) + H_2O(l)$$

The dark brown colour formed at first is caused by the combination of nitrogen monoxide with iron (II) ions to form complex ions:

$$Fe^{2+}(aq) + NO(g) \rightarrow Fe(NO)^{2+}(aq)$$
$$\text{pale green} \qquad\qquad \text{dark brown}$$

These dissociate on heating releasing nitrogen monoxide gas. The formation of these complex ions also takes place in the brown ring test for nitrates (34.12) and is responsible for the brown ring which gives the test its name.

2. Reactions with sulphur and its compounds

If concentrated nitric acid is diluted with about half its own volume of water and hydrogen sulphide is bubbled through it, a dense precipitate of sulphur is formed and brown fumes are given off:

$$H_2S(g) + 2HNO_3(aq) \rightarrow S(s) + 2NO_2(g) + 2H_2O(l)$$
$$\text{50}\%$$

The nitrogen dioxide is formed by the reduction of the nitric acid and the sulphur is the usual product formed when hydrogen sulphide is oxidised.

If concentrated nitric acid is used, very little sulphur is formed as the hydrogen sulphide is oxidised a stage further to sulphate ions, thus making a solution of sulphuric acid:

$$H_2S(g) + 8HNO_3(aq) \rightarrow H_2SO_4(aq) + 8NO_2(g) + 4H_2O(l)$$

The concentrated acid is thus a more powerful oxidising agent than the more dilute acid.

The element sulphur and the gas sulphur dioxide can both be oxidised in the same way to sulphate ions by concentrated nitric acid.

3. Reactions with organic materials

Concentrated nitric acid is highly corrosive to organic matter, such as paper, wood, fabrics and flesh, and this corrosive action is largely due to its oxidising power.

34.10 Oxides of nitrogen

Nitrogen dioxide

In the reactions of nitric acid described above, different oxides of nitrogen appear as reduction products. Nitrogen forms several oxides, but only the three most common ones are to be mentioned here.

Nitrogen dioxide, which has the formula NO_2, is the poisonous brown gas which is formed when copper reacts with concentrated nitric acid (34.7) or when lead(II) nitrate is heated (34.13). It is denser than air and has a pungent smell, rather similar to that of chlorine.

The brown gas actually contains an equilibrium mixture of molecules of nitrogen dioxide, which is brown, and dinitrogen tetroxide (N_2O_4), which is colourless (22.15):

$$2NO_2(g) \rightleftharpoons N_2O_4(g)$$
$$\text{brown} \qquad \text{colourless}$$

Nitrogen dioxide turns moist blue litmus paper red and it decolourises paper soaked in an acidified solution of potassium manganate(VII). These changes take place because the gas is very soluble in water, forming a colourless solution which is both acidic and a reducing agent. As it dissolves in the water, nitrogen dioxide reacts with it, producing a mixture of two acids, nitric acid and nitrous acid:

$$2NO_2(g) + H_2O(l) \rightarrow HNO_3(aq) + HNO_2(aq)$$
$$\text{nitric} \quad\quad \text{nitrous}$$
$$\text{acid} \quad\quad\quad \text{acid}$$

and it is because of this that the gas is sometimes referred to as a **mixed anhydride**. The nitrous acid (HNO_2) is responsible for the reducing property of the solution:

$$HNO_2(aq) + [O] \rightarrow HNO_3(aq)$$

This substance is unstable and, like sulphurous acid (27.16), it is known only in solution. Its salts are called nitrites (34.14) and sodium nitrite is one of the products formed when nitrogen dioxide is absorbed in sodium hydroxide solution:

$$2NO_2(g) + 2NaOH(aq) \rightarrow NaNO_3(aq) + NaNO_2(aq) + H_2O(l)$$
$$\text{sodium} \quad\quad \text{sodium}$$
$$\text{nitrate} \quad\quad \text{nitrite}$$

Nitrogen dioxide will not support the combustion of a wooden splint, but, like many gaseous oxides, it will allow magnesium ribbon to continue burning in it, the colour of the gas having disappeared when the flame is extinguished:

$$4Mg(s) + 2NO_2(g) \rightarrow 4MgO(s) + N_2(g)$$

The brown nitrogen dioxide has been reduced to colourless nitrogen.

A similar reaction takes place when the gas is passed over heated copper turnings and the nitrogen can then be collected over water in the usual way:

$$4Cu(s) + 2NO_2(g) \rightarrow 4CuO(s) + N_2(g)$$

Nitrogen monoxide

Nitrogen monoxide, formerly known as nitric oxide, was first obtained by J. B. van Helmont in 1620, but was more carefully studied by Joseph Priestley in 1772. It is still prepared by the method which Priestley used, the action of a mixture of equal volumes of water and concentrated nitric acid on copper:

$$3Cu(s) + 8HNO_3(aq) \rightarrow 3Cu(NO_3)_2(aq) + 4H_2O(l) + 2NO(g)$$

It is a colourless gas and has approximately the same density as air. The gas has one notable property which influences its others to an extent which makes a study of them difficult. When exposed to the air, it immediately combines with the oxygen, forming brown nitrogen dioxide:

$$2NO(g) + O_2(g) \rightarrow 2NO_2(g)$$

Air therefore has to be excluded from the gas if its properties are to be investigated.

Nitrogen monoxide is almost insoluble in water and since, like carbon monoxide, it does not react with water to form an acid, it is considered to be a neutral oxide. Collection of the gas over water will therefore free it from nitrogen dioxide which may have been formed at the same time and which is soluble in the water.

As is the case with nitrogen dioxide, nitrogen monoxide does not allow a burning splint to continue to burn in it, but it does support the combustion of burning magnesium ribbon, which is oxidised to its oxide while the gas is reduced to nitrogen:

$$2Mg(s) + 2NO(g) \rightarrow 2MgO(s) + N_2(g)$$

Heated copper will also reduce the gas, as it does nitrogen dioxide.

374

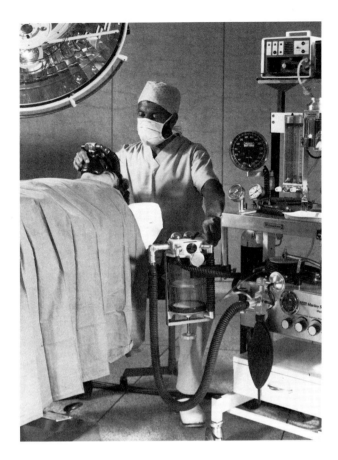

Fig. 34.3 A mixture of dinitrogen oxide and oxygen being used as an anaesthetic in an operating theatre. (Courtesy British Oxygen Company Ltd)

Dinitrogen oxide

Dinitrogen oxide (N_2O) is the gas formerly known as nitrous oxide and was also discovered in 1772 by the Yorkshire minister, Joseph Priestley, to whom is credited the discovery of oxygen five years later. It was another famous chemist, Sir Humphry Davy, who round about 1799 carried out a thorough investigation of the properties of dinitrogen oxide and discovered that it has anaesthetic properties. This led to the extensive use of the gas in surgical operations of short duration, e.g. in dentistry, in which it is administered mixed with oxygen, the pure gas being poisonous. Because of the effect it sometimes has on patients first feeling its effects or on regaining consciousness after its use, it is generally known as laughing gas.

The gas is produced on the industrial scale in the same way as in the laboratory, by the action of heat on ammonium nitrate:

$$NH_4NO_3(s) \rightarrow N_2O(g) + 2H_2O(g)$$

It is a colourless gas with a sweet smell and is denser than air. It has an appreciable solubility in water, but its solution is neutral. Because of its solubility, when it is prepared by heating ammonium nitrate, it has to be collected over warm water.

Dinitrogen oxide will support the combustion of a wood splint and, indeed, it may, like oxygen, rekindle one which is brightly glowing. There is, however, little chance of confusing the two gases because of the smell and solubility of dinitrogen oxide and because of the fact that there are very few reactions in which this oxide of nitrogen is released.

375

34.11
Nitrates

Nitrates are salts derived from nitric acid and containing the nitrate ion (NO_3^-). They are all solids and are all soluble in water. A number of them, for example potassium, sodium, ammonium, barium, lead and silver nitrates, crystallise in the anhydrous condition, while others, like copper(II) nitrate ($Cu(NO_3)_2.3H_2O$), contain water of crystallisation.

They give off nitric acid vapour when warmed with concentrated sulphuric acid,

e.g. $$KNO_3(s) + H_2SO_4(l) \rightarrow KHSO_4(s) + HNO_3(g)$$

Some of the vapour is decomposed by heat to nitrogen dioxide and hence slight brown fumes will be seen. The evolution of brown fumes can be increased by the addition of two or three copper turnings which are readily oxidised by the nitric acid:

$$Cu(s) + 4HNO_3(aq) \rightarrow Cu(NO_3)_2(aq) + 2NO_2(g) + 2H_2O(l)$$

34.12
Tests for nitrates

Since all nitrates are soluble in water, the nitrate ion cannot be detected by a precipitation test, like a sulphate or a chloride, and so an unusual test, making use of the reaction between nitric acid and iron(II) ions, has to be used. A few crystals of iron(II) sulphate are dissolved in dilute sulphuric acid and a solution of the suspected nitrate is added. The tube is held at $45°$ and concentrated sulphuric acid is carefully poured down the side so that it forms a dense layer beneath the aqueous solutions. If a brown ring is seen at the boundary between the two layers (increased by gentle tapping of the tube, or sometimes by cooling it), the presence of the nitrate ion is confirmed.

The nitrate ions react with the concentrated sulphuric acid at the boundary, forming nitric acid:

$$H_2SO_4(l) + NO_3^-(aq) \rightarrow HNO_3(aq) + HSO_4^-(aq)$$

This then reacts with the iron(II) ions, as described earlier (34.9), giving a brown colouration due to the formation of the complex ion $Fe(NO)^{2+}(aq)$.

A test which avoids the use of concentrated acid involves adding Devarda's alloy (an alloy of aluminium, zinc and copper) to an alkaline solution of a nitrate. The nitrate is reduced to ammonia. The distinctive smell of ammonia gas indicates the presence of a nitrate.

34.13
Action of heat on nitrates

All nitrates are decomposed by the action of heat, and in most cases the reaction is very similar to the decomposition by heat of carbonates in which the oxy-salt is decomposed to its basic and acidic oxides,

e.g. $$PbCO_3(s) \rightarrow PbO(s) + CO_2(g)$$
$$\text{basic} \qquad \text{acidic}$$

Lead(II) nitrate, decomposing in the same way, would produce dinitrogen pentoxide (N_2O_5) as the gaseous acidic oxide, but this is unstable and further decomposes to nitrogen dioxide and oxygen:

$$2Pb(NO_3)_2(s) \rightarrow 2PbO(s) + 4NO_2(g) + O_2(g)$$

Nitrates can be divided into three groups by their behaviour on heating. The first group consists of those which are similar to the lead(II) nitrate, described above, and which decompose to the **metal oxide**, **nitrogen dioxide** and **oxygen**. This class contains all nitrates except sodium, potassium and ammonium nitrates, and includes concentrated nitric acid itself which breaks down on heating (Fig. 34.4), forming water, nitrogen dioxide and oxygen:

$$4HNO_3(aq) \rightarrow 2H_2O(l) + 4NO_2(g) + O_2(g)$$

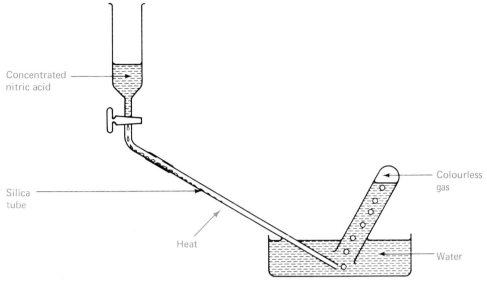

Fig. 34.4 The thermal decomposition of nitric acid

Concentrated nitric acid

Silica tube

Heat

Colourless gas

Water

Calcium, magnesium, aluminium, zinc, iron(III) and copper(II) nitrates in this class are hydrated, and decomposition does not take place until the water of crystallisation has been driven off. Lead(II) nitrate differs in being anhydrous and, on heating, the crystals start to decompose before the melting point is reached, flying apart with a sharp crackling noise. This behaviour is called **decrepitation**.

The action of heat on these nitrates is the basis of an efficient method of converting a metal into its oxide, the metal being first dissolved in concentrated nitric acid to give a solution of the nitrate which is then heated gently to evaporate the solution and then more strongly to decompose the residue.

The second group contains sodium nitrate and potassium nitrate and, being compounds of very reactive metals, they are decomposed to a lesser extent than the first group of nitrates. Both nitrates are anhydrous and have to be heated to a high temperature before they melt and then show a steady effervescence, evolving only **oxygen**. The residue is a **nitrite** which is a salt of nitrous acid,

e.g. $$2KNO_3(s) \rightarrow 2KNO_2(s) + O_2(g)$$

The third group contains only ammonium nitrate. When the anhydrous solid is heated, **steam** is given off, together with a sweet-smelling gas which can be collected over warm water and which then may rekindle a brightly-glowing wooden splint. When decomposition leaves only a small amount of the solid, a small explosion may take place. The gas produced is **dinitrogen oxide**, formed by what seems to be an internal redox reaction with the ammonium ion (NH_4^+) being oxidised and the nitrate ion (NO_3^-) being reduced:

$$NH_4NO_3(s) \rightarrow N_2O(g) + 2H_2O(g)$$

Since all metal nitrates break down on heating, giving oxygen, they are all oxidising agents in the solid state and will therefore promote the combustion of other substances. For this reason potassium nitrate is used as a constituent of gunpowder and many fireworks and explosives. Sodium nitrate is not used in this way since it is delinquescent and would cause the mixture to become wet before use.

One of the explosives used for the blasting of iron ore is a mixture of ammonium nitrate and fuel oil.

377

34.14
Nitrites

Nitrites are salts of the nitrous acid which is one of the products formed when nitrogen dioxide is dissolved in water. Sodium and potassium nitrites are the only two to exist in the solid state and they are best prepared by heating the corresponding nitrate (34.13).

As might be expected, the nitrite ion shows some similarity in behaviour to the sulphite ion, which it resembles in containing one atom of oxygen less than a comparatively stable and unreactive anion. The sulphite ion (SO_3^{2-}) needs to gain one atom of oxygen to turn it into a sulphate ion (SO_4^{2-}), while the nitrite ion (NO_2^-) could be similarly changed into a nitrate ion (NO_3^-). It is not surprising, therefore, that in aqueous solution nitrites and sulphites are reducing agents.

$$SO_3^{2-}(aq) + [O] \rightarrow SO_4^{2-}(aq)$$

$$NO_2^-(aq) + [O] \rightarrow NO_3^-(aq)$$

or $$NO_2^-(aq) + H_2O(l) \rightarrow NO_3^-(aq) + 2H^+(aq) + 2e^-$$

Thus, a solution both of a sulphite and of a nitrite will decolourise an acidified solution of potassium manganate(VII).

What is surprising is that a nitrite, when in acid solution, is also a good oxidising agent, since it can be easily reduced to nitrogen monoxide gas:

$$NO_2^-(aq) + 2H^+(aq) + e^- \rightarrow NO(g) + H_2O(l)$$

One of the tests by which a substance can be recognised as an oxidising agent is to see whether it will liberate iodine from an acidified solution of potassium iodide. If potassium nitrite solution is added to this reagent, the mixture turns black and brown fumes are evolved as nitrogen monoxide reacts with the air to form nitrogen dioxide (34.10):

$$I^-(aq) + NO_2^-(aq) + 2H^+(aq) \rightarrow \tfrac{1}{2}I_2(s) + NO(g) + H_2O(l)$$

Ammonium nitrite behaves on heating in a similar manner to ammonium nitrate (34.13). This compound is too unstable to be able to exist in the solid state in a bottle on the laboratory shelf, but it can be made on the spot by mixing together ammonium chloride and sodium nitrite in a little water, the mixture containing both ammonium ions (NH_4^+) and nitrite ions (NO_2^-) and therefore behaving as ammonium nitrite. On warming the solution effervesces, producing nitrogen:

$$NH_4^+(aq) + NO_2^-(aq) \rightarrow N_2(g) + 2H_2O(g)$$

Like the action of heat on ammonium nitrate, this is an example of an internal redox reaction where the cation (NH_4^+) is oxidised and the anion (NO_2^-) is reduced. This reaction provides a simple laboratory method of preparing a sample of nitrogen, the gas then being collected over water in the usual way.

34.15
The oxidation states of nitrogen

A large number of nitrogen compounds have been mentioned in this chapter and it may be helpful to explore the relationships between them in Table 34.1.

OXIDATION STATE OF NITROGEN	OXIDATION $\longrightarrow$				
	+1	+2	+3	+4	+5
Oxides	N_2O	NO	$[N_2O_3]$	$\left\{ \begin{matrix} NO_2 \\ N_2O_4 \end{matrix} \right\}$	$[N_2O_5]$
Acids	—	—	HNO_2		HNO_3
Salts	—	—	$NaNO_2$	—	$NaNO_3$
Ions	—	—	NO_2^-	—	NO_3^-

TABLE 34.1. *The oxidation states of nitrogen*

378

It can be seen that much of the complication of nitrogen chemistry arises from the fact that the two oxy-acids involve odd oxidation states of nitrogen (3 and 5), whereas the two most common oxides of nitrogen have the even oxidation states. The oxides N_2O_3 and N_2O_5 are shown in brackets since they are never met under normal conditions. The disproportionation of the mixed anhydride NO_2 on reaction with water is indicated by the arrows.

The trend in the acidic character of the oxides is typical of an element with variable valency, the acidity increasing as the oxidation state increases. The same trend was met in the case of sulphur (27.27). The lowest oxides N_2O and NO are neutral, N_2O_3 is weakly acidic (nitrous acid being a weak acid), while N_2O_5 is strongly acidic (nitric acid being a strong acid). Nitrogen(IV) oxide behaves, as was stated above, as if it were a mixture of the two anhydrides.

**34.16
Summary**

1. Nitric acid is manufactured by the catalytic oxidation of ammonia. It is used in the manufacture of explosives.
2. Nitric acid may be prepared by heating a nitrate with concentrated sulphuric acid.
3. Nitric acid, especially when concentrated, is a powerful oxidising agent. It oxidises most metals, iron(II) salts, and sulphur and some of its compounds. The nitric acid is reduced usually to oxides of nitrogen.
4. A number of oxides of nitrogen are known, all being somewhat unstable. The most common ones are nitrogen dioxide (which is a brown gas), nitrogen monoxide (which reacts with oxygen in air to form brown nitrogen dioxide) and dinitrogen oxide (commonly called laughing gas).
5. All nitrates are soluble in water and may be detected by the brown ring test and the Devarda's alloy test. They are all decomposed by heat, most of them yielding nitrogen dioxide and oxygen. Sodium and potassium nitrates evolve only oxygen and ammonium nitrate produces dinitrogen oxide.
6. Nitrites are the salts of the unstable acid, nitrous acid, and are both oxidising and reducing agents.

Data Section

The elements

Name	Symbol	Atomic Number	Relative Atomic Mass	Name	Symbol	Atomic Number	Relative Atomic Mass
Hydrogen	H	1	1·008	Iodine	I	53	126·9
Helium	He	2	4·003	Xenon	Xe	54	131·3
Lithium	Li	3	6·941	Caesium	Cs	55	132·9
Beryllium	Be	4	9·012	Barium	Ba	56	137·3
Boron	B	5	10·81	Lanthanum	La	57	138·9
Carbon	C	6	12·01	Cerium	Ce	58	140·1
Nitrogen	N	7	14·01	Praseodymium	Pr	59	140·9
Oxygen	O	8	16·00	Neodymium	Nd	60	144·2
Fluorine	F	9	19·00	Promethium	Pm	61	(145)
Neon	Ne	10	20·18	Samarium	Sm	62	150·4
Sodium	Na	11	22·99	Europium	Eu	63	152·0
Magnesium	Mg	12	24·31	Gadolinium	Gd	64	157·3
Aluminium	Al	13	26·98	Terbium	Tb	65	158·9
Silicon	Si	14	28·09	Dysprosium	Dy	66	162·5
Phosphorus	P	15	30·97	Holmium	Ho	67	164·9
Sulphur	S	16	32·06	Erbium	Er	68	167·3
Chlorine	Cl	17	35·45	Thulium	Tm	69	168·9
Argon	Ar	18	39·95	Ytterbium	Yb	70	173·0
Potassium	K	19	39·10	Lutetium	Lu	71	175·0
Calcium	Ca	20	40·08	Hafnium	Hf	72	178·5
Scandium	Sc	21	44·96	Tantalum	Ta	73	180·9
Titanium	Ti	22	47·90	Tungsten	W	74	183·9
Vanadium	V	23	50·94	Rhenium	Re	75	186·2
Chromium	Cr	24	52·00	Osmium	Os	76	190·2
Manganese	Mn	25	54·94	Iridium	Ir	77	192·2
Iron	Fe	26	55·85	Platinum	Pt	78	195·1
Cobalt	Co	27	58·93	Gold	Au	79	197·0
Nickel	Ni	28	58·70	Mercury	Hg	80	200·6
Copper	Cu	29	63·55	Thallium	Tl	81	204·4
Zinc	Zn	30	65·38	Lead	Pb	82	207·2
Gallium	Ga	31	69·72	Bismuth	Bi	83	209·0
Germanium	Ge	32	72·59	Polonium	Po	84	(209)
Arsenic	As	33	74·92	Astatine	At	85	(210)
Selenium	Se	34	78·96	Radon	Rn	86	(222)
Bromine	Br	35	79·90	Francium	Fr	87	(223)
Krypton	Kr	36	83·80	Radium	Ra	88	(226)
Rubidium	Rb	37	85·47	Actinium	Ac	89	(227)
Strontium	Sr	38	87·62	Thorium	Th	90	232·0
Yttrium	Y	39	88·91	Protactinium	Pa	91	(231)
Zirconium	Zr	40	91·22	Uranium	U	92	238·0
Niobium	Nb	41	92·91	Neptunium	Np	93	(237
Molybdenum	Mo	42	95·94	Plutonium	Pu	94	(244)
Technetium	Tc	43	(97)	Americium	Am	95	(243)
Ruthenium	Ru	44	101·1	Curium	Cm	96	(247)
Rhodium	Rh	45	102·9	Berkelium	Bk	97	(247)
Palladium	Pd	46	106·4	Californium	Cf	98	(251)
Silver	Ag	47	107·9	Einsteinium	Es	99	(254)
Cadmium	Cd	48	112·4	Fermium	Fm	100	(257)
Indium	In	49	114·8	Mendelevium	Md	101	(258)
Tin	Sn	50	118·7	Nobelium	No	102	(259)
Antimony	Sb	51	121·8	Lawrencium	Lr	103	(260)
Tellurium	Te	52	127·6	Unnilquadium	Unq	104	(261)
				Unnilpentium	Unp	105	(262)

These values are scaled to the relative atomic mass of $^{12}C = 12$. Numbers in brackets are the mass numbers of the most stable isotopes. Approximate relative atomic masses of selected elements are given in Table 1.

The Periodic Table

Groups 1 2   3 4 5 6 7 8

Period	1	2										3	4	5	6	7	8
1							H 1										He 2
2	Li 3	Be 4										B 5	C 6	N 7	O 8	F 9	Ne 10
3	Na 11	Mg 12					Transition metals					Al 13	Si 14	P 15	S 16	Cl 17	Ar 18
4	K 19	Ca 20	Sc 21	Ti 22	V 23	Cr 24	Mn 25	Fe 26	Co 27	Ni 28	Cu 29	Zn 30	Ga 31	Ge 32	As 33	Se 34	Br 35	Kr 36
5	Rb 37	Sr 38	Y 39	Zr 40	Nb 41	Mo 42	Tc 43	Ru 44	Rh 45	Pd 46	Ag 47	Cd 48	In 49	Sn 50	Sb 51	Te 52	I 53	Xe 54
6	Cs 55	Ba 56	La 57 *	Hf 72	Ta 73	W 74	Re 75	Os 76	Ir 77	Pt 78	Au 79	Hg 80	Tl 81	Pb 82	Bi 83	Po 84	At 85	Rn 86
7	Fr 87	Ra 88	Ac 89 †	Unq 104	Unp 105												

Periods

Alkali metals

Halogens   Noble gases

	Ce 58	Pr 59	Nd 60	Pm 61	Sm 62	Eu 63	Gd 64	Tb 65	Dy 66	Ho 67	Er 68	Tm 69	Yb 70	Lu 71	Lanthanides
†	Th 90	Pa 91	U 92	Np 93	Pu 94	Am 95	Cm 96	Bk 97	Cf 98	Es 99	Fm 100	Md 101	No 102	Lr 103	Actinides

Table 1 Physical properties of selected elements

Name	Symbol	Atomic number	Approximate relative atomic mass	Melting point/°C	†Boiling point/°C	*Density Solid or liquid/ g cm⁻³	Gas g dm⁻³	Atomic volume cm³ mol⁻¹	
Aluminium	Al	13	27	660	2470	2·70	–	10·0	
Argon	Ar	18	40	−189	−186	1·40	1·78	28·5	
Arsenic	As	33	75	–	613(s)	5·72	–	13·1	
Barium	Ba	56	137·5	714	1640	3·51	–	39·1	
Beryllium	Be	4	9	1280	2477	1·85	–	4·87	
Boron	B	5	11	2300	3930	2·34	–	4·62	
Bromine	Br	35	80	−7·2	58·8	3·12	–	25·6	
Calcium	Ca	20	40	850	1487	1·54	–	26·0	
Carbon	C	6	12	–	4000(s)	2·25	–	5·33	(graphite)‡
						3·51	–	3·42	(diamond)
Chlorine	Cl	17	35·5	−101	−34·7	1·56	3·17	22·8	
Chromium	Cr	24	52	1890	2482	7·19	–	7·23	
Cobalt	Co	27	59	1492	2900	8·90	–	6·62	
Copper	Cu	29	63·5	1083	2595	8·92	–	7·12	
Fluorine	F	9	19	−220	−188	1·11	1·70	17·1	
Gallium	Ga	31	70	29·8	2400	5·91	–	11·8	
Germanium	Ge	32	72·5	937	2830	5·35	–	13·6	
Gold	Au	79	197	1063	2970	19·3	–	10·2	
Helium	He	2	4	−270	−269	0·147	0·357	27·2	
Hydrogen	H	1	1	−259	−252	0·070	0·090	14·4	
Iodine	I	53	127	114	184	4·93	–	25·7	
Iron	Fe	26	56	1535	3000	7·86	–	7·10	
Krypton	Kr	36	84	−157	−152	2·16	3·74	38·8	
Lead	Pb	82	207	327	1744	11·3	–	18·3	
Lithium	Li	3	7	180	1330	0·53	–	13·1	
Magnesium	Mg	12	24	650	1110	1·74	–	14·0	
Manganese	Mn	25	55	1240	2100	7·20	–	7·63	
Mercury	Hg	80	200·5	−38·9	357	13·6	–	14·8	
Neon	Ne	10	20	−249	−246	1·20	0·902	16·8	
Nickel	Ni	28	59	1453	2730	8·90	–	6·60	
Nitrogen	N	7	14	−210	−196	0·808	1·25	17·3	
Oxygen	O	8	16	−218	−183	1·115	1·43	13·9	
Phosphorus	P	15	31	44·2	280	1·82	–	17·0	(white)
					400(s)	2·34	–	13·2	(red)‡
Platinum	Pt	78	195	1769	4530	21·4	–	9·12	
Potassium	K	19	39	63·7	774	0·86	–	45·5	
Rubidium	Rb	37	85·5	38·9	688	1·53	–	55·9	
Scandium	Sc	21	46	1540	2730	2·99	–	15·4	
Selenium	Se	34	79	217	685	4·81	–	16·4	
Silicon	Si	14	28	1410	2360	2·33	–	12·1	
Silver	Ag	47	108	961	2210	10·5	–	10·3	
Sodium	Na	11	23	97·8	890	0·97	–	23·7	
Strontium	Sr	38	87·5	768	1380	2·62	–	33·4	
Sulphur	S	16	32	113	445	2·07	–	15·5	(rhombic)‡
				119	445	1·96	–	16·4	(monoclinic)
Tin	Sn	50	118·5	232	2270	7·28	–	16·3	
Titanium	Ti	22	48	1675	3260	4·54	–	10·6	
Uranium	U	92	238	1130	3820	19·1	–	12·5	
Vanadium	V	23	51	1900	3000	5·96	–	8·54	
Xenon	Xe	54	131	−112	−108	3·52	5·86	37·3	
Zinc	Zn	30	65·5	420	907	7·14	–	9·16	

†(s) indicates sublimation point.
*For most elements the listed density is at room temperature. For an element which is a gas at room temperature the first density listed is that of the liquid at its boiling point, the second is that of the gas at s.t.p.
⊕Atomic volume is the volume containing 1 mole of atoms. For an element which is a gas at room temperature this is given for the liquid at its boiling point.
‡Stable allotrope at room temperature.

Table 2 Physical properties of selected compounds which do not contain any metal

Name	Formula	Melting point/°C	†Boiling point/°C	Density* solid or liquid /g cm^{-3}	Density* gas /g dm^{-3}
Hydrogen fluoride	HF	−83	20	0·95	
Hydrogen chloride	HCl	−115	−85		1·63
Hydrogen bromide	HBr	−88	−67		3·62
Hydrogen iodide	HI	−51	−35		5·71
Water	H_2O	0	100	1·00	
Hydrogen sulphide	H_2S	−85	−61		1·52
Ammonia	NH_3	−77	−33		0·76
Methane	CH_4	−182	−161		0·714
Tetrachloromethane	CCl_4	−23	77	1·59	
Carbon disulphide	CS_2	−111	47	1·26	
Carbon dioxide	CO_2	–	−78(s)		1·96
Carbon monoxide	CO	−205	−190		1·25
Nitrogen monoxide	NO	−164	−152		1·34
Dinitrogen monoxide	N_2O	−91	−88		1·96
Phosphorus(V) oxide	P_4O_{10}	–	360(s)	2·39	
Phosphorus trichloride	PCl_3	−91	76	1·58	
Sulphur dioxide	SO_2	−73	−10		2·86
Sulphur trioxide	SO_3	17	45	2·75	
Sulphuric acid	H_2SO_4	10	ca.317 (dec)	1·83	
Nitric acid	HNO_3	−41	ca.86 (dec)	1·52	
Silicon dioxide	SiO_2	ca.1700	v. high	2·66	
Ethanol	C_2H_5OH	−117	78	0·79	

† (s) indicates sublimation point.
(dec) means that some decomposition occurs at the boiling point.
* For compounds which are gases at normal temperatures the listed density is that of the gas at s.t.p. For other compounds the density given is that at room temperature.

Table 3 Physical properties of selected compounds which contain a metal

Name	Formula	†Melting point/°C	†‡Boiling point/°C	Density /g cm^{-3}
Potassium fluoride	KF	847	1530	2·57
Lithium chloride	LiCl	607	1350	2·07
Sodium chloride	NaCl	801	1470	2·17
Potassium chloride	KCl	767	1410	1·98
Potassium bromide	KBr	727	1440	2·75
Potassium iodide	KI	677	1330	3·13
Sodium hydroxide	NaOH	318	1380	2·13
Potassium hydroxide	KOH	360	1330	2·04
Magnesium chloride*	$MgCl_2$	714	1410	2·32
Calcium chloride*	$CaCl_2$	782	1600	2·51
Magnesium oxide	MgO	2900	3600	3·58
Calcium oxide	CaO	2600	2850	3·35
Aluminium oxide	Al_2O_3	2040	2980	3·97
Iron(III) oxide	Fe_2O_3	1450	1560	5·24
Calcium hydroxide	$Ca(OH)_2$	ca.600 (dec)	–	2·34
Calcium carbonate	$CaCO_3$	ca.800 (dec)	–	2·93
Sodium nitrate	$NaNO_3$	316	ca.550 (dec) (high)	2·26
Sodium sulphate*	Na_2SO_4	883	(dec) (high)	2·70
Sodium carbonate*	Na_2CO_3	854	(dec)	2·51

*Anhydrous
†(dec) means that some decomposition occurs before the melting or boiling point is reached.
‡All the values given for the boiling points are approximate.

Questions

Chapters 1–7

1. (a) Name the method which can be used to separate the pigments in ink.

(b) Describe how you would separate oil from a mixture of oil and water.

(c) Describe clearly how you would separate chalk from a mixture of chalk and sugar.

(Y.R.E.B., 78)

2. These curves show the variation of solubility with temperature of the crystalline salts: sodium nitrate, potassium nitrate, potassium chloride and sodium chloride.

(a) Name the solid which is
   (i) the most soluble at 0°C.
   (ii) the least soluble at 0°C.

(b) What is the solubility of each solid at 50°C?
   (i) Potassium nitrate.
   (ii) Sodium nitrate.
   (iii) Potassium chloride.
   (iv) Sodium chloride.

(c) At what temperature are the solubilities of sodium nitrate and potassium nitrate the same?

(d) How much potassium nitrate would be precipitated if a saturated solution containing 100 grammes of water at 70°C is cooled to 10°C?

(e) What is the minimum mass of water necessary to dissolve 100 grammes of potassium chloride at 60°C?

(f) How much sodium chloride will dissolve in 25 grammes of water at 90°C?

(g) Potassium nitrate may be manufactured by mixing solutions of potassium chloride and sodium nitrate and allowing crystallisation to take place firstly at 75°C and then at about 15°C. Explain why this procedure is used.

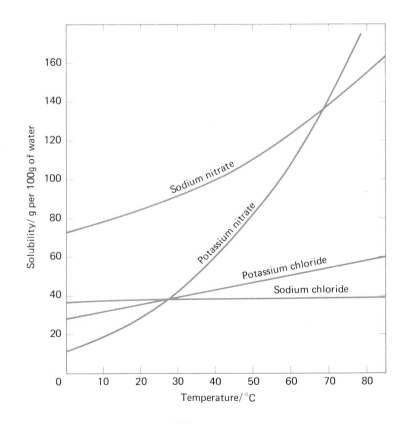

3. (a) How does (i) the motion of the molecules,
   and (ii) the spacing between the molecules
   differ in the three states of matter, viz. solid, liquid and vapour?
   (b) How and why does atmospheric pressure affect the boiling-point of a liquid?

   (*J.M.B.*)

4. (a) Describe an experiment to demonstrate the diffusion of gases. How may the diffusion of gases be explained by the kinetic theory?
   (b) Describe, using appropriate examples, how the following processes may be used in the separation of the components of a mixture:
   (i) fractional distillation, (ii) sublimation.

   (*W.J.E.C.*)

5. (a) (i) Explain why a gas exerts pressure on the walls of a closed vessel containing it.
   (ii) How and why is this pressure affected by increasing the temperature of the gas?
   (b) 30 cm³ of a gaseous oxide of carbon, measured at room temperature and pressure, have a mass of 0·035 g.
   (i) What is the mass of 1 mole of this oxide?
   (ii) Use this information to suggest the most probable formula of the oxide.
   (c) At room temperature and pressure, 1 g of hydrogen occupies 12 000 cm³. Use this information to determine the atomicity of hydrogen.

   (*L.*)

6. (a) 5 g of an oxide (*A*) of chromium were found on reduction to give 2·6 g of chromium. Find the simplest (or empirical) formula of the oxide.
   (b) Oxide (*A*) on strong heating forms oxygen and a second oxide of chromium, (*B*), which is found to contain 24 g of oxygen combined with the molar mass of chromium. Find the simplest formula of oxide (*B*).
   (c) Using the simplest formulae for the two oxides, write an equation for the formation of oxide (*B*) from oxide (*A*).

   (*L.*)

7. A colourless liquid *Q* boils at 46 °C and burns readily in air. It contains only the elements carbon and sulphur, in the proportions 15·8% C and 84·2% S by mass.
   (a) What is the ratio of moles of C atoms to moles of S atoms in *Q*?
   (b) If 1 mole of *Q* has a mass of 76 g, what is the molecular formula of *Q*?

   (*L.*)

8. (a) Write down the relative atomic masses of:– (i) Fe, (ii) S, (iii) O.
   (b) Calculate the mass of 1 mole of each of the following compounds:– (i) $FeSO_4$, (ii) Fe_2O_3, (iii) SO_3.
   (c) Calculate the mass of (i) iron(III) oxide and (ii) sulphur trioxide obtained from 15.2 grams of iron(II) sulphate.
   $$2FeSO_4 \rightarrow Fe_2O_3 + SO_2 + SO_3$$

9. (a) Give the names of the elements which have the following symbols:– Cu, K, Mg, Sn, He, Si.
   (b) The formula of Gallium (III) oxide is Ga_2O_3. Write down the formula for each of the following compounds:– Gallium (III) chloride, Gallium (III) sulphate, Gallium (III) hydroxide
   (c) Complete and balance the equations below:–
   (i)   Mg   +   HCl   → $MgCl_2 + H_2$
   (ii)   NO   + O_2 → NO_2
   (iii) NH_4OH + H_2SO_4 →   ... $+ 2H_2O$
   (iv) $FeCl_2$ ...   → $FeCl_3$
   (d) Calculate the percentage by mass of oxygen in Gallium (III) oxide.

Chapters 8–11

1. (a) Give two important processes which generate or release carbon dioxide into the atmosphere, and two which remove carbon dioxide from the atmosphere.
   (b) Of the gases in the air, name one which is
   (i) most abundant by volume,
   (ii) extracted and used industrially on a large scale.
   For (ii), give three examples of its use.
   (c) Name one noble (rare or inert) gas.
   (d) Name two gases notorious for causing atmospheric pollution.

   (*O.*)

2. Describe, giving essential practical details, a laboratory preparation of reasonably pure samples of the following salts: (a) sodium chloride, (b) copper(II) sulphate, (c) lead(II) chloride from lead.

   (*W.J.E.C.*)

3. In carrying out an experiment with hydrated magnesium sulphate, $MgSO_4.xH_2O$, the following results were obtained:
   (i) Mass of crucible plus lid = 14·636 g
   (ii) Mass of crucible plus lid plus hydrated magnesium sulphate = 15·374 g
   (iii) Mass of crucible plus lid plus crystals after heating = 14·996 g
   (iv) Mass of crucible plus lid plus crystals after further heating = 14·996 g
   (a) What was the mass of water of crystallisation in the sample of hydrated crystals used? (b) What was the mass of anhydrous magnesium sulphate formed? (c) What is the value of *x* in the formula $MgSO_4.xH_2O$? (d) Why was step (iv) necessary?

   (*O.*)

4. (a) Name one compound in each case which can cause (i) temporary hardness and (ii) permanent hardness in water.
   (b) Show, by equations, how both temporary and permanent hard water react with soap and state what you would see during the process. Contrast this with the action of a soapless detergent on a sample of hard water.
   (c) Describe how these two types of hardness in water arise in nature.
   (d) Why is it necessary, in hard water areas, to check domestic hot water and central heating pipes regularly?

   (*J.M.B.*)

5. (a) Briefly describe three types of treatment which may be necessary to convert river water into water which is suitable for domestic use.
   (b) Name three sources of pollution of natural waterways.
   (c) Stretches of polluted waterway are sometimes treated with oxygen.
   What is the purpose of the oxygen?

Chapters 12–16

1. (a) Complete the table below:

Element	Mass Number	Atomic Number	Number of Protons	Number of Neutrons
Sodium	23	11		
Calcium			20	20
$^{35}_{17}Cl$				
$^{37}_{17}Cl$				

(b) What name is given to a pair of atoms such as $^{35}_{17}Cl$ and $^{37}_{17}Cl$?
(*O.*)

2. (a) Give the arrangement of the electrons in (i) one atom of sodium (atomic number 11). (ii) one atom of chlorine (atomic number 17).

(b) Explain how one atom of sodium combines with one of chlorine and name the type of bond formed.

(c) By means of a diagram, show the arrangement of electrons in one molecule of chlorine.
(*Y.R.E.B., 78*)

3. (a) Write down the electronic structures of the following elements, whose atomic numbers are given in brackets: S(16); Cl(17); Ar(18).

(b) With reference to the electronic structures of the elements concerned, account for the fact that chlorine is a diatomic gas, whereas argon is a monatomic gas.

(c) Sulphur and chlorine combine to form a covalent liquid SCl_2. By means of a diagram, show the electronic structure of this compound.

(d) How do you account for the fact that simple covalent substances are usually quite stable to heat, but have relatively low boiling points?
(*W.J.E.C.*)

4. What are the constituent units of crystals of (i) iodine, (ii) sodium chloride? Compare and account for the ease with which these substances can be vaporised.
(*O.*)

5. 'The type of bond present in a crystal can influence its chemical and physical properties.'
(a) Explain what is meant by the above statement in the case of crystals of (i) iodine, (ii) diamond, (iii) sodium chloride. In your answer, in addition to the type of bond present, you should refer where relevant to the following:
the particles (i.e. atoms, ions or molecules) present in the crystal,
the action of heat on the crystal,
whether the melting point is low or high,
the conditions under which the substance will conduct electricity.

(b) A compound, *X*, is a white crystalline solid. It has a melting point of 767°C and is soluble in water.
State, with a reason, what type of bonding is present in the crystal.
Would you expect an aqueous solution of *X* to conduct electricity?
(*J.M.B.*)

6. Magnesium (atomic number 12), aluminium (atomic number 13) and chlorine (atomic number 17) are in the same short period.
(a) Describe the physical state and appearance of these elements at room temperature and pressure. Do they conduct electricity? Are they metals or non-metals? What trends in the properties of the elements can be observed across the period?

(b) Draw diagrams to show the arrangement of electrons in the atoms of these elements. Which element would you expect to have the highest first ionization energy and why?

(c) What are the charges on the ions formed by these elements? How can these charges be explained in terms of the number and arrangement of electrons in the atoms?
(*L.*)

7. By reference to the work of Döbereiner (1817), Newlands (1863), Mendeleev (1869) and Moseley (twentieth century) trace briefly the development of ideas for the classification of elements up to the Periodic Table as we know it today.
If all the successive ionisation energies of a particular element are known, how might the electron configuration be deduced?
(*L. Nuff.*)

8. (a) (i) Indicate the nature of α-rays, β-rays, γ-rays.
(ii) Write down these types of radiation in the order of their penetrating power, starting with the most powerful radiation.

(b) (i) The half-life period of uranium-238 is $4 \cdot 5 \times 10^9$ years. Briefly explain what is meant by this statement.
(ii) What implications does the possession of large half-life periods have for the disposal of radioactive waste?

(c) Quote two examples of the use of radioactive isotopes other than to produce energy. Give brief explanations of these applications (the isotopes need not be identified).
(*L. Chem in Soc.*)

Chapters 17–21

1. When iron is dissolved in dilute sulphuric acid, a gas is released and a salt is formed in the solution (which will be referred to below as solution *A*).
(a) Give the equation for the reaction and name the substances formed.

(b) What volume of gas (measured at s.t.p.) would have been released if 112 g iron had been dissolved?

(c) The crystalline salt obtained by evaporating a portion of solution *A* is strongly heated until a reddish brown residue is obtained. Name this residue and give an equation for the reaction.

(d) Hydrogen peroxide and dilute sulphuric acid are added to another portion of solution *A* and the mixture is warmed for several minutes. What colour change occurs and what chemical change has taken place?

(e) Compare the effect of sodium hydroxide solution on solution *A* with its effect on the solution obtained in (d).
(*O.*)

2. (a) Place the following metals in the order in which they appear in the activity series, starting with the most reactive: lead, sodium, iron, zinc.

(b) A fifth metal *M* reacts slowly with cold water but vigorously when heated with water vapour. It reacts with dilute acids according to the equation $M + 2H^+ \rightarrow M^{2+} + H_2$.
(i) Where would you place *M* in the activity list of the four metals in (a)?
(ii) Write an equation for the thermal decomposition of the carbonate of *M*.
(iii) Write an equation for the thermal decomposition of the nitrate of *M*.
(iv) In which group of the Periodic Table would you place *M*?
(*O.*)

3. How do you account for the fact that when an aqueous solution containing both sodium sulphate and copper sulphate is electrolysed, the platinum cathode changes colour and no gas is evolved at this electrode? What reaction occurs at the platinum anode?
(*0.*)

4. What is metallic corrosion and how may it be prevented? How does the chemist explain that the corrosion of a particular metal may be increased when it is in contact with some metals and decreased when it is in contact with other metals?
(*L. Nuff.*)

5. Explain briefly why each of the following reactions is termed an oxidation: (a) the combustion of carbon in oxygen; (b) the reaction of zinc with dilute sulphuric acid; (c) the reaction of sodium with chlorine.

(*O.*)

Chapters 22–23

1. The following equation represents a reversible reaction:

$$Ag^+(aq) + Fe^{2+}(aq) \rightleftharpoons Ag(s) + Fe^{3+}(aq)$$

(a) (i) Name two substances that you could mix to make the reaction go from left to right.
   (ii) What would you expect to see if the substances do react?
   (iii) Describe a chemical test to prove that some Fe^{3+} ions have been formed.

(b) (i) Name two substances that you could mix to make the reaction go from right to left.
   (ii) What would you expect to see if the substances do react?
   (iii) Describe a chemical test to prove that some Fe^{2+} ions have been formed.

(*L.*)

2. In many industrial processes the use of one or more of the following is essential for a satisfactory yield of the desired product: (a) a high temperature, (b) a high pressure, (c) a catalyst, (d) an electric current.
Take each of these factors in turn and describe the chemistry of an industrial process where it is used. You must choose a different process each time. Technical details are not required.

(*L.*)

3. (a) The heat of neutralization of acqueous sodium hydroxide by hydrochloric acid is the same as the heat of neutralization of aqueous potassium hydroxide by nitric acid. Give a detailed explanation of why this is so.

(b) Describe how you would carry out an experiment to find the heat of neutralization in one of the above cases. You should state clearly what you would do, what you would see and how you would work out your results.

(*L.*)

4. Marble chippings (calcium carbonate) were added to an excess of dilute hydrochloric acid and the gas which evolved was collected. The volume of the gas collected was noted every half minute and recorded as shown in the table below.

Time/min	$\frac{1}{2}$	1	$1\frac{1}{2}$	2	$2\frac{1}{2}$	3	$3\frac{1}{2}$	4
Volume/cm³	80	150	200	240	270	290	300	300

(a) (i) Using the data given in the table, plot a graph to show the results.
   (ii) How long did it take for 120 cm³ of gas to be collected?
   (iii) Why does the line of the graph remain horizontal after $3\frac{1}{2}$ minutes?
   (iv) Sketch on the graph the curve you would expect if smaller pieces of marble, but of the same total mass, were used.

(b) What effect, if any, would each of the following changes have on the final volume of gas collected? (i) Using more acid. (ii) Using less marble.

(c) Write a balanced chemical equation to represent the reaction which takes place between the marble and dilute hydrochloric acid.

(d) Draw a diagram of the apparatus which you would use to carry out the experiment described at the beginning of the question.

(*Y.R.E.B., 78*)

5. Describe how you would carry out an experiment to measure the volume of hydrogen which would be set free from dilute hydrochloric acid, at room temperature and pressure, by 1 gram of zinc.
The addition of a few drops of aqueous copper(II) sulphate speeds up the reaction between the acid and the zinc. What tests would you carry out to show that it is the Cu^{2+} ions and not the SO_4^{2-} ions or the water present in the aqueous copper(II) sulphate that cause the increase in speed?

(*L.*)

6. The curves below show the results of two experiments which were carried out using magnesium ribbon and a dilute acid. In each case, the same mass of magnesium ribbon was taken and the same volume of acid of the same concentration was used. The acids were 1 M hydrochloric acid and 1 M ethanoic acid.

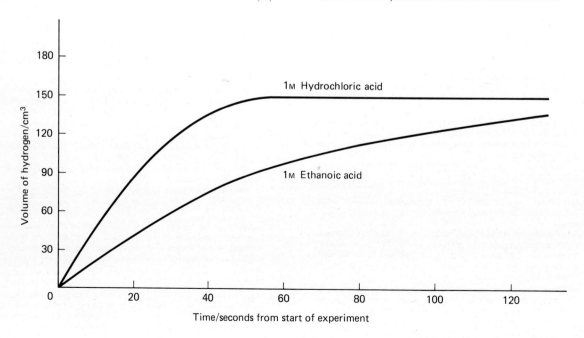

(a) An equation which represents *both* reactions is
$$Mg(s) + 2H^+(aq) \rightarrow Mg^{2+}(aq) + H_2(g)$$
   (i) What do the curves tell you about the concentration of H^+ ions in each acid solution?
   (ii) If left long enough, the volume of hydrogen collected at room temperature and pressure is the same in each case. Why is this so?
(b) How many seconds would elapse before half of the magnesium was used up (i) with hydrochloric acid, (ii) with ethanoic acid?
(c) What shape would the curve be if the same mass of powdered magnesium were used with the same volume of 1 M hydrochloric acid under the same conditions?

(L.)

Chapter 24

1. State concisely the chemical principles which determine methods of extracting metals from their compounds on the industrial scale (but without technical detail of industrial plant), referring to sodium or magnesium and to zinc or iron.
(a) Suggest a method for the large-scale production of a metal M found in large quantities as the carbonate MCO_3; M is known to displace hydrogen from dilute acids rather slowly.
(b) What do you understand by the statement that zinc oxide (or zinc hydroxide) is amphoteric? Describe experiments which you could perform to illustrate your explanation, starting with a solution of zinc sulphate.
(c) What mass of iron is obtainable from 3560 tonnes of the ore limonite $Fe_2O_3H_2O$?

(O.)

2. (a) Give, with a diagram, an account of the manufacture of aluminium from its oxide (purified bauxite).
(b) Give two important uses of this metal and explain the property on which each use depends.
(c) Describe one experiment to show that aluminium is a more reactive metal than copper.

(O. & C.)

3. Read the paragraph below, which concerns the metal aluminium, then answer the questions which follow.
'Purified bauxite (Al_2O_3) is electrolysed in a steel vessel with a carbon lining as a cathode and carbon anodes dipping into the electrolyte. Aluminium ions are discharged at the cathode, forming molten aluminium which is collected.'
(a) (i) What is liberated at the anode?
   (ii) Name the common alkali used in the purification of bauxite.
   (iii) Which compound is mixed with purified bauxite to make the electrolyte?
(b) (i) The reaction at the anode is such that the carbon anodes are being continually worn away. Give a possible explanation.
   (ii) Give two uses of aluminium.
   (iii) Why is aluminium a relatively expensive element?

(E.M.R.E.B.)

4. (a) The blast furnace process is used to obtain pig iron (cast iron) from iron ore. (i) What materials are put into the blast furnace? (ii) What chemical reactions occur in the blast furnace? (iii) What are the gases which emerge from the top of the furnace used for?
(b) Describe the experiments you would carry out in order to find out if air or water, or both, must be present for iron to rust.
(c) Describe three methods by which the rusting of iron can be prevented.

Chapters 25, 26

1. A pure white solid (A) was found to be soluble in water. This solution when acidified with dilute nitric acid gave a white precipitate with silver nitrate. The solution of (A) also gave a brilliant yellow colour in a flame test.
(a) Name the compound A.

When solid (A) was treated with concentrated sulphuric acid a gas was formed that 'steamed' in air and dissolved easily in water, giving a solution which turned blue litmus red.
(b) (i) Name the gas. (ii) Name the solution.

The solution of (A) in water when conducting electricity produced a green gas at the anode and a colourless gas at the cathode. When tested with a burning splint the colourless gas went 'pop'.
(c) (i) Name the green gas. (ii) Name the colourless gas.
(d) Write down equations to illustrate the chemical changes at (i) the cathode. (ii) the anode.

2. Give three physical properties of chlorine.
Under what conditions and with what results does chlorine react with (i) hydrogen sulphide, (ii) iron, (iii) sulphur dioxide?
The elements chlorine, bromine and iodine appear in the same group of the Periodic Table. Write a brief comment on their classification, referring to the manner in which their atoms combine with those of other substances, and mentioning two series of similar compounds.
Describe two simple experiments which you could carry out to demonstrate the relative activity of chlorine, bromine and iodine.

(O.)

3. Chlorine, bromine and iodine are in Group VII of the Periodic Table and are members of the same family.
(a) Explain the characteristics of the family by reference to
   (i) the physical state of each element,
   (ii) the similarities and differences in the electronic configuration of their atoms,
   (iii) the gradation in the sizes of their atoms and ions.
   (iv) the ease with which they combine with iron.
Give an equation for the reaction of iron with one of the elements.
(b) What is seen when an excess of chlorine is slowly bubbled through a solution of potassium iodide?
Write an equation for the reaction.
(c) Dichlorodifluoromethane, CCl_2F_2, is used as a refrigerant called 'Freon'
Write its structural formula and suggest what shape its molecule might be.

(J.M.B.)

Chapter 27

1. A yellow, non-metallic solid (A) burns in oxygen to give a gas (B) which turns damp blue litmus red. The gas (B) is oxidised in the presence of a catalyst to form another substance (C) which turns damp blue litmus red.
(a) Name the solid A
(b) Name the gas B
(c) Write the equation for the oxidation of A to B
(d) What compound is produced when B dissolves in water?
(e) What catalyses the change from B to C?
(f) Describe the appearance of C at room temperature
(g) Write the equation for the oxidation of B to C

2. (a) What do you understand by the term 'allotropy'?
(b) Name two crystalline allotropes of sulphur.
(c) Which is the stable allotrope at room temperature?

(d) Describe briefly how you would prepare a sample of the allotrope which is unstable at room temperature.

(e) What would happen to the crystals formed in (d) if they were left at room temperature?

(f) If 3·2 g of allotrope (i) were completely oxidised to sulphur dioxide, what mass of sulphur dioxide would be formed? (Relative atomic masses: $O = 16$, $S = 32$).

(*J.M.B.*)

3. (a) Give an outline account of the manufacture of sulphuric acid by stating the chemistry of the following steps:

   (i) the production of sulphur dioxide;
   (ii) the oxidation of sulphur dioxide;
   (iii) the absorption of sulphur trioxide.

   (b) Briefly describe one reaction in which sulphuric acid behaves as an oxidising agent, and one reaction in which the acid behaves as a dehydrating agent.

   (c) If x cm³ of a solution of sulphuric acid is exactly neutralised by y cm³ of a solution of sodium hydroxide, in what proportions by volume must the acid and the alkali be mixed to give a solution which will deposit sodium hydrogen sulphate on evaporation?

   (d) Mention three large-scale uses of sulphuric acid.

(*O.*)

4. Over 90% of the sulphur dioxide which is released into the atmosphere is in the northern hemisphere of the world. What are the possible explanations of this marked difference between the hemispheres and explain whether or not we ought to be concerned about the difference?
Describe two harmful effects of the pollution of the atmosphere by sulphur dioxide.

Chapter 28

1. (a) Give equations for three quite different reactions as a result of which carbon dioxide is released.

   (b) What is the chemistry of the test for carbon dioxide?

(*O.*)

2. When oxalic acid (ethanedioc acid) crystals are heated with concentrated sulphuric acid, a mixture of carbon monoxide and carbon dioxide is produced.

   (a) How would you demonstrate that both carbon monoxide and carbon dioxide are present in the gas evolved? You should state what you would do, what you would see and how you would interpret your results.

   (b) Suppose that you had collected 100 cm³ of the mixture in a gas syringe. How would you determine the percentage by volume of carbon dioxide in the mixture? You should state what you would do, what you would see and how you would interpret your results.

   (c) Suggest two safety precautions that should be taken when carrying out the above experiments. Why would these be necessary?

(*L.*)

Chapter 29

1. (a) Limestone is roasted in a kiln using producer gas as the fuel.
   (i) Explain how producer gas is made.
   (ii) Give the chemical name and formula of limestone.
   (iii) Name two other naturally occurring materials which have the same chemical composition as limestone.

   (b) Calcium oxide and calcium hydroxide are both white powders which can be made from limestone. Name a substance that will distinguish between them and state what happens when it is added to each powder.

   (c) Calcium hydroxide is used in the manufacture of bleaching

powder. With what is calcium hydroxide treated to convert it into bleaching powder?

(d) For what purpose is limestone added to the blast furnace?

(*E.M.R.E.B.*)

2. (a) Give, with equations, an account of the manufacture of sodium carbonate by the Solvay process. Include in your account

   (i) the names of the raw materials,
   (ii) the reactions by which the sodium carbonate is produced from the raw materials.

   (b) State two factors which make the Solvay process efficient and name the by-product formed.

   (c) How does sodium carbonate solution react with

   (i) dilute nitric acid,
   (ii) a solution of calcium chloride?

   For each reaction state what would be seen and either name the products or write an equation.

(*J.M.B.*)

3. (a) Sodium is manufactured from sodium chloride. For this process

   (i) state whether the electrolyte is molten or in solution,
   (ii) name the materials of which the electrodes are made.
   (iii) write an equation for the reaction in which the sodium is formed.

   (b) How is the process in (a) modified to obtain sodium hydroxide as the product? Explain how the sodium hydroxide is formed.

   (c) Name the other product common to both processes and outline the reactions of this substance with

   (i) water,
   (ii) hydrogen.

   In each case, give the conditions and either an equation or the name(s) of the product(s).

(*J.M.B.*)

Chapter 30–32

1. (a) Name and give the structural formula of an organic compound that is formed by the oxidation of ethanol.

   (b) Ethanol vapour, when passed over a heated catalyst, produces an unsaturated hydrocarbon X with relative molecular mass (molecular weight) of 28. (i) Name a catalyst which can be used. (ii) Write down the equation for the reaction. (iii) What is the empirical formula for X? (iv) Write down in full the structural formula for X. (v) Describe two tests by which you could show that X is unsaturated.

(*O.*)

2. Two saturated organic compounds have the same molecular formula C_2H_6O.

   (a) Write the structural formulae of the two compounds and name them.

   (b) How could sodium metal be used to distinguish between the two compounds?

   (c) Write an equation for the preparation of an ester starting with one of the two compounds. Name both the second reactant and the ester produced.

   (d) Name a suitable catalyst for the esterification process.

(*O.*)

3. (a) What is the major constituent, other than water, of wine? Give its name and formula. (b) What reaction occurs when the wine becomes sour? (c) What reagent would you use in the laboratory to bring reaction (b) about more quickly? (A balanced equation using this reagent is not expected.)

(*O.*)

4. (a) Name four fractions obtained by the fractional distillation of petroleum and give one major use for each fraction.

(b) At present, most of our ethanol is made from ethene. How is this done on a large scale?

(c) In future, it may be necessary to make ethene from ethanol. How could this be done industrially?

(d) Ethene and bromine were allowed to react together at a temperature high enough to ensure that all the substances involved in the reaction were gaseous.
  (i) Write the equation for the reaction.
  (ii) Give the structural formula for the product.
  (iii) State what would be seen during the course of the reaction.
  (iv) If 50 cm³ of ethene and 50 cm³ of bromine vapour were used, what volume of gaseous product would be formed?

(J.M.B.)

5. (a) What reaction occurs between ethanol and (i) sodium, (ii) acidified potassium dichromate(VI) solution?

(b) Describe how an ester can be made from ethanoic (acetic) acid. Give the structural formula of this ester.

(c) Describe the hydrolysis of an ester and give one example of a large-scale application of this hydrolysis.

(O. & C.)

6. (a) Outline the factors which are likely to be taken into account when choosing a site for the construction of a new oil refinery. (b) Explain one chemical method for the dispersal of oil slicks.

7. Oil is used (i) as a fuel for heating, (ii) as a fuel for transportation and (iii) as a source of chemicals which can be converted into products such as plastics man-made fibres, medicines and herbicides. Giving reasons, explain which of these uses you think should have prior claim on the world's diminishing oil supplies.

8. Name three examples of fossil fuels.
Compare the advantages and disadvantages of these fuels with respect to (i) the costs of transporting the fuels, (ii) the pollution of the atmosphere by sulphur dioxide and (iii) the need to use them sparingly in order to conserve the world's supply.

9. (a) Name two plastics which are made from ethene.
(b) Give one use for each plastic and explain why it is suitable for this purpose.
(c) Explain why the disposal of plastic objects is a problem.
(d) In the U.K. ethene is made from oil. An alternative source is to make it from ethanol which has been obtained by fermentation processes. Why do you think that this is not a feasible alternative in the U.K. at the moment?

Chapters 33, 34

1. Nitric acid is manufactured by the catalytic oxidation of ammonia. The reactions which take place are represented by the equations A and B below. (N.B. Equation A is *not* balanced).
A. $4NH_3 + 5O_2 \rightarrow 4NO + H_2O$
B. $4NO + 3O_2 + 2H_2O \rightarrow 4HNO_3$

(a) How many molecules of water would have to be formed in order to balance equation A?

(b) Name the catalyst used in the manufacture of nitric acid.

(c) Is the catalyst essential for the reaction represented by equation A or that represented by equation B, or for both of these reactions?

(d) Describe fully, one test for a nitrate.

(Y.R.E.B., 77)

2. Ammonia is manufactured by passing a mixture of nitrogen and hydrogen, at a pressure of about 250 atmospheres, over a catalyst at 500°C. Under these conditions, about 15% of the gases are converted to ammonia.

(a) Write an equation for the reaction and state whether it is exothermic or endothermic.

(b) Why is a catalyst used?

(c) What happens to the gases which are not converted to ammonia?

(d) The yield of ammonia is increased if the process is carried out at (i) even higher pressures and (ii) lower temperatures. Explain why these conditions are not used in industry.

(e) Explain why industrial plants for manufacturing ammonia used to be situated near coalfields but the plants built more recently are situated near oil refineries.

(f) If ammonia is passed over heated copper(II) oxide, the oxide is reduced to copper and the other products of the reaction are nitrogen and water.
Starting with dry ammonia coming through a delivery tube, draw a diagram of apparatus which could be used to carry out the above experiment and collect the three products.

(J.M.B.)

3. (i) When copper reacts with nitric acid, a gas *A* is released and a solution *B* is formed. When the solution *B* is evaporated to dryness and the residue heated strongly a solid *C* is produced together with gas *A* and another gas *D* which supports combustion. Name the substances *A*, *B*, *C* and *D*, and describe the appearance of *A*, *B* and *C*.
(ii) Write an equation for the decomposition of nitric acid by heat.

(O.)

4. (a) Briefly outline the production of an NPK (nitrogen, phosphorus, potassium) fertiliser from naturally occurring substances.

(b) What are some of the indications that the soil in which plants are growing is deficient in (i) nitrogen, (ii) phosphorus, (iii) potassium?

(c) In what ways can the large-scale use of fertilisers be harmful to the environment?

Revision Questions I

1. Metal M, which forms an ion M^{2+}, has a relative atomic mass of 59. The metal forms a basic oxide which is insoluble in water. The sulphate and chloride of M are both water soluble: the chloride exists as hexahydrate crystals, i.e. with 6 molecules of water of crystallisation, and the sulphate as a heptahydrate, i.e. with 7 molecules of water of crystallisation. The metal is just below zinc in the activity series (electrochemical series).
   (a) Write the formulae of (i) the hydrated chloride, (ii) the hydrated sulphate, of M.
   (b) Describe, giving all essential practical details, how you would prepare a dry sample of one of the compounds in (a), starting from the oxide of M.
   (c) Name two metals which you would expect M to displace from solutions of their salts. Write an equation for one of the displacement reactions.
   (d) What is the relative molecular mass of the nitrate $M(NO_3)_2$?
   (e) Calculate the maximum mass of this nitrate obtained in solution by reaction of the oxide of M with 100 cm³ of 0·050M nitric acid.

   (C.)

2. Make use of the following information about silicon (Si) and its compounds to answer the questions below.
   Silicon, atomic number 14, is the element immediately below carbon in Group IV of the Periodic Table. It does not react with water nor with dilute acids. It can be obtained by heating sand with an excess of magnesium. Sand is an oxide of silicon. When sand is heated with carbon at a high temperature, the products formed are carbon monoxide and carborundum which is a compound of silicon and carbon only and is a very hard substance.

   (a) State the characteristic valency of silicon and hence write down the chemical formulae for (i) sand, (ii) sodium silicate.
   (b) Write the equation for the reaction between sand and magnesium.
   (c) Describe how you would obtain pure dry silicon from the products of reaction (b).
   (d) How are the electrons arranged in a silicon atom?
   (e) Give (i) two physical differences, (ii) one chemical similarity, between carbon dioxide and sand.
   (f) Suggest a formula for carborundum.
   (g) Write the equation for the reaction of carbon with sand.
   (h) Name another substance which you would expect to have the same crystal structure as carborundum.

   (C.)

3. In two titration experiments it was found that
   I 20 cm³ of 0·1M potassium chromate(VI) (K_2CrO_4) required 20 cm³ of 0·2M silver nitrate for complete reaction.
   II 20 cm³ of 0·1M potassium chromate(VI) required 10 cm³ of 0·2M barium chloride for complete reaction.
   Silver chromate(VI) and barium chromate(VI) are both insoluble in water.

   (a) How many moles of silver nitrate are required to react completely with one mole of potassium chromate(VI)?
   (b) How many moles of barium chloride are required to react completely with one mole of potassium chromate(VI)?
   (c) Write equations for the reactions taking place in Experiments I and II.
   (d) Calculate the mass of silver chromate(VI) precipitated in Experiment I and the mass of barium chromate(VI) precipitated in Experiment II.
   (e) Write the equation for the reaction between solutions of barium chloride and silver nitrate and calculate the volume of 0·2M silver nitrate required to react completely with 20 cm³ of 0·2M barium chloride.
   (f) Give the formulae of three halides of silver which are insoluble in water.

   (C.)

4. Aluminium (valency 3) forms a compound Al_4C_3 with carbon. This compound reacts with water according to the equation

   $$Al_4C_3 + 12H_2O \rightarrow 4Al(OH)_3 + 3CH_4.$$

   Aluminium reacts with other non-metals to form compounds which react with water in a similar way to Al_4C_3.
   (a) Write the formulae of the compounds formed when aluminium combines separately with chlorine (valency 1), sulphur (valency 2) and nitrogen (valency 3).
   (b) Give the formulae of the compounds, other than aluminium hydroxide, which you would expect to be formed when the three compounds in (a) react with water.
   (c) Write equations for the reactions of the three compounds in (a) with water.
   (d) Without referring to their effect on indicators, describe briefly and write equations for four different reactions (one in each case) in which
      (i) one of the compounds in (b) reacts as an acid,
      (ii) one of the compounds in (b) reacts as a base,
      (iii) one of the compounds in (b) reacts as a reducing agent,
      (iv) two of the compounds in (b) react together to give a solid product.

   (C.)

5. At 30°C and 756 mm of mercury
   one mole of any gas occupies 25·0 dm³,
   one dm³ of a gaseous compound P has a mass of 1·20 g,
   one dm³ of a gaseous compound Q has a mass of 1·08 g.
   The empirical (simplest) formula of P is CH_2O.
   Q burns in oxygen to give water, carbon dioxide and nitrogen as the only products.
   25·0 dm³ of Q give, on complete combustion, 9·00 g of water and 12·5 dm³ of nitrogen. (Both volumes at 30°C and 756 mm of mercury.)
   Oxidation of P gives an organic acid R of molecular formula CH_2O_2.
   Q reacts slowly with water in the presence of sodium hydroxide to give ammonia and the sodium salt of acid R as the only products.
   (a) Calculate the relative molecular masses of P and Q.
   (b) Write the molecular formula of P.
   (c) What elements must be present in Q?
   (d) What is the mass of hydrogen in 9·00 g of water?
   (e) What is the mass of 12·5 dm³ of nitrogen at 30°C and 756 mm of mercury?

(f) Use your answers and the information above to deduce the molecular formula of Q.

(g) Write the equation for the combustion of Q in oxygen.

(h) Name the organic acid R and give the formula for its sodium salt.

(i) Write the equation for the reaction of Q with water in the presence of sodium hydroxide.

(j) Use your knowledge of valencies to write structural formulae for P and R, using lines to represent covalent bonds.

(C.)

6. Use the information given in (i) to (viii) to identify the substances A to H, selecting your answers only from substances in the list below. Write equations for all the chemical reactions mentioned in (i) to (viii).

calcium	calcium oxide	ammonium chloride
carbon	copper(II) oxide	copper(II) nitrate
chlorine	lead(II) oxide	hydrogen chloride
oxygen	calcium hydroxide	lead(II) nitrate

(i) A is a white solid. When heated, it does not melt but produces a dense white smoke.

(ii) An alkaline gas is formed when a mixture of A and C is warmed.

(iii) When water is added to E, heat is evolved and C is formed.

(iv) B burns brightly in air to form E.

(v) When D is heated, it gives off brown fumes and leaves a black residue of F.

(vi) A solution of D is formed by warming F with dilute nitric acid.

(vii) G is a gaseous, non-metallic element which reacts with hydrogen to form H.

(viii) A solution of H will neutralise a solution of C.

(C.)

7. Make use of the information in the table below to answer the following questions about strontium and bismuth. Either write equations or name the products of any definite chemical reactions to which you refer.

Element	Strontium	Bismuth
Symbol	Sr	Bi
Relative Atomic Mass	88	209
Valency	2	3
Position in the activity series (electrochemical series)	Between sodium and calcium	Between lead and copper

(a) Write formulae for the oxides of strontium and bismuth and state how these oxides would differ
  i) when treated with water,
  (ii) when heated in hydrogen.

(b) How would you expect the elements strontium and bismuth to differ in their behaviour when added to cold water?

(c) Write formulae for the chlorides of strontium and bismuth.

(d) Name the substances you would expect to be liberated at platinum cathodes (negative electrodes) when solutions of
  (i) strontium chloride,
  (ii) bismuth chloride,
  are electrolysed separately.

(e) What masses of the products in (d) would be liberated by the passage of 96 500 C (coulombs) of electricity?

(C.)

8. Nitrogen dioxide (NO_2) can be obtained by the action of heat on lead(II) nitrate. Nitrogen dioxide reacts with cold water forming nitrous acid (HNO_2) and nitric acid (HNO_3) as the only products.

(a) Describe what you would observe if lead(II) nitrate were heated until the reaction had finished and the residue left to cool.

(b) (i) Write the equation for the action of heat on lead(II) nitrate.

  (ii) Construct the equation for the reaction between nitrogen dioxide and a cold aqueous solution of sodium hydroxide.

(c) Suggest why each of the following nitrates is not used for the preparation of nitrogen dioxide:
  (i) sodium nitrate ($NaNO_3$),
  (ii) copper(II) nitrate ($Cu(NO_3)_2.3H_2O$),
  (iii) ammonium nitrate (NH_4NO_3).

(d) How is nitrogen dioxide formed during the manufacture of nitric acid from ammonia?

(C.)

9. Describe briefly and write equations for reactions by which you could convert

(a) oxygen gas into a compound containing oxide ions,

(b) hydrogen gas into a solution containing hydrogen ions,

(c) a solution containing calcium ions into an insoluble compound of calcium,

(d) metallic iron into a solution containing Fe^{3+} ions,

(e) sulphur into a solution containing sulphate ions,

(f) carbon dioxide into a solution containing carbonate ions.

(C.)

10. (a) Outline the manufacture of sulphuric acid from sulphur.

(b) Describe one reaction in which sulphuric acid behaves as a dehydrating agent.

(c) Three reactions of dilute sulphuric acid can be represented by the ionic equations:
  (i) $Mg(s) + 2H^+(aq) \rightarrow Mg^{2+}(aq) + H_2(g)$
  (ii) $2H^+(aq) + CO_3^{2-}(aq) \rightarrow H_2O(l) + CO_2(g)$
  (iii) $Ca^{2+}(aq) + SO_4^{2-}(aq) \rightarrow CaSO_4(s)$.

In which one of these reactions has oxidation and reduction taken place?

11. For each of the following products give one reason why they are useful: (i) nitrate fertilisers, (ii) lead compounds in petrol, (iii) detergents.
Give one important disadvantage of each product.

12. For producing chemicals on a large scale, cheap and abundant supplies of raw materials are required. Some examples of such raw materials are:

Air, fats, metal oxides, petroleum, sea water.

(a) Select ONE chemical in the production of which one or more of the above raw materials plays an essential part and describe the production of this chemical (technical details are not required). The raw material(s) must be converted into other substances.

(b) Comment briefly on the importance to society of the chemical you have selected and suggest what would be likely to happen if the raw material(s) in the list required for production of this chemical became unobtainable.

(L.)

13. Briefly indicate the factors which are likely to be taken into account when selecting a site for the construction of:
  (i) an aluminium-producing plant,
  (ii) an iron and steel works,
  (iii) a chlorine-producing plant.
Which of these elements are recycled after use? For each element which is recycled name one everyday object which is used as a source of the element for recycling.

Revision Questions II

1. Malachite is an ore of copper. In an attempt to obtain copper from the ore a pupil tried heating it in a test-tube. A gas which turned lime water milky was given off and a black powder was left in the test-tube.

   (a) The pupil thought that the black powder might be copper metal. What simple test could be carried out to check this?

   (b) Name the gas given off when malachite is heated.

   (c) The pupil added dilute hydrochloric acid to the black powder which on warming dissolved in the acid. What is the name of the compound in the resulting solution?

   (d) The pupil then passed an electric current through the solution, obtained in part (c), and copper was deposited on one of the electrodes. Which electrode would that be?

   (e) At the other electrode a gas was evolved and was found to bleach damp litmus paper. Name the gas.

   (f) The pupil could have obtained copper from the black powder. Describe how this could have been done.

   (g) Could he have used the same method to obtain zinc from zinc oxide and lead from lead oxide?

   (h) What is the mass of one mole of the black powder?

   (i) What mass of copper(II) sulphate crystals ($CuSO_4.5H_2O$) could be obtained from 8 g of the black powder by reacting with dilute sulphuric acid?

   (j) Write the ionic equation for the reaction which takes place at the electrode at which copper is produced in part (d).
   A current of 0·482 amp was passed through the solution for 1000 seconds. What mass of copper was deposited on the electrode?

   (k) If electrolysis of the solution of the black powder in hydrochloric acid was continued until no further change took place, what changes in
   (i)   the appearance,
   (ii) the electrical conductance
   of the solution would be observed? Explain the changes you describe.

   (l) Outline how, starting with malachite in each case, you could obtain samples of the following substances:
   (i)   copper(II) hydroxide,
   (ii) copper(II) sulphide.

2. The graph shows the results of an experiment on the reaction between 1 g of powdered chalk and 20 cm³ of dilute hydrochloric acid (2M hydrochloric acid) at room temperature (20 °C). The time when the reaction started is shown as t_0 and the finishing time as t_4.

   (a) What gas would you expect to be evolved?

   (b) At which time was the gas being evolved at the highest rate?

   (c) By which time was the reaction complete?

   (d) What change would you expect in the shape of the graph from t_0 to t_3 if 1 g of small pieces of marble had been used instead of powdered chalk?

   (e) Draw a diagram of a suitable apparatus you could use to obtain the results of this experiment.

   (f) What shape would the graph have had if 40 cm³ of M hydrochloric acid has been reacted with the 1 g of powdered calcium carbonate in this experiment? Explain your answer.

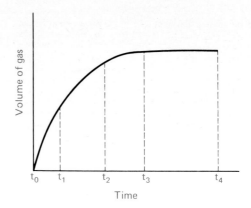

   (g) What shape would the graph have had if 20 cm³ of 2M hydrochloric acid had been reacted with 1 g of powdered calcium carbonate at 50 °C? Explain your answer.

   (h) If 1 g of powdered calcium carbonate was reacted with 20 cm³ of 2M hydrochloric acid, which reactant would be left over when the reaction stopped?

   (i) If 1 g of powdered calcium carbonate was reacted with 20 cm³ of 2M hydrochloric acid, what volume of gas, measured at s.t.p., would be given off?

   (j) You think that a powder might act as a catalyst in the reaction between calcium carbonate and dilute hydrochloric acid. Describe what you would do to find whether this is so and what would happen if it is.

3. (a) Crude oil is a mixture containing alkanes. In a refinery it can be fractionally distilled to produce a number of useful fractions.
   (i)   Write the names and molecular formulae of the first three members of the alkane series.
   (ii) As the relative molecular mass of the alkanes increases, each alkane will differ from the next in a number of ways. State two ways in which an alkane of a higher relative molecular mass will differ from an alkane of a lower relative molecular mass.
   (iii) Name three fractions obtained from the fractional distillation of crude oil.
   (iv) One of the fractions obtained by distilling crude oil is 'cracked'. Explain why the 'cracking' process is necessary and what 'cracking' means.

   (b) Butane, C_4H_{10}, is an example of a saturated compound. Ethene, C_2H_4, is an alkene and is an example of an unsaturated compound.
   (i)   Draw the structural formulae of butane and ethene.
   (ii) Name the products formed if each of these compounds is burned completely in a plentiful supply of air.
   (iii) Ethene quickly decolourises bromine water but butane has no effect on this reagent. Explain the reason for this, using your answer to (b) (i) to help.
   (iv) In the manufacture of margarine, vegetable oils are treated with hydrogen to make them into fats.

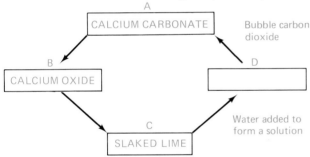

The periodic table extract shows: Li (3), Mg (12), K (19), Ca (20), Cu (29), N (7), F (9), Ne (10), Si (14), S (16), Ar (18), Br (35), with numbered boxes 1-36.

Vegetable oils contain unsaturated compounds, while those in fats are saturated. Describe the reaction which takes place during the hardening of oils.

(v) Unsaturated compounds, such as ethene, can be polymerised. Show, using structural formulae, what happens when ethene polymerises.

State and explain what you would expect to happen when polyethene is:
  (a) burned,
  (b) treated with bromine water,
  (c) heated to a very high temperature with no air present.

(vi) Oil and products made from it can cause pollution problems. Explain why each of the following causes pollution of our environment:
  (a) the incomplete combustion of petrol in a car engine,
  (b) the throwing away of plastic bags in fields and ditches,
  (c) the discharge of detergents into a river.

(*J.M.B./W.M.E.B.*)

Diagram: A — CALCIUM CARBONATE → B — CALCIUM OXIDE → C — SLAKED LIME → D. "Bubble carbon dioxide", "Water added to form a solution".

4. (a) Name THREE forms in which calcium carbonate is found in rocks.

(b) How may a solid lump of A, calcium carbonate, be converted to B, calcium oxide?

(c) When a few drops of water are added to a piece of calcium oxide, slaked lime, C, is produced.
  (i) Describe what you see happen in this reaction.
  (ii) The reaction is said to be 'exothermic'. What does this mean?

(d) If slaked lime is added to litmus solution, the litmus solution turns blue.
  (i) What sort of substance is slaked lime?
  (ii) Why does a farmer put slaked lime on the soil in his fields?

(e) What is solution D?

(f) When substance A (calcium carbonate) is converted into substance B (calcium oxide), a gas, E, is given off. What is this gas and what would you see when E is passed into D?

(g) What is the mass of one mole of (i) calcium carbonate (ii) gas E?

(h) What mass of pure calcium carbonate would need to be heated to produce 2·2 g of gas E?

(i) Gas E can also be produced from calcium carbonate by adding dilute hydrochloric acid. What volume of M hydrochloric acid is required to react with 10 g of pure calcium carbonate?

(j) Outline how you would prepare the following compounds, starting with calcium carbonate in each case:
  (i) calcium nitrate,
  (ii) calcium sulphate,
  (iii) a solution of calcium hydrogencarbonate,
  (iv) carbon monoxide.

(k) How would you show that the three substances you named in (a) are forms of the same substance, calcium carbonate?

5.

(a) Using only the elements whose symbols are given in the above extract of the Periodic Table, answer the following questions.
  (i) Give the symbols of two elements present in the same group in the classification.
  (ii) Give the symbols of two elements present in the same period in the classification.
  (iii) Name the element which has the electronic structure 2.8.4.
  (iv) Name TWO elements, the ions of which have the electronic structure 2.8.8.
    Give the symbol of each ion.
  (v) Name the element which is most likely to have a valency of 3.
  (vi) Name a transition element.
  (vii) Name the element, 1 mole of atoms of which would have the greatest mass.
  (viii) Which element would have the highest number of atoms in a 1 g sample?
  (ix) Name a noble gas.
    Describe the characteristic features of the electronic structure of a noble gas.
  (x) Potassium is the most reactive metal listed. Explain this statement in terms of the structure of a potassium atom.
  (xi) Fluorine is the most reactive non-metallic element listed. Explain this statement in terms of the structure of the fluorine atom.
  (xii) Explain in terms of electronic structure why you would expect the elements magnesium and calcium to show some similarity in chemical reactions.
    State a reaction which is common to both magnesium and calcium. In each case name the product(s) of the reaction you have stated.

(b) In a complete Periodic Table the element barium (Ba) appears below calcium in the same vertical column.
  (i) Will the element be more likely to form an ionic chloride or a covalent chloride?
  (ii) What is the formula of barium chloride?

(c) In a complete Periodic Table the element selenium (Se) appears below sulphur in the same vertical column.
  (i) Will the element be more likely to form an ionic hydride or a covalent hydride?
  (ii) What is the formula of selenium hydride?

(*J.M.B./W.M.E.B.*)

6. Metals can be obtained from their ores in a number of ways. The following two methods are frequently used.
Method 1 Electrolysis of a molten compound of the metal, e.g. in the extraction of sodium, magnesium and aluminium.
Method 2 Reduction of an oxide of the metal using carbon, e.g. in the extraction of zinc, iron and lead.

(a) (i) Explain why the methods are used for the metals named.

(ii) Iron can be formed at an electrode during electrolysis. Explain why it is preferable to reduce iron(III) oxide in a blast furnace.

(b) Iron ore, coke and calcium carbonate are fed into the top of a furnace and a blast of hot air is forced in through pipes at the bottom of the furnace. The calcium carbonate is used to remove acidic impurities which are present in the iron ore.
  (i) Name the gas which acts as the reducing agent in the blast furnace.
  (ii) State the reactions which occur leading to the formation of the reducing agent.
  (iii) Write the equation for the reaction between the reducing agent and iron(III) oxide.
  (iv) Name one acidic impurity present in iron ore.
  (v) What products are formed when calcium carbonate is heated in the blast furnace?
  (vi) State, giving an explanation for your answer, which one of the decomposition products of calcium carbonate will react with the acidic impurity. Give the chemical name of the product of the reaction.

(c) Magnesium is produced by the electrolysis of molten magnesium chloride.
  (i) Write the ionic equation for the formation of the magnesium during the electrolysis.
  (ii) Name the other product of this electrolysis.
  (iii) Magnesium chloride solution undergoes electrolysis at room temperature. Explain why this method cannot be used for the extraction of the magnesium.

(d) The element titanium is a metal which occurs naturally as titanium(IV) oxide, TiO_2. In the extraction of this metal, the oxide is mixed with carbon and heated in a stream of chlorine, producing titanium(IV) chloride and carbon monoxide.
The titanium(IV) chloride is converted into titanium by reacting it with sodium in an atmosphere of argon.
  (i) Write the equation for the reaction between titanium(IV) oxide, carbon and chlorine.
  (ii) Why is an atmosphere of argon necessary in order to carry out the reaction between sodium and titanium(IV) chloride?
  (iii) Name another metal which could be used in place of sodium.
  (iv) Calculate the mass of sodium which is theoretically required to produce 96 tonnes of titanium.
    Equation $TiCl_4 + 4Na \rightarrow Ti + 4NaCl$
    Relative atomic masses   Na = 23   Ti = 48.
    (J.M.B./W.M.E.B.)

7. (a) Bromine dissolves in water forming a reddish-brown solution. A dynamic equilibrium which can be represented by the following equation is obtained:

$$\underbrace{Br_2(aq) + H_2O(l)}_{\text{reddish-brown}} \rightleftharpoons \underbrace{H^+(aq) + Br^-(aq) + HOBr(aq)}_{\text{colourless}}$$

  (i) What is the meaning of the sign $\rightleftharpoons$?
  (ii) Explain what is meant by the term 'dynamic equilibrium'.

(iii) What would happen to the colour of the equilibrium mixture if a little more water was added to it?
(iv) Explain why the addition of an alkali to the equilibrium mixture would produce a colourless solution.
(v) What change would occur, if any, when excess dilute hydrochloric acid is added to the colourless alkaline solution produced in (a) (iv)?

(b) The gas methylamine (CH_3NH_2) dissolves in water forming an equilibrium:
     $CH_3NH_2(g) + H_2O(l) \rightleftharpoons CH_3NH_3^+(aq) + OH^-(aq)$
  (i) Describe one simple test you would carry out to show that methylamine which is dissolved in water behaves in this way.
  (ii) The solution obtained is a very poor conductor of electricity. What does this indicate about the composition of the equilibrium mixture?
  (iii) Name a substance which, when added to the mixture, would cause the position of equilibrium to move towards the left.
    Give an explanation for your answer.
  (iv) In view of your answers to (b) (iii), suggest how methylamine may be obtained, most efficiently, from the salt methylammonium chloride $CH_3NH_3^+Cl^-$.

(c) Methanol (CH_3OH) is manufactured by the reaction between carbon monoxide and hydrogen at a temperature of 300 °C and a pressure of 300 atmospheres:
     $CO(g) + 2H_2(g) \rightleftharpoons CH_3OH$ (g) $\Delta H = -90\ kJ\ mol^{-1}$
  (i) Explain why a high pressure increases both the percentage of methanol in the equilibrium mixture and the rate of reaching equilibrium.
  (ii) What two main factors are likely to determine the choice of 300 °C as the most suitable temperature?
  (iii) State one other way in which the yield of methanol can be increased at a constant temperature and pressure.

    (J.M.B./W.M.E.B.)

8. The following is a list of metals in order of reactivity, the most reactive being first in the list:
sodium, calcium, magnesium, aluminium, zinc, iron, tin, lead, copper, silver.
Use this series to help you answer the questions asked.

(a) Which metal will react quickly with cold water to produce a solution with a pH value greater than 7?
Name the gas given off during the reaction.

(b) Magnesium reacts vigorously with dilute hydrochloric acid producing a gas, but copper does not react with the acid.
  (i) Write the equation for the reaction of magnesium with hydrochloric acid.
  (ii) In the equation in (b) (i) hydrogen ions are reduced. Explain what this means.

(c) Some metal oxides react when heated with carbon at the temperature of a Bunsen flame.
Choose from copper(II) oxide, lead(II) oxide and magnesium oxide in answering the following questions.
  (i) Which oxide would be reduced to a reddish-brown powder?

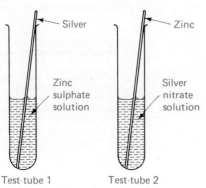

Test-tube 1   Test-tube 2

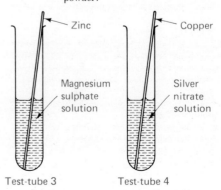

Test-tube 3   Test-tube 4

(ii) With which oxide would there be no reaction?
(iii) Write an equation for the reduction of one of the oxides listed by a gaseous non-metallic element.
(d) Study the diagrams:

In which test-tube(s) do the following occur?
(i) Zinc atoms become zinc ions.
(ii) Silver ions (Ag^+) become silver atoms.
(iii) There is no chemical change.
(iv) Metal atoms are oxidised.
(v) Write an ionic equation for one reaction from any tube.

(e) A piece of zinc briefly dipped into copper(II) sulphate solution reacts more quickly with dilute sulphuric acid than when the zinc is pure. The surface of the zinc after dipping into copper(II) sulphate solution and then placing in the dilute acid can be pictured as:

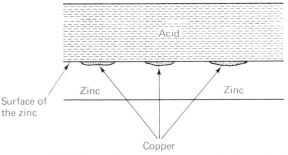

Surface of the zinc

(i) Why are there areas of copper on the zinc surface?
(ii) The zinc corrodes and bubbles of hydrogen are given off at the surface of the copper.
Use equations involving electrons to explain what happens to zinc atoms in the chemical change and what happens to the hydrogen ions undergoing chemical change in the acid.

(f) Two metal plates, one of tin and the other zinc, were connected to a voltmeter and then placed in a solution of dilute hydrochloric acid. Hydrogen was evolved from the tin and the zinc passed into solution as ions. Initially a voltage of 0·5 V was shown on the meter, but this value rapidly decreased.
(i) Explain why the zinc, rather than the tin, passes into solution as ions.
(ii) Describe the reaction taking place at the surface of the tin, resulting in the formation of hydrogen gas.
(iii) Explain why the voltage rapidly decreases.
(iv) Describe one method by which the rate of decrease of the voltage could be reduced.
(v) Name one metallic element, in each case, which could be used to produce an initial voltage bigger than 0·5 V by replacing the tin plate only, and then by replacing the zinc plate only.

(g) Aluminium can be used to replace a metal, chromium, from its oxide. The equation for the reaction is

$$2Al + Cr_2O_3 \rightarrow Al_2O_3 + 2Cr.$$

Chromium when hot will react with steam.
(i) Explain why, in spite of the information given, one of the chemical properties of aluminium makes it a better metal than chromium to use in the manufacture of saucepans.
(ii) Name the products formed when hot chromium reacts with steam.

(*J.M.B./W.M.E.B.*)

9. (a) (i) Complete the following table concerning the family of elements named the halogens. There are five omissions.

NAME OF HALOGEN	STATE AT ROOM TEMPERATURE	ATOMIC NUMBER	ARRANGEMENT OF ELECTRONS
Chlorine		17	
	liquid		2.8.18.7
Iodine		53	2.8.18.18.7

Use the table to help you answer the following questions.
(i) How many protons are there in a chloride ion (Cl^-)?
(ii) How many electrons are there in an iodide ion (I^-)?

(b) Dutch metal is a very thin sheet of an alloy of zinc and copper. It bursts into flame when dropped into chlorine.
(i) Name the two products of the reaction.
(ii) Under the same conditions, would you expect the reaction between Dutch metal and bromine vapour to be more, less or equally vigorous?
(iii) Which of the two halogens, chlorine and bromine, is the more powerful oxidising agent?

(c) Read through the passage and then answer the questions asked.
An oily colourless liquid (A) was added to some sodium chloride. A misty gas (B) was given off which turned damp litmus paper (or Universal Indicator paper) red. When gas B is dissolved in water and treated with silver nitrate solution, a white precipitate (C) is produced.
(i) Name the gas B.
(ii) Name the white precipitate C.
Write an equation for the reaction between the solution of gas B and silver nitrate solution.
(iii) Name the colourless liquid A.
(iv) Name the gas which would be given off if a piece of magnesium ribbon were placed in a solution of gas B. Name the salt which would be left in solution after the reaction.
(v) If an oxidising agent is mixed with liquid A and sodium chloride, a gas is given off which bleaches damp indicator paper. (Heat may be necessary.) Name the gas and a suitable oxidising agent. Explain why the substance is said to be acting as an oxidising agent.

(d) A pupil was asked to grind some sodium chloride crystals (NaCl) to a fine powder, weigh out one tenth of a mole and introduce it into a length of glass tubing, closed at one end. He repeated the procedure with one tenth of a mole of sodium iodide (NaI), using glass tubing of the same cross-sectional area. The height of the sodium iodide powder was greater than the height of the sodium chloride powder after each tube had been tapped to settle the powder.
(i) What fraction of a mole of sodium ions is there in the tube containing sodium chloride?
(ii) What fraction of a mole of iodide ions is there in the tube containing sodium iodide?
(iii) What does the experiment suggest about the relative sizes of a chloride and an iodide ion? Explain how you deduce your answer.

(*J.M.B./W.M.E.B.*)

10. (a) Nitrogen can be obtained in the laboratory by removing some of the gases from the air (a suitable apparatus is shown in the diagram), or by heating laboratory chemicals.

(i) Give the chemical name of solution A which removes carbon dioxide from the air.
(ii) Name the salt formed when carbon dioxide reacts with solution A.
(iii) Name the element which is removed from the air by hot copper.
(iv) Write the equation for the reaction between copper and the element you named in (iii).
(v) The gas collected in the gas jar will not be pure nitrogen. Assuming that all the carbon dioxide and all

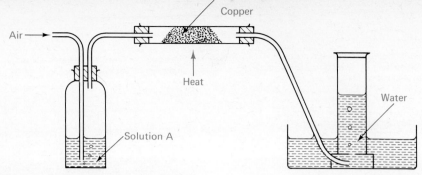

the gas you have named in (iii) are removed, name one element and one compound which will be present in small quantities in the gas jar with nitrogen.

(vi) Name a laboratory chemical or mixture of chemicals which will produce nitrogen when heated.

(b) Some nitrogenous fertilisers contain ammonium nitrate, NH_4NO_3. Nitrogen compounds are required for the healthy development of plants and animals.

  (i) Into what type of compound is the nitrogen converted within the plant?

 (ii) Describe the natural processes by which the nitrogen in the plants may be eventually returned to the soil.

(iii) In view of your answer to (ii), why is it necessary to use artificial fertilisers such as ammonium nitrate?

(iv) If some of the fertiliser containing ammonium nitrate is warmed with a solution P, a gas Q is given off which turns damp red litmus paper blue.
Name the gas Q and the substance contained in solution P.

 (v) Why would it not be advisable to lime soil immediately after using ammonium nitrate fertiliser?

(vi) Which of the following would contain the largest amount of nitrogen, 1000 kg of ammonium nitrate, NH_4NO_3, or 1000 kg of urea, $CO(NH_2)_2$? Show how you deduce your answer.

(c) Nitrogen is produced when ammonia is passed over heated copper(II) oxide.

  (i) Name two other products of the reaction.

 (ii) Which reactant is the reducing agent? Give a reason for your answer.

(d) By means of a simple diagram show the arrangement of the bonding electrons in a molecule of ammonia. The atomic number of nitrogen is 7 and of hydrogen is 1.

(J.M.B./W.M.E.B.)

11. (a) Study the following diagram of a laboratory experiment:

  (i) What is the purpose of the concentrated sulphuric acid?

 (ii) Name the white needle-like crystals formed.
Write an equation for the reaction producing the crystals.

(iii) Name substance (X).
What is the reason for using substance (X)?

(iv) What is the anhydrous calcium chloride for?

 (v) What is the purpose of the freezing mixture?

(b) The damage done in the United Kingdom as a result of pollution of the atmosphere by oxides of sulphur is over £100 million per year. Much of the pollution is liberated from power stations and sulphuric acid plants.

  (i) Give two reasons why oxides of sulphur are pollutants.

 (ii) Why do some power stations produce sulphur dioxide?

(iii) Why could a process to recover the sulphur dioxide be of economic value as well as a means of reducing pollution?

(c) Read the following passage and then answer the questions asked.
Sulphur has two crystalline allotropes with densities $2\cdot07$ g cm^{-3} and $1\cdot96$ g cm^{-3}. Both allotropes consist of S_8 molecules.
The transition temperature of these two allotropes is 96 °C.
The melting points of the allotropes are fairly low.

  (i) Name the two crystalline allotropes of sulphur.

 (ii) If each of the allotropes contains S_8 molecules, why are the crystals different shapes and why are their densities different?

(iii) Why are the melting points of the allotropes low?

(iv) What is meant by the statement that the transition temperature of the allotropes is 96 °C?

(d) Another form of sulphur is known as plastic sulphur.

  (i) Describe how you would prepare a sample of plastic sulphur.

 (ii) Why is this form called plastic sulphur?

(iii) How are the sulphur atoms arranged in plastic sulphur?

(iv) How do the properties of plastic sulphur change on standing? What is the reason for the change?

(J.M.B./W.M.E.B.)

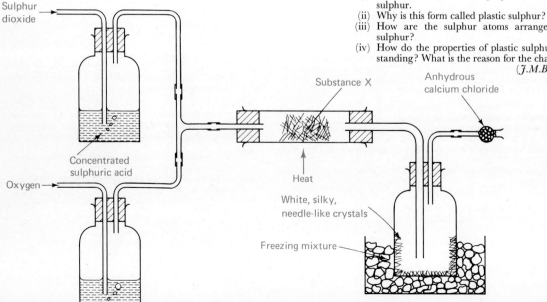

Index